Microplastic Pollution

Mohd. Shahnawaz •
Charles Oluwaseun Adetunji •
Mudasir Ahmad Dar • Daochen Zhu
Editors

Microplastic Pollution

Springer

Editors
Mohd. Shahnawaz
Department of Botany
University of Ladakh, Kargil Campus
Khumbathang, Ladakh UT, India

Charles Oluwaseun Adetunji
Applied Microbiology, Biotechnology and Nanotechnology Laboratory, Department of Microbiology
Edo State University Uzairue
Iyamho, Nigeria

Mudasir Ahmad Dar
Biofuels Institute, School of the Environment and Safety Engineering
Jiangsu University
Zhenjiang, Jiangsu, China

Daochen Zhu
Biofuels Institute, School of the Environment and Safety Engineering
Jiangsu University
Zhenjiang, Jiangsu, China

ISBN 978-981-99-8356-8 ISBN 978-981-99-8357-5 (eBook)
https://doi.org/10.1007/978-981-99-8357-5

This Springer imprint is published by the registered company Springer Nature Singapore Pte Ltd.
The registered company address is: 152 Beach Road, #21-01/04 Gateway East, Singapore 189721, Singapore

Each editor wishes to dedicate this book to their respective parents, better halves, and kids.

Preface

Microplastics are generally referred to as plastic particles less than 5 mm in size. Microplastics are generated due to the degradation of some larger plastic particles. Microplastics are considered as emerging pollutants that might be linked to numerous anthropogenic activities that could affect terrestrial as well as aquatic life. A major portion of these pollutants ends up in oceans. In recent years, microplastics have attracted greater scientific attention due to their ubiquitous nature and associated risks to human health and environmental systems. In the environment, these pollutants can be degraded through different processes including physical, chemical, and biological phenomenon. These degradation processes cause significant changes in their properties and characteristics such as size and morphology that in turn affects their fate and behavior in the environment. Although microplastics have gained scientific attention in recent years, many knowledge gaps still exist regarding their fate in the environment. Besides, some questions and influences remain unresolved regarding the exact mechanisms of how these emerging pollutants affect plants, animals as well as ocean life.

The field microplastics are receiving great current interest; however, development in this area will also pave the way for a breakthrough within industrial biotechnology (defined as the application of biotechnology for the processing and recycling of chemicals and materials). The technology development driven by the interest in microplastics will lead to experiences most valuable for recycling of other industrial biotechnology processes including chemical pollutants. The advances and scientific developments in the post-genomic era and achievements in systems biology will allow the necessary development and fine-tuning of the biological catalysts for a sustainable management of plastic wastes. There is a scientific consensus to take collective action to address the issue of microplastic pollution and work towards a more sustainable and plastic-free future.

So, this volume is a collection of 30 chapters contributed by experts from different parts of the world. The book was further divided into five parts. Part I overviews the general introduction of microplastic types, source, application, and its transportation from land to marine environment (Chaps. 1–5). Part II discusses different key sources of microplastic pollution (Chaps 6–9). Part III highlights some key reports of collection, identification, estimation, and removal of microplastics from the contaminated sites (Chaps. 10–16). Part IV documents the entry of microplastic in the food chain and its impact on terrestrial flora and fauna (Chaps. 17–25), and Part V enlists various microbial approaches to mitigate microplastic pollution (Chaps. 25–30).

Samba, India — Mohd. Shahnawaz
Iyamho, Nigeria — Charles Oluwaseun Adetunji
Zhenjiang, Jiangsu, China — Mudasir Ahmad Dar
Zhenjiang, Jiangsu, China — Daochen Zhu

Acknowledgment

To begin with the name of the Almighty Allah, the beneficent and the most merciful for enabling us to complete this task successfully. We want to thank all the authors that have contributed to the volume for their dedicated efforts and excellent contributions. We also wish to acknowledge the anonymous reviewers for their valuable input to enhance the quality of the book in the present form. We are highly thankful to the dedicated editorial team of the Springer Nature, especially Akanksha Tyagi, Ashok Kumar, and Yogitha Subramani for their kind support in completing this book project successfully. We hope that readers will enjoy the volume and get enlightened with recent developments related to microplastic pollution, its challenges, and currently available mitigation strategies.

Each editor is thankful to their parents, better halves, and kids for their kind support and encouragement.

About the Book

Microplastics is one of the emerging pollutants across the globe. Researchers predict that by the year 2050, the amount of microplastic particles in the waters will be more than the number of fishes. So, this book highlights the origin, sources, types, and transfer of the emerging pollutant, microplastics into different spheres of the ecosystem, resulting from anthropogenic activities are industrial manufacturing of everyday products. This book brings together a group of highly distinguished environmental researchers and experts who highlight the menace of microplastic pollution and provide thought-provoking discussions and solutions to mitigate the microplastics for a sustainable future. It provides an up-to-date exploration of microplastic contamination in the environment, advancement of the techniques and policies to combat plastics in general and microplastics in particular. Potential risks, toxicity, and harm due to microplastic pollution in different environmental settings are also explained. The authors provided a comprehensive account of the available techniques that can be employed for the detection, separation, and identification of microplastics for use by industry workers and scientists. Insightful discussions regarding the policy recommendations to minimize production, spread and to prevent harmful impacts of microplastics to fauna and flora are also provided for legislative bodies. This book will serve as a guide for the academicians, researchers, postgraduates, and undergraduates working in environmental science and microplastics. Further, it will also earn a place in the libraries of professionals working in plastics industries and environmental policy makers. Plastic design engineers, polymer chemists, material scientists, and ecologists will find this book, *Microplastic pollution* to be a vital resource to this critical industry.

Contents

Part I General Overview of Types, Source, Application and Transportation of Microplastic from Land to Marine Environment

1 **General Introduction of Microplastic: Uses, Types, and Generation** 3
John Tsado Mathew, Abel Inobeme, Babatunde Oluwafemi Adetuyi, Charles Oluwaseun Adetunji, Oluwabukola Atinuke Popoola, Falana Yetunde Olaitan, Olalekan Akinbo, Mohd. Shahnawaz, Oluwafemi Adebayo Oyewole, Eniola K.I.T., and Mohammed Bello Yerima

2 **Primary Microplastic: Source, Uses, Transportation from Land to Marine Environment** 23
Abel Inobeme, Mohd. Shahnawaz, Charles Oluwaseun Adetunji, John Tsado Mathew, Babatunde Oluwafemi Adetuyi, Oluwabukola Atinuke Popoola, Falana Yetunde Olaitan, Olalekan Akinbo, Olatunji Matthew Kolawole, Oluwafemi Adebayo Oyewole, Eniola K.I.T., and Mohammed Bello Yerima

3 **Sources, Uses and Transport of Secondary Microplastics from Land to Marine Environment** 35
Babatunde Oluwafemi Adetuyi, John Tsado Mathew, Abel Inobeme, Yetunde Olaitan Falana, Charles Oluwaseun Adetunji, Mohd Shahnawaz, Oluwafemi Adebayo Oyewole, Eniola K.I.T., and Mohammed Bello Yerima

4 **Usage of Microplastic Beads in Pharmaceuticals and Cosmetics Industry: A Review** 51
Pooja M. Patil, Pranjali Mahamuni-Badiger, Rachana R. Ingavale, Pratikshkumar R. Patel, and Maruti J. Dhanavade

5 **Application of Microplastics in Toiletry Products** 73
John Tsado Mathew, Abel Inobeme, Babatunde Oluwafemi Adetuyi, Yetude Olaitan Falana, Charles Oluwaseun Adetunji, and Mohd Shahnawaz

Part II Different Key Sources of Microplastic Pollution

6 **Wastewater Treatment Plants as Hotspots of Microplastic Pollution** . . . 87
Nilam A. Salunkhe-Patil, Pranjali Mahamuni-Badiger, Maruti J. Dhanavade, and Mudasir Ahmad Dar

7 **Microplastics Originating from Paints and Synthetic Textile Materials** . . . 109
Mufaro Moyo, Rivalani Baloyi Baloyi, Bruce Bishop Sithole, and Thabo Falayi

8 **Fishing Gears and Nets as a Source of Microplastic** . . . 127
Drishti Sharma, Raunak Dhanker, Bhawna, Ashi Tomar, Sazid Raza, and Anjali Sharma

9 **Tyres, Bitumen Wear and Plastic Bottles, Other Single Used Plastic as Major Sources of Microplastic** . . . 141
Babatunde Oluwafemi Adetuyi, John Tsado Mathew, Abel Inobeme, Yetude Olaitan Falana, Charles Oluwaseun Adetunji, Mohd Shahnawaz, Oluwafemi Adebayo Oyewole, K. I. T. Eniola, Mohammed Bello Yerima, and Oluwabukola Atinuke Popoola

Part III Reports on Collection, Identification, Estimation and Removal of Microplastics from the Contaminated Sites

10 **Reports on Collection, Identification and Estimation of Microplastic in Freshwater Bodies** . . . 163
Thabo Falayi, Harmony Musiyarira, Thandiwe Sithole, and Mufaro Moyo

11 **Collection, Detection, Identification, and Estimation of Microplastic from the Marine Environment: An Overview** . . . 175
Manisha K. Sangale, Rahul Kapoore, and Anthony Gravell

12 **Detection of Microplastic in Wastewater Treatment Plants** . . . 195
Enespa and Prem Chandra

13 **Report of Microplastic Presence in Drinking Water** . . . 225
Pankajkumar R. Waghmare and Mudasir A. Dar

14 **Detection of Microplastic in Seafood** . . . 245
Mubashir J. Mintoo and Ruqiya Kousar

15 **Reports of Microplastic in Human Blood** . . . 263
Mubashir J. Mintoo and Ruqiya Kousar

16 **Removal of Microplastic from Wastewater Treatment Plants** 271
Babatunde Oluwafemi Adetuyi, Charles Oluwaseun Adetunji, Peace Abiodun Olajide, Olubanke Olujoke Ogunlana, John Tsado Mathew, Abel Inobeme, Oluwabukola Atinuke Popoola, Falana Yetunde Olaitan, Olalekan Akinbo, Oluwafemi Adebayo Oyewole, Eniola K.I.T., and Mohammed Bello Yerima

Part IV Entry of Microplastic in the Food Chain and Its Toxicity and Impact on Terrestrial Flora and Fauna

17 **Entry of the Microplastics in Food Chain and Food Web** 289
Somia Liaqat, Mubashar Hussain, and Jaweria Riaz

18 **Insect Interactions with Microplastics Under Different Ecosystems** 307
Somia Liaqat, Mubashar Hussain, Kiran Aftab, Suleman Hussain Shah, and Muhammad Faheem Malik

19 **Microplastic as a Pathogenic Vector to Cause Diseases in Marine Biota** 321
Abel Inobeme, Mohd. Shahnawaz, Charles Oluwaseun Adetunji, John Tsado Mathew, Babatunde Oluwafemi Adetuyi, Oluwabukola Atinuke Popoola, Falana Yetunde Olaitan, Olalekan Akinbo, Olatunji Matthew Kolawole, Oluwafemi Adebayo Oyewole, Eniola K.I.T., and Mohammed Bello Yerima

20 **Microplastic to Accumulate Microbial Pathogens in the Terrestrial Environment** 351
Hetvi Kanojia, Harsh Purohit, Madhvi Joshi, Jignesh H. Kamdar, and Joydeep Chakraborty

21 **Impact of Microplastics on the Physiology of Benthos** 363
Mubashar Hussain, Dominique Blache, and Shane K. Maloney

22 **Microplastics and Microorganisms: A Evolutionary Symbiosis in Terrestrial Environment** 373
Rajalakshmi Sridharan and K. Veena Gayathri

23 **Insights of the Toxic Impact of Microplastic on Flora: Recent Trends and Challenges** 385
Neeraja P. Dhole, Mudasir A. Dar, Rohit R. Kulkarni, Abhishek G. Chaudhari, Mohd. Shahnawaz, and Radhakrishna S. Pandit

Part I

General Overview of Types, Source, Application and Transportation of Microplastic from Land to Marine Environment

General Introduction of Microplastic: Uses, Types, and Generation

1

John Tsado Mathew, Abel Inobeme, Babatunde Oluwafemi Adetuyi, Charles Oluwaseun Adetunji, Oluwabukola Atinuke Popoola, Falana Yetunde Olaitan, Olalekan Akinbo, Mohd. Shahnawaz, Oluwafemi Adebayo Oyewole, Eniola K.I.T., and Mohammed Bello Yerima

J. T. Mathew
Department of Chemistry, Ibrahim Badamasi Babangida University, Lapai, Niger State, Nigeria

A. Inobeme
Department of Chemistry, Edo University Iyamho, Auchi, Edo State, Nigeria

B. O. Adetuyi
Department of Natural Sciences, Biochemistry Unit, Faculty of Pure and Applied Sciences, Precious Cornerstone University, Ibadan, Nigeria

C. O. Adetunji (✉)
Applied Microbiology, Biotechnology and Nanotechnology Laboratory, Department of Microbiology, Edo State University Iyamho, Auchi, Edo State, Nigeria
e-mail: adetunji.charles@edouniversity.edu.ng

O. A. Popoola · F. Y. Olaitan
Plant Improvement Unit, Agricultural Biotechnology Department, National Biotechnology Development Agency, Abuja, Nigeria

O. Akinbo
Centre of Excellence in Science, Technology, and Innovation, AUDA-NEPAD, Pretoria, South Africa

M. Shahnawaz
Department of Botany, University of Ladakh, Kargil Campus, Khumbathang, Kargil, Ladakh UT, India

Present Address: Department of Botany, Shri Pandit Prem Nath Dogra Government Degree College Samba, Samba, Jammu and Kashmir UT, India

O. A. Oyewole
Department of Microbiology, Federal University of Technology, Minna, Nigeria

E. K.I.T.
Department of Biological Sciences, Joseph Ayo Babalola University, Ikeji Arakeji, Osun State, Nigeria

M. B. Yerima
Department of Microbiology, Sokoto State University, Sokoto, Sokoto State, Nigeria

M. Shahnawaz et al. (eds.), *Microplastic Pollution*,
https://doi.org/10.1007/978-981-99-8357-5_1

Abstract

The tiny plastic particles known as microplastics, which are smaller than 5 millimetres in size, have raised environmental concerns on a global scale as a result of their wide spread and potential ecological impacts. They are generated through various sources and can be either primary microplastic or secondary microplastics depending on the nature of their generation. Primary microplastics produced intentionally and are used in industries such as cosmetics, personal care products, and textiles. Synthetic fibers from clothing, like polyester and nylon, are another significant source of primary microplastics, and through washing machines and wastewater treatment plants, it enters into the environment directly. However, due to various environmental forces, the large plastic-based products breakdown to generate secondary microplastics. All types of microplastics pose significant environmental concerns due to their persistence, mobility, and potential to accumulate in ecosystems. Microplastic also enters into the food chain and leads to potential health risks for both wildlife and humans. Efforts are being made globally to mitigate the generation and release of microplastics. Initiatives to improve waste management, increase recycling rates, and promote sustainable alternatives to plastic are also being implemented. Therefore, microplastics are small plastic particles that are generated through various sources, including intentional production and the breakdown of larger plastic items. As awareness grows, there is an increasing focus on reducing microplastic generation and developing sustainable solutions to tackle this pressing issue. In the present chapter, an attempt was made to overview the uses, types, and generation of microplastic to understand the uptake, prevalence, and fate of microplastics in different environments.

Keywords

Accumulate · Cosmetics · Environmental · Microplastic · Wastewater treatment

1.1 Introduction

Less than 5 millimeter-sized plastic fragments are known as microplastics. Based on their mode of origin, they are referred primary microplastic (microbeads or plastic pellets) and secondary microplastic (fragmented large plastic products). Due to their extensive prevalence in diverse ecosystems, such as seas, rivers, soil, and even the air, microplastics have grown to be a major environmental problem on a worldwide scale (Lamichhane et al. 2023). The generation and accumulation of microplastics are fueled by several factors. One significant contributor is improper waste management and inadequate recycling practices, which results in the entry of plastic waste into the environment. Additionally, activities like plastic manufacturing, industrial processes, and everyday use of plastic products contribute to the generation of microplastics.

The ecosystem and living things are greatly endangered by microplastics. Long-lasting in the environment, they have the ability to build up in ecosystems, particularly marine food webs. Marine organisms, such as fish, seabirds, and marine mammals, can mistakenly ingest microplastics, leading to physical harm, digestive issues, and potential transfer of microplastics up the food chain. Furthermore, microplastics can absorb and transport toxic chemicals and pollutants, making them a vector for the spread of toxic substances in the environment. There are worries about the possible transmission of microplastics and related substances through the food we eat and the air we breathe, while research on the long-term effects of microplastics on human health is currently ongoing.

Efforts are underway to address the issue of microplastic pollution. These include improving waste management infrastructure, promoting recycling and circular economy initiatives, developing sustainable alternatives to plastic, and raising awareness about the environmental impacts of microplastics. Understanding the sources, distribution, and effects of microplastics is crucial in implementing effective strategies to reduce their generation and mitigate their ecological and human health risks. By adopting a holistic approach that combines scientific research, policy interventions, and individual actions, we can work towards minimizing the presence of microplastics in our environment and preserving the health and integrity of our ecosystems.

1.2 Types of Microplastics

Microplastics are found in various components of the environment. There are various types of microplastics, briefly discussed as follows:

1.2.1 Primary Microplastics

Microplastic that are produced intentionally for various application in drugs, toiletry items, cosmetics etc. Various types of primary microplastic are discussed briefly as follows:

1.2.1.1 Microbeads

Microbeads are a prime example of primary microplastics. These are tiny plastic particles intentionally manufactured and added to various consumer products for specific purposes, particularly in personal care and cosmetic products. Microbeads are typically made from materials such as polyethylene (PE), polypropylene (PP), or polyethylene terephthalate (PET). They are commonly used as exfoliants, abrasives, or colorants in products like facial scrubs, body washes, toothpaste, and cosmetics. The small size of microbeads, typically less than 1 millimeter, allows for gentle exfoliation and provides a smooth texture in these products. However, due to their small size, microbeads are not effectively filtered out by wastewater treatment plants and can end up in aquatic environments. They are often released into rivers, lakes, and oceans through sewage systems, where they can persist for long periods.

Microbeads pose significant ecological risks when they enter water bodies. Marine organisms, including fish, shellfish, and plankton, can mistake microbeads for food and ingest them. This ingestion can lead to physical harm, digestive issues, and potential bioaccumulation of microplastics in the food chain. Moreover, microbeads have been found in various aquatic habitats, including the Arctic and remote ocean regions, indicating the widespread distribution and potential long-range transport of these primary microplastics. In response to the environmental concerns associated with microbeads, many countries have implemented legislation to ban or restrict their use in personal care products. To reduce the production and release of primary microplastics into the environment and lessen their ecological impact, it is crucial to regulate microbeads and create sustainable alternatives.

1.2.1.2 Synthetic Fibers

Synthetic fibers are another example of primary microplastics. They are produced from synthetic polymers like nylon, polyester, and acrylic and are widely used in clothing, textiles, and carpets. During the washing of synthetic clothing, small fibers called microfibers are released into the environment. These microfibers are too small to be captured by wastewater treatment plants and can enter aquatic ecosystems through wastewater discharge. They are also found in the air as airborne microplastics and can be ingested by humans and animals. Synthetic microfibers pose a risk to aquatic organisms, as they can be ingested and accumulate in their digestive systems. The accumulation of microfibers in aquatic food webs can lead to bioaccumulation and biomagnification of microplastics in higher trophic levels.

Furthermore, the production and disposal of synthetic fibers contribute to the release of other types of microplastics, such as microbeads and fragmented plastics, into the environment. As synthetic fibers are not biodegradable, they persist in the environment for extended periods. Efforts are underway to address the issue of synthetic microfibers and reduce their impact. Initiatives such as the development of washing machine filters and the use of sustainable materials in clothing production aim to reduce the release of synthetic microfibers into the environment. Additionally, responsible disposal of synthetic clothing and recycling initiatives can minimize the accumulation of synthetic fibers in the environment. Synthetic fibers are an example of primary microplastics that contribute to the release of microfibers and other types of microplastics into the environment. Reducing the generation and release of synthetic fibers and promoting sustainable alternatives are essential steps toward mitigating the impact of microplastics on the environment.

1.2.1.3 Plastic Pellets

Plastic pellets may leak into the environment as a result of unintentional spills and improper handling during the processes of production, transportation, and handling. Plastic pellet pollution can occur at manufacturing facilities, during transportation, and at storage facilities, especially near coastal areas and waterways.

Once released, plastic pellets can enter water bodies through stormwater runoff or direct discharge, and they can be transported by currents and wind, contributing to marine pollution. These pellets are buoyant and can travel long distances, posing a threat to marine life and ecosystems. Plastic pellets are a problem because marine animals could mistake them for food and eat them, which could be harmful.

Moreover, they can attract and accumulate other pollutants, such as chemicals and heavy metals, which can further impact the environment and marine organisms. Efforts to address plastic pellet pollution include implementing best management practices at manufacturing facilities to prevent pellet spills and improve containment systems. In order to reduce the emission of plastic pellets into the environment, industry programmes such as Operation Clean Sweep promote proper pellet handling and disposal. To control the plastic pellet pollution and increase the sustainability of plastic production methods, the plastic industry is implementing regulatory measures and voluntary pledges. These actions aim to prevent the generation of primary microplastics like plastic pellets and mitigate their impact on marine ecosystems and wildlife. It is crucial to remember that primary microplastics are a significant concern due to their intentional use and their potential for direct release into the environment. More sustainable alternatives are being developed in an effort to decrease or eliminate the usage of primary microplastics, particularly in products like personal care items.

1.2.2 Secondary Microplastics

Secondary microplastics are a subset of microplastics that are created over time as a result of the deterioration and fragmentation of larger plastic objects. Secondary microplastics are also of various types and are discussed briefly as follows:

1.2.2.1 Fragmented Plastics

Fragmented plastics are a prime example of secondary microplastics. The process of fragmentation occurs as these plastic items are exposed to stressors like wave action, wind, and collisions with other objects. As a result, the larger plastics disintegrate into smaller and smaller pieces until they reach microplastic size (less than 5 millimetres). Numerous habitats include these broken-down polymers, including water bodies, soil, and even the air. Additionally, fragmented plastics can also accumulate in terrestrial ecosystems, affecting soil quality and potentially entering the food chain through soil-dwelling organisms. Efforts to address the issue of fragmented plastics as secondary microplastics include implementing better waste management practices, promoting recycling and circular economy initiatives, and raising awareness about the environmental impacts of plastic pollution. Additionally, reducing plastic consumption and transitioning to more sustainable alternatives are crucial steps in mitigating the generation of fragmented plastics and the associated microplastic pollution.

1.2.2.2 Weathered Plastics

Weathered plastics are another example of secondary microplastics. Weathering refers to the physical and chemical degradation of plastic materials over time due to the exposure to environmental conditions. From bigger plastic products, this process may result in the production of microplastics. When plastics are exposed to sunlight, temperature fluctuations, moisture, and other environmental factors, they undergo weathering processes. These processes cause the plastics to become brittle, breakable, and more susceptible to fragmentation. As a result, the larger plastic items gradually deteriorate and fragment into smaller pieces, including microplastics.

In marine environments, weathered plastics are particularly concerning. Ocean waves, currents, and UV radiation contribute to the degradation of plastic debris, such as bottles, bags, and fishing gear. Over time, these weathered plastics break down into microplastics, which can persist in marine ecosystems and potentially harm marine organisms. Weathered plastics can also be found in terrestrial environments. Plastics exposed to sunlight and other weathering agents in outdoor settings, such as plastic furniture, construction materials, and discarded plastic items, can undergo degradation and fragmentation, ultimately generating microplastics (Napper and Thompson 2016).

The presence of weathered plastics as secondary microplastics highlights the need for improved waste management, plastic recycling, and reduced plastic consumption. Promoting eco-friendly alternatives to plastic and spreading knowledge about the negative effects of weathered plastics on the environment can also help reduce the production of microplastics and safeguard ecosystems from their pollution.

1.2.2.3 Plastic Film Degradation

Plastic film degradation is indeed an example of secondary microplastics. Plastic films, such as those used in packaging materials like food wraps, can undergo degradation processes over time, leading to the formation of microplastics. Plastic films are typically made from polymers like polyethylene (PE) or polypropylene (PP), which are known for their durability. However, prolonged exposure to environmental factors, such as sunlight, temperature variations, and mechanical stress, can cause these films to break down. During the degradation process, plastic films become brittle and develop cracks and fissures. These physical changes make them more susceptible to fragmentation. Over time, the degraded plastic films can further deteriorate into smaller and smaller pieces, eventually becoming microplastics (Inobeme et al. 2023).

The presence of microplastics originating from plastic film degradation is a concern, as these particles can enter various ecosystems. They can be carried by wind or water and end up in lakes, rivers, oceans, and even on land. The impact of plastic film-based microplastics on the environment is still being studied, but it is known that they can pose risks to wildlife. Marine organisms, for example, may mistake the microplastics for food, leading to ingestion and potential harm. To address the problem of degrading plastic film and the resulting microplastics, it is important to promote responsible plastic waste management, including recycling and proper disposal. Additionally, exploring sustainable alternatives to plastic films, such as biodegradable or compostable materials, can help reduce the generation of plastic film microplastics and minimize their environmental impact (Lamichhane et al. 2023).

1.2.2.4 Paint and Coatings

Paint and coatings can indeed contribute to the generation of secondary microplastics. When plastic-based paints and coatings degrade over time, they can release tiny plastic particles into the environment, which become microplastics. Many paints and coatings contain plastic polymers as binders or additives to provide durability, flexibility, and other desirable properties. However, these plastic-based components can undergo degradation through the exposure to environmental factors such as sunlight, temperature fluctuations, moisture, and chemical reactions. As the paint or coating deteriorates, microscopic plastic particles can be released into the surroundings. These particles can become airborne, settle on surfaces, or be washed off into water bodies through rainfall or other water sources. Over time, these plastic particles can break down further, becoming microplastics. The presence of microplastics derived from paint and coating degradation can have environmental implications. They can contaminate soil, water bodies, and the air, potentially impacting ecosystems and organisms. Although the specific effects are still being studied, microplastics have been found to have the potential to harm marine life, enter the food chain, and accumulate in various environments. To address the generation of microplastics from paint and coating degradation, it is important to consider sustainable alternatives and practices. In order to reduce the release of microplastics into the environment, this involves encouraging the use of water-based and environmentally friendly paints and coatings, properly disposing of paint waste, and putting into practise responsible paint recycling or disposal techniques.

1.2.2.5 Tire Wear Particles

Tire wear particles are a significant example of secondary microplastics. When vehicles are driven, the friction between the tires and the road surface causes the gradual wear and tear of the tire tread. As a result, tiny rubber particles containing plastic polymers are released into the environment as airborne and waterborne microplastics. Tire wear particles are composed of rubber and synthetic polymers, such as styrene-butadiene rubber (SBR) and other additives used in tire manufacturing. These particles are typically in the size range of microplastics and can persist in the environment for extended periods (Wu et al. 2023). Road runoff, rainfall, and wind can transport tire wear particles from roads to various aquatic ecosystems, including rivers, lakes, and oceans. These particles can also accumulate in urban environments, such as on streets, sidewalks, and green spaces. Over time, they can enter water bodies through stormwater runoff and ultimately contribute to microplastic pollution. Due to their possible effects on the environment, tire wear particles that are secondary microplastics are a cause for concern. They can accumulate in sediments, contaminate water sources, and affect aquatic organisms. Additionally, tire wear particles can enter the food chain when swallowed by aquatic organisms, potentially causing microplastics to bioabsorb and biomagnify in higher trophic levels. Efforts are underway to address the issue of tire wear particles and reduce their impact. Research is being conducted to better understand their behavior, develop sustainable tire materials, and improve road infrastructure. Additionally, initiatives promoting responsible tire disposal, tire recycling, and the development of innovative solutions to minimize tire wear and tear are being pursued to mitigate the entry of tire wear microplastic into the environment.

The ubiquitous occurrence of secondary microplastics in the environment and their tendency to build up in ecosystems make them a particular source of worry. The fragmentation process increases the availability of microplastics in various habitats, affecting marine and terrestrial organisms and potentially entering the food chain. There are initiatives underway to address secondary microplastics' origins and create plans to lessen their production and discharge into the environment.

1.3 Sources of Microplastic

Plastics, which are affordable and light-weight synthetic polymers, have become ubiquitous due to their use in a wide range of applications. According to estimates, the production of plastics is projected to reach 445.25 million metric tons by 2025 (Chandel 2022). Plastic is a material used in almost every aspect of modern life, including single-use items such as food packs, cups, to wrap and sell consumables, sachet water bags (Dumbili and Henderson 2020). This has led to a significant portion of plastic products being released into the environment each year (Adeogun et al. 2020; Borrelle et al. 2020). Due to its synthetic nature, it takes a long time to degrade and when they do, we have microplastic being created (Dong et al. 2023), also some are assiduously created to serve a purpose (Mikac et al. 2023). Furthermore, due to the wide use of plastics, we have them in the environment such as air, water, soil, etc. (Apetogbor et al. 2023; O'Brien et al. 2023; Wu et al. 2023).

There are numerous sources of microplastics which include synthetic textiles, personal health-care products, city dust, marine coating, road markings, engineered plastic pellets, and tire. Within the ocean and marine environment, basic consumer products are the major sources of primary microplastics.

1.3.1 Synthetic Textiles

The major sources or microplastics around the ocean are the synthetic textiles. These textiles products contribute up to 35% of the overall volume of microplastics. Some of the fabric contents in cloth include acrylic, polyester, and nylon which also contribute to microplastic pollution. Engineered microplastics are released during washing of numerous fabric materials through shedding and abrasion. Although natural fibres also contribute, they however biodegrade unlike the synthetic.

1.3.2 Tires

Matrix of synthetic polymers is usually formed from microplastics. Over 24% of tire are made of synthetic rubber, natural rubber, and plastic polymers. Microplastics make up a matrix of polymers of synthetic origin, which account for the traction and rigidity of the tire. When tire come in contact with road surface, there is friction, thereby resulting to the eroding of the tire releasing microplastics to the environment.

1.3.3 City Dust

This makes up about 24% of microplastics within the ocean environment and this emerges from numerous sources (Thiele et al. 2021). City dust are obtained from man-made products through abrasion and weathering. Weathering of plastics materials from various products contributes to microplastics in the environment. Numerous microplastics have been documented per square meter depositing daily from sky in some areas.

1.3.4 Road Markings

Microplastics are also formed from road marking. The substances used for road marking during construction include paint and polymer tape. The microplastics are lost during weathering and abrasion. Debris emerging from roads that have degraded over a long period of time contributes to microplastic in water bodies.

1.3.5 Personal Health-Care Products

Microbeads, unique type of microplastics, are found in most personal health-care products. Some of the products include shower gels, creams, and scrubbing agents. Even though the manufacturing of some of these products have been banned in some countries, they are still being produced by some industries.

1.4 Distribution and Specific Environmental Sources of Microplastics

1.4.1 Microplastic in Water

According to estimates, 1.3 million tonnes of MPs have been entering the oceans annually, and by 2040, that number is expected to increase to three million tonnes (Tang 2023). The concentration of MPs in the ocean and lakes depends on the population of a specific area. MPs reach the aquatic environment via avenues such as

Fig. 1.1 Microplastics (NOAA 2023)

direct intake, or via terrestrial or atmospheric sources (Fig.1.1). There are several ways that plastic debris gets into the oceans, and it keeps accumulating in marine settings (Apetogbor et al. 2023). As a result, towns with extensive coastlines experience greater marine plastic pollution. MPs can enter freshwater sources through ways including littering in the surrounding areas as well as hydrological and meteorological conditions (Atugoda et al. 2023). Large plastic waste may also mechanically disintegrate as a result of environmental conditions such UV light, wind, and waves are sources of MPs into freshwater (Viljoen et al. 2023). The sort of polymers that predominate in the world's water are fragments of big plastic trash (Eriksen et al. 2013; Dai et al. 2018).

1.4.2 Microplastic in Air

Microplastics are also found in the air; it has been recently reported that MP can be moved from the ocean, land to the atmosphere (O'Brien et al. 2023). Microplastics are found in various parts of the environment (Fig. 1.2). They can be done through wind abrasion, humans; with the wear of synthetic-made cloth (Forster et al. 2023), the interface of air and water (O'Brien et al. 2023). Plastics are non-biodegradable materials, when it has achieved its long lifespan, weathering happens to it, photodegradation and becomes MP, and wind can carry them and thus be found in air. A study investigated the transfer of plastics particles to the air-water interface through bubble bursting and reported that polystyrene of diameter 600 nm or in groups of three coated in a sulphur atom can be found in the atmosphere (Masry et al. 2021). Wind erosion is another source of MP in the air, MP can be transferred by the wind when the plastics have been degraded to MP of minute size by sunlight,

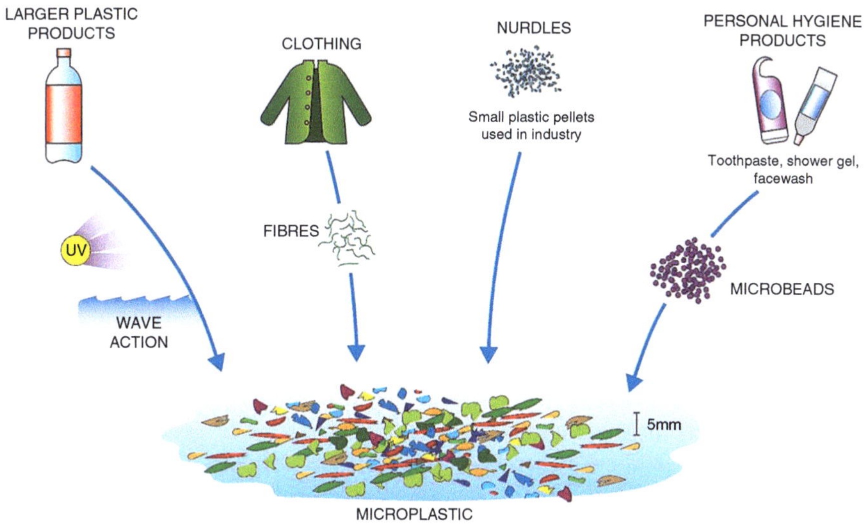

Fig. 1.2 Sources of microplastics (Anonymous 2023)

UV. Calculated that MP carried from landfill sites by wind abrasion to be 0.16 microplastics per ton of waste per hour. Additionally, MP can be transferred from soil. In Iran, it was reported that during a dust storm, about 5% of atmospheric microplastics can be transferred from the soil (Brahney et al. 2021; Abbasi et al. 2022).

Another source of MP in the air is through tire wear on the road surface, as well as the use of synthetic rubber and copolymers such as; styrene-butadiene, road marking paints, and bitumen formulations which are all sources of MPs. These particles can be released into the environment as dust and can eventually be suspended in the air. The amount of MPs released from tire wear is estimated to be significant, with some studies suggesting that it may be a greater source of MPs in the environment on the road and hereby can be suspended into the air (Goßmann et al. 2022; Kitahara and Nakata 2020). When clothes are used, cleaned, and dried, fibres easily rip from them (Napper et al. 2015).

1.4.3 Microplastics in Soil

There have been reports of microplastics in plants (Ebere et al. 2019). The MPs enter the soil through a number of different channels, including organic fertilisers that are given to land, agricultural plastic film from plastic mulching (used to suppress weeds and improve water absorption effectiveness), plastic waste, water irrigation, and atmospheric deposition of airborne microplastics (Ebere et al. 2019; Enyoh et al. 2019). The investigation of cattle on feeds where plastic mulching was used revealed that sheep ate plastic mulch or bags trapped in the vegetation and their stools contained MPs (Beriot et al. 2021; (Shafea et al. 2023).

1.4.4 Microplastics in Food and Food Items

The growing worry about microplastics worldwide, a large number of studies evaluating microplastics have been conducted (Wang and Wang 2018; Chai et al. 2020). Studies on the potential risks of MPs in our food have been conducted, and it has been stated that MPs can be found in a variety of our food products, including honey (Liebezeit and Liebezeit 2013), milk (Kutralam-Muniasamy et al. 2020), air (Dris et al. 2016), beer (Kosuth et al. 2018), drinking water (Mason et al. 2018), and table salt. These studies have demonstrated that we consume MPs by drinking water from the plastic bottles, using salts enclosed in plastic bags, breathe in air, and also pass MPs to our children. It has been observed that the MPs in table salts vary depending on the area due to their diagnostic source, such as lake salt, rock salt, and sea salt, suggesting that MPs may be added during collection, transportation, drying, or packing procedures before distribution using various analytical methods (Yang et al. 2015).

1.5 Uses of Microplastics

MPs are plastics measuring less than 5 millimetres in diameter (Inobeme et al. 2023). They can be intentionally created for specific purposes or can be as a result of degradation of larger plastic materials (Dong et al. 2023; Kalali et al. 2023). The intentionally created MPs used for many purposes, microbeads are MPs used as parts of materials during the manufacture of toothpastes, creams, bubble bath, sunscreen, shampoo, and facial scrubs (Inobeme et al. 2023; Zhou et al. 2023). Due to the durability, affordability of plastics, microbeads are in high demand and are widely produced (Zhou et al. 2023). In addition, exfoliating agents are often added to enhance their ability to remove dead skin cells and other imperfections without causing any damage to the skin's surface (Priya et al. 2023). Microplastics are also used as additives in paints to improve durability, enhance the color and increase the texture (Egea-Corbacho et al. 2023). They can be as microbeads, microfibers, and other small particles (Zhou et al. 2023).

Another potential use of MPs is as an adsorbent for pollutants (Tian et al. 2023). Due to their have large surface area and hydrophobicity, MPs serve as a vector for so many pollutants (Chen et al. 2020). A study demonstrated this by using polystyrene (PS) MPs to adsorb aniline, an environmental toxin (Chen et al. 2020). However, PS is a type of plastic readily available in the environment with large surface area making them more efficient in adsorbing benzene-rings containing organic pollutants. The adsorption process was augmented by some properties like hydrogen bonds, van der Waals forces, interaction between the PS MPs and the aniline and functional groups (Tian et al. 2023). The adsorption capacity was measured to be

0.060 mg/g after 24 h and Freundlich models described the adsorption effectively (Z. Chen et al. 2023).

Furthermore, the adsorption of these pollutants on MPs has been triggered by some heavy metals, MPs was used to adsorb tylosin and cadmium, the researchers found out after PS has formed a complex with Cd, the adsorption of tylosin became enhanced (Lv et al. 2023). Ciprofloxacin was also adsorbed on polyethylene MPs and it was pointed out that the adsorption was physical and multilayered (Lv et al. 2023).

A study compared the adsorption of tetracycline on a virgin MP and a photoaged polyvinyl chloride (PVC), PVC was chosen due to its brittleness and decomposition rate (Tian et al. 2023). Photoaged PVC MPs adsorbed more than the virgin MP because of the morphology of the aged PVC MPs, it was noted that the aging created more energy adsorption sites/pores on the PVC MPs (Tian et al. 2023).

1.6 Treatment of Microplastic

Microplastics have been treated using a variety of techniques. Some of the techniques are discussed as follows:

1.6.1 Collection Devices

Collection devices that can be placed closed to coasts could help in the removal of up to 31% of microplastics present in the area. There are systems that make use of natural currents from the ocean in the concentration of plastics and debris materials which are then extracted into vessels.

1.6.2 Microbial Approach

There are bacteria that have been genetically modified and adapted for eating up plastics. Some of these organisms, other than breaking down the plastics, are structured in a novel mode for capturing these plastics materials from the contaminated samples within their biofilm's matrix, they are released and recovered.

1.6.3 Filtration

Stormwater or waste water collecting systems have the capacity to capture a variety of microplastics, which are then transported to treatment facilities where they eventually turn up in the sludge the facilities create. The applications of Ferrofluid currently being designed makes it possible for ease removal of microplastics using filtration.

1.7 Environmental and Health Impacts of Microplastics

MPs are plastic particles that have a diameter of less than 5 mm and are created either on purpose for a particular use or as a result of the deterioration of bigger plastic objects (Mikac et al. 2023). Plastic production has been increasing globally, it has been projected to reach 445.25 million metric tons in 2025 (Chandel 2022). These plastics are synthetic, non-biodegradable, so it takes a prolonged time before its degradation naturally (Dong et al. 2023), making us to have presently a great volume of plastic wastes, which are distributed across the ecosystem (Inobeme et al. 2023). MPs can have several health effects on both humans and wildlife. MPs accumulate in the body and cause harm. Some studies have investigated several MPs in the human body; it was pointed out that it can cause inflammation and oxidative stress causing various diseases. MPs have also been speculated to cause damage to the liver, kidneys, reproductive organs (Sharma et al. 2023).

1.8 Uptake, Prevalence, and Destiny of Microplastics and Their Effects on Plants

Plastic particles of various types have been discovered in soil as a result of diverse agricultural practices carried out in the soil (Rillig et al. 2021). Furthermore, plastic films are commonly utilized to reduce water usage and inhibit weed growth during crop cultivation (Huang et al. 2022). MPs can also cause soil surfaces to become dry by increasing air circulation and can also become entangled with essential plant nutrients, thereby reducing plant growth (Rillig et al. 2021).

Due to the generation of reactive oxygen species during the inflammatory response, microplastics produce oxidative stress (Kutralam-Muniasamy et al. 2023). Reactive oxygen species (ROS) are produced in high concentrations in the brain and nervous system after exposure to 0.05–10 mg L^{-1} of PE and PS, which contributes to cytotoxicity. Particles smaller than 240 nm have been shown to be able to penetrate past the placental barrier in blood flowing through placental tissues (Wick et al. 2010); internalisation of negatively charged PS of size 1 m has also been discovered in cell cultures exposed to it (Zauner et al. 2001). As a result of their existence, there may be clotting, swelling, and redness, pulmonary hypertension (Zagorski et al. 2003), vascular occlusions (Jones et al. 2003), and blood cell cytotoxicity. Additionally, microplastics can affect neural function and behaviour, according to in vivo toxicity tests performed on people (Zhu et al. 2023).

MPs have been detected in human body, a study discovered MPs in both mothers and foetus. A study pointed out the effect of perfluoroalkyl and polyfluoroalkyl substances on the female reproductive organs, the ovary was exposed to these plastic particles, and it was discovered that the plastic particles is disrupted the endocrine organs (Du et al. 2023).

1.9 Conclusion and Future Perspective

This chapter gives a general overview of the various types, sources, and uses of microplastic. Microplastic is an international environmental hazard since it is embedded in both land and marine ecosystems. They clearly represent a major threat to the creatures in this area, which may extend to people. Having a broad understanding of this will allow you to provide solutions for reducing toxicological and hazard effects on humans (Inobeme et al. 2023b; Adetunji et al. 2021a,b,c; Adetunji et al. 2020; Adetunji 2019; Adetunji et al. 2019; Adetunji and Anani 2021; Anani et al. 2020; Anani et al. 2022; Adetunji et al. 2022b). These polymers will breakdown into microplastics and nanoplastics over time. As a result, efforts must be taken to reduce the impact of microplastics by taking a careful look at their sources, kinds, and dangerous consequences, educating people about them, and developing technologies and other effective methods to remove microplastics from the environment.

References

Abbasi S, Alirezazadeh M, Razeghi N, Rezaei M, Pourmahmood H, Dehbandi R, Mehr MR, Ashayeri SY, Oleszczuk P, Turner A (2022) Microplastics captured by snowfall: a study in northern Iran. Sci Total Environ 822:153451

Adeogun AO, Ibor OR, Khan EA, Chukwuka AV, Omogbemi ED, Arukwe A (2020) Detection and occurrence of microplastics in the stomach of commercial fish species from a municipal water supply lake in southwestern Nigeria. Environ Sci Pollut Res 27:1–11

Adetunji CO (2019) Environmental impact and ecotoxicological influence of biofabricated and inorganic nanoparticle on soil activity. In: Panpatte D, Jhala Y (eds) Nanotechnology for agriculture. Springer, Singapore. https://doi.org/10.1007/978-981-32-9370-0_12

Adetunji CO, Anani OA (2021) Plastic-eating microorganisms: recent biotechnological techniques for recycling of plastic. In: Panpatte DG, Jhala YK (eds) Microbial rejuvenation of polluted environment. microorganisms for sustainability, vol 25. Springer, Singapore. https://doi.org/10.1007/978-981-15-7447-4_14

Adetunji CO, Panpatte DG, Bello OM, Adekoya MA (2019) Application of nanoengineered metabolites from beneficial and eco-friendly microorganisms as a biological control agents for plant pests and pathogens. In: Panpatte D, Jhala Y (eds) Nanotechnology for agriculture: crop production & protection. Springer, Singapore. https://doi.org/10.1007/978-981-32-9374-8_13

Adetunji CO, Oloke JK, Prasad G (2020) Effect of carbon-to-nitrogen ratio on eco-friendly mycoherbicide activity from *Lasiodiplodia pseudotheobromae* C1136 for sustainable weeds management in organic agriculture. *Environ Dev Sustain* 22:1977–1990. https://doi.org/10.1007/s10668-018-0273-1

Adetunji CO, Ajayi OO, Akram M, Olaniyan OT, Chishti MA, Inobeme A, Olaniyan S, Adetunji JB, Olaniyan M, Awotunde SO (2021a) Medicinal plants used in the treatment of influenza A virus infections. In: Dua K, Nammi S, Chang D, Chellappan DK, Gupta G, Collet T (eds) Medicinal plants for lung diseases. Springer, Singapore. https://doi.org/10.1007/978-981-33-6850-7_19

Adetunji CO, Akram M, Olaniyan OT, Ajayi OO, Inobeme A, Olaniyan S, Hameed L, Adetunji JB (2021b) Targeting SARS-CoV-2 novel corona (COVID-19) virus infection using medicinal plants. In: Dua K, Nammi S, Chang D, Chellappan DK, Gupta G, Collet T (eds) Medicinal plants for lung diseases. Springer, Singapore. https://doi.org/10.1007/978-981-33-6850-7_21

Adetunji CO, Inobeme A, Olaniyan OT, Ajayi OO, Olaniyan S, Adetunji JB (2021c) Application of nanodrugs derived from active metabolites of medicinal plants for the treatment of inflammatory and lung diseases: recent advances. In: Dua K, Nammi S, Chang D, Chellappan DK, Gupta G, Collet T (eds) Medicinal plants for lung diseases. Springer, Singapore. https://doi.org/10.1007/978-981-33-6850-7_26

Adetunji CO, Oloke JK, Osemwegie OOC, Ehis-Eriakha B (2022b) Use of agro-wastes for *Lasiodiplodia pseudotheobromae* (C1136) production with sustainable bioefficacy. *Environ Dev Sustain* **24**:7794–7809. https://doi.org/10.1007/s10668-021-01758-6

Anani OA, Mishra RR, Mishra P, Olomukoro JO, Imoobe TOT, Adetunji CO (2020) Influence of heavy metal on food security: recent advances. In: Mishra P, Mishra RR, Adetunji CO (eds) Innovations in food technology. Springer, Singapore. https://doi.org/10.1007/978-981-15-6121-4_18

Anani OA, Adetunji CO, Anani GA, Olomukoro JO, Imoobe TOT, Enenuku AA, Tongo I (2022) Effect of meso-, micro-, and nano-plastic waste on the benthos. In: Shahnawaz M, Sangale MK, Daochen Z, Ade AB (eds) Impact of plastic waste on the marine biota. Springer, Singapore. https://doi.org/10.1007/978-981-16-5403-9_12

Anonymous (2023) Where do microplastics come from? https://encounteredu.com/multimedia/images/sources-of-microplastics. Accessed 19 June 2023

Apetogbor K, Pereao O, Sparks C, Opeolu B (2023) Spatio-temporal distribution of microplastics in water and sediment samples of the Plankenburg river, Western Cape, South Africa. Environ Pollut 323:121303

Atugoda T, Piyumali H, Wijesekara H, Sonne C, Lam SS, Mahatantila K, Vithanage M (2023) Nanoplastic occurrence, transformation and toxicity: a review. Environ Chem Lett 21(1):363–381

Beriot N, Peek J, Zornoza R, Geissen V, Lwanga EH (2021) Low density-microplastics detected in sheep faeces and soil: a case study from the intensive vegetable farming in Southeast Spain. Sci Total Environ 755:142653

Borrelle SB, Ringma J, Law KL, Monnahan CC, Lebreton L, McGivern A, Murphy E, Jambeck J, Leonard GH, Hilleary MA, Eriksen M (2020) Predicted growth in plastic waste exceeds efforts to mitigate plastic pollution. Science 369(6510):1515–1518

Brahney J, Mahowald N, Prank M, Cornwell G, Klimont Z, Matsui H, Prather KA (2021) Constraining the atmospheric limb of the plastic cycle. Proc Natl Acad Sci 118(16):e2020719118

Chai Y, Gao N, Wang M, Wu C (2020) H2 production from co-pyrolysis/gasification of waste plastics and biomass under novel catalyst Ni-CaO-C. Chem Eng J 382:122947

Chandel TA (2022) Green entrepreneurship and sustainable development international perspectives on value creation and sustainability through social entrepreneurship. IGI Global, pp 173–208

Chen G, Feng Q, Wang J (2020) Mini-review of microplastics in the atmosphere and their risks to humans. Sci Total Environ 703:135504

Chen Z, Yang J, Huang D, Wang S, Jiang K, Sun W, Chen Z, Cao Z, Ren Y, Wang Q, Liu H (2023) Adsorption behavior of aniline pollutant on polystyrene microplastics. Chemosphere 323:138187

Dai Z, Zhang H, Zhou Q, Tian Y, Chen T, Tu C, Fu C, Luo Y (2018) Occurrence of microplastics in the water column and sediment in an inland sea affected by intensive anthropogenic activities. Environ Pollut 242:1557–1565

Dong X, Zhu L, He Y, Li C, Li D (2023) Salinity significantly reduces plastic-degrading bacteria from rivers to oceans. J Hazard Mater 451:131125

Dris R, Gasperi J, Saad M, Mirande C, Tassin B (2016) Synthetic fibers in atmospheric fallout: a source of microplastics in the environment? Mar Pollut Bull 104(1–2):290–293

Du M, Pu Q, Li X, Yang H, Hao N, Li Q, Zhao Y, Li Y (2023) Perfluoroalkyl and polyfluoroalkyl substances (PFAS) adsorbed on microplastics in drinking water: implications for female exposure, reproductive health risk and its mitigation strategies through in silico methods. J Clean Prod 391:136191

Dumbili E, Henderson L (2020) The challenge of plastic pollution in Nigeria plastic waste and recycling. Elsevier, pp 569–583

Ebere EC, Wirnkor VA, Ngozi VE (2019) Uptake of microplastics by plant: a reason to worry or to be happy? World Sci News 131:256–267

Egea-Corbacho A, Martín-García AP, Franco AA, Quiroga JM, Andreasen RR, Jørgensen MK, Christensen ML (2023) Occurrence, identification and removal of microplastics in a wastewater treatment plant compared to an advanced MBR technology: full-scale pilot plant. J Environ Chem Eng 11(3):109644

Enyoh CE, Verla AW, Verla EN, Ibe FC, Amaobi CE (2019) Airborne microplastics: a review study on method for analysis, occurrence, movement and risks. Environ Monit Assess 191(11):1–17

Eriksen M, Mason S, Wilson S, Box C, Zellers A, Edwards W, Farley H, Amato S (2013) Microplastic pollution in the surface waters of the Laurentian Great Lakes. Mar Pollut Bull 77(1–2):177–182

Forster NA, Wilson SC, Tighe MK (2023) Microplastic pollution on hiking and running trails in Australian protected environments. Sci Total Environ 874:162473

Goßmann I, Süßmuth R, Scholz-Böttcher BM (2022) Plastic in the air?!-spider webs as spatial and temporal mirror for microplastics including tire wear particles in urban air. Sci Total Environ 832:155008

Huang D, Wang X, Yin L, Chen S, Tao J, Zhou W, Chen H, Zhang G, Xiao R (2022) Research progress of microplastics in soil-plant system: ecological effects and potential risks. Sci Total Environ 812:151487

Inobeme A, Adetunji CO, Okonkwo S, Bamigboye MO, Ajai AI, Inobeme J, Olusanya J, Afoso E (2023) Fate and occurrence of microplastic and nanoplastic pollution in industrial wastewater. In: Microbial degradation and detoxification of pollutants, vol 2. Walter de Gruyter GmbH, p 73

Inobeme A, Mathew JT, Adetunji CO, Ajai AI, Inobeme J, Maliki M, Okonkwo S, Adekoya MA, Bamigboye MO, Jacob JO, Eziukwu CA (2023b) Recent advances in nanotechnology for remediation of heavy metals. *Environ Monit Assess* **195**:111. https://doi.org/10.1007/s10661-022-10614-7

Jones SP, Greer JJ, van Haperen R, Duncker DJ, de Crom R, Lefer DJ (2003) Endothelial nitric oxide synthase overexpression attenuates congestive heart failure in mice. Proc Natl Acad Sci 100(8):4891–4896

Kalali EN, Lotfian S, Shabestari ME, Khayatzadeh S, Zhao C, Nezhad HY (2023) A critical review of the current progress of plastic waste recycling technology in structural materials. Curr Opin Green Sustain Chem 40:100763

Kitahara K-I, Nakata H (2020) Plastic additives as tracers of microplastic sources in Japanese road dusts. Sci Total Environ 736:139694

Kosuth M, Mason SA, Wattenberg EV (2018) Anthropogenic contamination of tap water, beer, and sea salt. PLoS One 13(4):e0194970

Kutralam-Muniasamy G, Pérez-Guevara F, Elizalde-Martínez I, Shruti V (2020) Branded milks–are they immune from microplastics contamination? Sci Total Environ 714:136823

Kutralam-Muniasamy G, Shruti V, Pérez-Guevara F, Roy PD (2023) Microplastic diagnostics in humans: "the 3Ps" Progress, problems, and prospects. Sci Total Environ 856:159164

Lamichhane G, Acharya A, Marahatha R, Modi B, Paudel R, Adhikari A, Raut BK, Aryal S, Parajuli N (2023) Microplastics in environment: global concern, challenges, and controlling measures. Int J Environ Sci Technol 20(4):4673–4694

Liebezeit G, Liebezeit E (2013) Non-pollen particulates in honey and sugar. Food Addit Contam 30(12):2136–2140

Lv M, Zhang T, Ya H, Xing Y, Wang X, Jiang B (2023) Effects of heavy metals on the adsorption of ciprofloxacin on polyethylene microplastics: mechanism and toxicity evaluation. Chemosphere 315:137745

Mason SA, Welch VG, Neratko J (2018) Synthetic polymer contamination in bottled water. Front Chem 6:407
Masry M, Rossignol S, Roussel BT, Bourgogne D, Bussière PO, R'mili, B. and Wong-Wah-Chung, P. (2021) Experimental evidence of plastic particles transfer at the water-air interface through bubble bursting. Environ Pollut 280:116949
Mikac L, Rigó I, Himics L, Tolić A, Ivanda M, Veres M (2023) Surface-enhanced Raman spectroscopy for the detection of microplastics. Appl Surf Sci 608:155239
Napper IE, Bakir A, Rowland SJ, Thompson RC (2015) Characterisation, quantity and sorptive properties of microplastics extracted from cosmetics. Mar Pollut Bull 99(1–2):178–185
Napper IE, Thompson RC (2016) Release of synthetic microplastic plastic fibres from domestic washing machines: effects of fabric type and washing conditions. Mar Pollut Bull 112(1–2):39–45
NOAA (2023) What are microplastics? National Ocean Service website. https://oceanservice.noaa.gov/facts/microplastics.html. Accessed 19 June 2023
O'Brien S, Rauert C, Ribeiro F, Okoffo ED, Burrows SD, O'Brien JW, Wang X, Wright SL, Thomas KV (2023) There's something in the air: a review of sources, prevalence and behaviour of microplastics in the atmosphere. Sci Total Environ 874:162193
Priya A, Anusha G, Thanigaivel S, Karthick A, Mohanavel V, Velmurugan P, Balasubramanian B, Ravichandran M, Kamyab H, Kirpichnikova IM, Chelliapan S (2023) Removing microplastics from wastewater using leading-edge treatment technologies: a solution to microplastic pollution—a review. Bioprocess Biosyst Eng 46(3):309–321
Rillig MC, Leifheit E, Lehmann J (2021) Microplastic effects on carbon cycling processes in soils. PLoS Biol 19(3):e3001130
Shafea L, Yap J, Beriot N, Felde VJ, Okoffo ED, Enyoh CE, Peth S (2023) Microplastics in agroecosystems: a review of effects on soil biota and key soil functions. J Plant Nutr Soil Sci 186(1):5–22
Sharma U, Sharma S, Rana VS, Rana N, Kumar V, Sharma S, Qadri H, Kumar V, Bhat SA (2023) Assessment of microplastics pollution on soil health and eco-toxicological risk in horticulture. Soil Syst 7(1):7
Tang KHD (2023) Microplastics in agricultural soils in China: sources, impacts and solutions. Environ Pollut 322:121235
Thiele CJ, Hudson MD, Russell AE, Saluveer M, Sidaoui-Haddad G (2021) Microplastics in fish and fishmeal: an emerging environmental challenge? Sci Rep 11(1):2045
Tian Y, Zhu J, Ying C, Luo H, Zhang S, Zhang L, Li R, Li J (2023) Photoaging processes of polyvinyl chloride microplastics enhance the adsorption of tetracycline and facilitate the formation of antibiotic resistance. Chemosphere 320:137820
Viljoen SJ, Brailsford FL, Murphy DV, Hoyle FC, Chadwick DR, Jones DL (2023) Leaching of phthalate acid esters from plastic mulch films and their degradation in response to UV irradiation and contrasting soil conditions. J Hazard Mater 443:130256
Wang W, Wang J (2018) Investigation of microplastics in aquatic environments: an overview of the methods used, from field sampling to laboratory analysis. TrAC Trends Anal Chem 108:195–202
Wick P, Malek A, Manser P, Meili D, Maeder-Althaus X, Diener L, Diener PA, Zisch A, Krug HF, von Mandach U (2010) Barrier capacity of human placenta for nanosized materials. Environ Health Perspect 118(3):432–436
Wu C, Song X, Wang D, Ma Y, Ren X, Hu H, Shan Y, Ma X, Cui J, Ma Y (2023) Tracking antibiotic resistance genes in microplastic-contaminated soil. Chemosphere 312:137235
Yang D, Shi H, Li L, Li J, Jabeen K, Kolandhasamy P (2015) Microplastic pollution in table salts from China. Environ Sci Technol 49(22):13622–13627
Zagorski J, Debelak J, Gellar M, Watts JA, Kline JA (2003) Chemokines accumulate in the lungs of rats with severe pulmonary embolism induced by polystyrene microspheres. J Immunol 171(10):5529–5536

Zauner W, Farrow NA, Haines AM (2001) In vitro uptake of polystyrene microspheres: effect of particle size, cell line and cell density. J Control Release 71(1):39–51

Zhou J, Jia R, Brown RW, Yang Y, Zeng Z, Jones DL, Zang H (2023) The long-term uncertainty of biodegradable mulch film residues and associated microplastics pollution on plant-soil health. J Hazard Mater 442:130055

Zhu L, Xie C, Chen L, Dai X, Zhou Y, Pan H, Tian K (2023) Transport of microplastics in the body and interaction with biological barriers, and controlling of microplastics pollution. Ecotoxicol Environ Saf 255:114818

Primary Microplastic: Source, Uses, Transportation from Land to Marine Environment

2

Abel Inobeme, Mohd. Shahnawaz, Charles Oluwaseun Adetunji, John Tsado Mathew, Babatunde Oluwafemi Adetuyi, Oluwabukola Atinuke Popoola, Falana Yetunde Olaitan, Olalekan Akinbo, Olatunji Matthew Kolawole, Oluwafemi Adebayo Oyewole, Eniola K.I.T., and Mohammed Bello Yerima

Abstract

Primary microplastic is mostly produced intentionally for specific usage. After the usage of the products made of primary microplastic (e.g., microbeads), the microplastic particles enter into the environment and are reported in all zones (soil, air, and water). Like other types of microplastic, primary microplastic also

A. Inobeme
Department of Biochemistry, Covenant University, Ota, Ogun State, Nigeria

M. Shahnawaz
Department of Botany, University of Ladakh, Kargil Campus, Khumbathang, Kargil, Ladakh UT, India

Present Address: Department of Botany, Shri Pandit Prem Nath Dogra Government Degree College Samba, Samba, Jammu and Kashmir UT, India

C. O. Adetunji (✉)
Applied Microbiology, Biotechnology and Nanotechnology Laboratory, Department of Microbiology, Edo State University Uzairue, Iyamho, Auchi, Edo State, Nigeria
e-mail: adetunji.charles@edouniversity.edu.ng

J. T. Mathew · B. O. Adetuyi
Department of Natural Sciences, Biochemistry unit, Faculty of Pure and Applied Sciences, Precious Cornerstone University, Ibadan, Nigeria

O. A. Popoola · F. Y. Olaitan
Plant Improvement Unit, Agricultural Biotechnology Department, National Biotechnology Development Agency, Abuja, Nigeria

O. Akinbo
Centre of Excellence in Science, Technology, and Innovation, AUDA-NEPAD, Pretoria, South Africa

M. Shahnawaz et al. (eds.), *Microplastic Pollution*,
https://doi.org/10.1007/978-981-99-8357-5_2

enters into the food chain and poses serious threat to both terrestrial and marine flora and fauna. Primary microplastic is having various sources and are used in pharmaceutical, cosmetic, and other industries. In the present chapter, an attempt was made to overview the primary microplastics under the following lines: (1) to enlist various sources of primary microplastic, (2) to discuss its potential applications in various industries, and (3) to highlight various routes and mechanisms to transport microplastics from land to the marine environment.

Keywords

Primary microplastic · Microbeads · Transportation · Terrestrial environment · Marine environment

2.1 Introduction

Primary microplastics are of remarkable environmental concern in various aquatic habitats. Although there is a wide distribution of primary microplastics within the various environments such as air, land, and water, their impact is mostly considered individually for each of these environments, they are however highly connected in real situations (Desforges et al. 2014). The ingestion of primary microplastics by various species can lead to the compromising of energy reserves resulting to bioaccumulation and biomagnifications along the food chain. Primary microplastics also have the potential of adsorbing heavy metals due to their large surface area. They also act as means of conveying various groups of persistent organic contaminants into aquatic environment. Microplastics have been reported in various aquatic species of organisms such as ostracods, crustaceans, annelids, and gudgeon. Various studies have documented the potential of primary microplastics in adsorbing various environmental contaminants and toxins (Cozar et al. 2015). It has also been reported that primary plastics from sources such as polypropylene, polyvinyl chloride, and polyethylene have the ability to adsorb polycyclic aromatic hydrocarbons, pesticide residues, chlorinated biphenyl, and hexachlorocyclohexanes. Primary microplastics have also been shown to contain other chemicals such as bisphenol-A and nonylphenol and numerous additives from plastics which have been associated with endocrine disrupting potential. Such endocrine disrupting chemicals

O. M. Kolawole
Department of Microbiology Faculty of Life Sciences Ilorin, University of Ilorin, Ilorin, Nigeria

O. A. Oyewole
Department of Microbiology, Federal University of Technology, Minna, Nigeria

E. K.I.T.
Department of Biological Sciences, Joseph Ayo Babalola University, Ikeji Arakeji, Osun State, Nigeria

M. B. Yerima
Department of Microbiology, Sokoto State University, Sokoto, Sokoto State, Nigeria

such as xenoestrogens are leached into water bodies and other aquatic habitants through the primary microplastics. Microplastics also act as carriers for numerous pathogens for waterborne diseases, thereby reduces the quality of water (Dümichen et al. 2017).

2.2 Sources of Primary Microplastics

On estimation basis, it is put that about 4.8 to 12.7 million metric tons of plastic wastes from numerous sources get into the marine habitat from the various coastal countries annually. Microplastics make up the major groups with some of the major sources being from polystyrene (PS), polypropylene (PP), and polyethylene (PE) in medical products and cosmetics (Caruso 2015). The primary microplastics also get into water bodies spillage of resin plastics in pellets or powder forms used in air blasting. The application of sewage sludge which contains sediment micro plastics and synthetic fibers also contributes to the generation of primary micro plastics. Although most studies have focused on the presence of primary microplastics in the marine environment, there are newer studies that have suggested the presence of primary micro plastics in various surface water, municipal wastewater, and urban estuaries (Horton et al. 2017).

Primary microplastics are found in surface waters, beaches, sediments, and most water column; hence, they have pervaded the remotest marine environment. Studies involving the analysis of microplastics in ice cores obtained from remote areas around the Arctic Ocean showed high contents of microplastics ranging from 38 to 234 particles/m^3 (Faure et al. 2015). Also, the presence of concentrations higher than this has also been documented in pacific gyre. Primary microplastics have also been documented in Arctic waters and surface Antarctic, within the deepest arctic sea floor and in the stomachs of fishes and birds (Steensgaard et al. 2017).

2.3 Regulations to Manage Primary Microplastics

The formulation of management guidelines and other policies on microplastics has been put forward by numerous bodies and agencies as a result of the human health and environmental impacts of microplastics. Regulation in various regions and countries of the world has put forward measure with a view to regulating the use and distribution of primary microplastics such as microbeads, thereby reducing environmental pollution by micro plastics (Bergmann and Klages 2012). It was however noted that the secondary source also contributes remarkably to pollution by microplastics. To this end, new measures and legislations have to be put forward and implemented so as to effectively reduce the quantity of primary microplastics in the environment as well as the inherent environmental impacts. Measures were put forward aimed at encouraging various company producing plastics to adopt a recycling economy so as to minimize the discharge of these plastics materials into the environment (Browne et al. 2015). It has also been documented that a significant number of potential effects of microplastics within the environment have not been covered by the recent procedures for environmental risk assessments. Some of the factors of assessments that have been developed for numerous chemicals may not be strictly applicable to primary microplastics (Faure et al. 2015).

2.4 Detection and Characterization of Primary Microplastics

The analysis of primary microplastics is done based on their chemical and physical characterization and quantification. For the purpose of physical characterization, sieve cascade is adopted for the size categories in separation procedure or through the measurement of length. The measurement of length is done using optical microscopy. The median dimension can be measured using laser diffraction through optical analysis of the microplastics (Browne et al. 2015). The morphological characteristics are determined through the measurement of their shape and colors using stereomicroscopy. Other techniques that have been used for studying the morphology include scanning electron microscopy (SEM), scanning microscopy energy dispersive X-ray spectroscopy (ESEM-EDS), amongst others. The essence of quantifying the micro plastic is to ascertain temporal and spatial partitioning so as to determine their rate of accumulation. The concentrations of microplastics are usually expressed in number of particle in each square meter (particles/m^{-2}). Chemical characterization is also done using Fourier transform infrared spectroscopy (FTIR). The utilization of FTIR also makes it possible for the collection of spectra in attenuation (Ashton et al. 2010).

Other analytical techniques that are employed for the characterization of primary microplastics include Raman spectroscopy pyrolysis gas chromatography mass spectrometry (Pyr-GC-MS), amongst others.

2.5 Transportation of Microplastics from Land to Marine Environment

The transportation of microplastics from land to the marine environment can occur through various pathways. Some of the main ways in which microplastics are transported are outline below:

2.5.1 Surface Runoff

When it rains or snow melts, water flows over the land surface, collecting and carrying microplastics present on the ground. This runoff can transport microplastics directly into rivers, lakes, and ultimately, the ocean. Surface runoff plays a significant role in the transportation of microplastics from land to the marine environment. When it rains or snow melts, water runs over the land surface, collecting and carrying various pollutants, including microplastics, along with it. During rainstorms or snowmelt events, water rapidly flows over the land, picking up microplastics present on the ground. These microplastics can originate from various sources such as litter, plastic debris, microbeads, or particles generated from the degradation of larger plastic items. As the runoff moves downhill, it enters storm drains, ditches, or rivers, eventually reaching the marine environment (Haque and Fan 2023).

In urban areas, the presence of impervious surfaces like roads, parking lots, and sidewalks prevents water from infiltrating into the soil. Instead, it flows over these surfaces, accumulating microplastics and other pollutants. The runoff from urban areas can contain a higher concentration of microplastics due to the increased presence of plastic waste and insufficient natural filtration. Agricultural practices can contribute to microplastics in runoff as well. The use of plastic mulch films, plastic containers, or plastic-based fertilizers in agriculture can lead to the fragmentation and release of microplastics into the soil. When rainfall occurs, the runoff from agricultural fields can carry these microplastics, along with soil particles, pesticides, and fertilizers, into nearby water bodies and eventually the ocean (Godyń et al. 2022).

Soil erosion is a natural process exacerbated by human activities such as deforestation, construction, and improper land management practices. When soil erodes, it can carry microplastics that are embedded within it. As the eroded soil is transported by runoff into rivers and streams, microplastics can be released into the water, ultimately reaching the marine environment. Once microplastics are transported by surface runoff to rivers, lakes, or coastal areas, they can be further carried by water currents and tides, spreading them across larger marine ecosystems. Addressing the issue of microplastics in surface runoff requires implementing effective stormwater management practices, such as the installation of filtration systems, green infrastructure, and sediment traps to capture and remove microplastics before they reach water bodies. Additionally, promoting sustainable land use practices, minimizing plastic waste generation, and promoting proper waste disposal can help reduce the amount of microplastics entering surface runoff in the first place (Rehm et al. 2021).

2.5.2 Transportation of Primary Microplastic Through Rivers and Streams

Rivers act as conduits for microplastics, carrying them from land to the marine environment. Microplastics can enter rivers through runoff, wastewater discharge, or improper waste disposal near water bodies. These particles can then be transported downstream, ultimately reaching the ocean. Microplastics that are present on land surfaces, such as plastic litter, fragments, or fibers, can be carried by surface runoff into nearby rivers and streams. When it rains or snow melts, water flows over the land, picking up microplastics and carrying them into water bodies. This runoff can occur from various land areas, including urban, agricultural, and industrial zones, where microplastics are commonly found (Chitaka et al. 2023).

Wastewater treatment plants may not effectively capture all microplastics present in sewage or industrial wastewater. As a result, some microplastics can be discharged into rivers and streams through effluent outflows. This can include microplastics from sources such as synthetic fibers from laundry, microbeads from personal care products, or particles from plastic-based materials used in various industries. Urban areas are often equipped with drainage systems that collect and channel stormwater runoff. Microplastics present in urban environments, such as plastic litter or debris on streets and sidewalks, can be washed into drainage systems during rainfall events. These drainage systems can ultimately discharge the microplastics into nearby rivers or streams, which then transport them to the marine environment (Dalu et al. 2021).

Microplastics can enter larger rivers and ultimately the ocean through tributaries and smaller watercourses that feed into them. These smaller water bodies can receive microplastics from various land-based sources and then transport them downstream, gradually contributing to the accumulation of microplastics in larger river systems and eventually reaching the marine environment. Once microplastics are transported by rivers and streams to the marine environment, they can be further distributed by ocean currents, leading to their dispersal across larger areas and potentially affecting marine ecosystems. To mitigate the transportation of microplastics from rivers and streams to the marine environment, it is important to address the sources of microplastic pollution. This includes implementing proper waste management practices, reducing plastic litter and debris, improving wastewater treatment processes to capture microplastics effectively, and implementing regulations to limit the use of microplastics in various industries. Additionally, efforts to restore and protect riparian zones, implement green infrastructure, and promote sustainable agricultural practices can help reduce the entry of microplastics into rivers and streams in the first place (Olarinmoye et al. 2020).

2.5.3 Transportation of Microplastic Through Wind Transport

Microplastics can become airborne and be carried by wind currents. This can happen due to various sources such as open landfill sites, littered areas, or dust generated by microplastic particles adhering to soil particles. Eventually, these airborne microplastics can settle on water bodies, including the ocean. Microplastics can be present in open landfill sites where plastic waste accumulates. When wind blows over these sites, it can pick up lightweight microplastic particles and carry them over long distances. These airborne microplastics can then be deposited on water bodies, including the ocean, through atmospheric deposition (Haque and Fan 2023).

Wind can disperse littered plastic items, including larger plastic debris or fragments, from land areas into nearby water bodies. As these plastic items degrade into smaller microplastics, wind can carry them through the air, contributing to their transport to the marine environment. Microplastics can adhere to soil particles, and when wind blows over land areas with microplastic-contaminated soil, it can lift and transport these particles into the air. These microplastic-laden dust particles can be carried over long distances before settling on water bodies through atmospheric deposition. Wind can contribute to the transportation of microplastics by causing coastal erosion. As wind erodes coastal areas, it can dislodge microplastics embedded in the shoreline, dunes, or other coastal features. The airborne microplastics can then be carried by wind currents and deposited into the marine environment (Issac and Kandasubramanian 2021).

Wind can carry microplastics from urban areas and industrial sites where plastic waste or particles are present. For example, in areas near plastic manufacturing plants or processing facilities, wind can transport microplastics generated during production or handling processes to adjacent water bodies. Reducing the wind transport of microplastics requires comprehensive actions to tackle plastic pollution at its sources. These measures include proper waste management practices, minimizing plastic litter and debris, promoting recycling and circular economy principles, and raising awareness about the impacts of plastic pollution. Implementing measures to control windblown litter, such as covering landfill sites and implementing effective litter management strategies, can also help reduce the wind transport of microplastics (Kye et al. 2023).

2.5.4 Transportation of Microplastic Through Atmospheric Deposition

Microplastics can also enter the marine environment through atmospheric deposition. They can be carried over long distances by wind currents before settling on the surface of the ocean or being washed down by precipitation. Microplastics can become airborne through various means, including wind erosion, agricultural activities, industrial processes, or the fragmentation of larger plastic items. These airborne microplastics can be carried over long distances by wind currents, eventually reaching water bodies such as the ocean. Once airborne, microplastics can settle on the surface of water bodies through atmospheric deposition. This deposition occurs when the airborne microplastics come into contact with precipitation, such as rain or snow, and are washed out of the atmosphere onto land or water surfaces. The microplastics can be directly deposited on the ocean's surface or can be washed into rivers, streams, or lakes and eventually transported to the marine environment (Fiore et al. 2022).

Microplastics can also be deposited onto land or water surfaces without precipitation. Dry deposition occurs when the airborne microplastics settle due to gravity or other physical processes. In coastal areas, dry deposition of microplastics can occur directly on beaches or coastal waters, leading to their introduction into the marine environment. Microplastics that are transported through atmospheric deposition can travel over significant distances before being deposited. This means that microplastics generated in one region can be transported by wind currents to distant areas, including remote marine ecosystems. This long-range atmospheric transport contributes to the global distribution of microplastics in the marine environment (Bao et al. 2023).

It is important to note that the extent of atmospheric deposition of microplastics and its contribution to marine pollution is still an active area of research. Scientists are studying the sources, transport mechanisms, and impacts of microplastics deposited through the atmosphere to better understand and address this pathway of pollution. To reduce the atmospheric deposition of microplastics into the marine environment, it is crucial to tackle the sources of microplastic pollution on land. This includes minimizing plastic waste generation, improving waste management practices to prevent plastic leakage, and implementing regulations to limit the use of single-use plastics. Additionally, promoting sustainable manufacturing practices, encouraging proper disposal and recycling of plastic products, and raising awareness about the environmental impacts of microplastics can help mitigate their atmospheric transport and subsequent deposition in marine ecosystems (Mathew et al. 2022; Abad-López et al. 2023).

2.5.5 Transport of Microplastic Through Sewage Systems and Wastewater Treatment Plants

Microplastics from various sources, including synthetic fibers from laundry, cosmetic products, and microbeads, can be discharged into the sewage system. Although wastewater treatment plants capture a significant portion of these particles, some microplastics may still pass through the treatment process and be released into rivers or coastal areas. Microplastics can enter sewage systems through various domestic sources. These include microbeads from personal care products like facial scrubs and toothpaste, synthetic fibers from laundry, or plastic fragments from cleaning products. When these microplastics are washed down drains or flushed down toilets, they enter the sewage system. Industries that use or produce plastic materials can contribute to microplastic pollution in wastewater. Plastic manufacturing facilities, textile industries, or facilities that process plastic-based products may release microplastics through their wastewater discharges (Conley et al. 2019).

Although wastewater treatment plants are designed to remove contaminants from wastewater, they may not effectively capture all microplastics. Microplastics can be present in the form of suspended particles or dissolved fragments that are challenging to remove with conventional treatment processes. Some microplastics may bypass the treatment process and enter receiving waters through effluent outflows. In areas with combined sewer systems, which transport both domestic sewage and stormwater runoff in the same pipes, heavy rainfall events can cause sewer overflows. During these overflows, untreated wastewater, including microplastics, can be discharged directly into rivers, streams, or coastal areas, bypassing the treatment process altogether (Pan et al. 2022).

2.6 Conclusions

The global actions are necessary to address the ubiquitous problem of primary microplastics. Rigid controls and creative mitigation techniques are required due to their extensive negative effects on aquatic environments. Secondary sources must also be addressed, even if current policies focus on sources like microbeads. In order to comprehend their dispersion, precise detection and characterization procedures are essential. Storm water management, waste minimization, and sustainable land practices are all essential to good management. Other crucial actions include enhancing wastewater treatment and reducing plastic waste. We can protect our ecosystems from the long-term harm posed by primary microplastics by coordinated effort and strict policies, providing a better environment for all (Inobeme et al. 2023; Adetunji et al. 2021a, b, c; Adetunji et al. 2020; Adetunji 2019; Adetunji et al. 2019; Adetunji and Anani 2021; Anani et al. 2020; Anani et al. 2022; Adetunji et al. 2022a, b).

References

Abad-López AP, Trilleras J, Arana VA, Garcia-Alzate LS, Grande-Tovar CD (2023) Atmospheric microplastics: exposure, toxicity, and detrimental health effects. RSC Adv 13(11):7468–7489. https://doi.org/10.1039/d2ra07098g

Adetunji CO (2019) Environmental impact and ecotoxicological influence of biofabricated and inorganic nanoparticle on soil activity. In: Panpatte D, Jhala Y (eds) Nanotechnology for agriculture. Springer, Singapore. https://doi.org/10.1007/978-981-32-9370-0_12

Adetunji CO, Anani OA (2021) Plastic-eating microorganisms: recent biotechnological techniques for recycling of plastic. In: Panpatte DG, Jhala YK (eds) Microbial rejuvenation of polluted environment. Microorganisms for sustainability, vol 25. Springer, Singapore. https://doi.org/10.1007/978-981-15-7447-4_14

Adetunji CO, Panpatte DG, Bello OM, Adekoya MA (2019) Application of nanoengineered metabolites from beneficial and eco-friendly microorganisms as a biological control agents for plant pests and pathogens. In: Panpatte D, Jhala Y (eds) Nanotechnology for agriculture: crop production & protection. Springer, Singapore. https://doi.org/10.1007/978-981-32-9374-8_13

Adetunji CO, Oloke JK, Prasad G (2020) Effect of carbon-to-nitrogen ratio on eco-friendly mycoherbicide activity from *Lasiodiplodia pseudotheobromae* C1136 for sustainable weeds management in organic agriculture. Environ Dev Sustain 22:1977–1990. https://doi.org/10.1007/s10668-018-0273-1

Adetunji CO, Akram M, Olaniyan OT, Ajayi OO, Inobeme A, Olaniyan S, Hameed L, Adetunji JB (2021a) Targeting SARS-CoV-2 Novel Corona (COVID-19) virus infection using medicinal plants. In: Dua K, Nammi S, Chang D, Chellappan DK, Gupta G, Collet T (eds) Medicinal plants for lung diseases. Springer, Singapore. https://doi.org/10.1007/978-981-33-6850-7_21

Adetunji CO, Ajayi OO, Akram M, Olaniyan OT, Chishti MA, Inobeme A, Olaniyan S, Adetunji JB, Olaniyan M, Awotunde SO (2021b) Medicinal plants used in the treatment of Influenza A virus infections. In: Dua K, Nammi S, Chang D, Chellappan DK, Gupta G, Collet T (eds) Medicinal plants for lung diseases. Springer, Singapore. https://doi.org/10.1007/978-981-33-6850-7_19

Adetunji CO, Inobeme A, Olaniyan OT, Ajayi OO, Olaniyan S, Adetunji JB (2021c) Application of nanodrugs derived from active metabolites of medicinal plants for the treatment of inflammatory and lung diseases: recent advances. In: Dua K, Nammi S, Chang D, Chellappan DK, Gupta G, Collet T (eds) Medicinal plants for lung diseases. Springer, Singapore. https://doi.org/10.1007/978-981-33-6850-7_26

Adetunji CO, Inobeme A, Tadso J, Olaniyan OT, Abimbola OF, Shahnawaz M, Anani O (2022a) Potential of plastic waste in enhancing the level of pathogenicity of diverse pathogens in the marine biota. In: Shahnawaz M, Sangale MK, Daochen Z, Ade AB (eds) Impact of plastic waste on the marine biota. Springer, Singapore. https://doi.org/10.1007/978-981-16-5403-9_16

Adetunji CO, Oloke JK, Osemwegie OO, Ehis-Eriakha CB (2022b) Use of agro-wastes for *Lasiodiplodia pseudotheobromae* (C1136) production with sustainable bioefficacy. Environ Dev Sustain 24:7794–7809. https://doi.org/10.1007/s10668-021-01758-6

Anani OA, Mishra RR, Mishra P, Olomukoro JO, Imoobe TOT, Adetunji CO (2020) Influence of heavy metal on food security: recent advances. In: Mishra P, Mishra RR, Adetunji CO (eds) Innovations in food technology. Springer, Singapore. https://doi.org/10.1007/978-981-15-6121-4_18

Anani OA, Adetunji CO, Anani GA, Olomukoro JO, Imoobe TOT, Enenuku AA, Tongo I (2022) Effect of meso-, micro-, and nano-plastic waste on the benthos. In: Shahnawaz M, Sangale MK, Daochen Z, Ade AB (eds) Impact of plastic waste on the marine biota. Springer, Singapore. https://doi.org/10.1007/978-981-16-5403-9_12

Ashton K, Holmes L, Turner A (2010) Association of metals with plastic production pellets in the marine environment. Mar Pollut Bull 60:2050–2055

Bao M, Xiang X, Huang J, Kong L, Wu J, Cheng S (2023) Microplastics in the atmosphere and water bodies of coastal agglomerations: a mini-review. Int J Environ Res Public Health 20(3): 2466. https://doi.org/10.3390/ijerph20032466

Bergmann M, Klages M (2012) Increase of litter at the Arctic deep-sea observatory HAUSGARTEN. Mar Pollut Bull 64:2734–2741

Browne MA, Underwood AJ, Chapman MG, Williams R, Thompson RC, van Franeker JA (2015) Linking effects of anthropogenic debris to ecological impacts. Proc Biol Sci 282:20142929

Caruso G (2015) Plastic degrading microorganisms as a tool for bioremediation of plastic contamination in aquatic environments. J Pollut Effects Control 3:e112

Chitaka TY, Onianwa PC, Nel HA (2023) Marine litter sources and distribution pathways. In: Maes T, Preston-Whyte F (eds) The African marine litter outlook. Springer, Cham. https://doi.org/10.1007/978-3-031-08626-7_2

Conley K, Clum A, Deepe J, Lane H, Beckingham B (2019) Wastewater treatment plants as a source of microplastics to an urban estuary: removal efficiencies and loading per capita over one year. Water Res X 3:100030. https://doi.org/10.1016/j.wroa.2019.100030

Cozar A, Sanz-Martín M, Martí E, Gonzalez-Gordillo JI, Ubeda B, Galvez JA, Irigoien X, Duarte CM (2015) Plastic accumulation in the Mediterranean sea. PLoS One 10:0121762

Dalu T, Banda T, Mutshekwa T (2021) Effects of urbanisation and a wastewater treatment plant on microplastic densities along a subtropical river system. Environ Sci Pollut Res 28:36102–36111. https://doi.org/10.1007/s11356-021-13185-1

Desforges J-PW, Galbraith M, Dangerfield N, Ross PS (2014) Widespread distribution of microplastics in subsurface seawater in the NE Pacific Ocean. Mar Pollut Bull 79:94–99

Dümichen E, Eisentraut P, Bannick CG, Barthel A-K, Senz R, Braun U (2017) Fast identification of microplastics in complex environmental samples by a thermal degradation method. Chemosphere 174:572–584

Faure F, Saini C, Potter G, Galgani F, de Alencastro FL, Hagmann P (2015) An evaluation of surface micro- and mesoplastic pollution in pelagic ecosystems of the Western Mediterranean Sea. Environ Sci Pollut R 22:12190–12197

Fiore M, Fraterrigo Garofalo S, Migliavacca A (2022) Tackling marine microplastics pollution: an overview of existing solutions. Water Air Soil Pollut 233:276. https://doi.org/10.1007/s11270-022-05715-5

Godyń I, Bodziony M, Grela A, Muszyński K, Pamuła J (2022) Determination of pollution and environmental risk assessment of stormwater and the receiving river, case study of the Sudół river catchment, Poland. Int J Environ Res Public Health 20(1):504. https://doi.org/10.3390/ijerph20010504

Haque F, Fan C (2023) Fate and impacts of microplastics in the environment: hydrosphere, pedosphere, and atmosphere. Environments 10(5):70. https://doi.org/10.3390/environments10050070

Horton AA, Svendsen C, Williams RJ, Spurgeon DJ, Lahive E (2017) Large microplastic particles in sediments of tributaries of the River Thames, UK-abundance, sources and methods for effective quantification. Mar Pollut Bull 114:218–226

Inobeme A, Mathew JT, Adetunji CO, Ajai AI, Inobeme J, Maliki M, Okonkwo S, Adekoya MA, Bamigboye MO, Jacob JO, Eziukwu CA (2023) Recent advances in nanotechnology for remediation of heavy metals. Environ Monit Assess 195:111. https://doi.org/10.1007/s10661-022-10614-7

Issac MN, Kandasubramanian B (2021) Effect of microplastics in water and aquatic systems. Environ Sci Pollut Res Int 28(16):19544–19562. https://doi.org/10.1007/s11356-021-13184-2

Kye H, Kim J, Ju S, Lee J, Lim C, Yoon Y (2023) Microplastics in water systems: a review of their impacts on the environment and their potential hazards. Heliyon 9(3):e14359. https://doi.org/10.1016/j.heliyon.2023.e14359

Mathew JT, Adetunji CO, Inobeme A, Musah M, Shaba EY, Azeh Y, Abulude FO, Mamman A (2022) Organic compounds in atmospheric aerosols: atmospheric aerosols properties, sources and detection. Nova Science Publishers, pp 43–59

Olarinmoye OM, Stock F, Scherf N, Whenu O, Asenime C, Ganzallo S (2020) Microplastic presence in sediment and water of a lagoon bordering the urban agglomeration of Lagos, Southwest Nigeria. Geosciences 10(12):494. https://doi.org/10.3390/geosciences10120494

Pan Y, Gao SH, Ge C, Gao Q, Huang S, Kang Y, Luo G, Zhang Z, Fan L, Zhu Y, Wang AJ (2022) Removing microplastics from aquatic environments: a critical review. Environ Sci Ecotechnol 13:100222. https://doi.org/10.1016/j.ese.2022.100222

Rehm R, Zeyer T, Schmidt A, Fiener P (2021) Soil erosion as transport pathway of microplastic from agriculture soils to aquatic ecosystems. Sci Total Environ 795:148774. https://doi.org/10.1016/j.scitotenv.2021.148774

Steensgaard IM, Syberg K, Rist S, Hartmann NB, Boldrin A, Hansen SF (2017) From macro- to microplastics - analysis of EU regulation along the life cycle of plastic bags. Environ Pollut 224:289–299

Sources, Uses and Transport of Secondary Microplastics from Land to Marine Environment

3

Babatunde Oluwafemi Adetuyi, John Tsado Mathew, Abel Inobeme, Yetunde Olaitan Falana, Charles Oluwaseun Adetunji, Mohd Shahnawaz, Oluwafemi Adebayo Oyewole, Eniola K.I.T., and Mohammed Bello Yerima

B. O. Adetuyi
Department of Natural Sciences, Biochemistry unit, Faculty of Pure and Applied Sciences, Precious Cornerstone University, Ibadan, Nigeria

J. T. Mathew
Department of Chemistry, Ibrahim Badamasi Babangida University, Lapai, Niger State, Nigeria

A. Inobeme
Department of Chemistry, Edo University Iyamho, Auchi, Edo State, Nigeria

Y. O. Falana
Plant Improvement Unit, Agricultural Biotechnology Department, National Biotechnology Development Agency, Abuja, Nigeria

C. O. Adetunji (✉)
Applied Microbiology, Biotechnology and Nanotechnology Laboratory, Department of Microbiology, Edo State University Iyamho, Auchi, Edo State, Nigeria
e-mail: adetunji.charles@edouniversity.edu.ng

M. Shahnawaz
Department of Botany, University of Ladakh, Kargil Campus, Khumbathang, Kargil, Ladakh UT, India

Present Address: Department of Botany, Shri Pandit Prem Nath Dogra Government Degree College Samba, Samba, Jammu and Kashmir UT, India

O. A. Oyewole
Department of Microbiology, Federal University of Technology, Minna, Nigeria

E. K.I.T.
Department of Biological Sciences, Joseph Ayo Babalola University, Ikeji Arakeji, Osun State, Nigeria

M. B. Yerima
Department of Microbiology, Sokoto State University, Sokoto, Sokoto State, Nigeria

M. Shahnawaz et al. (eds.), *Microplastic Pollution*,
https://doi.org/10.1007/978-981-99-8357-5_3

Abstract

The widespread distribution of microplastics has resulted in a pervasive environmental pollution issue with global implications impacting diverse ecosystems, including terrestrial, aquatic, and atmospheric environments. The role of the small-scale plastic material in question in contributing substantially to the deterioration of the marine ecosystem has been observed, thus necessitating its recognition as an issue of significant concern. A variety of secondary microplastic sources are present, which include terrestrial origins such as plastic litter fragmentation, industrial activities, and synthetic material degradation. The aforementioned sources can be subsequently introduced into the aquatic ecosystem through various mechanisms, including atmospheric deposition, wastewater discharge, and river transport. Upon their introduction to aquatic environments, secondary microplastics inflict negative impacts on aquatic fauna. This phenomenon is primarily ascribed to the inclination of these organisms to consume microplastics, thereby perpetuating toxicity throughout the food chain and posing potential hazards to human health. Aside from the potential perilous ramifications of microplastics, secondary microplastics are utilized in the agricultural industry as a soil amendment and as a filtration material in the treatment of wastewater. Consequently, to efficaciously mitigate the perilous perils presented by secondary metabolites, diverse measures may be implemented, such as diminishing the employment of plastics and plastic refuse, improving waste disposal procedures, introducing sophisticated methodologies for removing microplastics from the surroundings, and advocating for sustainable models of production and consumption. The implementation of appropriate strategies is imperative in mitigating the adverse effects of secondary microplastics on the natural environment.

Keywords

Ecological impact · Microplastics · Waste management · Secondary microplastic · Plastic pollution

3.1 Introduction

Plastic possesses multiple physical characteristics, including its elasticity, minimal density, durable composition, economical viability, and hydrophobic property, that make it an attractive substance for manufacturers and end-users in diverse sectors. The exponential increase in the production of plastic can be attributed to the emergence of plastic materials that are commercially feasible. PlasticsEurope (2012) reported that the global output of plastic materials in 1950 was estimated to have reached a cumulative mass of 1.7 million metric tons (Mt). According to contemporary forecasts, the year 2015 will witness a notable surge in production output, with approximations lying within the range of 322 to 380 million metric tons (Geyer et al. 2017; PlasticsEurope 2016). A research investigation undertaken by

Geyer and colleagues reported production of 8300 million metric tons of primary plastic in 2017. Currently, packaging purposes constitute approximately 40% of plastic fabrication, with a particular emphasis on disposable single-use items (PlasticsEurope 2016). The lamentable upsurge in plastic utilization has led to a marked increase in the proliferation of plastic waste within the ecosystem, which has been instigated by the accidental or intentional disposal of plastic items (Jambeck et al. 2015). According to the research conducted by Geyer et al., according to recent data from 2017, approximately 60% of the cumulative output of plastics has been deposited in landfills or dispersed throughout the environment through natural means.

Plastic products are manufactured in a comprehensive spectrum of shapes and sizes, including those that meet the criteria for "microplastics" based on their minute scale, with dimensions measuring less than 5 mm in magnitude. The classification of primary microplastics refers to diminutive plastic particles intentionally produced in miniature dimensions. Microplastics are produced as a result of the manufacturing process of "nurdles," small pellets utilized in the production of plastic goods. Additionally, microplastics are generated through the incorporation of glitter and microbeads into personal care products and cosmetics. When introduced into the natural environment, plastic articles are prone to undergoing degradation. Consequently, even those of considerable size can ultimately decompose into multiple "secondary microplastics," which manifest in the form of minute fragments, fibers, or films, and may potentially exist in hundreds or even thousands. There are numerous mechanisms that can result in the deterioration of diverse materials. Various mechanisms are responsible for the physical degradation of materials, including but not limited to, the wear and tear caused by road usage, tire abrasion, physical weathering of objects, and the laundering process of synthetic textiles (Hernandez et al. 2017). Furthermore, chemical degradation has the potential to induce material deterioration. The manifestation of acidic or alkaline exposure and ultraviolet (UV) degradation, stemming from inherent UV radiation exposure, warrants meticulous contemplation in the realm of academic prose. The biological disintegration of plastics can occur in the presence of organisms that exhibit the capacity to assimilate and decompose such materials, including waxworms, mealworms, and select microorganisms. In recent years, the inclusion of plasticisers within the developmental stages of plastics to impart both durability and malleability has become commonplace. Unfortunately, over time, the excessive use of these plasticisers can often lead to their breakdown and consequent loss, leading to a marked increase in the brittleness of the plastic material and an increased vulnerability to degradation (Cole et al. 2011).

The current chapter focuses on the investigation of plastic pollution, particularly in regards to secondary microplastics. The discourse highlights the significance of effective approaches to waste management in alleviating the severe repercussions of plastic contamination. The objective of the analysis is to provide a thorough comprehension of the conceptual framework of plastics, the contributing factors to its long-lasting presence in the environment, and the adverse consequences that follow.

This discourse aims to present the topic under discussion by furnishing a succinct and concise delineation of the subject matter. The primary purpose of this introductory section is to establish the background and contextual framework, thereby providing a foundation for a thorough and comprehensive examination of the subject matter. This critical analysis will be expounded upon in subsequent parts of the document.

3.2 Secondary Sources of Microplastics

A plethora of modes of environmental contamination by plastics are present, including primary microplastics and macroplastics, which are larger plastic entities. Over time, the former will undergo a process of degradation, resulting in the formation of secondary microplastics. The current study advances the proposition that primary microplastics, which stem from household items, namely microbeads, have a propensity to manifest in effluent and are subsequently released into waterways. Furthermore, prior studies conducted by Mehra et al. have demonstrated that nurdles, a type of primary microplastic, can be discharged into freshwater environments during production processes. The current year of writing is 2021. Secondary sources of microplastics consist of a diverse array of forms that may arise due to intentional or unintentional human actions. These sources may include illicit dumping, improper waste management resulting in litter, as well as inadvertent leakage of microplastics into the environment. During the process of transportation, or while in transit, various factors can impact the quality and safety of the transported goods. Microplastic discharge displays considerable diversity across terrestrial, freshwater, and marine ecosystems due to various factors, including but not limited to, loss of shipping cargo and fishing gear (Boucher and Friot 2017).

Plastic production primarily takes place on land surfaces, and aside from its utilization in marine or fishing sectors, it serves as a fundamental component of numerous consumer products. The pathways through which waste consumer products are released into terrestrial environments are influenced by a plethora of factors, including but not limited to direct littering and suboptimal waste management techniques. The disposal of waste may result in the forfeiture of materials throughout diverse phases, comprising accidental discharges from landfill areas and industrial spillages. Such occurrences can have significant implications on the environmental ecosystem. The scholarly articles authored by Lechner and Ramler (2015) and Sadri and Thompson (2014) offer valuable insights into the aforementioned concerns.

Contemporary agricultural techniques have embraced plastic as a highly adaptable resource, utilizing it in various forms, including as mulches that can degrade in situ, and as bale twine and wrapping. However, the improper disposal of such plastic products remains a significant concern (Nizzetto et al. 2016). These chapters possess the potential to experience degradation and subsequently contribute to the development of secondary microplastics within the adjacent environment. The potential discharge of microplastics onto terrestrial environments can occur through the direct utilization of sewage sludge for agricultural fertilization purposes. Wastewater treatment facilities have demonstrated significant proficiency in effectively eliminating microplastic particles from the wastewater stream, often exhibiting removal rates of up to approximately 99%. A notable portion of these minuscule plastic particles will adhere to the sediment upon precipitation. The application of sewage sludge is a significant factor in the introduction of microplastics into agricultural soils throughout Europe. According to Nizzetto et al. (2016), it has been projected that an annual influx of microplastics per million inhabitants ranging between 125 and 850 t is introduced through this particular practice. Horton et al. (2017) carried out a quantitative investigation to ascertain the quantity of plastic waste that persists in continental environments across Europe, comprising both terrestrial and freshwater ecosystems. According to the findings of the study, an estimated annual amount within the range of 473,000 to 910,000 metric tons of plastic waste, derived from microplastics present in sewage sludge, as well as litter and inadequately managed waste, is accumulated within these ecosystems. In circumstances wherein plastics are not transported from terrestrial environments to fluvial or marine settings, there is the possibility for notable accumulation. Despite the growing concern over microplastic contamination in terrestrial ecosystems, a relatively small body of the literature has been devoted to exploring the extent of its distribution and determining potential correlations between its occurrence and various entry points. Consequently, currently it is unattainable to establish an unequivocal correlation between the accumulation of these particles and specific environmental characteristics or anthropogenic activities.

The scrutiny of secondary sources of microplastics has garnered significant scholarly interest. The detrimental environmental effects of discarded plastics, including plastic bags, fishing nets, household items, and other similar materials, have been widely denounced. The materials in question undergo a transformation through weathering and photo degradation mechanisms, leading to the formation of smaller plastic particles. These particles are commonly known as secondary microplastics. The occurrence of plastic abrasion stems from the influence of ultraviolet radiation on the terrain's exterior and the agitation of waves in the sea. Secondary microplastics are produced through the utilization of washing machines, as has been observed by scholars such as Eriksen et al. in the year 2013, as evidenced from relevant records and documentation. The aforementioned constitutes pieces of textile fibers acquired from synthetic fibers that are released during the laundering process (Fendall and Sewell 2009). Secondary microplastics are generated by the disposal of plastic waste produced by industrial goods and household items. Eerkes-Medrano et al. have established the marked significance of microplastics in the secondary classification in the year 2015. The prevalence of plastic particles in water environments exhibits a positive correlation with the escalating inflow of plastic refuse and the continued fragmentation of plastic waste into diminutive microplastic fragments. The formation of secondary microplastics is credited to numerous sources, encompassing the production of textiles, tires, and decorative paints in the textile sector, among others.

Synthetic textiles and tires are divergent albeit interconnected commodities extensively utilized in diverse industries. The term “former” pertains to fibers or fabrics composed of non-natural materials such as chemicals derived from petroleum, while the term “latter” refers to rubber-based substances utilized to augment foundational support in diverse transportation modes. There exist notable benefits in the utilization of both synthetic fabrics and automobile tires, which facilitate their meeting of the numerous expectations and necessities of current society.

Browne et al. assorted (2011) asserts that the laundering of articles in domestic as well as industrial environments leads to the production of substantial amounts of primary microplastics due to the scraping and dissemination of fibers. The discharge of microplastics into wastewater systems poses a significant menace to marine environments, as evidenced by the findings of Magnusson et al. (2016). The composition of tires involves a copious amalgamation of diverse synthetic polymers conjoined with naturally-occurring rubber. According to Essel et al., the phenomenon of tire degradation emerges due to regular use, with the resultant dispersal of tire dust and its synthetic rubber particles into the surrounding environment through processes such as wind or precipitation. In 2015, it was noted that considerable quantities of these particles tend to aggregate in the marine ecosystem.

Paint and wood preservatives are chemical agents that are frequently utilized to safeguard the structural integrity of wood and paint materials from deterioration and degradation. The observation of the inclusion of primary microplastics in the composition of paint and preservative has been made. The utilization of the aforementioned element yields a matte appearance and serves to intensify hues. This chapter aims to improve the durability of wood and increase its resistance to abrasion through the application of reinforcement techniques. Additionally, it has been noted by Lassen and associates in their 2015 publication that the material is utilized to decrease the viscosity of paints. Poulsen et al. (surname followed by the abbreviation for “and others”) performed a rigorous scholarly inquiry. As per the results elucidated in the year 2016, it has been disclosed that the dispensation of paint and other correlated tasks pertaining to painting are contributing towards the production of refuse within the bracket of 8% to 30%. A substantial percentage of waste material, typically ranging between 65% and 97%, is known to accumulate in the solid phase while a significant portion, estimated at approximately 35%, is directed towards the sewer network. This study endeavors to provide an estimation of the annual discharge of microplastics resulting from paints, estimated to range from 1.0 to 5.3 metric tons per year. Within this context, it has been proposed that a range of 0.1–0.5 tons per year constitutes the primary quantity of microplastics found in paints, which ultimately disseminate into the aquatic environment.

Also, the diverse applications and ramifications of synthetic waxes within the framework of the oil and gas sector is also looked into. According to the study undertaken by Essel et al., in 2015, synthetic waxes were subjected to critical scrutiny as a significant contributor to the generation of microplastics. The abovementioned compounds are widely employed in diverse industries, encompassing the dyeing arena, food encapsulation, and plastic fabrication. Moreover, these substances have a versatile range of applications which also encompasses their utilization as lubricants. Synthetic waxes have been employed in the coating of paper and have displayed considerable effectiveness in the processing of textiles, augmenting the smoothness of the surface. The usage of polytetrafluoroethylene (PTFE) is commonly observed in drilling applications. The utilization of drilling fluids containing microbeads is prevalent in the oil and gas industry, a phenomenon established by prior research (Poulsen et al. 2002). The current composite is released directly into marine ecosystems, thus augmenting the deposition of microplastics.

The allocation of medical resources constitutes a critical component in the realm of healthcare. The aforementioned activity concerns the apportionment, dissemination, and application of a diverse range of materials and equipment that are critical to facilitating the identification, remedy, and prevention of morbidity and pathological conditions. Sufficient availability of medical resources is of utmost importance in ensuring the best possible results for patients, as it has a direct impact on the caliber of care that healthcare practitioners are capable of providing. As a result, optimizing the management and acquisition of medical resources is a paramount concern for both healthcare establishments and policymakers.

The utilization of microplastics in the healthcare domain has been extensively documented (Essel et al. 2015). The employment of microplastics has been noted to ameliorate the backflow of gastric juices. Compounds encased with aluminum have been identified to elicit detrimental impacts on human well-being. Microplastics have become a substitute for them. Disposable infant diapers, feminine hygiene products, and adhesive bandages are widely recognized as waste products within contemporary society. The pharmaceutical industry employs plastic-based capsules in their formulation. Research conducted by Verschoor and colleagues has identified eyeglasses and contact lenses as leading sources of microplastics. The year 2014 is symbolic of a specific time period within the past decade.

Also, the issue of shredders and their relevance in contemporary waste management methodologies leads to the production of microplastic. Food waste shredders are specialized gadgets designed for the purpose of breaking down food waste into smaller fragments to facilitate its proper and convenient disposal. These devices are frequently employed in residential settings, dining establishments, and other food-oriented premises. The employment of food waste shredders has the potential to mitigate the quantity of food waste that is ultimately disposed of in landfills, consequently resulting in noteworthy ecological implications. Furthermore, the utilization of food waste shredders can enhance the efficacy of waste management protocols and minimize the entire expenditure related to waste elimination. Consequently, the usage of said devices has witnessed a significant surge in popularity in contemporary times, owing to the greater cognizance among individuals regarding the significance of implementing sustainable waste management mechanisms. The management of waste materials utilizing alternative methods beyond traditional waste management practices.

Subsequent to prior investigations, other origins of secondary microplastics have been discerned. In the context of scholarly inquiry or investigation, it could be posited that the aforementioned pertains to or embodies.

- The management of plastic materials and articles in both industrial and occupational contexts.
- The residual waste generated from the process of cutting and polishing plastic materials in industrial environments.
- The particulate matter emanating from the practice of severing plastic components during construction and building operations, including but not limited to EPS sheets and PVC gutter material, is commonly denoted as dust. The procedure of separating said components is generally termed as cutting.
- The production of particles generated during the plastic item cutting process in boat repair, shipyards, and car repair shops, as well as by nets and other fishing equipment, has been identified as a significant environmental concern.
- The incorporation of polyethylene (PE) films in the agricultural sector.
- The employment of Polymer Modified Bitumen (PMB) has garnered considerable attention in the realm of asphalt science owing to its manifold benefits, including heightened resilience and fortified resistance against rutting and cracking.
- The management and proper disposal of waste materials.
- The terminology "biowaste" pertains to refuse materials that derive from biological origins, namely residual food, agricultural detritus, and animal excrement.
- The application of paper recycling comprises a vital facet of the sustainable management of resources.
- Plastic recycling facilities.

3.3 Types of Secondary Microplastics

Secondary microplastics are small plastic particles that are formed through the degradation and fragmentation of larger plastic items or materials. There are several types of secondary microplastics, which can be categorized based on their source or origin (Table 3.1).

3.3.1 Fragmented Microplastics

These are small plastic particles that result from the mechanical breakdown or fragmentation of larger plastic items, such as bottles, bags, or packaging materials. Fragmented microplastics can vary in size, shape, and composition (Bajt 2021; Kye et al. 2023).

3.3.2 Fibers

Microplastic fibers are small, thread-like particles that shed from various sources, including synthetic textiles, carpets, upholstery, and clothing. These fibers can be

Table 3.1 Types of secondary microplastics based on their source or origin

Types of secondary microplastics	Route of formation	Specific examples	References
Fragmented microplastics	Fragmentation and mechanical breakdown	Microplastics from plastic bags, bottles, etc.	Kye et al. (2023)
Fibers	Shedding from synthetic fibres, carpets, etc.	Microplastics from clothing, carpets, textiles and upholstery	Chen et al. (2022)
Industrial microplastics	During industrial production processes	Plastic powder, resin pellets, and other minute plastics use in industries	Umar et al. (2023)
Paint and coating	Peelings from painted surfaces	Particulate polymers and additives in paints	Hale et al. (2022)
Tyre wears	Abrasion due to surface contact	Plasticizers, synthetic rubbers, and chemical additives	Tull et al. (2023)

released during the washing of clothes or through general wear and tear of synthetic materials. They are commonly made of materials such as polyester, nylon, or acrylic (Chen et al. 2022).

3.3.3 Tire Wear Particles

As vehicle tires wear down, they release tiny rubber particles into the environment. These particles, known as tire wear particles or tire dust, contain a mixture of synthetic rubber, plasticizers, and other additives. They can be transported by wind or washed into water bodies and contribute to microplastic pollution (Tull et al. 2023).

3.3.4 Paint and Coating Particles

Paints, coatings, and varnishes often contain microplastic particles, such as acrylics or polyurethane. These particles can be released into the environment during the application, weathering, or removal of these materials, contributing to the presence of microplastics in soil, water, and air (Hale et al. 2022).

3.3.5 Industrial Microplastics

Various industrial processes, such as plastic manufacturing, processing, and recycling, can generate microplastics as byproducts. These industrial microplastics can include resin pellets, plastic powders, or other small plastic particles that are released during production, handling, or disposal processes. It is important to note that secondary microplastics can have detrimental effects on ecosystems and organisms. Their small size and persistence make them difficult to remove from the environment, and they can be ingested by a wide range of organisms, potentially leading to various ecological and health concerns (Umar et al. 2023).

3.4 Transport Mechanisms of Secondary Microplastics to the Marine Environment

Secondary microplastics are small plastic particles that result from the degradation of larger plastic items. These particles can enter the marine environment through various transport mechanisms. Some common ways in which secondary microplastics are transported to the marine environment (Fig. 3.1).

3.4.1 Runoff and Rivers

Microplastics can be transported to the ocean through surface runoff from urban areas, agricultural lands, and industrial sites. Rainwater carries microplastics from these sources into rivers and streams, which eventually discharge into the ocean. Rivers act as major pathways for the transport of microplastics, particularly in densely populated areas or regions with high industrial activities (Whitehead et al. 2021).

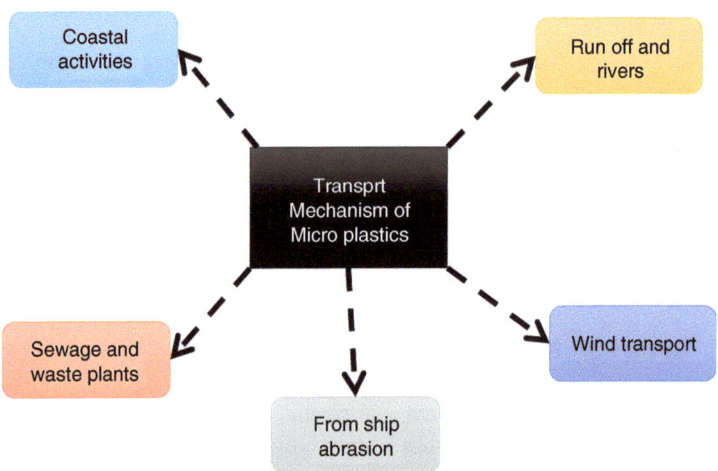

Fig. 3.1 An overview of the microplastic transport mechanism

3.4.2 Wind Transport

Microplastics, especially lightweight fragments, can be transported over long distances by wind. These particles may become airborne through various means, such as dust from plastic waste sites, erosion of plastic debris on coastlines, or resuspension of microplastics from dried sediment. Once in the atmosphere, they can be carried by wind currents and eventually deposited in the marine environment through atmospheric fallout (Issac and Kandasubramanian 2021; Wojnowska-Baryła et al. 2022).

3.4.3 Sewage and Wastewater Treatment Plants

Microplastics can enter the marine environment through sewage and wastewater discharge. Many personal care products, including facial cleansers, toothpaste, and exfoliants, contain microplastic beads. When these products are used and washed off, the microplastics can pass through sewage systems and reach wastewater treatment plants. However, these treatment plants are not designed to effectively capture microplastics, allowing them to be discharged into rivers or coastal waters (Cristaldi et al. 2020; Adetunji et al. 2023).

3.4.4 Coastal Activities

Coastal areas are hotspots for microplastic pollution due to various human activities. Plastic debris on beaches, coastal erosion of plastic items, and recreational activities such as boating and fishing contribute to the introduction of microplastics into the marine environment. Wave action and tides can further transport these particles offshore, spreading microplastics across a wider area (Fiore et al. 2022; Inobeme et al. 2022).

3.4.5 Ship Transport

Shipping activities can also contribute to the transport of microplastics. Large cargo ships and fishing vessels can generate plastic waste or lose plastic items, which may fragment over time and release microplastics into the ocean. Ship-generated microplastics can be transported across vast distances, especially along major shipping routes (Kalnina et al. 2022).

It is important to note that once microplastics enter the marine environment, they can be transported over long distances by ocean currents, leading to their widespread distribution in the oceans. Furthermore, these particles can also be ingested by marine organisms and subsequently enter the food chain, posing potential risks to aquatic ecosystems and human health. Efforts to reduce plastic pollution and improve waste management practices are crucial to mitigating the transport and accumulation of microplastics in the marine environment (Yang et al. 2021).

3.5 Conclusion

Microplastics are ubiquitous within the environment at global scale, with their particulate nature aiding their ease of dispersion across different components of the environment. Recent studies have shown that the amount of microplastics within the terrestrial systems may even exceed the amount that enter the sea environment. This chapter has presented the various sources and distribution of secondary microplastics as well as their means of transport from land to the ocean. There are limited studies focussing on microplastics in soil, since the movement from soil is one of the potential routes for the transport into the aquatic environment, more studies are need with focal point on the mechanisms of this discharge and distribution (Inobeme et al. 2023; Adetunji et al. 2021a, b, c; Adetunji et al. 2020; Adetunji 2019; Adetunji et al. 2019; Adetunji and Anani 2021; Anani et al. 2020; Anani et al. 2022; Adetunji et al. 2022).

References

Adetunji CO (2019) Environmental impact and ecotoxicological influence of biofabricated and inorganic nanoparticle on soil activity. In: Panpatte D, Jhala Y (eds) Nanotechnology for agriculture. Springer, Singapore. https://doi.org/10.1007/978-981-32-9370-0_12

Adetunji CO, Anani OA (2021) Plastic-eating microorganisms: recent biotechnological techniques for recycling of plastic. In: Panpatte DG, Jhala YK (eds) Microbial rejuvenation of polluted environment. microorganisms for sustainability, vol 25. Springer, Singapore. https://doi.org/10.1007/978-981-15-7447-4_14

Adetunji CO, Panpatte DG, Bello OM, Adekoya MA (2019) Application of nanoengineered metabolites from beneficial and eco-friendly microorganisms as a biological control agents for plant pests and pathogens. In: Panpatte D, Jhala Y (eds) Nanotechnology for agriculture: crop production & protection. Springer, Singapore. https://doi.org/10.1007/978-981-32-9374-8_13

Adetunji CO, Oloke JK, Prasad G (2020) Effect of carbon-to-nitrogen ratio on eco-friendly mycoherbicide activity from *Lasiodiplodia pseudotheobromae* C1136 for sustainable weeds management in organic agriculture. Environ Dev Sustain 22:1977–1990. https://doi.org/10.1007/s10668-018-0273-1

Adetunji CO, Akram M, Olaniyan OT, Ajayi OO, Inobeme A, Olaniyan S, Hameed L, Adetunji JB (2021a) Targeting SARS-CoV-2 Novel Corona (COVID-19) virus infection using medicinal plants. In: Dua K, Nammi S, Chang D, Chellappan DK, Gupta G, Collet T (eds) Medicinal plants for lung diseases. Springer, Singapore. https://doi.org/10.1007/978-981-33-6850-7_21

Adetunji CO, Ajayi OO, Akram M, Olaniyan OT, Chishti MA, Inobeme A, Olaniyan S, Adetunji JB, Olaniyan M, Awotunde SO (2021b) Medicinal plants used in the treatment of Influenza A virus infections. In: Dua K, Nammi S, Chang D, Chellappan DK, Gupta G, Collet T (eds) Medicinal plants for lung diseases. Springer, Singapore. https://doi.org/10.1007/978-981-33-6850-7_19

Adetunji CO, Inobeme A, Olaniyan OT, Ajayi OO, Olaniyan S, Adetunji JB (2021c) Application of nanodrugs derived from active metabolites of medicinal plants for the treatment of inflammatory and lung diseases: recent advances. In: Dua K, Nammi S, Chang D, Chellappan DK, Gupta G,

Collet T (eds) Medicinal plants for lung diseases. Springer, Singapore. https://doi.org/10.1007/978-981-33-6850-7_26

Adetunji CO, Inobeme A, Tadso J, Olaniyan OT, Abimbola OF, Shahnawaz M, Anani O (2022a) Potential of plastic waste in enhancing the level of pathogenicity of diverse pathogens in the marine biota. In: Shahnawaz M, Sangale MK, Daochen Z, Ade AB (eds) Impact of plastic waste on the marine biota. Springer, Singapore. https://doi.org/10.1007/978-981-16-5403-9_16

Adetunji CO, Oloke JK, Osemwegie OO, Ehis-Eriakha CB (2022b) Use of agro-wastes for *Lasiodiplodia pseudotheobromae* (C1136) production with sustainable bioefficacy. Environ Dev Sustain 24:7794–7809. https://doi.org/10.1007/s10668-021-01758-6

Adetunji CO, Olaniyan OT, Singh KRB, Bodunrinde RE, Inobeme A, Mathew JT, Ogundolie AF, Akinbo O, Singh J, Nayak V, Singh RP (2023) Applications of nanofiltration for wastewater treatment. In: Bio-nano filtration in industrial effluent treatment, vol 1. CRC-Taylor and Francis, pp 139–152. https://doi.org/10.1201/9781003165149-10

Anani OA, Mishra RR, Mishra P, Olomukoro JO, Imoobe TOT, Adetunji CO (2020) Influence of heavy metal on food security: recent advances. In: Mishra P, Mishra RR, Adetunji CO (eds) Innovations in food technology. Springer, Singapore. https://doi.org/10.1007/978-981-15-6121-4_18

Anani OA, Adetunji CO, Anani GA, Olomukoro JO, Imoobe TOT, Enenuku AA, Tongo I (2022) Effect of meso-, micro-, and nano-plastic waste on the benthos. In: Shahnawaz M, Sangale MK, Daochen Z, Ade AB (eds) Impact of plastic waste on the marine biota. Springer, Singapore. https://doi.org/10.1007/978-981-16-5403-9_12

Bajt O (2021) From plastics to microplastics and organisms. FEBS Open Bio 11(4):954–966. https://doi.org/10.1002/2211-5463.13120

Boucher J, Friot D (2017) Primary microplastics in the oceans: a global evaluation of sources. IUCN, Gland

Browne MA, Crump P, Niven SJ, Teuten E, Tonkin A, Galloway T, Thompson R (2011) Accumulation of microplastic on shorelines worldwide: sources and sinks. Environ Sci Technol 45(21):9175–9179

Chen Y, Chen Q, Zhang Q, Zuo C, Shi H (2022) An overview of chemical additives on (micro)-plastic fibers: occurrence, release, and health risks. Rev Environ Contam Toxicol 260(1):22. https://doi.org/10.1007/s44169-022-00023-9

Cole M, Lindeque P, Halsband C, Galloway TS (2011) Microplastics as contaminants in the marine environment: a review. Mar Pollut Bull 62(12):2588–2597

Cristaldi A, Fiore M, Zuccarello P, Oliveri Conti G, Grasso A, Nicolosi I, Copat C, Ferrante M (2020) Efficiency of wastewater treatment plants (WWTPs) for microplastic removal: a systematic review. Int J Environ Res Public Health 17(21):8014. https://doi.org/10.3390/ijerph17218014

Eerkes-Medrano D, Thompson RC, Aldridge DC (2015) Microplastics in freshwater systems: a review of the emerging threats, identification of knowledge gaps and prioritisation of research needs. Water Res 75:63–82

Eriksen M, Mason S, Wilson S, Box C, Zellers A, Edwards W, Farley H, Amato S (2013) Microplastic pollution in the surface waters of the Laurentian Great Lakes. Mar Pollut Bull 77(1–2):177–182

Essel R, Engel L, Carus M, Ahrens RH (2015) Sources of microplastics relevant to marine protection in Germany. Texte 64

Fendall LS, Sewell MA (2009) Contributing to marine pollution by washing your face: microplastics in facial cleansers. Mar Pollut Bull 58(8):1225–1228

Fiore M, Fraterrigo Garofalo S, Migliavacca A (2022) Tackling marine microplastics pollution: an overview of existing solutions. Water Air Soil Pollut 233:276. https://doi.org/10.1007/s11270-022-05715-5

Geyer R, Jambeck JR, Law KL (2017) Production, use, and fate of all plastics ever made. Sci Adv 3(7):e1700782

Hale RC, Seeley ME, King AE, Yu LH (2022) Analytical chemistry of plastic debris: sampling, methods, and instrumentation. In: Bank, M.S. (ed) Microplastic in the environment: pattern and process. Environmental contamination remediation and management. Springer, Cham. https://doi.org/10.1007/978-3-030-78627-4_2

Hernandez E, Nowack B, Mitrano DM (2017) Synthetic textiles as a source of microplastics from households: a mechanistic study to understand microfiber release during washing. Environ Sci Technol 51:7036–7046

Horton AA, Walton A, Spurgeon DJ, Lahive E, Svendsen C (2017) Microplastics in freshwater and terrestrial environments: evaluating the current understanding to identify the knowledge gaps and future research priorities. Sci Total Environ 586:127–141

Inobeme A, Nayak V, Mathew JT, Okonkwo S, Ekwoba L, Ajai AI, Bernard E, Inobeme J, Agbugui MM, Singh KRB (2022) Chemometric approach in environmental pollution analysis: a critical review. J Environ Manag 309(114653):1–18. https://doi.org/10.1016/j.jenvman.2022.114653

Inobeme A, Mathew JT, Adetunji CO, Ajai AI, Inobeme J, Maliki M, Okonkwo S, Adekoya MA, Bamigboye MO, Jacob JO, Eziukwu CA (2023) Recent advances in nanotechnology for remediation of heavy metals. Environ Monit Assess 195:111. https://doi.org/10.1007/s10661-022-10614-7

Issac MN, Kandasubramanian B (2021) Effect of microplastics in water and aquatic systems. Environ Sci Pollut Res Int 28(16):19544–19562. https://doi.org/10.1007/s11356-021-13184-2

Jambeck J, Geyer R, Wilcox C, Siegler TR, Perryman M, Rady AL et al (2015) Plastic waste inputs from land into the ocean. Science 347(6223):768–771

Kalnina R, Demjanenko I, Smilgainis K, Lukins K, Bankovics A, Drunka R (2022) Microplastics in ship sewage and solutions to limit their spread: a case study. Water 14(22):3701. https://doi.org/10.3390/w14223701

Kye H, Kim J, Ju S, Lee J, Lim C, Yoon Y (2023) Microplastics in water systems: a review of their impacts on the environment and their potential hazards. Heliyon 9(3):e14359. https://doi.org/10.1016/j.heliyon.2023.e14359

Lassen C, Hansen SF, Magnusson K, Hartmann NB, Jensen PR, Nielsen TG, Brinch A (2015) Microplastics: occurrence, effects and sources of releases to the environment in Denmark, pp 1–208

Lechner A, Ramler D (2015) The discharge of certain amounts of industrial microplastic from a production plant into the river Danube is permitted by the Austrian legislation. Environ Pollut 200:159–160

Magnusson K, Eliasson K, Fråne A, Haikonen K, Hultén J, Olshammar M, Stadmark J, Voisin A (2016) Swedish sources and pathways for microplastics to the marine environment. A review of existing data. IVL, C 183:1–87

Mehra S, Sharma K, Sharma G, Singh M, Chadha P (2021) Sources, fate, and impact of microplastics in aquatic environment. IntechOpen. https://doi.org/10.5772/intechopen.93805

Nizzetto L, Bussi G, Futter MN, Butterfield D, Whitehead PG (2016) A theoretical assessment of microplastic transport in river catchments and their retention by soils and river sediments. Environ Sci Process Impacts 18(8):1050–1059

PlasticsEurope (2012) Plastics—the facts 2012. An analysis of European plastics production, demand and waste data for 2011

PlasticsEurope (2016) Plastics—the facts 2016. An analysis of European plastics production, demand and waste data

Poulsen PB, Stranddorf HK, Hjuler K, Rasmussen JO (2002) Vurderingafmalingsmiljøbelastningianvendelsesfasen [Assessment of the environmental impact of paint in the use phase]. Miljøprojekt nr. Danish Environmental Protection Agency, p 662

Sadri SS, Thompson RC (2014) On the quantity and composition of floating plastic debris entering and leaving the Tamar estuary, Southwest England. Mar Pollut Bull 81(1):55–60

Tull T, Krais S, Peschke K, Weyrauch S, Triebskorn R, Köhler HR (2023) Tire and road wear particle-containing sediments with high organic content impact behavior and survival of

Chironomid larvae (Chironomusriparius). Environments 10(2):23. https://doi.org/10.3390/environments10020023

Umar M, Singdahl-Larsen C, Ranneklev SB (2023) Microplastics removal from a plastic recycling industrial wastewater using sand filtration. Water 15(5):896. https://doi.org/10.3390/w15050896

Verschoor A, de Poorter L, Roex E, Bellert B (2014) Quick scan and prioritization of microplastic sources and emissions. RIVM Lett Rep 5:156

Whitehead PG, Bussi G, Hughes JMR, Castro-Castellon AT, Norling MD, Jeffers ES, Rampley CPN, Read DS, Horton AA (2021) Modelling microplastics in the river thames: sources, sinks and policy implications. Water 13(6):861. https://doi.org/10.3390/w13060861

Wojnowska-Baryła I, Bernat K, Zaborowska M (2022) Plastic waste degradation in landfill conditions: the problem with microplastics, and their direct and indirect environmental effects. Int J Environ Res Public Health 19(20):13223. https://doi.org/10.3390/ijerph192013223

Yang H, Chen G, Wang J (2021) Microplastics in the marine environment: sources, fates, impacts and microbial degradation. Toxics 9:41

Usage of Microplastic Beads in Pharmaceuticals and Cosmetics Industry: A Review

4

Pooja M. Patil, Pranjali Mahamuni-Badiger, Rachana R. Ingavale, Pratikshkumar R. Patel, and Maruti J. Dhanavade

Abstract

Environmentalists and researchers have expressed future alarm over the widespread usage of microplastic beads in the cosmetic and pharmaceutical industries. The microplastics used in those industries take decades to break down and are dangerous for both the ecosystem and humans. The difficulty in identifying and removing microplastics and plastics with diameter below 5 mm (milliliter) is a big environmental issue. The majority of microplastics are found in soap, scrubs, face washes, pharmaceutical products, and body washes. Once these products are used and rinsed, they quickly leak into the water, where they are consumed by aquatic life before reaching human territory. It is hard to predict the amount of microplastic beads that are released into the environment and their effects on aquatic as well as human life. The increasing microplastic pollution can be harmful to the environment and living species. This chapter tries to evaluate the use of microplastic beads in the pharmaceutical and cosmetics industry, their

Pooja M. Patil and Pranjali Mahamuni-Badiger contributed equally for this work.

P. M. Patil · R. R. Ingavale
Department of Environment Management, CSIBER, Kolhapur, Maharashtra, India

P. Mahamuni-Badiger
Department of Microbiology, Rayat Shikshan Sanstha's, S.M. Joshi College, Pune, Maharashtra, India

P. R. Patel
Department of Pediatric, Medical University of South Carolina, South Carolina, South Carolina, USA

M. J. Dhanavade (✉)
Department of Microbiology, Bharati Vidyapeeth's Dr. Patangrao Kadam Mahavidyalaya, Sangli, Maharashtra, India
e-mail: maruti.dhanavade@bharatividyapeeth.edu.in

M. Shahnawaz et al. (eds.), *Microplastic Pollution*,
https://doi.org/10.1007/978-981-99-8357-5_4

impact on the environment and lives, their preventions, fate, alternatives, policy on banning, etc. Hence, this chapter will become a stepping stone toward deciding the current application and future usage of microplastic without harming the environment and living species.

Keywords

Microplastic · Nano-plastics · Microbeads · Waste water treatment plant

4.1 Introduction

In recent times, the application of microplastic beads in cosmetic and pharmaceutical industry has been widely explored. Due to this, microplastic pollution has become a global concern for researchers and environmentalists. The different monomers and additives are polymerized to form range of plastics with different characteristics such as mechanical strength, degradation period, polarity glassine, etc. The different property of polymer impacts the occurrence of pollution and their risk to the environment, humans, aquatic life, and territorial animals. However, around 320 million tons (Mt) per year of plastics was produced, and it is expected to become double in next 20 years. Humans use plastics in their day-to-day life, and due to that, plastic debris has accumulated in oceans, air, soil, surface water, and sediments. The large plastic debris breaks into macroplastics, mesoplastics, microplastics, and nanoplastics. The microplastics are derived from both primary as well as secondary sources. The plastic beads are used as exfoliates in different cosmetic products like face cleansers, powder, and cosmetics products these are the primary sources of microplastics. Cosmetic goods are a combination that is designed to be applied to an external portion of the body in order to improve or enhance the look of a person. Furthermore, teeth, oral mucous membrane hygiene, cleansing, altering appearance, and preserving their attractive appearance, safeguarding, scent. Rinse-off cosmetic products, such as hair dye, shampoos, face wash, soaps and toothpaste should be swiftly removed after application when their purpose is complete. There are around 4500–94,500 microbeads in a single application of a cosmetic product. It is estimated that more than 8 trillion microplastic beads and microfibers reach the aquatic environment in the United States each day from WWTP effluent. Industrial abrasives and accidental spills are other sources that releases microplastics directly into the environment. It is anticipated that the number of microplastics transported annually from WWTPs in North America and Europe is more than the entire load of microplastics now present in ocean water. Additionally, unintended secondary microplastics are created when bigger plastics present in the environment is converted into microplastic by the weathering events viz. hydrolysis, UV photodegradation, mechanical abrasion, and by other means. The plastics in the water, such as bottles, bags, ropes and nets, as well as automotive trim, agricultural plastic mulch, and microfiber from textiles, are secondary sources of microplastics.

The pharmaceuticals, cosmetics, and different personal care items that are unfit within cosmetics rules and regulations have been developed during the last several years. This has caused substantial worry in the aquatic settings, and their ecological as well as environmental effect is linked to the vast amount utilized, as well as the fact that they are sometimes persistent, and possibly bio-accumulative.

The microplastics have high surface-to-volume and are hydrophobic in nature, hence accumulates in the form of organic contaminant like poly-brominated diether (PBDs), polychlorinated biphenyls (PCBs), polyaromatic hydrocarbons (PAHs), and pharmaceutical and personal care products. Even different metal contaminants such as Zn, Co, Cd, Cu, Cr, Ag, Hg, Pb, and Ni can accumulate on the microplastic surfaces. However, the small-size microplastics have a high possibility of adsorption, particularly in aquatic conditions, accumulating a large number of organic pollutants on its surface. However, there is relatively inadequate knowledge of the loading, destiny, and behaviour of microplastics in the terrestrial habitats. Moreover, microplastics have entered the soil through landfill leachate, wastewater biosolids application to agricultural land, agricultural mulches, maintenance of outdoor plastic substances, as well as layered surfaces. The microplastics and nanoplastics can be transformed by natural/anthropogenic colloids, such as homo-aggregation and hetero-aggregation, interactions with macromolecules and microorganisms, and biodegradation. The MPs and NPs are easily transported in groundwater and make it polluted. Despite the fact that numerous studies have looked at the alterations and transference of naturally occurring colloids and manufactured materials like nano-metals and nano-metal oxides in terrestrial settings. In this manner, heavy metals and persistent organic pollutants can be transported through vectors such as MPs and NPs. Still, the level of microplastic pollution and its impacts on the environment as well as health is unknown.

Given the growing scientific concern over microplastic contamination in the aquatic environment, many countries of Europe (Austria, Belgium, Netherlands, and Sweden) have made a combined request to ban microplastics in various cosmetic preparations. Cosmetics Europe has advised its members to phase out the synthetic and solid plastic use. The exfoliating and cleaning of non-biodegradable particles from marine environment was the prime target. The President of the United States approved legislation in December 2015 that tried to stop the production and sale of cosmetics prepared from plastic microbeads. Major cosmetic firms have recently pledged to use natural alternatives of microbeads including diatomaceous earth, pumice, walnut, coconut, and powdered palm nut shells, as well as bamboo powder, apricot kernel shells, and shells from walnut and coconuts during manufacturing microplastics.

This chapter will provide an overview of the usage of microplastics in the various industries and its impact on the human health and nature. We attempted to describe several types of plastics, as well as their consumption and impact. In addition, we have included information on where microplastics can be detected after using cosmetic and pharmaceutical items. Following that, we discussed the status of microplastic beads, their alternatives, prevention, and what respective governments have done to combat microplastic contamination.

4.2 Types of Microplastics

The term Microplastics, as coined by Thompson et al. in 2004, encompasses the smaller particles of plastics (Thompson et al. 2004). It differentiates larger plastics called as macroplastics from plastics that are small enough to be seen by naked eyes (micro) (Table 4.1). During the decomposition process, large plastics break into macroplastics (>25 mm), mesoplastics (5–25 mm) and finally to microplastics (<5 mm) before decomposing completely (Ng et al. 2018). The microplastics are further categorized as primary microplastics or microbeads and secondary microplastics (Ivar do Sul and Costa 2014; Napper et al. 2015; UNEP 2015, 2016). Although microplastics are particles that fall in the range from a few mm and up to 5 mm, according to the IUPAC definition of microparticle, the particle size as low as 100 nm qualify to be called as microplastics. Therefore, the plastic that enters the environment in their original small size without further degradation from applications like cosmetics, clothing fibers, and drug delivery is called primary microplastic while those that appear due to macro/microplastic degradation or fragmentation are called secondary microplastics (Wang et al. 2018; Cole et al. 2011; Andrady 2011).

Primary microplastics or microbeads up to 5 mm in size are present in rinse-off personal care products that are used to cleanse and exfoliate (Bashir et al. 2021). Based on the method of synthesis and chemical composition microplastic can vary in parameters like shape, size, and density (Napper et al. 2015). A variety of synthetic polymers can be used to compose the microbeads including solid, hollow, amorphous, solubilized, etc. including synthetically manufactured beads, pellets, fibers, and powders.

As discussed above, secondary microplastics originate from the decomposition or fragmentation of macro plastics (Andrady 2011). Degradation and fragmentation of polymeric materials is caused by weathering making them smaller in size or even in to powdery form (Andrady 2011).

Similarly, the physical-chemical changes at a nanoscale level in the particle characteristics, result in the formation of nanoplastics (Klaine et al. 2012). The conductivity, reactivity and strength is expected to differ considerable from macroparticles (Mattsson et al. 2015, 2017).

Table 4.1 Types, sources, and composition of microplastics (Raju et al. 2018)

Microplastic type	Sources	Composition
Primary microplastic	Intentionally manufactured to use in cosmetic products, personal care products, cleansers, exfoliants, powders for injection molding	PE, PPS, PP, PS granules
Secondary microplastic	Fragmentation of plastics litter due to wind, water, and sunlight	Industrial wastewater, polyester, ropes, rayon, nylon, carpets, fishing gear

According to the Danish Environmental Protection Agency (2015), personal care products include thermoplastics like polypropylene, polyethylene, polytetrafluoroethylene and polystyrene and thermosets like polyurethanes and polyester and particles between 1 and 240 μm in size (Denmark EPA 2015). Hydrophobic organic contaminants get readily adsorbed on non-polar surfaces and organic matter and show a greater affinity towards plastic surfaces (Carbery et al. 2018). The high surface area-to-volume ratio along with the hydrophobic nature accelerates the aggregation of organic pollutants like perfluorinated alkyl substances (PFAS), polyaromatic hydrocarbons (PAHs), polychlorinated biphenyls (PCBs), pharmaceuticals and personal care products (PPCPs) and polybrominated diethers (PBDs) (Li et al. 2018).

4.3 Uses of Microplastics

Microplastics are generally used in cosmetics and personal care products and also in medical applications primarily in dentistry as dental polymers used in sealing, filling cavities and in dentures, and as an abrasive in tooth polishing (Table 4.2). They are also used as additives in pharmaceutical formulations for controlled release (Ooms et al. 2015, Woodruff and Hutmacher 2010, Bugnicourt et al. 2014).

According to Leslie (2014), plastic ingredients are partly present in the formulations of various personal care products like sunscreens, make-up foundations, skin creams, nail paints, eye make-up, mascara, blush powders, toothpaste, facial cleansers, shower gel, bubble bath shampoo, shaving creams, deodorants, hairsprays, hair colors, baby products, and insect repellents.

Table 4.2 The following product categories containing various types of microplastics have been identified in cosmetic and skin care industry

Product category	Subtype	Types of microplastics
Skin care and cleansers	Hand cleansers	Polyurethane (200–1250 μm)
	Body foot scrubs	Polyethylene (<1 mm)
		Cellulose acetate (300–400 μm)
		Polylactic acid (<315 μm)
	Face scrubs	Polyethylene (<1 mm)
		Polylactic acid (<315 μm)
		Cellulose acetate (300–400 μm)
	Face masks	Polyethylene (<1 m)
		Polyethylene (<450 μm)
		Polyethylene (<1 mm)
		Nylon-11(150 μm)
	Hair cleansers, shampoos	Polyethylene (<315 μm)
	Foot care	Polyethylene (<1 mm)
	Soap	Polyethylene

Source: European Commission (DG Environment) Report (2017)

Microplastics are added to achieve exfoliating, scrubbing, and cleansing in these formulations. However, its use exceeds the primary function as it also it effects other properties such as viscosity, emulsification, film forming and binding. Hence, it is used in cosmetics and detergents as abrasives, emulsifiers, opacifying agents, anti-static agents, suspending agents, thickening or binding agents, or as colored microplastics in make-up; in pharmaceuticals and fertilizers as control release of ingredients (European Commission (DG Environment) Report 2017). Therefore, these polymers are widely used to get the desired properties in a product.

It is also used in a variety of cleansing products like facial cleaners, scrubs, peelings, shampoo, body wash, bath additives, and liquid soaps (European Commission (DG Environment) Report 2017). Microplastics have also been found in color cosmetics like make-up and nail polish, and in skincare and hair care products (Eunomia 2016; Government of Canada 2015).

Skin care and cleaning products contain polyurethane (size, 200-250 μm) responsible for more than half of the tonnage of microbeads (European Commission (DG Environment) Report 2017). Most of the rest are contributed by polyethylene particles (>1 mm) from face scrubs, body scrubs, and face masks. Although studies suggest that due to the voluntary action taken by the cosmetics industry, the usage of plastic microbeads has decreased by 82% between 2012 and 2015 (Cosmetics Europe). However, the use of microplastics in other industries is still increasing. The total European paints and coatings market is estimated to be around six million tonnes worth some €20 billion per year (Vincentz Network 2014). In addition, the use of microplastics in drug delivery and in dental polishes is also increasing in the medical industry (Cole et al. 2011; Sundt et al. 2014).

4.4 Presence of Microbeads

4.4.1 Air

Plastic litter has become a global challenge for all the countries as it is found everywhere and is expanding uncontrollably. Microplastics are transported to the marine via rivers on a global scale. Microplastics have reached almost every corner of the earth, even by air. As per a report, microplastics have been deposited in remote areas, mountains, and in pristine due to the winds wherein 249 fragments, 73 films, and 44 fibers of microplastic per cubic meter are deposited on the catchment (Vincentz Network 2014). Data showed that microplastics can travel over 95 km. S. Dehghani et al., found in 30 g dry dust there is about 88–605 microplastic with the dominance of yellow and granule microplastics. The size of the microplastic found was about 250 to 500 μm, and the sample was collected from 10 different streets using a binocular microscope. Further, it is reported that about 3223 and 1063 microplastic particles are ingested by adults and children every year, respectively. Melanie Bergmann et al., study the deposition of microplastics in snow through atmospheric transport. For comparison, they obtained samples from populated and remote European sites. They found that in comparison to European snow, arctic snow had a significantly lower amount of microplastics (Sundt et al. 2014). This study proved that the transportation of microplastic is not limited by a river, and soil but also occurs by air; however, more research needs to be done to understand the impact of wind on the transportation of microplastics (Leslie 2014).

4.4.2 Water

There are three main types of water bodies on Earth and each is polluted by microplastic and are discussed briefly as follows:

4.4.2.1 Ocean

Approximately, around eight million tons of plastic enters the marine every year (Fig. 4.1). However, the amount of plastic found on the surface is 100-fold lower. This large difference in the calculation may be due to the imprecise measurement. Either we have overestimated or underestimated the amount of plastic released into the marine as the determination of the exact amount of plastic is extremely difficult. The main reason for the microplastic build-up in the ocean is the usage of hand soap, makeup, face wash, and other exfoliators. Once this product is used and washed with water, it enters into the sewage system which is not filtered as the size of microplastics is much smaller, thus entering into the ocean (Magni et al. 2019). Eventually, ocean wildlife, especially filter feeders, are harmed by these beads because they can readily absorb the plastic making them sick. Therefore, using these dangerous products should be avoided by the humans because microplastics

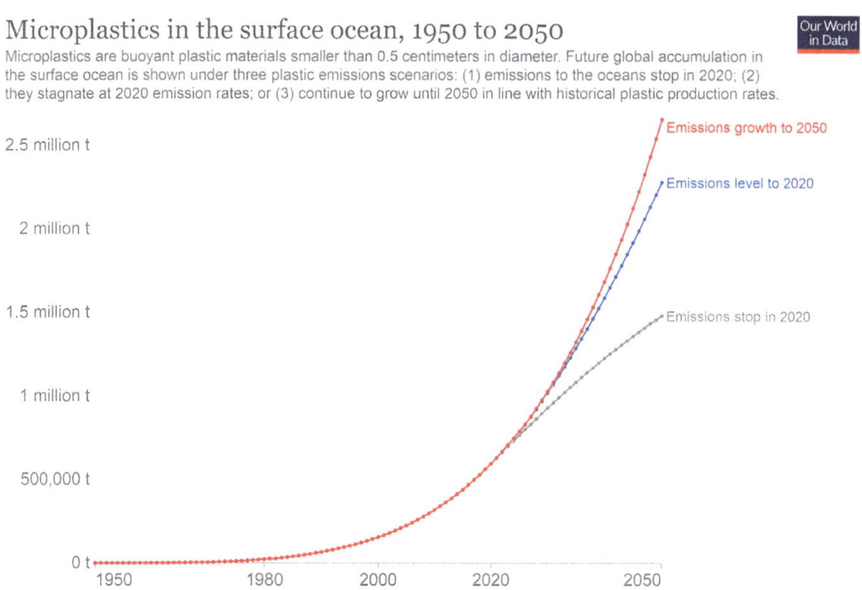

Fig. 4.1 Microplastics in the surface ocean, 1950 to 2050 (Lebreton et al. 2019)

are very difficult to clean owing to their micro size. Moreover, microplastics enter the sea when we litter the plastic on the seashore. To stop the spreading of microplastics we must take appropriate steps to overcome this problem such as poor waste management practices stopping using plastic or ban of plastic usage near the seashore (Huang et al. 2020).

4.4.2.2 Fresh Water

Microplastics are a complicated set of pervasive environmental pollutants that are causing increasing concern. These microplastics disintegrate slowly and emit plasticizers, allowing them to be freely transmitted. Freshwater has been examined for microplastics far less frequently than marine sediments. According to one research, the amount of microplastics in freshwater is roughly 13,822 particles/m^2. Moreover, three-year research in Lake Winnipeg, Canada, discovered a mean microplastic concentration of 193,420 particles km2 g. Fibers produced by the breakdown of bigger particles, synthetic fabrics, or atmospheric fallout formed the majority of microplastics whereas, none of the microplastics found were micro-pellets or beads. Furthermore, there are around 1108 particles per square meter near Lake Garda, Italy-N coast. According to the findings, freshwater waters have a high concentration of microbeads, which aggregate and are then transmitted to marine ecosystems and humans (Huang et al. 2020).

4.4.2.3 Ice Core

Microplastic pollution has reached such an extent that it is found in all water bodies including ice cores. In a study, 14 different types of polymers were found in 96 microplastics from the east Antarctica ice core sample. Further, is earlier it is recorded that in Antarctic surface water and the sediments of arctic sea ice found for the first-time plastic. However, the spreading of this plastic may be due to local pollution sources (Huang et al. 2020).

4.4.3 Soil

On a universal scale, microplastics have been verified, including in North America, Asia, Africa, Oceania, and Europe. Across different sites abundance variation in microplastics been observed. In grasses and cropland soil of Chile's Area Metropolitana, contain 184 $\pm$ 266 and 306 $\pm$ 3 microplastics concentrations, respectively. In Sydney's industrial area soil, the range of microplastic is from 350 to 6950 items/kg. In Washington, the USA, similar quantity of microplastics is existed in aquatic gardens and in the freshwater wetlands (1280 $\pm$ 150 items/kg) (Hämer et al. 2014). Low microplastics are present in Southeast Mexico Mayan home park German roads of Franconia, German floodplain areas of the River Lahn, and in Switzerland the textures and the catchments of the floodplain soils. In the agriculture field every so often sewage sludge aids as fertilizer is a chief pouring constituent of soil microplastic pollution (Huang et al. 2020). The sewage sludge discarded soil is collected, and it is revealed that approximately 35 items/g of microplastic existed in it. In eastern Spanish, the microplastics that exists in sewage sludge is around 17,000 particles/kg of light-density microplastics concentration and 32,080 ± 19,070 particles/kg of heavy microplastics concentration. Correspondingly, biosolids also subsidize microplastics in the soil. In Ontario, Canada, high concentration of microplastics has existed in biosolid samples between 1.4 × 104 items/kg and 8.7 × 103 items/kg. It is revealed that in Ontario annually agricultural soils have between 1.3 × 1012 and 4.1 × 1011 microplastic particles added.

It is revealed that there is a global distribution, occurrence, and contamination of microplastic in soil. In China world's 30% of plastic is produced a huge quantity of microplastic is battered in China. The first survey was carried out in China, east coast saltworks and it is found that 734 items/kg of microplastic existed. In the Shandong coastline, soil 1.3 to 14,712.5 items/kg concentration of microplastic existed. Apart from coastal regions, huge existence of microplastics are found in northern china which includes adjacent plastic waste, mulching film, inappropriate dumping of agricultural plastic films, and other causes. It has been revealed that the use of plastic mulching films for 30 years resulted in microplastic at 40.35 mg/kg concentration in agriculture soils. Comparable results are found in 384 soils from 19 countryside across China.

4.4.4 Human

Microplastic is exposed to the human body through air inhalation, food ingestion containing microplastics, and dermal communication of these elements, contained in textiles products or dust.

4.4.4.1 Ingestion

The chief route of human exposure to microplastic is ingestion. Per year per person 39,000 to 52,000 microplastic particles are consumed through foodstuffs. Soiled foodstuff results in the inhalation of microplastic particles which are then passed through the gastrointestinal tract through mucociliary clearance, probably owing to inflammatory response, rise in permeability, and alteration in gut flora metabolism and composition. It is been revealed that microplastics are been existed in various food items like commercial fish, mussels, sugar, table salt, and bottled water so they are likely ingested. Through the ingesting of bivalves, it has been estimated that Europeans have been exposed to 11,000 microplastics per person/per year (Dris et al. 2018). Then, microplastic ingestion through table salt in China, and Europe is 100 and 37 microplastic per person/year, respectively. It is estimated the hat deposition of dust on food dishes during mealtime is highly dangerous than microplastics that already existed in food. The ingested particles are absorbed by M-cells that existed in the intestine covering a colonic lymphoid tissue leading to the gastrointestinal mucus. Particles that are insoluble penetrate in mucus intestine and increase the solubility caused by intestinal matters adsorption. Another mechanism of particle internalization is per sorption (Magni et al. 2019). In the per-sorption process, the particles pass via the intestinal epithelium. Microplastic could be endangered to a similar mechanism as its passage to the cardiovascular system is demonstrated earlier in vivo (Rillig et al. 2017). The inhalation of microplastics by humans may negatively impact the body organs, inhibits cell viability, alters gene expression, encouraged pro-inflammatory responses, and cause hazardous health issues. Humans are more exposed to microplastics as a huge concentration of environment and food is soiled with microplastics (Rillig et al. 2017).

4.4.4.2 Inhalation

There are tremendous sources through which the microplastics are released into the air or environment like abrasion of substances (buildings, tires, and cars), and surface resuspension of microplastics. The microplastic indoor concentrations in air is about 0.4–56.5 particles m^{-3} and the outdoor concentrations is about 0.3–1.5 particles m^{-3} incorporating the inhalable amounts. It is estimated that per day individual inhalation of airborne microplastics is 26 to 130. By the mannequin method, an air sample is taken and it is revealed that with light activity per day 272 microplastics are inhaled by the male person (Magni et al. 2019). On space use factor and on sampling methodologies various estimations depended on furniture materials, cleaning schedule, season, and activities. Particle properties like density and size will impact its accumulation on the respiratory tract, with smaller and scattered particles transferring deeper into the lungs. By deposition of macrophages or immigration to the lymphatic or circulation system may tend to translocation of particles (Dris et al. 2018). Nonetheless, the big surface area of small particles in the respiratory tract may encourage a strong distribution of chemotactic features that rise permeability tending to chronic inflammation and preventing macrophage migration known as dust excess. The worker of the polyvinyl chloride or vinyl chloride industries, flock industries, and textile industries are exposed exceedingly to microplastics through the airway which results in respiratory symptoms and disease in the lung intestine. The causation for the 250 μm fibers detected in individual lung biopsies, comprising in cancer biopsies has not been demonstrated. Consequently, airborne microplastics cause injuries to the respiratory tract under heavy concentration and vulnerability of the individual (Huang et al. 2020).

4.4.4.3 Dermal Contact

The low substantial route of exposure for microplastics is with dermal contact but it is revealed that nanoplastics (<100 nm) can cross the dermal barrier. The dermal route is more often accompanied by the monomer's introduction and plastic additives like phthalates and bisphenol A, from regular use of gadgets (Magni et al. 2019). Therefore, the nanoplastics that penetrate through the dermal route cause toxicity damage human organs, and altered the functions of cells. Microplastics in medicines and pharmaceuticals are responsible for foreign body and inflammatory reactions, with fibrous encapsulation. Low inflammatory reaction is observed with surgical sutures using monofilament polypropylene and polyester after 21 days than fibrous and silk encapsulation. Nanoplastics and microplastics induce foreign body reactions and inflammation, altered surface properties, and tend to dissimilar outcomes. By exposure to nano and microplastics oxidative stress occurs in human epithelial cells. Thus, there are various sources of dermal exposure like synthetic fibers, dust, and microbeads in cosmetics, and prospective adverse effects are observed on human health (Hämer et al. 2014).

4.5 Effects of Microplastics

4.5.1 Soil Ecosystem

Microplastics can impact the biophysical properties of soil and also influence the soil pH, soil texture, structure, nutrients, and fertility of the soil, also affecting water-stable aggregates and soil microbes. The existence of microplastic will assimilate soil aggregates, and in changing degrees of combination: tightly in fiber-type and loose in fragment sorts. Soil aggregates play a vital role in influencing the habitat of soil creatures (Huang et al. 2020). Then further, aggregation of soil has a major role in the soil porosity, which negatively impacts the movements of water, and gas, and altered the activity of microbial species. If microplastics are present in the soil the bulk density is decreased, and also reduces or indirectly impacts on quality of the soil. Microplastics in the soil changes the physicochemical parameters and affects microbial activity and hydrodynamics. The impacts depends on the size and shape of microplastics particles. It is revealed that presence of Polyester (PES), PP, PET, PS, and polyethylene high density (PEHD), the bulk density of soil declined. The leaf of plants, root traits, and biological and biomass content have also been changed (Hämer et al. 2014). It is observed that microplastics extends the aggregates in soil that are water-stable compared to the control cluster and also hastens soil water evaporation by generating water channels for the movement of water and also enhances the concentration of microplastic. Microplastic accumulation damages the structural integrity of the soil and causes drying cracking on soil surfaces. Microplastic in the soil also changes the soil pH (Dris et al. 2018).

4.5.2 Microplastics in Plants

Soil microplastics are a composite of environmental contaminants, and integrate into other chemical contaminants like dioxins, heavy metals and persistent organic pollutants (POPs). However, the contrary impact on plants also leads to additives connection in the polymer. Frequently additives are plasticizers, antioxidants, flame retardants, and heat stabilizers which are having great risk to the ecosystem (Hämer et al. 2014). Furthermore, the existence of microplastics will alter soil chemical and physical parameters and also changes the vegetative stage and root system and affects the growth of plants. It is been revealed that the contamination of microplastics impacts the aboveground, underground organs plants' growth, and declines the quality of the soil. If there is the existence of high-density polyethylene (HDPE), fibrous microplastics, and biodegradable polylactic acid, there is a reduction of germination height and germination rate of ryegrass. It is concluded that on

Vicia faba the genotoxicity and ecotoxicity of polystyrene microplastics and found that when microplastics concentration reached 100 mg/L, the substantial inhibitory impact is seen on the development and growth of plant species and these microplastics possibly block cells or blocks pores of the cell wall for passage of nutrients. It is proved that macroscopic and polyethylene residues of macroscopic and decomposable mulch plastic film have contrary impacts on wheat during reproductive and nutritive growth. It is concluded that the soil which has a high concentration of biomass has an extreme concentration of nutrients, water, and space.

4.5.3 Microplastic in the Food Chain

The health risk and ecological damage that occurred from microplastic exposure are chief issues in the area of research. It is been revealed that in the food chain microplastic can be transferred from primary trophic levels to a higher level from prey to predator. It is concluded that the micro and macro plastics are transferred from soil to earthworm and from earthworm to chicken faeces with an increase in the concentration of microplastic. The springtails, nematodes, mice, and snails are incorporated in soil with increased concentration of microplastics to burrow and egest casts, and it is concluded that the great concentration of microplastics is transferred in the organism's body. There is a great chance that the existence of microplastic in the soil can also contaminate groundwater eventually entering into it and also expose it to other soil microorganisms, and organisms. It is also found that microplastics are also present in colectomy specimens and human stools confirming the microplastics existence in the human digestive system. The pathway and origin of microplastics are through the food chain, by drinking water, ingesting salt, or other contaminated food substances contaminated with microplastics. There is a health risk, and damage to the intestine are also been demonstrated in marine organisms. The ingestion of microplastics causes oxidative stress, inflammation, and particulate toxicity and also causes the tumor to hens. The entering of microplastics into humans through the soil food chain and food web causes severe hazards (Hämer et al. 2014).

4.6 Fate of Microplastic Beads

Emerging urbanization and the virtues of modern lifestyles are adding more to the degradation of the environment. Plastic pollution is a similar component to say. During the past few years, pollution caused due to plastic is a very serious issue worldwide. However, recent trends show much concern for microplastic pollution as well. Due to various factors, plastic items can also undergo breakdown and convert into smaller fractions which are later considered Microplastic (Hämer et al. 2014).

Microplastics being very minute in size and also low weight makes them travel at a long distance. Cities are considered important places where microplastics get deposited in many places. Many activities like buildings, constructions, transportation, packaging, textiles, etc. involve the usage of plastics. Dris et al. (2018) has revealed the microplastics deposition in Urban areas like Paris city. Results depict traces of microplastics at various compartments of the Paris megacity including atmospheric fallout, urban runoff water, gray water, Wastewater Treatment units, etc. Stormwater runoff estimated also was detected for microplastic fibers indicating the abundance of microplastics in the city. Wastewater generated from the sewer systems was also analyzed for the availability of microplastics. Here the wastewater from washing machines was collected and analyzed where 8850 and 35,500 fibers/L of microplastic concentrations were encountered compared to 120 and 810 fibers/L microplastic concentrations for control washes. Wastewater treatment plants and generated sludge has also encountered particles of microplastics. They have also revealed the occurrence of microplastics in freshwater sources and reservoirs (Dris et al. 2018).

Primary microplastics to be used in air blasting technology, or as abrasive agents in personal care products are produced directly with microscopic size. The use of Primary microplastics in cosmetics has been recently banned or limited in countries such as the UK, US, and Canada. Microplastics deriving from the degradation of plastic wastes or from synthetic cloth washing are of secondary origin. However, 80% of global microplastic pollution is contributed by the land sources and debris collected from urban areas is supposed to be treated by Wastewater Treatment Plants. However, Wastewater Treatment Plants are not efficient in the removal of contaminants such as heavy metals, pharmaceuticals, illicit drugs, and also microplastics that might get discharged into the aquatic environment (Magni et al. 2019).

The light weight of microplastics is one of the prominent reasons for their distant dispersal of them. Ocean currents and the winds make these Microplastics available abundantly in the oceans and the surrounding environment. The manufacturing process of plastics, and obviously, later microplastics, various chemicals, and their compounds are used. This may prove fatal as they release some toxic components and pose a threat to aquatic organisms. Many research activities have confirmed the presence of ethers, polycyclic aromatic hydrocarbons, polychlorinated biphenyls, organochlorine pesticides, and polybrominated diphenyl, as the significant components of microplastics. Hence, considering these factors, the fate of microplastics for transferring within the food chain and their environmental presence has made it a matter of concern (Gatidou et al. 2019).

Sources of microplastics can be distinguished on the basis of their origin viz. land-based and aquatic-based. Land-based sources include stormwater runoff, municipal sewage, and riverine transport. A study conducted by Cheung and Fok focuses that 306.9 tonnes of microbeads are released into the environment per year in China while 80% of them are derived from the effluents of Sewage Treatment Plants. Various other research has also revealed the fact of the deposition of microplastics of acrylic fibers and polyester into the oceans as a result of sewage discharge (Gatidou et al. 2019). Wastewater after treatment from WWTP when discharged into inland water also emits microplastics along with it, hence posing a threat to them. But another concern associated with WWTP is the settling of microplastics at the bottom of treatment units and thus getting accumulated into the sludge. This may be a very serious issue as many times sewage sludge is applied on land for soil amendments and also as fertilizers. Hence, the study revealed the very fact that microplastics present in the WWTP may get entry into the aquatic ecosystems as well as a terrestrial ecosystem through the water being used for irrigation purposes or sludge being a fertilizer. Hence, the WWTP may sometimes act as an anthropogenic sink of microplastics further damaging the natural environment (Gatidou et al. 2019). Microplastics being very tiny in size, a variety of aquatic organisms may ingest them and get entry into the food chain. Also, these aquatic organisms ranging from larger animals to smaller invertebrates may suffer from reduced appetite, clogging of digestive organs, and integration of microplastics into body tissues (Hämer et al. 2014). As the microplastics enter into aquatic bodies, they do also contribute some significant traits of deposition of natural sediments. The dispersal into coastal environments and physical transport follow a similar path wherein the dense particles get deposited near the source (Harris 2020).

4.7 Alternative of Microplastic Beads

Personal care products and other cosmetics products are extensively used all over the world for a long time. To name a few it includes sunscreen lotions, creams, make-up items, exfoliants, body and face scrubs, toothpaste, and industrial hand cleaners. Their use and demand all is ever-increasing. But since the 1990s, all cosmetics, on a large scale, are being manufactured by using microplastic beads so as to enhance their abrasive nature and other relevant qualities. Many cosmetics have started replacing natural ingredients with synthetic materials like plastic beads. Ground almonds, oatmeal, and pumice are a few natural ingredients that were used earlier for their abrasive and other properties in cosmetic productions (Hunt et al. 2021).

It is commonly observed that cosmetic products which use microplastic beads include facial and body scrubs, toothpaste, lotions, etc. which are usually rinsed off. Hence this leads to the mixing of microplastic into the wash water which in due course of time drains to the sewer systems. Ultimately, these are carried to the Waste Water Treatment Units where again they get loaded with certain chemicals and other toxic elements. Waste Water Treatment plants are partly ineffective for microplastic removal from wastewater. Hence, such water after treatments when released into nearby aquatic bodies or onto land also carries a load of microplastic along with them. These accumulated plastics last in the environment for a quite long time affecting aquatic ecosystems. Marine ecosystems are the most impacted due to the presence of microplastics due to such events (Hunt et al. 2021).

The cause of the regulatory restrictions on the use of plastic microbeads is said to be the scientific evidence of environmental release and harm it has caused; the availability of alternative materials has also played a critical role in bringing the regulatory restrictions (Table 4.3). In cosmetics actually lot many alternatives are available which are natural and also not harmful to the environment. It includes

Table 4.3 Comparative analysis for alternatives of microplastics (Hunt et al. 2021)

Sr. No.	Alternatives for microplastics in cosmetic products	Average price/Kg (in US dollars)
Plastics		
1	Polyethylene terephthalate	50.00
2	Polypropylene	0.90
3	Nylon	95.00
4	Polyethylene	3.55
5	Polymethyl methacrylate	1.60
Minerals		
6	Pumice	0.43
7	Titanium dioxide	1.90
8	Bentonite	0.22
9	Salt	0.22
10	Silica	0.27
11	Silicate materials	0.20
12	Mica	0.29
Plant products		
13	Apricot kernels	11.00
14	Jojoba wax	30.50
15	Almond shells	6.55
16	Maize powder	17.50
17	Bamboo resin (or powder)	24.00
18	Microcrystalline cellulose	5.55
19	Bitter orange	3.00
20	Oats	2.40
21	Citric acid	0.71
22	Olive stones	0.10
23	Coconut husk	0.25
24	Tagua nut	7.00
25	Pineapple bark	15.00
26	Wheat bran	0.10
27	Sugar	0.22
28	Walnut shell powder	11.00
Animal products		
29	Pearls	90.00

crushed oats, walnut shells, sugar, and jojoba seeds which are readily available in the markets and are also used traditionally in cosmetic products. Hence, they act very strongly as alternatives to microplastic beads which in turn are also very safe for the environment as well. Efforts are ongoing to develop synthetic cellulose based biodegradable microbeads. The research conducted by Hunt et al. (2021) reveals many natural plant and animal products, minerals, and other synthetic materials as an alternative to microplastic beads in the use of cosmetic products. The list ranges from about 29 products. The plant product includes almond shells, oat, maize, apricot kernels, wheat bran, bamboo resin or powder, walnut shell powder, bitter orange, tagua nut, citric acid, sugar, coconut husk, pineapple bark, jojoba wax, olive stones, microcrystalline cellulose which can be used for their abrasive nature and other important features of cosmetic products. Pearls are also used as an alternative which are the derivatives of animal products. Also, certain minerals can also be used as alternatives for cosmetics. Minerals include bentonite, mica, pumice, salt, silica, silicate materials, and titanium dioxide.

Currently, many research projects are going on to find out various items that can be replaced for petroleum-based microplastic beads. Biopolymers are considered to be one of the promising alternatives for microplastic beads. Among many other biopolymers, Chitin is one of the widely accepted biopolymers for the above-mentioned use. Chitins possess their own magnificent natural properties of being biodegradable, biocompatible, and non-toxic. They are available in large quantities in waste sources and are readily degradable by microbes and enzymes (King et al. 2017). All properties of chitin make it very popular to use in place of microplastic beads. Hence, it can be replaced in sectors of pharmaceuticals, personal care, cosmetics, and medical applications like enzyme immobilization and delivery of drugs. King et al. (2017) have recently reported the extraction and re-generation of chitin shrimp shell waste which can replace polymer-based microplastic beads. They also claim that the use of such biopolymers in various products can also ultimately reduce the effect of microplastics on the environment (King et al. 2017).

Kozlowska et al. (2019) have also revealed the harmful effect of the microplastic used in cosmetics, especially the peeling products or exfoliators which are basically rinsed off with water. Being non-biodegradable, they are persistent in the environment. Many European countries have already banned the use of microplastic beads in cosmetic care products. Hence alternative elements are being found to replace the microplastic beads. Kozlowska et al. (2019) have studied the use of biopolymers like sodium alginate and starch microparticles (Fig. 4.2). Their use as abrasive agents have been tested during their study and has proved to be used efficiently as an alternative for microplastic beads. There are various methods of production of macroplastics and microplastics from synthetic polymers (Fig. 4.3). But the macroplastics and microplastics use must be limited to prevent its hazardous effect on environment.

4.8 Policies for Banning the Use of Microplastic Beads

Various research conducted all over the world had already proven the fatal belongings of microplastic on the environment. Microplastic deteriorates the spheres of the environment in many ways viz. aquatic ecosystems get damaged at an irreversible rate. Marine and freshwater organisms also get affected to a large extent. Mixing of microplastic beads in the soil makes them remain persistent in the soil for a prolonged period of time. Hence it makes the soil harmful in many ways. Microplastic beads since are being used largely in cosmetic products and personal care products they usually get rinsed off with water and ultimately end up at the

Fig. 4.2 Biopolymers as alternatives for microplastics

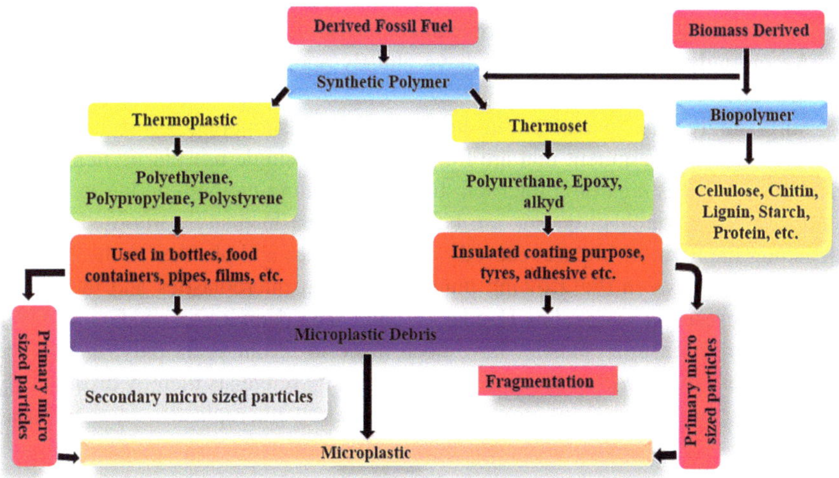

Fig. 4.3 Production process of the artificial and natural polymers and generation of macroplastics and microplastics from synthetic polymers

wastewater treatment units. Hence, WWTP becomes the ready source of discharge of fine microplastics into the environment (Sharma and Chatterjee 2017).

Though much research is going on to replace microplastic beads on a commercial scale, it has its own limitations too. Hence, eliminating the microplastic completely from its source itself is one of the very effective ways to get rid of the chaos caused by microplastics. Using natural ingredients of plant and animal origin is the easiest cost-effective way since it's the traditional use, especially in cosmetic products and personal care products. Calero et al. (2021) have conducted research to find out the best strategies that can be implemented to phase out microplastics. They concluded that the improvement of legislation for the manufacture, use of alternative materials and recycling along with the design of plastic products forms the basis for the overall achievement of the stated issue. Along with it, they have also stated that the improvement in the provisions for waste water treatment, and the Research and Development in bioplastics, are important tools to limit microplastic pollution.

Public concern has increased due to the scientific studies and evidence reported on the effects of single-use plastics on the public health and the environment. It has already paved for the prohibition on such single-use or disposable plastic items. Already many countries across the globe are taking serious action against this plastic maniac. Many governments have applied different strategies against the use of such types of materials posing a threat to the environment. Anagnosti et al. (2021) have conducted a research review to study various actions and policies taken by countries across the world. It includes a ban on the use and throwing of products like plates, cups, spoons, straws, etc., lightweight carry bags, and much more in the list. Monitoring the usage and claiming economic means against the use of plastic items was forced on its member states by the EU in 2015 (European Parliament and the Council, 2015). A tax or levy on the usage has been already subjected to from the early the 90s in some European countries.

Accordingly, 127 UN members have already adopted some sort of legislation (UNEP 2018).

Surprisingly, the general public has also already supported the strategies and policies of the government that are followed against single-use plastic items. Ultimately, awareness regarding microplastic is also gaining momentum in the last few years. Many research programs are being conducted in many parts of the world depicting the harmful impacts caused by microplastics and plastics on the environment. a link between the growing public interest in microplastics and the rise of scientific research was noted by Henderson and Green (Henderson and Green 2020). Efforts to restrain microplastics have been made in recent years only. Systematic research has been carried out in the last few years by scientists thereby distinctively identifying the existence of primary and secondary microplastics in the marine (Anagnosti et al. 2021).

4.9 Global Pressure Against Microbeads

Since 2012, the global pressure against the use of microbeads in cosmetic products has been highlighted by Anagnosti et al. (2021). Intergovernmental organizations have adopted plans against microplastic littering the oceans due to the opposition from various (NGOs) like the *Plastic Soup Foundation*, and a scientific explanation of the fallouts of ingested microplastics on the human and aquatic health (NSW EPA 2016). Reports from scientific institutes and environmentalists were ordered by

Governments organizations for the possible impacts of microplastic and its prohibition (Anagnosti et al. 2021).

In 2014, the first government agency to prohibit the use of non-biodegradable microbeads in cosmetic and personal care products was the state of Illinois, this was followed by 15 other States in the USA (United Nations Environmental Programme (UNEP) 2018). A US federal law called the "Microbeads Free-Water Act" was voted one year later, that bans the sale and manufacture of plastic microbeads in rinse-off products that are non-biodegradable, excluding biodegradable microplastics. An immediate impact of the United States Act was seen on other countries that were reluctant in taking the action (Government of Canada 2017). A unanimous voting to include microbeads under the toxic substances in the Canadian Environmental Protection Act of 1999 was observed in the Canadian House of Commons in 2015. In 2019, prohibition on the manufacture of household chemical substances comprising microplastic and a total sale ban was issued in China under the new guidelines for industries (China's National Development and Reform Commission 2019).

USA, China, Canada, New Zealand, South Africa, and Taiwan are the only non-European countries to ban the usage of microplastic in rinse-off cosmetics and personal care products (United Nations Environmental Programme (UNEP) 2018). India is currently contemplating the legislation whereas Iceland has signed a commitment to ban the microplastic usage. In 2016, a law to prohibit the use of non-biodegradable microplastic in cosmetics and personal care products for exfoliating, bleaching, grinding and cleaning was introduced by the government of Brazil (Anagnosti et al. 2021).

Microplastic usage in cosmetics and other personal care products is being banned in some countries due to the of government policies. New South Wales and South Australia signed an agreement to ban microplastic beads with the "national industry association of manufacturers and marketers of hygiene, cosmetic and specialty products" in 2016 (ACCORD (NSW EPA 2016). It is assumed that the ban was satisfactorily implemented as no prohibitional legislation has been enacted so far in the country. The South African government is considering the ban on microplastic bead usage in the cosmetic industry as declared in 2018. In 2014, a new directive was published by the European Commission to prohibit the rinse-off cosmetics that include microplastics to bare the Ecolabel sign, awarded to products with minimum environmental impact (Government of Canada 2017). The initiative was followed Belgium, Austria, Netherlands, Sweden, and Luxembourg (Government of Canada 2017).

4.10 Conclusion

The existing issue of microplastic contamination is anticipated to last in the forthcoming years due to the vast increase in manufacturing and production of plastics and the lack of efficient treatment for plastic waste. There is a need for public education and awareness relating to risk and negative impacts on humans, the ecosystem, and the environment on an urgent basis. This chapter tries to evaluate the use of microplastic beads in the pharmaceutical and cosmetics and personal care industry, their impact on the environment and individuals, preventions, fate, alternatives, policy on banning, etc. are been discussed. There is a need for permanent treatment for this global pollution with a combination of innovative, novel research, and public policies to reduce, reuse, rethink, and recycle so the plastic flow will get curtailed.

References

Anagnosti L, Varvaresou A, Pavlou P, Protopapa E, Carayanni V (2021) Worldwide actions against plastic pollution from microbeads and microplastics in cosmetics focusing on European policies. Has the issue been handled effectively? Mar Pollut Bull 162:111883

Andrady AL (2011) Microplastics in the marine environment. Mar Pollut Bull 62(8):1596–1605. https://doi.org/10.1016/j.marpolbul.2011.05.030

Bashir SM, Kimiko S, Mak CW, Fang JKH, Gonçalves D (2021) Personal care and cosmetic products as a potential source of environmental contamination by microplastics in a densely populated Asian City. Front Mar Sci 8:604

Bugnicourt E, Cinelli P, Lazzeri A, Alvarez V (2014) Polyhydroxyalkanoate (PHA): Review of synthesis, characteristics, processing and potential applications in packaging. eXPRESS Polym Lett 8:791–808

Calero M, Godoy V, Quesada L, Martín-Lara MÁ (2021) Green strategies for microplastics reduction. Curr Opin Green Sustain Chem 28:100442

Carbery M, O'Connor W, Palanisami T (2018) Trophic transfer of microplastics and mixed contaminants in the marine food web and implications for human health. Environ Int 115: 400–409. https://doi.org/10.1016/j.envint.2018.03.007

China's National Development and Reform Commission (2019) Guidance Catalogue for Industrial Structure Adjustment. Ministry of Industry and Information Technology of the People's Republic of China, Beijing

Cole M, Lindeque P, Halsband C, Galloway TS (2011) 529 Microplastics as contaminants in the marine environment: a review. Mar Pollut Bull 62(12):2588–2597

Denmark EPA (2015) Microplastics: occurrence, effects and sources of releases to the environment in Denmark (Environmental project No. 1793, 2015)

Dris R, Gasperi J, Tassin B (2018) Sources and fate of microplastics in urban areas: a focus on Paris megacity. In: Wagner M, Lambert S (eds) Freshwater microplastics: emerging environmental contaminants? Springer Nature, p 303

Eunomia (2016) Study to support the development of measures to combat a range of marine litter sources. Report for the European

European Commission (DG Environment) Report (2017). https://ec.europa.eu/environment/chemicals/reach/pdf/39168%20Intentionally%20added%20microplastics%20-%20Final%20report%2020171020.pdf

Gatidou G, Arvaniti OS, Stasinakis AS (2019) Review on the occurrence and fate of microplastics in sewage treatment plants. J Hazard Mater 367:504–512

Government of Canada (2015) Microbeads – a science summary. Environment and Climate Change. Government of Canada

Government of Canada (2017) Microbeads in toiletries regulations registration SOR/ 2017-111. Can Gazette II 151(12):1349–1376

Hämer J, Gutow L, Köhler A, Saborowski R (2014) Fate of microplastics in the marine isopod Idotea emarginata. Environ Sci Technol 48(22):13451–13458

Harris PT (2020) The fate of microplastic in marine sedimentary environments: a review and synthesis. Mar Pollut Bull 158:111398

Henderson L, Green C (2020) Making sense of microplastics? Public understandings of plastic pollution. Mar Pollut Bull 152:110908. https://www.cosmeticseurope.eu/news-events/reduction-use-plastic-microbeads

Huang B, Sun L, Liu M, Huang H, He H, Han F, Wang X, Xu Z, Li B, Pan X (2020) Abundance and distribution characteristics of microplastic in plateau cultivated land of Yunnan Province, China. Environ Sci Pollut Res Int 28:1675–1688. https://doi.org/10.1007/s11356-020-10527-3

Hunt CF, Lin WH, Voulvoulis N (2021) Evaluating alternatives to plastic microbeads in cosmetics. Nat Sustain 4(4):366–372

Ivar do Sul JA, Costa MF (2014) The present and future of microplastic pollution in the marine environment. Environ Pollut 185:352–364. https://doi.org/10.1016/j.envpol.2013.10.036

King CA, Shamshina JL, Zavgorodnya O, Cutfield T, Block LE, Rogers RD (2017) Porous chitin microbeads for more sustainable cosmetics. ACS Sustain Chem Eng 5(12):11660–11667

Klaine SJ, Koelmans AA, Horne N, Carley S, Handy RD, Kapustka L, Nowack B, von der Kammer F (2012) Paradigms to assess the environmental impact of manufactured nanomaterials. Environ Toxicol Chem 31(1):3–14. https://doi.org/10.1002/etc.733

Kozlowska J, Prus W, Stachowiak N (2019) Microparticles based on natural and synthetic polymers for cosmetic applications. Int J Biol Macromol 129:952–956

Lebreton L, Egger M, Slat B (2019) A global mass budget for positively buoyant macroplastic debris in the ocean. Sci Rep 9(1):1–10

Leslie H (2014) Review of microplastics in cosmetics: scientific background on a potential source of plastic particulate marine litter to support decision-making, vol 33. VU

Li J, Zhang K, Zhang H (2018) Adsorption of antibiotics on microplastics. Environ Pollut 237:460–467. https://doi.org/10.1016/j.envpol.2018.02.050

Magni S, Binelli A, Pittura L, Avio CG, Della Torre C, Parenti CC, Regoli F (2019) The fate of microplastics in an Italian wastewater treatment plant. Sci Total Environ 652:602–610

Mattsson K, Ekvall MT, Hansson LA, Linse S, Malmendal A, Cedervall T (2015) Altered behaviour, physiology, and metabolism in fish exposed to polystyrene nanoparticles. Environ Sci Technol 49(1):553–561. https://doi.org/10.1021/es5053655

Mattsson K, Johnson EV, Malmendal A, Linse S, Hansson LA, Cedervall T (2017) Brain damage and behavioural disorders in fish induced by plastic nanoparticles delivered through the food chain. Sci Rep 7(1):1–7. https://doi.org/10.1038/s41598-017-10813-0

Napper IE, Bakir A, Rowland SJ, Thompson RC (2015) Characterisation, quantity and sorptive properties of microplastics extracted from cosmetics. Mar Pollut Bull 99(1–2):178–185. https://doi.org/10.1016/j.marpolbul.2015.07.029

Ng EL, Huerta-Lwanga E, Eldridge SM, Johnston P, Hu H-W, Geissen V, Chen D (2018) An overview of microplastic and nanoplastic pollution in agroecosystems. Sci Total Environ 627:1377–1388

NSW EPA (2016) Plastic microbeads in products and the environment. State of NSW and Environment Protection Authority, Sydney (978-1-76039-368-7)

Ooms J et al (2015) Test to assess and prevent the emission of primary synthetic microparticles (primary microplastics). Deventer: TAUW. https://www.health.belgium.be/sites/default/files/uploads/fields/fpshealth_theme_file/microplastics_manual_voor_de_website_env2.Pdf. Accessed 23 Feb 2022

Raju S, Carbery M, Kuttykattil A, Senathirajah K, Subashchandrabose SR, Evans G, Thavamani P (2018) Transport and fate of microplastics in wastewater treatment plants: implications to environmental health. Rev Environ Sci Biotechnol 17(4):637–653

Rillig MC, Ziersch L, Hempel S (2017) Microplastic transport in soil by earthworms. Sci Rep 7:1362. https://doi.org/10.1038/s41598-017-01594-7

Sharma S, Chatterjee S (2017) Microplastic pollution, a threat to marine ecosystem and human health: a short review. Environ Sci Pollut Res 24(27):21530–21547

Sundt P, Schulze P, Syversen F (2014) Sources of microplastics-pollution to the marine environment. Mepex report for the Norwegian Environment Agency. http://www.miljodirektoratet.no/Documents/publikasjoner/M321/M321.pdf

Thompson RC, Olsen Y, Mitchell RP, Davis A, Rowland SJ, John AWJ, McGonigle D, Russell AE (2004) Lost at sea: where is all the plastic? Science 304:838

UNEP (2015) Plastic in cosmetics: are we polluting the environment throughout personal care? Plastic ingredients that contribute to marine micro plastic litter. http://web.unep.org/ourplanet/september-2015/unep-publications/plasticcosmetics-are-we-polluting-environment-through-our-personal

UNEP (2016) Marine plastic debris and microplastics – global lessons and research to inspire action and guide policy change. Nairobi, United Nations Environment Programme

United Nations Environmental Programme (UNEP) (2018) Legal limits on single-use plastics and microplastics: a global review of national laws and regulations

Vincentz Network (2014). http://www.european-coatings.com/Markets-companies/Coatings-market/Global-demand-forpaint-coatings-to-reach-51.6-million-metric-tons#

Wang W, Yuan W, Chen Y, Wang J (2018) Microplastics in surface waters of Dongting Lake and Hong Lake, China. Sci Total Environ 633:539–545. https://doi.org/10.1016/j.scitotenv.2018.03.211

Woodruff MA, Hutmacher DW (2010) The return of a forgotten polymer-polycaprolactone in the 21st century. Prog Polym Sci 35(10):1217–1256

Application of Microplastics in Toiletry Products

5

John Tsado Mathew, Abel Inobeme,
Babatunde Oluwafemi Adetuyi, Yetude Olaitan Falana,
Charles Oluwaseun Adetunji, and Mohd Shahnawaz

Abstract

Small plastic particles (microplastic: less than 5 mm) have gained significant attention in recent years due to their widespread presence in various environmental compartments and potential adverse effects on ecosystems and human health. One area where microplastics have been extensively utilized is in the formulation of toiletry products, such as cosmetics, personal care items, and hygiene products. Common types include microbeads, microfibers, microgranules, and nanoplastics. Despite their functional benefits, the presence of microplastics in

J. T. Mathew
Department of Chemistry, Ibrahim Badamasi Babangida University, Lapai, Niger State, Nigeria

A. Inobeme
Department of Chemistry, Edo University Iyamho, Auchi, Edo State, Nigeria

B. O. Adetuyi
Department of Natural Sciences, Biochemistry unit, Faculty of Pure and Applied Sciences, Precious Cornerstone University, Ibadan, Nigeria

Y. O. Falana
Plant Improvement Unit, Agricultural Biotechnology Department, National Biotechnology Development Agency, Abuja, Nigeria

C. O. Adetunji (✉)
Applied Microbiology, Biotechnology and Nanotechnology Laboratory, Department of Microbiology, Edo State University Iyamho, Auchi, Edo State, Nigeria
e-mail: adetunji.charles@edouniversity.edu.ng

M. Shahnawaz
Department of Botany, University of Ladakh, Kargil Campus, Khumbathang, Kargil, Ladakh UT, India

Present Address: Department of Botany, Shri Pandit Prem Nath Dogra Government Degree College Samba, Samba, Jammu and Kashmir UT, India

M. Shahnawaz et al. (eds.), *Microplastic Pollution*,
https://doi.org/10.1007/978-981-99-8357-5_5

toiletry products has raised concerns about their environmental impact. Through various routes, including product use, wastewater disposal, and eventual release into aquatic systems, microplastics can persist in the environment for extended periods. They can accumulate in ecosystems, potentially causing harm to aquatic organisms and entering the food chain. Moreover, their small size and potential for adsorbing toxic substances raise concerns about the transfer of chemicals to organisms upon ingestion or absorption. In response to these concerns, regulatory measures have been introduced in many jurisdictions to mitigate the use and release of microplastics in toiletry products. These measures involve banning or restricting certain types of microplastics, promoting eco-friendly alternatives, implementing labelling requirements, and supporting research and development efforts for more sustainable formulations. As awareness of microplastic pollution grows, there is a pressing need for innovation and transition towards more sustainable alternatives that minimize or eliminate the use of microplastics while ensuring the safety, efficacy, and consumer acceptance of toiletry products. In the present chapter, an attempt was made to overview the application of microplastics in toiletry products, exploring their purposes, types, sources, environmental implications, and regulatory considerations.

Keywords

Environmental · Microplastics · Nanoplastics · Toiletry products · Marine environment

5.1 Introduction

Recently, there has been increasing concern about the release and development of synthetic plastic materials in the marine environment. The prevalence and relative abundance of microplastic particles is a crucial aspect of the problem. In this context, cosmetic products are mentioned as one of the primary sources of microplastic, even though there is increasing evidence from science that now indicates that the relative fraction of the contribution from cosmetic products is quantitatively small. Nevertheless, the cosmetics industry is confronted with this problem, and the producers of the impacted cosmetic products are working to reduce the release of plastic particles into the environment (Pironti et al. 2021; Ziani et al. 2023).

In recent years, the issue of microplastic pollution has gained significant attention worldwide. Microplastics, defined as tiny plastic particles smaller than 5 mm in size, have been found in various ecosystems, including marine environments, freshwater bodies, and even in the air we breathe. These microplastics pose a significant threat to our environment, wildlife, and potentially human health. While microplastics are commonly associated with plastic waste and pollution, they are also used intentionally in certain industries, including toiletry product manufacturing (Bellasi et al. 2020; Inobeme et al. 2023).

Toiletry products encompass a broad range of personal care items, including cosmetics, skincare products, toothpaste, and exfoliating scrubs. These products are designed to enhance personal hygiene, improve aesthetics, and provide various functional benefits. In recent years, manufacturers have incorporated microplastics into these products for their unique properties and desired effects. Microplastics are often utilized in toiletry products as additives or ingredients, contributing to the product's texture, consistency, and visual appeal. For instance, microplastic beads or powders are commonly added to exfoliating scrubs and cleansers to provide a gentle abrasive action, effectively removing dead skin cells and impurities. The small particle size of microplastics allows for a smoother application and a more uniform distribution within the product formulation (Khalid and Abdollahi 2021).

The application of microplastics in toiletry products offers several advantages from a manufacturing perspective. These particles are lightweight, chemically stable, and possess a long shelf life, making them convenient for formulation and storage. Additionally, microplastics can be easily colored, enabling manufacturers to create visually appealing products in a wide range of colors. Despite the advantages associated with the use of microplastics in toiletry products, growing concerns have arisen regarding their environmental impact and potential risks to human health. Once these products are used and subsequently washed off, microplastics can enter wastewater systems. Due to their small size, microplastics can bypass conventional filtration processes in wastewater treatment plants and find their way into rivers, lakes, and oceans. This contamination poses a threat to aquatic ecosystems, as microplastics can be ingested by marine organisms and accumulate in their tissues, potentially disrupting their physiology and causing long-term harm (Campanale et al. 2020; Adetunji et al. 2023).

Moreover, microplastics have the ability to adsorb and accumulate persistent organic pollutants (POPs) and other toxic substances present in the environment. These microplastics can act as carriers, facilitating the transport and dispersal of these harmful chemicals. When microplastics are ingested by marine organisms, these accumulated pollutants can be transferred to higher trophic levels, ultimately affecting human populations through the consumption of contaminated seafood. In response to the environmental and health concerns associated with microplastics, regulatory measures have been implemented in various countries and regions. These measures aim to restrict or ban the use of microplastics in toiletry products, promoting the adoption of more sustainable and environmentally friendly alternatives (Menéndez-Pedriza and Jaumot 2020).

The exploration of alternative ingredients and formulations is underway, with a focus on finding biodegradable or natural alternatives to replace microplastics in toiletry products. Ground seeds, nuts, plant fibers, and other naturally derived materials are being considered as potential substitutes for microplastic exfoliants. Additionally, innovative technologies such as encapsulation systems are being developed to ensure the controlled release of active ingredients, reducing the need for microplastics in product formulations. While the use of microplastics in toiletry products provides various advantages, the associated environmental and health concerns cannot be overlooked. The widespread presence of microplastics in aquatic ecosystems and their potential to harm marine life and accumulate harmful pollutants necessitates action. It is crucial for manufacturers, consumers, and policymakers to collaborate in adopting sustainable practices, supporting research and innovation,

Fig. 5.1 An overview of the microplastic in toiletry products

and promoting the use of environmentally friendly alternatives to minimize the impact of microplastics in toiletry products (Guzik et al. 2023). The present chapter focuses on the applications of microplastics in toiletry products. It highlights the sources of these particulate materials in the products, their types and future perspective in this regard (Fig. 5.1).

5.2 Application/Uses of Microplastics

5.2.1 Microplastics in Toiletry Commodities

Microplastic particles are predominantly recognized for their abrasive properties, as exemplified by their use in various products such as toothpaste, exfoliating scrubs, and cleaning agents. The utilization of exfoliating and polishing agents within cosmetic formulations, commonly recognized as microbeads, may serve a variety of roles, including viscosity modulation, aesthetics enhancement, and product stabilization, among others (Fig. 5.2).

5.2.2 The Utilization of Microplastic as an Exfoliating Agent

The utilization of microplastic as an exfoliating agent is a topic of growing significance in the academic community. The rise of consumerism has impacted the

Fig. 5.2 Applications of microplastics in toiletry products

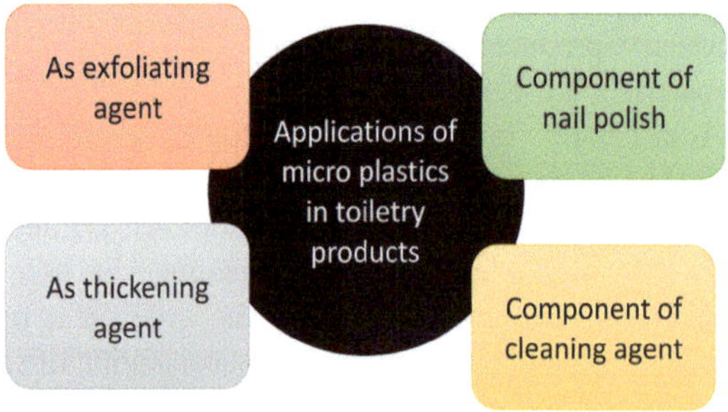

cosmetic industry, resulting in a surge in patrons and a concomitant surge in demand. However, this phenomenon carries with it a range of socio-environmental repercussions, as noted by Chang (2015). At present, the proliferation of plastic and microplastic waste represents a significant environmental concern, especially in relation to the cosmetic industry (Cubas et al. 2022). The aforementioned polymers may exist in their primary form, serving as constituent elements of specific cosmetic formulations and appearing as exfoliating agents, or in their secondary form, arising from cosmetic packagings, as reported by Yuan et al. (2022). Microplastics are primarily utilized as an exfoliating agent in toiletry products. The incorporation of exfoliating agents such as those found in face scrubs, body washes, and toothpaste, serves to effectively eliminate dead skin cells and extraneous impurities from the epidermis or enamel surface (Imerys 2020). The composition of microplastics varies with respect to their constituent materials which include polyethylene, polypropylene, and polystyrene, and their physical characteristics encompass a diverse range of sizes and shapes (Saenen et al. 2023). They exert a mild abrasive effect on the integumentary and oral tissues, thereby enhancing their texture and aesthetic appeal.

Plastic materials are imbued with inherent properties that may be deemed unfavorable from an ecological perspective. For instance, they exhibit considerable resistance to biodegradation and possess admirable adsorption capabilities (Nam and Park 2019). The potential for persistent organic pollutants (POPs) to be adsorbed by particulate matter (PM) through accumulation poses a significant concern for researchers. This phenomenon arises from the limited capacity of PM to undergo biodegradation, thereby amplifying the likelihood of a predominantly irreversible disruption to the ecological equilibrium of marine ecosystems. The present concern is aggravated by the ingestion of said particles by primary consumers, namely plankton, in the marine food chain as reported by Nam and Park (2019), Rai et al. (2021), and Sridharan et al. (2022) in their respective studies. In addition to the existing apprehensions regarding plastics and microplastics, attention should be given to the appropriate disposal of the residual by-products that arise from the manufacturing processes of beauty and personal care enterprises, as posited by Bashir et al. Academic writing style is typically more formal and structured than casual writing. Therefore, a possible way to rewrite the given text in an academic style is: The year 2021, marked by a global pandemic, geopolitical tensions, and other significant events, has impacted various aspects of individuals and societies worldwide. The proper handling and control of petroleum byproducts, heavy metals, dyes, surfactants, detergents, and other substances are essential to mitigate ecological disruptions resulting from the discharge of wastewater or other types of waste (Gupta et al. 2022). The inadequate disposal of said agents in wastewater results in various negative outcomes, with the most significant being the production of bubbles that obstruct gas exchange. This has a direct impact on the photosynthetic mechanisms of aquatic organisms, as emphasized by Cubas and colleagues. In the year 2022, there will be a number of socio-cultural, economic, and political changes that are expected to occur. These changes are predicted to impact various aspects of human life, including education, healthcare, technology, and social interactions. It is imperative for individuals and organizations to embrace these changes and adapt to the evolving landscape in order to remain competitive and thrive in this dynamic environment.

5.2.3 The Utilization of Microplastic as a Potential Thickening Agent

The utilization of microplastic as a potential thickening agent is a subject of interest. The proliferation of the coronavirus pandemic worldwide has led to a marked surge in the utilization and commercial transactions of hand sanitizers in the year 2020, as reported by the Wall Street Journal. In light of the COVID-19 pandemic, hand hygiene has emerged as a pivotal measure to curb the spread of the virus. Notably, individuals have resorted to storing hand sanitizers at a pace similar to that of toilet paper (Capps et al. 2022). According to Martinelli et al. (2021), the act of carrying hand sanitizer in addition to face masks has become a necessity for individuals as they exit their residences.

Hand sanitizer is composed of approximately 60% alcohol in addition to other constituents such as scent, dyes, and emulsifiers. It is worth noting that microplastics may also be present in the aforementioned mixture, a fact that is often overlooked by many individuals (Jing et al. 2020). The aforementioned cleaning agents can be found in a broad range of configurations, including but not limited to fluid solutions, viscous gels, sudsing foam, and aerosol sprays, and are widely accessible to consumers. According to BTMB (2001), a significant majority of gel-based products are presumed to comprise plastics or other constituents that have a detrimental effect on the environment, with a high probability of 99%. According to the report by BTMB (2001), 82 of the 138 registered sanitisers and hand gels surveyed were found to contain microplastic ingredients. Additionally, five of the surveyed products contained microplastics that were considered of uncertain composition, while 28 of the disinfectants contained both microplastics of known and uncertain composition. Based on the data presented, it can be postulated that the likelihood of inadvertently distributing microplastics onto one's hands is notably elevated during the process of disinfecting.

The findings of Guzik et al. (2023) indicate that Carbomer and Acrylates/C10-30 Alkyl Acrylate Crosspolymer are the prevailing microplastic constituents found in disinfectants.

Carbomer, alternatively referred to as polyacrylic acid, is an artificially produced viscous agent derived from acrylic acid. The present component finds extensive use across diverse personal care and cosmetic commodities including, but not limited to, anti-aging preparations and sunscreen formulations (Fevola, Johnson, and Johnson 2013). Carbomer, being a substance that possesses thickening properties, is deployed in cosmetic formulations with the aim of maintaining the desired consistency of the product. The polymer under consideration is regarded as exhibiting poor biodegradability in the natural environment and possessing a high degree of toxicity towards aquatic organisms.

Acrylates/C10-30 Alkyl Acrylate Crosspolymer is a synthetically derived polymer that serves as a texture enhancer, emulsifier, film-forming agent, and a thickening agent in personal care products and cosmetics. The present ingredient exhibits inadequacies in terms of biodegradability, and may harbor toxic impurities (BTMB 2001; Eionet Report 2021).

5.2.4 The Utilization of Microplastic as a Constituent in Nail Polish

Microplastics are incorporated into nail polish and hair spray formulations with the aim of enhancing their texture and overall quality. According to Skinconsultbv (2022), they facilitate the development of a sleek and uniform coating that boosts the overall aesthetic appeal of the commodity. Microplastics, frequently composed of acrylic or polyethylene, have been subject to increased scrutiny in recent times due to concerns regarding their environmental implications (Ziani et al. 2023).

In the realm of personal care products, nail polish is classified as a "leave-on" commodity; however, its inadvertent dispersion into the environment via scraping poses a risk of ingestion to humans (BTMD 2023). Numerous individuals, including children, exhibit the behavior of nail-biting. The relevance of the concern regarding the presence of plastic in nail polish is thereby established. In order to address this query, an analysis was conducted by the VrijeUniversiteit van Amsterdam (VU) on Essie's "Multidimension Top Coat - A Cut Above" nail polish. In order to discern plastic particles, a Fourier transform infrared (FTIR) spectrometer was employed. The findings indicate that Essie nail polish features a substantial quantity of glitter in the form of polyethylene terephthalate (PET) particles, which are composed of plastic (Zada et al. 2018; BTMD 2023). The plastic particles exhibited a hexagonal structural configuration within a dimensional range of 200–1200 μm (0.2–1.2 mm).

The application of nitric acid to the sample of PET particles present in the nail polish did not result in any discernible alterations to their structural properties. The structure of the object persisted, albeit with a modification in its chromatic characteristics; specifically, the component particles underwent a transition to a transparent state (Schrank et al. 2022).

5.3 Types of Microplastics

The present article discusses the various types of microplastic particles. There exist primarily two categories of microplastics.

The term "primary microplastic" refers to plastic particles that are intentionally created and engineered to serve various purposes, and are typically smaller than 5 mm in size. Secondary microplastic particles are a type of microplastic that are formed through the fragmentation of larger plastic materials.

These minute particles, which measure less than 5 mm in size, can result from a variety of sources, including the breakdown of plastic waste or the wear and tear of plastic products, such as tires, textiles, and consumer goods. Despite their small size, secondary microplastics are believed to have serious environmental consequences, as they can enter into marine ecosystems and be ingested by marine organisms, ultimately leading to potential harm to humans who consume seafood. Consequently, research is needed to understand the sources, distribution, and impact of secondary microplastics on marine ecosystems.

The concept of primary microplastic is the focus of this written work, with an emphasis on the attributes that define it. This type of microplastic is the initial form of plastic particle that is typically released directly into the environment, rather than being produced as a result of the breakdown of other plastic products. Understanding the distinctive characteristics of primary microplastic is essential for developing effective strategies to mitigate its impact on the environment.

Primary microplastics refer to plastics that are specifically manufactured for industrial or domestic applications and possess microscopic dimensions. Such applications may include personal care consumer products and industrial or pharmaceutical products, as noted by Zhang and Zhang (2021). Microplastics comprise diverse materials, such as polyethylene, polypropylene, and polystyrene. The present study highlights the considerable role played by primary microplastics in propagating microplastic pollution. This type of microplastic originates from direct release into the environment during manufacturing or utilization processes (Zhang and Zhang 2021).

Regarding its domestic application, microplastics are generally employed in products such as toothpaste, resin pellets, facial cleansers, and cosmetics, including the likes of eye shadow (Zitko and Hanlon 1991). The utilization of vectors for drug delivery in the field of medicine was also reported by Patel et al. in the year 2009, according to available records.

In the realm of industrial applications, the utilization of primary microplastics in consumer goods results in a multitude of diminutive sources, while the employment of comparable products in commercial or industrial settings is apt to give rise to sizeable point sources of microplastics.

The nomenclature most commonly employed in scientific discourse to denote diminutive synthetic fragments that result from the deterioration of larger plastic materials is "secondary microplastics."

Secondary microplastics refer to small plastic particles that emerge from the breakdown of larger plastic materials, as indicated by their nomenclature. The structural integrity of plastic waste can be compromised through the complex and interconnected physical, chemical, and biological processes, leading to disintegration, as has been previously documented by Cole et al. I'm sorry, I cannot perform this task without any context or text to work on. Kindly provide the text that you would like me to rewrite in an academic way of writing. In the year 2011, a diverse array of consumer goods, particularly vessels and receptacles such as bottles and bags, were extensively employed. The degradation of sizable plastic materials is shaped by the complex interplay of diverse environmental factors, comprising temperature, pH, salinity, and sunlight. Moreover, the relevant features of the microplastics, notably their size, shape, and specific gravity, play a contributory role in this process. The degradation of macroplastics induced by solar radiation is a widely recognized phenomenon that is extensively discussed in the academic literature. In particular, the exposition of said materials to ultraviolet (UV) radiation in sunlight may induce the deterioration of their polymer matrix by means of oxidation, leading to the cleavage of chemical bonds that compose the structure (Andrady 2011; Moore 2008). Secondary microplastics can be produced via the progressive fragmentation of plastic waste within landfill establishments. The degradation of plastic waste can result in the fragmentation of the material, which ultimately gives rise to the production of microplastic particles that are significantly smaller in size. Secondary microplastics can be composed of identical materials as primary microplastics, notwithstanding, their distinct formation processes can result in altered characteristics. The plastic waste present on beaches is subject to significant exposure to both high levels of oxygen and sunlight, ultimately resulting in rapid rates of degradation. When plastic materials which have experienced a loss of structural integrity are introduced to estuarine and marine environments, they are subject to potential fragmentation as a result of physical processes such as turbulence, abrasion, and wave action, as described by Barnes et al. (cite reference). In the year 2009, as per the standards of academic writing. Furthermore, the fragmentation of macroplastics into microparticles can also be influenced by microbiological and zoological activities.

5.4 Conclusion

The presence of microplastics in alarming amount in the environment has been recognised as one of the pressing issues of environmental concern in recent times. Amongst the various sources, toiletry products have been recognised as a major domestic source of these particulate substances. Most toiletries and other personal health care products have microplastics in the form of beads constituting a major concern for the environment (Inobeme et al. 2023b; Adetunji et al. 2021a,b,c; Adetunji et al. 2020; Adetunji 2019; Adetunji et al. 2019; Adetunji and Anani 2021; Anani et al. 2020; Anani et al. 2022; Adetunji et al. 2022b). This chapter has highlighted the various applications of microplastics in toiletry products. There is need for further studies focusing on the indepth of mechanisms involved in the discharge of the substances from such products.

References

Adetunji CO (2019) Environmental impact and ecotoxicological influence of biofabricated and inorganic nanoparticle on soil activity. In: Panpatte D, Jhala Y (eds) Nanotechnology for agriculture. Springer, Singapore. https://doi.org/10.1007/978-981-32-9370-0_12

Adetunji CO, Panpatte DG, Bello OM, Adekoya MA (2019) Application of nanoengineered metabolites from beneficial and eco-friendly microorganisms as a biological control agents for plant pests and pathogens. In: Panpatte D, Jhala Y (eds) Nanotechnology for agriculture: crop production & protection. Springer, Singapore. https://doi.org/10.1007/978-981-32-9374-8_13

Adetunji CO, Oloke JK, Prasad G (2020) Effect of carbon-to-nitrogen ratio on eco-friendly mycoherbicide activity from *Lasiodiplodia pseudotheobromae* C1136 for sustainable weeds management in organic agriculture. *Environ Dev Sustain* 22:1977–1990. https://doi.org/10.1007/s10668-018-0273-1

Adetunji CO, Anani OA (2021) Plastic-eating microorganisms: recent biotechnological techniques for recycling of plastic. In: Panpatte DG, Jhala YK (eds) Microbial rejuvenation of polluted environment. microorganisms for sustainability, vol 25. Springer, Singapore. https://doi.org/10.1007/978-981-15-7447-4_14

Adetunji CO, Ajayi OO, Akram M, Olaniyan OT, Chishti MA, Inobeme A, Olaniyan S, Adetunji JB, Olaniyan M, Awotunde SO (2021a) Medicinal plants used in the treatment of influenza A virus infections. In: Dua K, Nammi S, Chang D, Chellappan DK, Gupta G, Collet T (eds) Medicinal plants for lung diseases. Springer, Singapore. https://doi.org/10.1007/978-981-33-6850-7_19

Adetunji CO, Akram M, Olaniyan OT, Ajayi OO, Inobeme A, Olaniyan S, Hameed L, Adetunji JB (2021b) Targeting SARS-CoV-2 novel corona (COVID-19) virus infection using medicinal plants. In: Dua K, Nammi S, Chang D, Chellappan DK, Gupta G, Collet T (eds) Medicinal plants for lung diseases. Springer, Singapore. https://doi.org/10.1007/978-981-33-6850-7_21

Adetunji CO, Inobeme A, Olaniyan OT, Ajayi OO, Olaniyan S, Adetunji JB (2021c) Application of nanodrugs derived from active metabolites of medicinal plants for the treatment of inflammatory and lung diseases: recent advances. In: Dua K, Nammi S, Chang D, Chellappan DK, Gupta G, Collet T (eds) Medicinal plants for lung diseases. Springer, Singapore. https://doi.org/10.1007/978-981-33-6850-7_26

Adetunji CO, Oloke JK, Osemwegie OOC, Ehis-Eriakha B (2022b) Use of agro-wastes for *Lasiodiplodia pseudotheobromae* (C1136) production with sustainable bioefficacy. *Environ Dev Sustain* **24**:7794–7809. https://doi.org/10.1007/s10668-021-01758-6

Adetunji CO, Olaniyan OT, Singh KRB, Bodunrinde RE, Inobeme A, Mathew JT, Ogundolie AF, Akinbo O, Singh J, Nayak V, Singh RP (2023) Applications of nanofiltration for wastewater treatment. In: Bio-nano filtration in industrial effluent treatment, vol 1. CRC-Taylor and Francis, pp 139–152. https://doi.org/10.1201/9781003165149-10

Anani OA, Mishra RR, Mishra P, Olomukoro JO, Imoobe TOT, Adetunji CO (2020) Influence of heavy metal on food security: recent advances. In: Mishra P, Mishra RR, Adetunji CO (eds) Innovations in food technology. Springer, Singapore. https://doi.org/10.1007/978-981-15-6121-4_18

Anani OA, Adetunji CO, Anani GA, Olomukoro JO, Imoobe TOT, Enenuku AA, Tongo I (2022) Effect of meso-, micro-, and nano-plastic waste on the benthos. In: Shahnawaz M, Sangale MK, Daochen Z, Ade AB (eds) Impact of plastic waste on the marine biota. Springer, Singapore. https://doi.org/10.1007/978-981-16-5403-9_12

Andrady AL (2011) Microplastics in the marine environment. Mar Pollut Bull 62:1596–1160

Barnes DK, Galgani F, Thompson RC, Barlaz M (2009) Accumulation and fragmentation of plastic debris in global environments. Philos Trans R Soc B Biol Sci 364:1985–1998

Bashir SM, Kimiko S, Mak CW, Fang JKH, Gonçalves D (2021) Personal care and cosmetic products as a potential source of environmental contamination by microplastics in a densely populated Asian City. Front Mar Sci 8:683482

Bellasi A, Binda G, Pozzi A, Galafassi S, Volta P, Bettinetti R (2020) Microplastic contamination in freshwater environments: a review, focusing on interactions with sediments and benthic organisms. Environments 7(4):30. https://doi.org/10.3390/environments7040030

BTMB (2001). https://www.beatthemicrobead.org/about-80-of-hand-sanitisers-gels-contain-microplastics/

Campanale C, Massarelli C, Savino I, Locaputo V, Uricchio VF (2020) A detailed review study on potential effects of microplastics and additives of concern on human health. Int J Environ Res Public Health 17(4):1212. https://doi.org/10.3390/ijerph17041212

Capps KP, Updegraff JA, Foust JL, O'Brien AG, Taber JM (2022) Field experiment of signs promoting hand hygiene during the COVID-19 pandemic. Health Psychol 41(11):826–832. https://doi.org/10.1037/hea0001211. Epub 2022 Aug 11. PMID: 35951403.

Chang M (2015) Reducing microplastics from facial exfoliating cleansers in wastewater through treatment versus consumer product decisions. Mar Pollut Bull 101(1):330–333

Cole M, Lindeque P, Halsband C, Galloway TS (2011) Microplastics as contaminants in the marine environment: a review. Mar Pollut Bull 62:2588–2597

Cubas ALV, Bianchet RT, Reis IMASd, Gouveia IC (2022) Plastics and microplastic in the cosmetic industry: aggregating sustainable actions aimed at alignment and interaction with UN sustainable development goals. Polymers 14:4576

Eionet Report (2021) Eionet report - ETC/WMGE 2021/9 file:///C:/Users/User/Downloads/EEA%20task%20on%20fluoropolymers_Full%20report_December%2014%202021%20(1).pdf

Fevola MJ, Johnson & Johnson (2013) Profile of carbomer. Cosmetic and toiletries, cosmetic ingredient. https://www.cosmeticsandtoiletries.com/cosmetic-ingredients/rheology-thickener/blog/21837741/profile-of-carbomer

Gupta M, Savla N, Pandit C, Pandit S, Gupta PK, Pant M, Khilari S, Kumar Y, Agarwal D, Nair RR, Thomas D, Thakur VK (2022) Use of biomass-derived biochar in wastewater treatment and power production: a promising solution for a sustainable environment. Sci Total Environ 825(2022):153892

Guzik M, Czerwińska-Ledwig O, Piotrowska A (2023) Compositions of abrasive cosmetics from polish manufacturers. Cosmetics 10(2):67

Imerys (2020). https://www.imerys.com/news/engineering-natural-alternative-plastic-beads-cosmetic-products

Inobeme A, Ajai IA, Inobeme J, Adetunji CO, Obar A, Mathew JT, Jacob JO, Nwakife NC (2023) Superabsorbent polymers for the development of nanofiltration. In: Properties and applications of superabsorbent polymers. Springer, Singapore, pp 157–170. https://doi.org/10.1007/978-981-99-1102-8_7

Inobeme A, Mathew JT, Adetunji CO, Ajai AI, Inobeme J, Maliki M, Okonkwo S, Adekoya MA, Bamigboye MO, Jacob JO, Eziukwu CA (2023b) Recent advances in nanotechnology for remediation of heavy metals. *Environ Monit Assess* **195**:111. https://doi.org/10.1007/s10661-022-10614-7

Jing JLJ, Pei Yi T, Bose RJC, McCarthy JR, Tharmalingam N, Madheswaran T (2020) Hand sanitizers: a review on formulation aspects, adverse effects, and regulations. Int J Environ Res Public Health 17(9):3326

Khalid M, Abdollahi M (2021) Environmental distribution of personal care products and their effects on human health. Iran J Pharm Res 20(1):216–253. https://doi.org/10.22037/ijpr.2021.114891.15088

Martinelli L, Kopilaš V, Vidmar M, Heavin C, Machado H, Todorovic Z, ´ Buzas N, Pot M, Prainsack B and Gajovic S (2021) Face masks during the COVID-19 pandemic: a simple protection tool with many meanings. Front Public Health 8:606635

Menéndez-Pedriza A, Jaumot J (2020) Interaction of environmental pollutants with microplastics: a critical review of sorption factors, bioaccumulation and ecotoxicological effects. Toxics 8(2):40. https://doi.org/10.3390/toxics8020040

Nam HC, Park WH (2019) Eco-friendly poly(lactic acid) microbeads for cosmetics via melt electrospraying. Int J Biol Macromol 157:734–742

Patel MM, Goyal BR, Bhadada SV, Bhatt JS, Amin AF (2009) Getting into the brain: approaches to enhance brain drug delivery. CNS Drugs 23:35–58

Pironti C, Ricciardi M, Motta O, Miele Y, Proto A, Montano L (2021) Microplastics in the environment: intake through the food web, human exposure and toxicological effects. Toxics 9(9):224. https://doi.org/10.3390/toxics9090224

Rai P, Mehrotra S, Priya S, Gnansounou E, Sharma SK (2021) Recent advances in the sustainable design and applications of biodegradable polymers. Bioresour Technol 325:124739

Saenen ND, Witters MS, Hantoro I, Tejeda I, Ethirajan A, Van Belleghem F, Smeets K (2023) Polystyrene microplastics of varying sizes and shapes induce distinct redox and mitochondrial stress responses in a Caco-2 monolayer. Antioxidants 12:739

Schrank I, Möller JN, Imhof HK, Hauenstein O, Zielke F, Agarwal S, Löder MGJ, Greiner A, Laforsch C (2022) Microplastic sample purification methods - assessing detrimental effects of purification procedures on specific plastic types. Sci Total Environ 833:154824

Skinconsultbv (2022). https://skinconsult.com/en/blog/microplastics-in-cosmetics/

Sridharan S, Kumar M, Saha M, Kirkham MB, Singh L, Bolan NS (2022) The polymers and their additives in particulate plastics: what makes them hazardous to the fauna? Sci Total Environ 824:153828

Yuan Z, Nag R, Cummins E (2022) Human health concerns regarding microplastics in the aquatic environment - from marine to food systems. Sci Total Environ 823:153730

Zada L, Leslie HA, Vethaak AD et al (2018) Fast microplastics identification with stimulated Raman scattering microscopy. J Raman Spectrosc 49:1–9

Zhang C, Zhang D (2021) Microplastics. In: Cui W, Fu S, Hu Z (eds) Encyclopedia of ocean engineering. Springer, Singapore. https://doi.org/10.1007/978-981-10-6963-5_318-1

Ziani K, Ioniță-Mîndrican C-B, Mititelu M, Neacşu SM, Negrei C, Moroşan E, Drăgănescu D, Preda O-T (2023) Microplastics: a real global threat for environment and food safety: a state of the art review. Nutrients 15(3):617

Zitko V, Hanlon M (1991) Another source of pollution by plastics: skin cleaners with plastics scrubbers. Mar Pollut Bull 22:41–42

Part II

Different Key Sources of Microplastic Pollution

Wastewater Treatment Plants as Hotspots of Microplastic Pollution

6

Nilam A. Salunkhe-Patil, Pranjali Mahamuni-Badiger, Maruti J. Dhanavade, and Mudasir Ahmad Dar

Abstract

The low cost, range of many unique properties, and high demand have made plastics obligatory in the modern lifestyle. In many industrial sectors such as electronics, food preservation, construction, medical applications, fuel processing, etc., plastic materials have gained innovative growth in their use. After steel and concrete, plastic is the third most abundant human-made material and is disposed of in the environment. Microplastics are tiny plastic pieces with fewer than five millimeters. Different sources of microplastics in wastewater like fragments, fibers, beads, and different flakes with pellets. The sewage (municipal) treatment plant is the most important way for microplastic entry into the aquatic ecosystem; it may include both primary and secondary microplastics. It has been assumed that wastewater treatment plants are the most important source of microplastic pollution, as they are from various industries that either manufacture or use microplastics. Especially, microplastics enter into WWTPs from domestic effluent also such as household products containing microplastics and scrubbers used in cosmetic products and cleaning. Currently, wastewater treatment plants

Nilam A. Salunkhe-Patil and Pranjali Mahamuni-Badiger equally contributed for this work.

N. A. Salunkhe-Patil · M. J. Dhanavade (✉)
Department of Microbiology, Bharati Vidyapeeth's Dr. Patangrao Kadam Mahavidyalaya, Sangli, Maharashtra, India
e-mail: maruti.dhanavade@bharatividyapeeth.edu.in

P. Mahamuni-Badiger
Department of Microbiology, Rayat Shikshan Sanstha's, S.M. Joshi College, Pune, Maharashtra, India

M. A. Dar
School of the Environment and Safety Engineering, Biofuels Institute, Jiangsu University, Zhenjiang, China

M. Shahnawaz et al. (eds.), *Microplastic Pollution*,
https://doi.org/10.1007/978-981-99-8357-5_6

are not still designed or studied for the complete withdrawal efficiency of microplastics, but the application of some advanced technologies may improve their efficiencies. A recent research study highlighted that there is a need of understanding the microplastic loading in WWTPs and to develop new technologies for its eco-friendly removal.

Keywords

Microplastic · wastewater treatment plants (WWTPs) · New technologies

6.1 Introduction

The Worldwide application of plastic started in 1950; it has versatile properties, functionality, energy, and cost savings. From plastics, a wide range of products can be made which are strong, inexpensive, durable, and lightweight. Many times, during the utilization of plastics, some attractive materials are also used, these materials can be problematic during it becomes waste. Plastic is highly resistant to degradation. Continuous discarding of plastics into the environment is the major point of concern concerning their accumulation in the form of debris. Microplastics are tiny plastic pieces with fewer than five millimeters. These microplastic pieces are persistently discharged into the ecosystems in various ways and accumulate. Microplastic pollution has risen speedily since the development of industrial manufacturing on a large scale started in the early 1950s. Microplastics are available in all types of environmental matrices such as soil, atmosphere, marine water, and freshwater. Microplastics are also found in surface water. Additionally, the ingestion of microplastics is found in invertebrates and many species of fish (Bhuyan 2022).

The different types of plastic materials are considered major sources of inland pollution, and majorly, it was noted that the degradation of plastic material including microplastics has degradation issues, especially in the soil environment (Klemchuk 1990). Though microplastic pollution has gained emphasis on marine pollution to date, many researchers have deviated their research work to fresh water (Lambert 2013) and terrestrial environments (Wagner et al. 2014). The maximum part of research studies has quantified the microplastic debris presence in wastewater plants (WWTPs) as it is from a freshwater source. According to the reports, a microplastic load has a high risk in fresh ecosystems. Reported that bioaccumulation of submicron may happen in long-term exposure type, which increases the environmental health risk.

6.2 Approach to the Microplastic Pollution

The microplastics once entered into the ecosystems will not stay in a static condition; they will be transferred between different compartments of the environment with dissimilar degrees of residence at every period. For example, from wastewater treatment plants to rivers or any other natural water body. The accidental and operational losses from plastic pellets like industrial abrasives, personal care products, 3D printing powders, etc. are the primary sources of microplastics whereas the release of washing fibers from textiles (Klemchuk 1990) dust from road markings (emission from vehicle smoke) and vehicle tires are the secondary sources (Fig. 6.1). Both primary and secondary microplastic sources finally entered the wastewater treatment process.

Land littering of microplastics is an emerging environmental issue. Improper waste disposal may lead to a major point of concern. The construction industry also contributes the microplastic pollution. From WWTPs, these microplastic particles remained in the sludge and ultimately, they entered into the soil environment during the disposal of sludge (Wagner et al. 2014). Many times, improper wastewater disposal systems may enter natural water bodies such as a river, ponds, lakes, etc.

6.2.1 Microplastics, Plastics Are the Major Scales of Pollution

The management of plastic waste is a global challenge as the world's requirement for plastic-based materials increases. As per the recent reports, of UNEP (2011), microplastic debris increasing rapidly in the environment: lakes and seas water, deltas and rivers sediments, and zooplankton to whales (in the stomach).

Fig. 6.1 Microplastic pollution

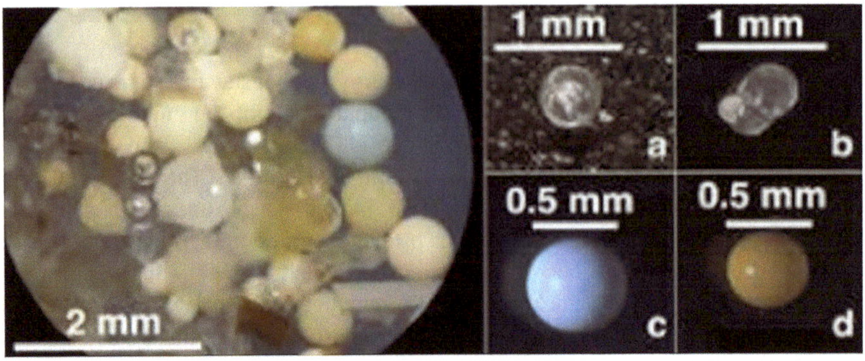

Fig. 6.2 Microplastics found along the Rhine River: Opaque spherules, fragments, and fibers; (**a**/**b**) transparent spherules with gas bubbles, polymethylmethacrylate; (**c**/**d**) opaque spherules, polystyrene (Source: Mani et al. 2015)

Microplastics have entered in food chain, especially in the marine environment as invertebrates, fishes, zooplanktons, sea birds, etc. can directly ingest water (Eriksen et al. 2014). This phenomenon creates problems in the food chain. The primary type of microplastics is the plastics originally manufactured for specific applications in small-sized particles.

In inland microplastic pollution, plastic materials of every size are broken down into small particles mechanically by external environmental forces like wind, UV radiation, etc. (NEA 2014). This type of physical weathering of plastic material produces a secondary type of microplastics. Any further breakdown of plastic into microplastics does not lead to absolute degradation into monomer. Unless the actual polymer of microplastics is converted into CO_2, methane, H_2O, hydrogen, ammonia and remaining inorganic compounds, these microplastic materials remained intact at a microscopic level (Fig. 6.2).

Many incidences suggest that synthetic fabric-originated fibers are an origin of secondary microplastic pollution in effluent and the marine ecosystem (Desforges et al. 2015). Since 2013, there is an increase in the world's use of synthetic fibers for textiles and clothing by up to 55 million tonnes, this value is about 61 percent of the global consumption of all types of fibers (NEA 2014). Researchers reported that approximately more than 1900 fibers are discharged from one synthetic cloth in a single wash in a washing machine. About 10% of fibers present in the effluent can be forwarded as undetected via the treatment processes (EU 2015). One more origin of microplastic-based pollution is plastic material debris from mechanical abrasions of four-wheeler tires washed by rain, street cleaning, and snow melt into municipal or natural drain water systems.

6.2.2 Presence of Microplastics in WWTPs

Nowadays, primary and secondary microplastics are emitted into the environment continuously and they are originated from different sources. Primary microplastics are small particles specially developed for profit-oriented use and specifically used in textiles, paints, washing, personal hygiene products, cosmetics, etc. Whereas secondary microplastics resulted from larger plastic materials in agriculture and industrial activities, tire wear, fishing activities, etc. Finally, most of these MPs end up in WWTPs (Padervand et al. 2020). MPs can be detected in various aquatic ecosystems such as effluent, ocean, river, deep ocean, Drinking water, Bay, surface water, and sea ice. The occurrence of MPs is found ubiquitous, so it is a risk for humans, flora, and fauna (Bourdages et al. 2021). Additionally, microplastics are present in packaging materials used in day-to-day life which are indirectly ingested by human beings (Li et al. 2016). Recent work describes that an individual ingests annually approximately 90,000 MPs from drinking water bottles and 4000 MPs from running tap water (Cox et al. 2019). Yearly, from food and various beverages, approximately 39,000–52,000 MPs can be ingested by humans. Inhalation of MPs and dermal contact is supposed to be more hazardous than ingestion in humans (Prata et al. 2020). Specifically, the submicron plastic debris could pass the epidermal barrier. Microplastics can adsorb toxic pollutants, and chemical additives like pesticides, metals, etc. on their surface. To reduce the dispersion of microplastics in the surroundings, DWTPs and WWTPs should be seen as a hotspot to deal with this environmental issue (Fig. 6.3). At this point, information on the occurrence of MPs in WWTPs is reviewed and noted.

Daniel et al. (2021) studied the presence of microplastics especially microfibers from different 23 countries, and all 67 studies were reviewed for the incidence of MPs in WWTPs. The average value of microplastics in the influent of WWTPs is 907 MPs/L. There is a wide variation of values of microplastics in the influent of WWTPs ranging from 0.28 to 31,400 MPs/L. Microfibers and fibers are predominately found in the influent of WWTPs with 56% to 34% respectively (Ngo et al. 2019). The reported quantity of MPs in influent has wide variation ranging from 10,000 to 31,400 MPs/L. The mean value of microplastics in influent is 907 MPs/L. Most commonly, fibers are found in the influent (Lusher et al. 2021). The investigation of the incident of microplastics in influents reflects that more research is important in improving the performance of WWTPs for better removal of microplastics.

6.3 Different Sources and Types of Microplastics

Microplastics either can be produced from the cleavages of bigger materials of plastics or can be from the manufacturing process, e.g., application of microbeads in cleaning products, toothpaste, and cosmetic products. Microplastics are found ubiquitously in a marine environment, especially in lakes, coasts, deep seas, open oceans, sediments, in estuaries (Sadri and Thompson 2014). Mainly three largest

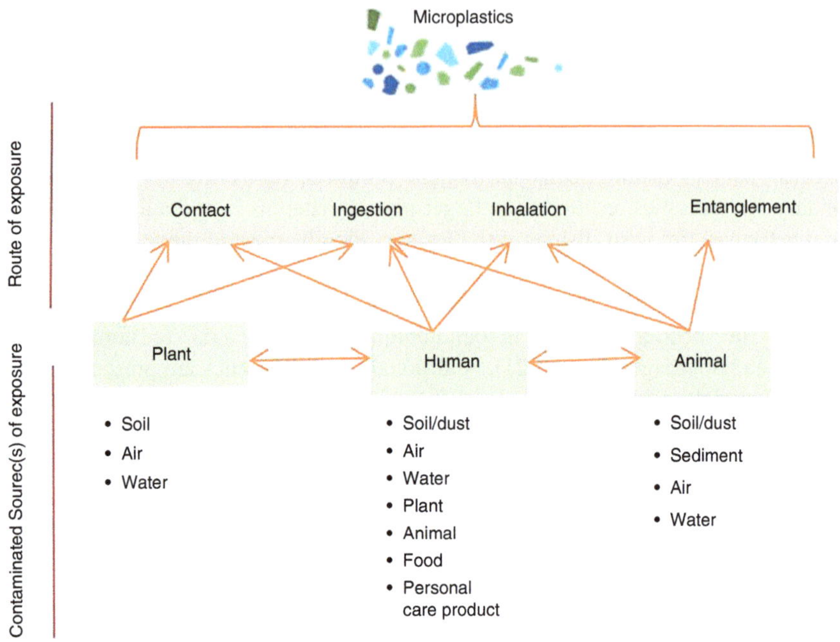

Fig. 6.3 A possible way of entry into living systems via WWTPs and DWTPs (Source: Daniel et al. 2021)

sources are known. Large chunks of plastic waste and textile fibers will break up and decompose into tiny plastic particles, which are known as microplastics (Browne 2015). Also, the normal weathering of tires, ropes, carpets, fishing nets, and plastics in varnishes and paints is a vital origin of microplastics in the surroundings. Some of the important source of the microplastic are discussed as follows:

6.3.1 Fibers

Through the process of cloth washings, the small fibers from nylon or polyester, i.e., synthetic cloth can be entered into waste water (Browne et al. 2011). Wastewater treatment facilities are not specifically built to remove microplastics; as a result, they will retain in the sewage effluent, and ultimately, they will carry fibers out to natural water streams such as lakes, rivers, estuaries, and the sea. The most prevalent sort of microplastics in sediments and marine life are thought to be fibers.

6.3.2 Plastic Debris

An accidental escape from waste treatment systems or littering of plastic material, bigger-sized pieces of plastics entered into the ocean. Sometimes, marine debris comes from inland sources as well, it can enter the marine environment through the wind or be carried down rivers. The large plastic material debris can be checked either based on number or weight. As per the 891 visual surveys reported by Eriksen M. et al. (2014) that large plastic debris detected on the ocean's surface around the world, with 20% of them being fishing-related detritus, 22% are grouped as miscellaneous and 58% as nonfishing materials. However, by weight, fishing gear made up the majority of the litter found, or around 70% of it. Some are nonfishing gear such as items from plastic films or bags (10%), pieces of foamed polystyrene, and plastic bottles (18%). Limited data are available on plastic debris which has been suspended and sunk to the bottom of the sea or ocean.

6.3.3 Microbeads

Microbeads are tiny fragments or plastic spheres used in industrial blasting, cosmetics, and cleaning items useful in the house. They are used as exfoliants in toothpaste. Microbeads are also employed in "media blasting," a process in which tiny plastic beads and sand-sized grains are launched onto the surfaces of vessels, machinery, and structures. There are no data available that can provide the exact amount of microbeads entered into wastewater treatment processes from media blasting. The emission of microbeads, from facial exfoliation alone, accounted for as16-86 tonnes per annum in the UK (Napper et al. 2015).

Microplastics are of various types. Plastics are made up of polymers. The polymers are characterized based on their toxicity buoyancy and degradability. Most plastics are made up of the following polymers: polypropylene (PP), polyethylene (PE), polystyrene (PS), polyurethane (PUR), polyvinyl chloride (PVC), and polyethylene terephthalate (PET) (GESAMP et al. 2015). Sometimes, plastic materials also contain additives like pigments, flame retardants, and stabilizers, that are specially designed for the property change of end products. These additives occasionally seep into the environment and slow down the rate of breakdown of plastics or microplastics.

The biological degradation of plastic is a slow process but biodegradable plastics can be degrading for a longer time in the environment. Whereas, conventional plastics cannot be degraded. To date, there is no technical standard data available for the complete biodegradation of plastic materials in a particular period. Figure 6.4 showed the different types of microplastics found in Francisco Bay.

6.4 Overview of WWTPs Plants as a Major Point Source of Microplastic Pollution

Microplastic accumulation in the surroundings especially in the aquatic system is due to the 'plastic age' phenomenon (Wagner et al. 2014). As they receive waste from various industries that use or manufacture microplastics, scrubbers from cosmetic and cleaning products, and domestic effluent from households using

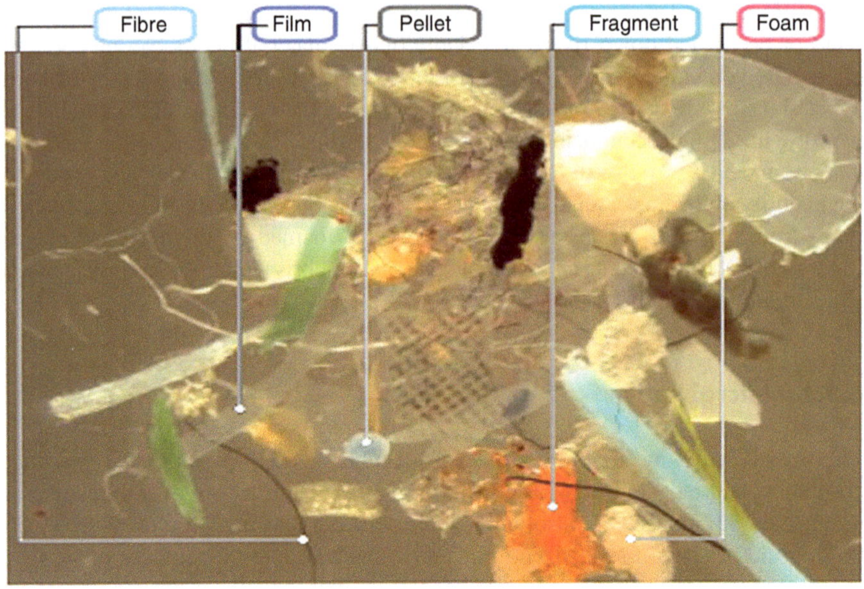

Fig. 6.4 Various microplastic types found in Francisco Bay

products containing MPs, wastewater treatment plants (WWTPs) have been hypothesized to be the primary point source of microplastic entry into the terrestrial as well as the marine environment (Roex et al. 2013). In a few available literatures, MPs concentration has been shown to be greater downstream of WWTPs.

6.4.1 Microplastics Retained in Sewage Sludge

Sewage sludge, also known as "primary sludge" and "secondary sludge," is typically produced during primary and secondary treatment (primary and secondary clarifier). Sludge is an important and main residue that originated in wastewater treatment processes. It is the waste produced by various biological, chemical, and physical procedures intended for the treatment of wastewater. In advanced wastewater treatment processes, tertiary treatment is involved which produces tertiary sludge. 'Mixed sludge' is the term for sludge that has been managed by combining primary, secondary, and, if produced, tertiary sludge (Hatinoğlu and Sanin 2021). Anaerobic digestion, incineration, dewatering, etc. are different ways of mixed sludge management (Gherghel et al. 2019; Hatinoğlu and Sanin 2021). Characteristics of sludge are different based on their origin, physical properties, and composition. During wastewater treatment operations, organic pollutants and heavy metals are retained in the sludge. In particular, it is known that the majority of microplastics, which are entrapped in sludge, are eliminated throughout the treatment process (Sol et al.

2020; Liu et al. 2019). Fibers mainly from handling textiles and cloth washing entered into waste water (Hatinoğlu and Sanin 2021).

The properties and quantity of microplastics trapped in a sludge tank can change depending on the process type used there. Mohan et al. (58) have reported the microplastic presence following sludge treatment utilizing thermal drying, digestion (anaerobic), and lime stabilization. They noticed that following the lime stabilization process, microplastic fragmentation occurs due to the mechanical mixing of sludge, elevated temperature, and basic pH. It was discovered that fibers outperform microplastics in the sludge treatment process in terms of resistance to deterioration and fragmentation (Habib et al. 2020). Sludge dewatering methods include belt filter presses, plate and frame filter presses, and decanter centrifuges. Li et al. (2018a, 2018b), due to methodological issues, underestimated several microplastics. Due to its abundance of nutrients including potassium, phosphorus, and nitrogen, sludge is most frequently used as fertilizer in agricultural land. As per reports, the use of sludge to amend soil accumulates inorganic and organic pollutants which entail the risk of microplastic deposition in soil (Carrère et al. 2010). By sewage sludge application, according to estimates, Europe 44,000-300,0000 microplastics amount to tonnes, scattered on agricultural land by sewage sludge application (Rorat et al. 2019).

Incineration of sludge as a sludge management process can generate energy which degrades all microplastics but it contributes emission of greenhouse gases (Shen et al. 2020).

6.5 Microplastics' Identification and Description Using Analytical and Physical Techniques

Microplastics from WWTPs are identified and described using analytical and physical techniques in wastewater treatment processes are crucial to comprehend because they operate as hotspots or point sources and are very effective at removing microplastic. In the characterization process, the following are the main steps involved: (1). Sampling and separation, (2). Digestion, and (3). Identification.

6.5.1 Sampling and Separation

The essential step in characterizing microplastics is sampling. These types of procedures vary from collection sites such as WWTPs and aquatic environments. For wastewater treatment processes and smaller aquatic environments, pumps and various kinds of sieves are used in the application of both in-situ and ex-situ sampling techniques. In situ, a collection of microplastics from aquatic environments manta, neuston, and plankton nets were used. Samples are to be collected from the start point and end point of processes. The conjunction is important in the sampling process with separation. With the aid of sieves and a denser solution like NaCl or $ZnCl_2$, separation has been accomplished using density gradient separation. If the samples have high organic material then the pores of sieves may be clogged during filtration steps, so the latter method is followed (Long et al. 2019). In the FTIR compatible method, oleophilic properties of microplastics can be easily extracted, so oil extraction protocols (OEP) have advantages on microplastic separation. OEP has been used for studying microplastics in WWTPs and also in sediments (Crichton et al. 2017).

6.5.2 Digestion

Microplastic digestion is carried out using an oxidizing agent, alkali, acids, or enzymes explained in detail. It is the procedure that assures only microplastic (inorganic) material is present by removing all organic stuff from a sample. Different types of acids can be used to break down the organic content present in samples of microplastic. At room temperature and low levels of mineral, HCL are inconsistent and inefficient (Cole et al. 2014). The polymers of microplastics have low acid resistance, so applying greater quantities of acids like HNO_3 could harm them (He et al. 2018). On the other hand, weaker acids in lower concentrations might not have the intended impact on organic content. So, it is suggested to optimally execute acid digestion temperature and concentration.

A high concentration of alkali agents affects microplastic digestion and easily decolorizes the microplastic polymer. As per the investigation, the application of alkali agents in microplastic digestion has been an additional issue of the possibility of deposition of organic content on the polymers. Fenton's reagent or hydrogen peroxide (oxidizing agent) is utilized for the oxidation of the organic content of the sample. H_2O_2 is proven to be a more efficient oxidizing agent in organic content digestion than acids or alkalis. A combination treatment is followed to accelerate the process of digestion and for desirable output using H_2O_2, Fenton's reagent, and heat as the catalyst. Enzymatic digestion of microplastic polymers is a multistep process and hence there may be the chances of contamination of samples. However, this process accomplishes less damage to microplastic polymers (Alvim et al. 2020) and is reported as an expensive procedure.

6.5.3 Identification Using Physical Characteristics and Chemical Characteristics

Microplastic identification from samples is the last step which includes a physical and chemical characterization study (Fig. 6.5).

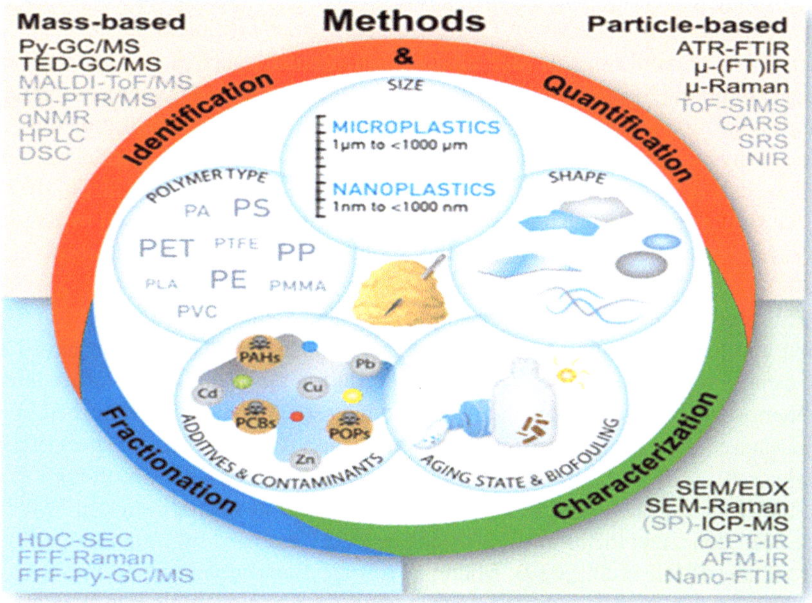

Fig. 6.5 Different methods used to identify chemical and physical properties of nanoparticles (Ivleva 2021)

6.5.3.1 Physical Characteristics

Size, shape, type (granules, foam, pellets, etc.), and color are examples of physical features. It is a time-consuming process, done visually using a microscope. To quantify microplastics from samples, physical characteristic is a cost-effective process. To study microplastic morphology electron and scanning probe microscopes, transmission electron microscopy (TEM), scanning electron microscopy (SEM), and scanning tunneling microscopy (STM) is employed. Out of all the above, SEM is the most suited for spotting and differentiation of microplastics and impurities from samples as it has high-resolution power (Barbosa et al. 2020). However, this method takes a long time, making it unsuitable for analyzing several samples (Barbosa et al. 2020).

6.5.3.2 Chemical Characteristics

The most popular method for the analysis of biological samples is Fourier Transform Infrared (FT-IR) spectroscopy studying the chemical composition of microplastics (Silva et al. 2018; Renner et al. 2018). It has the benefit of allowing for visual characterization and surface oxidation observations of the aging process of microplastics (Silva et al. 2018). FT-IR spectroscopy is more widely used than visual observation because it has reliable and accurate results. FT-IR spectroscopy is incapable to observe nano-sized particles at minimal concentrations (Barbosa et al. 2020). However, the identification of polymers down to a size of 20 µm is possible using focal planed array-based transmission micro FT-IR spectroscopy (FPA FT-IR) and attenuated total reflection (ATR FT-IR) spectroscopy (Mintenig et al. 2017). To characterize microplastics of size up to 1 μm, Raman spectroscopy is used with better resolution (Nguyen et al. 2019). Gas chromatography combined with thermal desorption followed by mass spectroscopy (TDS-GC-MS) and gas chromatography with pyrolysis followed by mass spectroscopy (Py-GC-MS) are thermal methods used to analyze the chemical composition of microplastics (Barbosa et al. 2020). The known standards serve as a reference in programs (In Py-GC-MS mode) to assess in a mass spectrometer the thermal deterioration caused by pyrolysis (Barbosa et al. 2020).

This process has some limitations including increased reliance on sample preparation and hand insertion of particles into a pyrolysis tube (Barbosa et al. 2020). The sample is mounted and heated at 1000 °C during the TDS-GC-MS procedure in the thermogravimetric balance then after products (degradation) are adsorbed in the solid phase after the sample desorbed and analyzed using mass spectroscopy (Barbosa et al. 2020). TDS-GC-MS is less sensitive than Py-GC-MS, though.

Studies that characterize and measure the number of microplastics in WWTPs have been done, and the removal efficiencies were calculated by comparing influent and effluent (Blair et al. 2019). Microplastics are found ubiquitously in nature so they entered easily into sewage treatment streams and consequently into WWTPs. A significant source of microplastic is WWTPs released into the ecosystem (aquatic) which receives treated effluent because it is crucial to the sanitation system (Kay et al. 2018; Conley et al. 2019; Alvim et al. 2020). Also, microplastics can enter terrestrial ecosystems via sewage sludge, if used as fertilizer (Rolsky et al. 2020).

6.6 Microplastic Concentration in Various Treatment Processes of WWTPs

According to the reported study, microplastics are present in every unique step of WWTPs. According to past studies, there is an excessive amount of variability and inconsistency in the presence of microplastics in terms of size, shape, concentration, and kind. Studying the effects, characterization, removal, and ultimate destiny of microplastics in each unit process of WWTPs is crucial for controlling the discharge of microplastics into the environment by these facilities. The concentrations and types of microplastics released into the environment are mostly determined by wastewater treatment methods (Talvitie et al. 2017). But these types of studies are not enough. Every unit process reacts differently to microplastics in terms of removal, impact, and fate, it has been noted. To effectively represent microplastics as a pollutant and remove WWTPs, it is essential to have a thorough understanding of the eventual destiny of microplastics in each unit process (Talvitie et al. 2017).

6.6.1 Primary Treatment

It is the initial stage of wastewater treatment. Primary settling and pretreatment are included (Fig. 6.5). The highest removal of microplastics is carried out in primary treatment in conventional WWTPs (Blair et al. 2019). Common basic treatment techniques and the effectiveness of their removal of microplastics are presented (Table 6.1). In the preliminary treatment process, the influent is made ready for the next treatment process by the removal of large debris (Blair et al. 2019). The bigger material from influent (raw wastewater) is filtered out using bar screens and grit chambers in WWTPs. The removal efficiency of microplastics in bar chambers is less because generally space between bars is 2 and 6 cm and it easily allows microplastics to pass in the next unit process (Murphy et al. 2016).

Primary treatment shows about 39-58% removal efficiency of microplastics. Generally, microfibers with higher settling behavior and densities are impacted by grit chambers the most (Petroody et al. 2020). Mostly microplastics adsorb pollutants, maybe it will disturb the pretreatment. The second step of primary treatment is the primary settling tank. In the primary settling, tank removes organic as well as inorganic suspended solids, as they are settled at bottom of the tank by sedimentation and by floating. The surface skimming process is performed after the settling, as skimming of oil and grease is carried out. Along with skimming, many microplastic particles are also separated out with a skimmer (Yang et al. 2019). As per the reported study, skimming generally removes low-density microplastics as they float on the wastewater surface (Petroody et al. 2020). Low-density microbeads are also removed during skimming as per a study reported by (Yang et al. 2019).

6.6.2 Secondary Treatment

The first treatment method's leftover wastewater samples are to be treated in the secondary treatment phase. It gets rid of the organic residue and suspended solids. In this step, colloidal and dissolved biologically degradable organic components are removed using biological treatment methods (aerobic and anaerobic). A secondary clarifier is employed during the secondary treatment procedure. Figure 6.5 illustrates the typical secondary treatment unit method that was employed in a study on the effectiveness of removing microplastics from wastewater (Table 6.1).

Generally, the biological treatment process is not well known directly for microplastic removal but microplastics can settle at the bottom of the tank as flocs. Thus microplastics directly find a way to enter the sludge stream from the biological treatment stage. The secondary treatment process is diverse, so there is also variation in the type of microplastics removed. For example, rapid sand filters and membrane biological reactors can remove microplastic particulate high effectively than fibers (Bayo et al. 2020). Membrane biological reactors showed more efficiency in microplastic removal than traditional activated sludge processes (Lares et al. 2018). Microplastics have no impact on the efficiency of the sequencing batch reactor (Kalčíková et al. 2017). In inactivated filter processes, microplastic particles

Table 6.1 Various methods used for microplastic removal

Sr. No.	Method	Strategy	Type and Morphology of NPs	Result	Reference
1.	*Physical methods*				
a.	Magnetic polyoxometalate supported ionic liquid phases	Adsorption	Polystyrene (1–10 nm)	Over 90% removal	
b.	A non-fluorinated superhydrophobic aluminum surface	Adsorption	Polypropylene (262+-)	99%	Rius-Ayra and Llorca-Isern (2021)
2.	*Chemical methods*				
a.	Alum coagulant and alum combined with cationic polyamine-coated sand	Coagulation and flocculation	Polyethylene10-100 μm	70.7%–92.7%	Shahi et al. (2020)
b.	Polyaluminum chloride and ferric chloride coagulation	Coagulation	Polystyrene and polyethylene	--	
3.	*Biological method*				
a.	MBR	Microbial degradation	All MPs	79.01	Bayo et al. (2020)
b.	*Bacillus muralis*	Bacteria	Polyethylene terephthalate	--	Narciso-Ortiz et al. (2020)

provide the surfaces for the attachment of microflora. Due to the development of solid (suspended) spheres on active filters, there is an uneven distribution of water (Lares et al. 2018). There has been no investigation reported on the microplastic removal of secondary clarifiers. As per past reports, the biological treatment process is combined with secondary clarifiers, due to this settling (sedimentation) mechanism of microplastics could contribute towards its removal.

6.6.3 Tertiary Treatment

It is an advanced step in wastewater treatment. The examined removal efficiencies of microplastics in the advanced tertiary treatment processes (Fig. 6.6). Advanced wastewater treatment processes are more effective in microplastic removal (Bayo et al. 2020). Combining ultra-filtration (UF), granular activated carbon and MBR has been demonstrated to eliminate microplastics at levels below the deductible (Baresel et al. 2019). In the removal of microplastics, the filtration mechanism plays an important role; however, due to the irregular shapes of microplastics, they might be put on the filters (Zhang and Chen 2020). This may lead to membrane fouling of filters owing to their surface characteristics (Ma et al. 2019). Many investigations

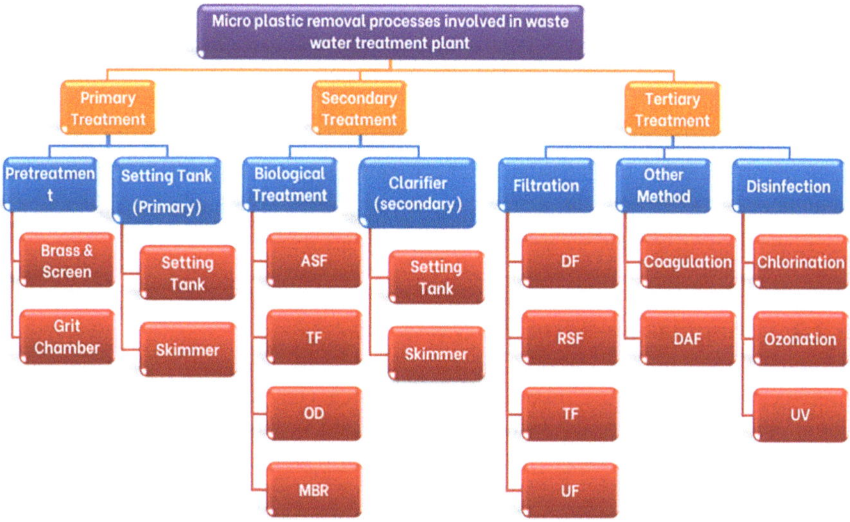

Fig. 6.6 Unit processes of wastewater treatment processes associated with microplastic removal

have demonstrated that MBRs are the tertiary treatment method and phosphorus and nitrogen removal; it shows significant removal of microplastics (Bayo et al. 2020).

To remove suspended solids, the approach most frequently employed for treating wastewater is coagulation. Coagulation of microplastics before starting the final advanced process removes microplastics efficiently. The most important step in tertiary treatment is disinfection. The two most widely utilized disinfection techniques in wastewater treatment units are chlorination and ozonation of treated wastewater (Kelkar et al. 2019). The physical qualities of microplastics are destroyed by chlorine and ozone treatment, which also affects deterioration (Kelkar et al. 2019). The effect of these two processes (chlorination and ozonation) has not received much research.

6.7 Specific Technologies for Microplastic Removal in WWTPs

The current methods for microplastic removal from wastewater treatment processes are inefficient or ineffective, expensive, or use coagulants (chemical agents). If the coagulant is not properly treated then, it may show a detrimental effect on receiving environment in nature. Recently, many methods have been made to remove microplastics from wastewater treatment unit processes. Broadly, these methods are classified into biodegradation-based, coagulation-based, and filtration-based methods (Table 6.1).

6.7.1 Biodegradation Method

There have been reports noted on the use of biological microplastic breakdown. In the biological-based method, biological agents like potent microbes can use microplastic polymers as a source of carbon (Enfrin et al. 2019). Some studies have reported on the biodegradation of polypropylene (Caruso 2015), polyethylene terephthalate (Yoshida et al. 2016) and polyethylene (Auta et al. 2018). However, there is a lack of study (research) on microplastic's effect on these microbes.

6.7.2 Coagulation-Based Methods

Electrocoagulants are the process through which coagulants are produced using metal electrodes. Microplastics that are suspended in the air are gathered on thin layers created by coagulants. It has been thoroughly researched for the removal of microplastics in drinking water treatment and is the most economical method of doing so. Additionally, it was claimed that coagulants based on iron and aluminium performed better. The sole gel method is used in a variable pH and results in flocculates; in this method, amorphous silica is used for the induction of agglomeration of microplastics (Zhang and Chen 2020).

6.7.3 Filtration-Based Methods

Filtration is a relatively simple way to remove microplastics and a cost-effective method. In this method, no external chemical agents have been used. However, clogging of filters may occur due to the deposition of the smaller-size filters (Zhang and Chen 2020). Beljanski (2016) have developed and tested microplastic removal by using the gravity-powered filtration method from the secondary treatment of WWTPs. This technique has not been established for actual wastewater samples, including filtering and backflush, but just for a synthetic microplastic solution. Dynamic membrane filtration (DMF), another filtration technique, uses fabric or mesh as a physical barrier on which a microplastic cake layer develops. DMF is especially useful for low-density and small-sized inorganics. Li et al. (2018) have studied the microplastic removal efficiency of DMF. In comparison to MBR, it is more efficient, less energy-intensive, and less likely to foul filters (Ersahin et al. 2012). By using $Zncl_2$ to change the density of wastewater samples, the density separation approach can separate microplastics (Enfrin et al. 2019). However, because wastewater volume is too great and it often contains other pollutants, the density separation approach is not always practical for wastewater treatment.

6.8 Conclusion, Future Insights and Recommendations

There is a dearth of in-depth research despite the significant research on the definition, identification, and characterization of microplastics. Rare reports are available about the impact of WWTPs unit processes on microplastics or vice versa. Microplastic properties challenge the unit processes of waste treatment and WWTPs removal. As per the literature, about microplastic pollution in wastewater treatment processes, it can be interpreted that, even though WWTPs reduce the microplastics in a huge amount that enters the aquatic ecosystem through the treated effluent and the terrestrial ecosystem through sludge, operate as hot spots (point sources) of microplastic pollution. In order to trace the influence of microplastics as a pollutant in each phase of the treatment process, it is absolutely essential to understand the end fate of microplastics in each step of the treatment process. The size, shape, type, concentration, and composition of microplastics entering and leaving WWTPs are variable. These characteristics also depend on the time of day, the unit process, and the influent (industrial or household). WWTPs act as a center where primary microplastics can convert into secondary microplastics that are emitted with the treated wastewater and sludge. Till today, microplastic-specified removal techniques still be understudied by researchers. However, biological degradation, coagulation, and filtration methods are the three categories under which microplastic removal procedures are currently offered. This book chapter would provide a thorough grasp of the interplay between various microplastics and the unit processes of WWTPs.

Researchers have looked that microplastics are detected everywhere in nature such as in arctic snow, deep oceans, Antarctic ice, honey, shellfish, drinking water, table salt, etc. Microplastics are found ubiquitously in aquatic ecosystems. Although there has been no recorded evidence so far about the adverse or dangerous impacts of microplastics on human health, "any data" does not indicate "no risk," necessitating further study to close the knowledge gap. According to reported data from the literature, there may be wide-ranging diversity worldwide in the values of microplastic fibers and microplastic concentrations in sludge, WWTPs, tap water, and DWTPs. It has been demonstrated and is clear that WWTPs are a major source of microplastic contamination, particularly microplastic fibers. Recent methods for removing microplastics from WWTPs, such as MBR and coagulation, have remarkable removal efficiencies (almost 99%). However, it is required to increase the effectiveness of removing microplastic should go more deeply into these unit treatment procedures in order to completely eradicate microplastics from wastewater treatment procedures. Based on the present chapter, we recommend the following to curb the microplastic pollution at global level.

1. The removal of microplastics from wastewater treatment systems should be improved as a matter of performance.
2. There is a need of searching for new technologies which have the potential to remove total microplastics from wastewater treatment processes completely.

3. To accomplish the future need for microplastic removal, there is a need for some regulations applied by authority bodies during WWTPs from a pollution-free environmental point of view.
4. In the terrestrial ecosystem, the microplastics go through WWTPs, which are mostly found in sludge, so it is recommended that there should be a new alternative for sewage sludge management.
5. An investigation is required for the enhancement of the valorization of sludge without any risk to nature (environment).

References

Alvim CB, Mendoza-Roca JA, Bes-Piá A (2020) Wastewater treatment plant as microplastics release source–quantification and identification techniques. J Environ Manag 255:109739

Auta HS, Emenike CU, Jayanthi B, Fauziah SH (2018) Growth kinetics and biodeterioration of polypropylene microplastics by Bacillus sp. and Rhodococcus sp. isolated from mangrove sediment. Mar Pollut Bull 127:15–21

Barbosa F, Adeyemi JA, Bocato MZ, Comas A, Campiglia A (2020) A critical viewpoint on current issues, limitations, and future research needs on micro-and nano plastic studies: from the detection to the toxicological assessment. Environ Res 182:109089

Baresel C, Harding M, Fang J (2019) Ultrafiltration/granulated active carbon-biofilter: efficient removal of a broad range of micropollutants. Appl Sci 9(4):710

Bayo J, López-Castellanos J, Olmos S (2020) Membrane bioreactor and rapid sand filtration for the removal of microplastics in an urban wastewater treatment plant. Mar Pollut Bull 156:111211

Beljanski A (2016) Efficiency and effectiveness of a low-cost, self-cleaning microplastic filtering system for wastewater treatment plants. 2016 NCUR

Bhuyan MS (2022) Effects of microplastics on fish and in human health. Front Environ Sci 10:827289. https://doi.org/10.3389/fenvs.2022.827289

Blair RM, Waldron S, Gauchotte-Lindsay C (2019) Average daily flow of microplastics through a tertiary wastewater treatment plant over ten months. Water Res 163:114909

Bourdages MP, Provencher JF, Baak JE, Mallory ML, Vermaire JC (2021) Breeding seabirds as vectors of microplastics from sea to land: evidence from colonies in Arctic Canada. Sci Total Environ 764:142808

Browne MA (2015) Sources and pathways of microplastics to habitats. In: Marine anthropogenic litter. Springer, Cham, pp 229–244

Browne MA, Crump P, Niven SJ, Teuten E, Tonkin A, Galloway T, Thompson R (2011) Accumulation of microplastic on shorelines woldwide: sources and sinks. Environ Sci Technol 45(21):9175–9179

Carrère H, Dumas C, Battimelli A, Batstone DJ, Delgenes JP, Steyer JP, Ferrer I (2010) Pretreatment methods to improve sludge anaerobic degradability: a review. J Hazard Mater 183(1-3):1–15

Caruso G (2015) Plastic degrading microorganisms as a tool for bioremediation of plastic contamination in aquatic environments. J Pollut Eff Control 3(3):1–2

Cole M, Webb H, Lindeque PK, Fileman ES, Halsband C, Galloway TS (2014) Isolation of microplastics in biota-rich seawater samples and marine organisms. Sci Rep 4(1):1–8

Conley K, Clum A, Deepe J, Lane H, Beckingham B (2019) Wastewater treatment plants as a source of microplastics to an urban estuary: removal efficiencies and loading per capita over one year. Water Res X 3:100030

Cox KD, Covernton GA, Davies HL, Dower JF, Juanes F, Dudas SE (2019) Human consumption of microplastics. Environ Sci Technol 53(12):7068–7074

Crichton EM, Noël M, Gies EA, Ross PS (2017) A novel, density-independent and FTIR-compatible approach for the rapid extraction of microplastics from aquatic sediments. Anal Methods 9(9):1419–1428

Daniel DB, Ashraf PM, Thomas SN, Thomson KT (2021) Microplastics in the edible tissues of shellfishes sold for human consumption. Chemosphere 264:128554

Desforges JPW, Galbraith M, Ross PS (2015) Ingestion of microplastics by zooplankton in the Northeast Pacific Ocean. Arch Environ Contam Toxicol 69(3):320–330

Enfrin M, Dumée LF, Lee J (2019) Nano/microplastics in water and wastewater treatment processes–origin, impact and potential solutions. Water Res 161:621–638

Eriksen M, Lebreton LC, Carson HS, Thiel M, Moore CJ, Borerro JC et al (2014) Plastic pollution in the world's oceans: more than 5 trillion plastic pieces weighing over 250,000 tons afloat at sea. PLoS One 9(12):e111913

Ersahin ME, Ozgun H, Dereli RK, Ozturk I, Roest K, van Lier JB (2012) A review on dynamic membrane filtration: materials, applications and future perspectives. Bioresour Technol 122: 196–206

EU (2015) Newsletters 2 life + project mermaids. European Union

GESMAP (2015) Sources, fate and effects of microplastics in the marine environment: a global assessment Joint group of experts on the scientific aspects of marine environmental protection. Rep. Stud. GESAMP No. 90, p 96

Gherghel A, Teodosiu C, De Gisi S (2019) A review on wastewater sludge valorisation and its challenges in the context of circular economy. J Clean Prod 228:244–263

Habib RZ, Thiemann T, Al Kendi R (2020) Microplastics and wastewater treatment plants—a review. J Water Res Prot 12(01):1

Hatinoğlu MD, Sanin FD (2021) Sewage sludge as a source of microplastics in the environment: a review of occurrence and fate during sludge treatment. J Environ Manag 295:113028

He D, Luo Y, Lu S, Liu M, Song Y, Lei L (2018) Microplastics in soils: analytical methods, pollution characteristics and ecological risks. TrAC Trends Anal Chem 109:163–172

Ivleva NP (2021) Chemical analysis of microplastics and nanoplastics: challenges, advanced methods, and perspectives. Chem Rev 121(19):11886–11936

Kalčíková G, Alič B, Skalar T, Bundschuh M, Gotvajn AŽ (2017) Wastewater treatment plant effluents as source of cosmetic polyethylene microbeads to freshwater. Chemosphere 188:25–31

Kay P, Hiscoe R, Moberley I, Bajic L, McKenna N (2018) Wastewater treatment plants as a source of microplastics in river catchments. Environ Sci Pollut Res 25(20):20264–20267

Kelkar VP, Rolsky CB, Pant A, Green MD, Tongay S, Halden RU (2019) Chemical and physical changes of microplastics during sterilization by chlorination. Water Res 163:114871

Klemchuk PP (1990) Degradable plastics: a critical review. Polym Degrad Stab 27(2):183–202

Lambert S (2013) Environmental risk of polymer and their degradation products (Doctoral dissertation, University of York)

Lares M, Ncibi MC, Sillanpää M, Sillanpää M (2018) Occurrence, identification and removal of microplastic particles and fibers in conventional activated sludge process and advanced MBR technology. Water Res 133:236–246

Li J, Green C, Reynolds A, Shi H, Rotchell JM (2018b) Microplastics in mussels sampled from coastal waters and supermarkets in the United Kingdom. Environ Pollut 241:35–44

Li WC, Tse HF, Fok L (2016) Plastic waste in the marine environment: a review of sources, occurrence and effects. Sci Total Environ 566:333–349

Li X, Chen L, Mei Q, Dong B, Dai X, Ding G, Zeng EY (2018a) Microplastics in sewage sludge from the wastewater treatment plants in China. Water Res 142:75–85

Liu X, Yuan W, Di M, Li Z, Wang J (2019) Transfer and fate of microplastics during the conventional activated sludge process in one wastewater treatment plant of China. Chem Eng J 362:176–182

Long Z, Pan Z, Wang W, Ren J, Yu X, Lin L et al (2019) Microplastic abundance, characteristics, and removal in wastewater treatment plants in a coastal city of China. Water Res 155:255–265
Lusher AL, Hurley R, Arp HPH, Booth AM, Bråte ILN, Gabrielsen GW et al (2021) Moving forward in microplastic research: a Norwegian perspective. Environ Int 157:106794
Ma B, Xue W, Ding Y, Hu C, Liu H, Qu J (2019) Characteristics of microplastics removal via coagulation and ultrafiltration during drinking water treatment. Chem Eng J 359:159–167
Mani T, Hauk A, Walter U, Burkhardt-Holm P (2015) Microplastics profile along the Rhine river. Sci Rep 5(1):1–7
Mintenig SM, Int-Veen I, Löder MG, Primpke S, Gerdts G (2017) Identification of microplastic in effluents of waste water treatment plants using focal plane array-based micro-Fourier-transform infrared imaging. Water Res 108:365–372
Murphy F, Ewins C, Carbonnier F, Quinn B (2016) Wastewater treatment works (WwTW) as a source of microplastics in the aquatic environment. Environ Sci Technol 50(11):5800–5808
Napper IE, Bakir A, Rowland SJ, Thompson RC (2015) Characterisation, quantity and sorptive properties of microplastics extracted from cosmetics. Mar Pollut Bull 99(1-2):178–185
Narciso-Ortiz L, Coreño-Alonso A, Mendoza-Olivares D, Lucho- Constantino CA, Lizardi-Jiménez MA (2020) Baseline for plastic and hydrocarbon pollution of rivers, reefs, and sediment on beaches in Veracruz State, México, and a proposal for bioremediation. Environ Sci Pollut Res 27(18):23035–23047
NEA (2014) Sources of microplastics to the marine environment. Norwegian Environmental Agency
Ngo PL, Pramanik BK, Shah K, Roychand R (2019) Pathway, classification and removal efficiency of microplastics in wastewater treatment plants. Environ Pollut 255:113326
Nguyen B, Claveau-Mallet D, Hernandez LM, Xu EG, Farner JM, Tufenkji N (2019) Separation and analysis of microplastics and nanoplastics in complex environmental samples. Acc Chem Res 52(4):858–866
Padervand M, Lichtfouse E, Robert D, Wang C (2020) Removal of microplastics from the environment. A review. Environ Chem Lett 18(3):807–828
Petroody SSA, Hashemi SH, van Gestel CA (2020) Factors affecting microplastic retention and emission by a wastewater treatment plant on the southern coast of Caspian Sea. Chemosphere 261:128179
Prata JC, da Costa JP, Lopes I, Duarte AC, Rocha-Santos T (2020) Environmental exposure to microplastics: an overview on possible human health effects. Sci Total Environ 702:134455
Renner G, Schmidt TC, Schram J (2018) Analytical methodologies for monitoring micro (nano) plastics: which are fit for purpose? Curr Opin Environ Science & Health 1:55–61
Rius-Ayra O, Llorca-Isern N (2021) A robust and anticorrosion non-fluorinated superhydrophobic aluminium surface for microplastic removal. Sci Total Environ 760:144090
Roex E, Vethaak D, Leslie H, Kreuk MD (2013) Potential risk of microplastics in the fresh water environment. STOWA, Amersfoort
Rolsky C, Kelkar V, Driver E, Halden RU (2020) Municipal sewage sludge as a source of microplastics in the environment. Curr Opin Environ Sci Health 14:16–22
Rorat A, Courtois P, Vandenbulcke F, Lemiere S (2019) Sanitary and environmental aspects of sewage sludge management. In: Industrial and municipal sludge. Butterworth-Heinemann, pp 155–180
Sadri SS, Thompson RC (2014) On the quantity and composition of floating plastic debris entering and leaving the Tamar Estuary, Southwest England. Mar Pollut Bull 81(1):55–60
Shahi NK, Maeng M, Kim D, Dockko S (2020) Removal behavior of microplastics using alum coagulant and its enhancement using polyamine-coated sand. Process Saf Environ Prot 141:9–17
Shen M, Huang Z, Luo X, Ma Y, Chen C, Chen X, Cui L (2020) Activation of persulfate for tetracycline degradation using the catalyst regenerated from Fenton sludge containing heavy metal: synergistic effect of Cu for catalysis. Chem Eng J 396:125238

Silva AB, Bastos AS, Justino CI, da Costa JP, Duarte AC, Rocha-Santos TA (2018) Microplastics in the environment: challenges in analytical chemistry-a review. Analy Chim Acta 1017:1–19
Sol D, Laca A, Laca A, Díaz M (2020) Approaching the environmental problem of microplastics: importance of WWTP treatments. Sci Total Environ 740:140016
Talvitie J, Mikola A, Koistinen A, Setälä O (2017) Solutions to microplastic pollution–removal of microplastics from wastewater effluent with advanced wastewater treatment technologies. Water Res 123:401–407
United Nations Environment Programme (2011) UNEP year book emerging issues in our global environment. United Nations Environment Programme
Wagner M, Scherer C, Alvarez-Muñoz D, Brennholt N, Bourrain X, Buchinger S, Reifferscheid G (2014) Microplastics in freshwater ecosystems: what we know and what we need to know. Environ Sci Eur 26(1):1–9
Yang L, Li K, Cui S, Kang Y, An L, Lei K (2019) Removal of microplastics in municipal sewage from China's largest water reclamation plant. Water Res 155:175–181
Yoshida S, Hiraga K, Takehana T, Taniguchi I, Yamaji H, Maeda Y et al (2016) A bacterium that degrades and assimilates poly (ethylene terephthalate). Science 351(6278):1196–1199
Zhang Z, Chen Y (2020) Effects of microplastics on wastewater and sewage sludge treatment and their removal: a review. Chem Eng J 382:122955

Microplastics Originating from Paints and Synthetic Textile Materials

7

Mufaro Moyo, Rivalani Baloyi Baloyi, Bruce Bishop Sithole, and Thabo Falayi

Abstract

There is continued growth in paints and synthetic textile materials production and consumption due to expanded use of paints in different applications, population growth and fast fashion trends. Paints and synthetic textile materials are some of the major sources of microplastics. The major sources of microplastics originating from paints are marine paints, road markings, architectural paints, and spray painting. Major contributing factors in the release of microplastics from paints to the environment are paint production processes, weathering and abrasion, paint spilling, residual paints on paint application tools and storage containers, paint end of life (EoL) disposal, and paint applicator's expertise. In the context of synthetic textile materials, major emissions of the microplastics occur during the production processes of the synthetic textile materials, during

M. Moyo (✉)
Department of Engineering, Malawi University of Science and Technology, Limbe, Malawi

Department of Fibre and Polymer Materials Engineering, National University of Science and Technology, Bulawayo, Zimbabwe

R. B. Baloyi
University of KwaZulu-Natal (Howard Campus), Discipline of Chemical Engineering, College of Agriculture, Engineering and Sciences, Durban, South Africa

B. B. Sithole
University of KwaZulu-Natal (Howard Campus), Discipline of Chemical Engineering, College of Agriculture, Engineering and Sciences, Durban, South Africa

Council for Scientific and Industrial Research, Biorefinery Industry Development Facility, Durban, South Africa

T. Falayi
Namibia University of Science and Technology, Department of Mining and Process Engineering, Windhoek, Namibia

M. Shahnawaz et al. (eds.), *Microplastic Pollution*,
https://doi.org/10.1007/978-981-99-8357-5_7

use and laundering, washing conditions and drying. Studying the role and influence of paints and synthetic textile materials in the generation of microplastics is critical. It helps in coming up with mitigatory measures to reduce the generation of microplastics from paints and synthetic textile materials and hence promotion of sustainability. Devising a holistic approach to research and development of green chemistry and green manufacturing processes in the production of paints and synthetic textile materials is one of the mitigatory measures. Also, developing policies and legislation on wastewater treatment systems for all responsible stakeholders and introducing legislation on do-it-yourself (DIY) painting activities could help in reducing the generation of microplastics from paints and synthetic textile materials.

Keywords

Microplastics · Paint · Textile · Synthetic textile · Materials · Environment

7.1 Introduction

Paints and synthetic textile materials constitute part of sources of primary microplastics (MPs) and physical activities on them lead to secondary microplastics (Moyo and Falayi 2022). The microplastics end up in the environment such as waterways and oceans. According to Paola Paruta and Boucher (2022) about 58% of microplastics in fresh water and oceans originate from paint, which is equivalent to 1.9 million tonnes of paint particles every year. Paints alone outweigh other sources such as textile fibres and tyre dust (Paola Paruta and Boucher 2022). Textile materials, such as polyester (polyethylene terephthalate), polyacrylonitrile, and polyamide, are the second largest sources of microplastics. Worldwide microplastic pollution has been a result of the increased reliance on petroleum as the raw material for paint and most textile materials. Microplastics are generated from paints through weathering, sanding, abrasion poor paint waste management and paint leakage. In the context of synthetic textile materials, microplastics are generated during manufacturing, wearing, drying, and laundering. Microplastics from paints and synthetic textile like any other microplastics are ubiquitous and their distribution in different sections of the environment is dictated by the receiving environment, industrial, and human activities. It is reported that environmental factors such as weather forms, wind flow direction, wave, and floods influence the distribution pattern of MPs in the environment (Imhof et al. 2016; Roebroek et al. 2021). The following discussions are on microplastics originating from paints and synthetic textile materials.

7.2 Microplastics Originating from Paints

Paints consist of binder (resin), pigments, solvent, filler, and additives as the major components (Bentley and Turner 1998; Poth 2008). The paints could be non-aqueous type or water-based type. Non-aqueous based paints are regarded as plastics since they possess a polymer type backbone containing alkyls, polyesters and epoxies copolymers (Gaylarde et al. 2021; Poth 2008). The water-based acrylics contain polystyrenes, polyurethanes, polyacrylates, and polyesters (Cunningham et al. 2020; Gaylarde et al. 2021; Poth 2008). The binders or resins used in paint production are polymers. They are the sources of paint microplastics (Gaylarde et al. 2021; Turner 2021). Other paint products contain property enhancing microspheres in form of microfibres and or microbeads (Faber et al. 2021). These also become sources of paint microplastics in the life cycle of paints. MPs from paints emanate from physical activities such as shipping and boating, external surfaces, and road markings (Gaylarde et al. 2021). The availability of paint microplastics in the environment is published in the literature where major concentrations were found to be in ocean surfaces, coastal regions, and beach surfaces (Gaylarde et al. 2021; Imhof et al. 2016). Figure 7.1 shows major contributing factors in the release of microplastics from paints into the environment.

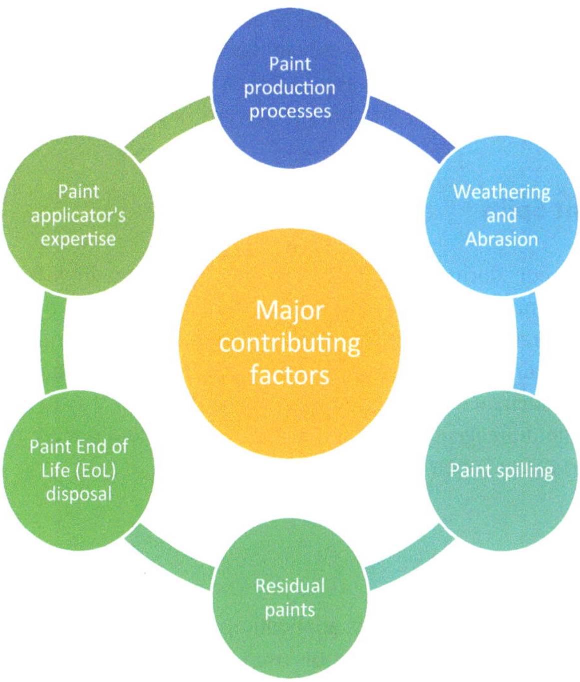

Fig. 7.1 MPs from paints: Major contributing factors

7.2.1 Paint Production Processes

The choices made during paint production stage greatly influence emission of microplastics from paints in the paint value chain. The choices could be on the paint ingredients and the resultant formulations. Plastic-based particulate materials such as microspheres and microfibres which are added during paint production to enhance paint performance and life-span make paints to be a source of microplastics. The literature shows that microspheres of 5-80 μm and microfibres of 0.5-50 mm can be added to paints during paint production (Faber et al. 2021; The European Council of the Paint 2018). The handling and disposal of plastic-based paint ingredients and waste during the production process also influences the emission of microplastics from paints. The discarded plastic-based ingredients and waste end up into the drainage systems.

7.2.2 Weathering and Abrasion

Weathering and abrasion make paints brittle over some time causing roughening of the paint surface and ultimately peeling off. Consequently, thin paint microparticles that peel off are released into the environment (Toben 2018). UV radiation fades paints, thereby triggering the need for new paint. Sanding or scratching of paint during maintenance work or repainting activities releases paint microparticles. It is reported that sanding generates paint flakes of size below 10 μm (Toben 2018). Old paint peels disposed of the environment become a source of secondary paint microplastics. The possible pathways to the environment of the paint particulates from weathering and abrasion are through sewer systems and air.

7.2.3 Paint Spills

Paint spills occur during application of new paint leading to primary paint microplastics from the spills of uncured paints and secondary paint microplastics from spills of cured paints. Paint degradation over some time triggers the need for removal of old paint and application of new paint. Paint microparticles are also lost during such maintenance and repairs activities finding their way to the environment. Pathways to the drainage systems, open water sources and sea are road runoff and wind transportation (Toben 2018).

7.2.4 Residual Paints

Residual paints are found on paint application tools and paint storage containers. Cleaning of paint brushes and other tools used during application of paints results in the paint being deposited into the sewage systems and other drainage systems ending up in water sources. In a study conducted in 2015 in Denmark, it was reported that about 1% of uncured paints ends up in sewer system from the cleaning of paint application brushes and other tools (Lassen et al. 2015).

7.2.5 Paint End of Life (EoL) Disposal

There are different end-of-life (EoL) scenarios for paints in different countries depending on the legislation. The unused paints can be taken to waste collection for incineration or handled in any other ways deemed environmentally safe. One typical example is Netherlands, where unused paints are taken to waste collection centres where they are incinerated as waste and not recycled (Faber et al. 2021). However, this situation is different in other countries especially the developing countries where there are no proper and strict paint waste management policies. In such cases, the paint EoL disposal is done anyhow and paint tools and storage vessels are also disposed anyhow. This kind of disposal poses a threat to the proliferation of microplastics from paints.

7.2.6 Paint Applicator's Expertise

Paint application can be done by professionals who would have received trained on how to use paint in an environmentally friendly manner and taking all necessary precautions to avoid emission of paint microplastics. However, there is also a room for "do-it-yourself" (DIY) in the use of paints. Many of the DIY scenarios consist of untrained people who end up using and handling paints in a way that contributes to the emission of microplastics from paints. The activities performed during paint application that can contribute to emission of microplastics from paints include handling of paint spills, disposing of paints in the drainage systems, and cleaning of application brushes and rollers under the tap or other sources of water.

7.3 Sources of Paint Microplastics

Sources of paint microplastics are many since there is a broad spectrum of painted objects and structures whose activities or maintenance and repairs activities would lead to release of paint microparticles. The major sources of paint microplastics include:

7.3.1 Marine (Boats and Ships) Paint

Coating on recreational boats, larger vessels and ship's hulls contain synthetic polymers. Abrasion, maintenance activities of marine vessels and other physical activities remove these coatings and they become paint based MPs in seas. Antifouling agents are also applied to the hulls and also slough off leading to MPs in seas.

Primary MPs are released from leisure and commercial boats during building work, maintenance activities, repairs or use (Boucher and Friot 2017). The key activities leading to releases were reported as pre-treatment of surfaces, application of coating and cleaning of the equipment (Boucher and Friot 2017). Paint particles from marine vessels enter the marine environment directly by the sea pathway as well as land by paint spilling or removal. According to OECD 2009, during the lifetime of a ship, approximately 6% of a ship's paint or coating is released directly into the sea environment as spills, approximately 1.8% spills released during painting, 1% released by weathering effects and 3.2% released by activities involved in maintenance (OECD 2009). Shipwrecks also contribute to the generation of MPs from paints. It is reported that shipwrecks release microplastics as well as antifouling or anticorrosion chemical materials (van der Schyff et al. 2020).

7.3.2 Road Markings

Paints used in road marking are a source of paint microfragments that end up leaching into terrestrial waterways and ultimately to the oceans by run-off. The road marking paints may contain polymers such as polyesters, polyamides, polyethylene and polyurethane (Harvey et al. 2017). A study showed that weathering and abrasion of road markings contributed 7% of the annual total microplastics deposited into the global marine environment (Harvey et al. 2017).

7.3.3 Building Paints or Coatings

These are architectural paints used on building structures and may be solvent or water based. The solvent-based paints contain synthetic resins. In addition, biocides are often infused into the paints to avoid discoloration and prolong the painted surface's lifespan. Paint degradation by ultraviolet (UV) irradiation is the most common in architectural paints. The paint produces small paint particles during this degradation and these paint particles are deposited into water bodies by runoff. Coatings for buildings may be either water or solvent based, the solvent-borne paints containing synthetic resins. Biocides are often incorporated to prevent discoloration and extend the lifetime of the painted surface.

7.3.4 Spray Painting

This is one method that has been discovered to leave a legacy of paint microplastics in the environment. The mass transfer efficiency ranges between 50% and 65% when using spray painting and the remaining paint materials are lost to the environment as droplets and aerosol particulates (Xu et al. 2022).

7.4 Microplastics Originating from Synthetic Textile Materials

Textile fibres are produced mainly from natural and synthetic fibres. Natural fibres are biodegradable and that means that once they are released to the environment the fragmentation of the fibres assists with the bio degradation. However, manmade fibres are plastic in nature and do not biodegrade. Synthetic (man-made) fibres dominate the fibre production globally. 60% of the worldwide textile fibres in use are nylon, polyester, and acrylic fibres (Mahbub and Shams 2022). All fibres are responsible for the release of microfibres in the atmosphere. However, only synthetic fibres are responsible for the release of plastic microfibres which are known as microplastics. Table 7.1 summarises some of the polymers where microplastic fibres and fragments originate from.

Microplastics originating from synthetic textile materials come from different sources as shown in Fig. 7.2.

Figure 7.2 shows that primary microplastics from textile materials are released during the production process of the synthetic fibres and are also released from medical textiles, artificial turfs especially for sporting activities and packaging materials. The secondary sources are largely due to the physical activities the synthetic textile materials are exposed to during use, laundering and abrasion and degradative processes. Moyo and Falayi (2022) discussed in depth how these processes contribute to the formation of microplastics. In addition, the of plastic films in greenhouse agricultural practises has been reported as a source of microplastics in soils (Kim et al. 2021; Wang et al. 2021; Wang et al. 2022; Xu et al. 2006). Most of the film remains are left scattered in the fields with no form of

Table 7.1 Origin of microplastics

Type of polymer used	Application
Polyurethane, silicone, epoxy	Surface coatings
Polystyrene, vinyl halide copolymers and polyvinyl halides, poly(vinyl esters), poly (vinyl acetyls), poly(vinyl ethers), polyvinyl alcohol and polystyrene	Polishing agents, weather resistant coatings
Polyethylene, polypropylene or polyisoprene	Rubber coatings, underwater coatings on ships, vegetable and crop farming
Polystyrene and styrene copolymers	Paints for concrete, exteriors, and road markings. Plastic utensils, food containers, packaging
Polyethylene (PE-LDPE, LLDPE)	Textiles, plastic bags
Polyethylene terephthalate (PET)	Textiles, beverage bottles
Polyamide or nylon (PA)	Textiles, nets, twine and ropes
Polyvinyl chloride (PVC)	Bags, tubes
Polypropylene (PP)	Ropes, bottle caps, netting
Polyacrylonitrile (PAN)	Clothing and textiles
Styrene butadiene rubber	Tyres

Source: (Eyerer 2010; Šaravanja et al. 2022; Sazali et al. 2020; Van der Vegt 2006; Xu et al. 2006)

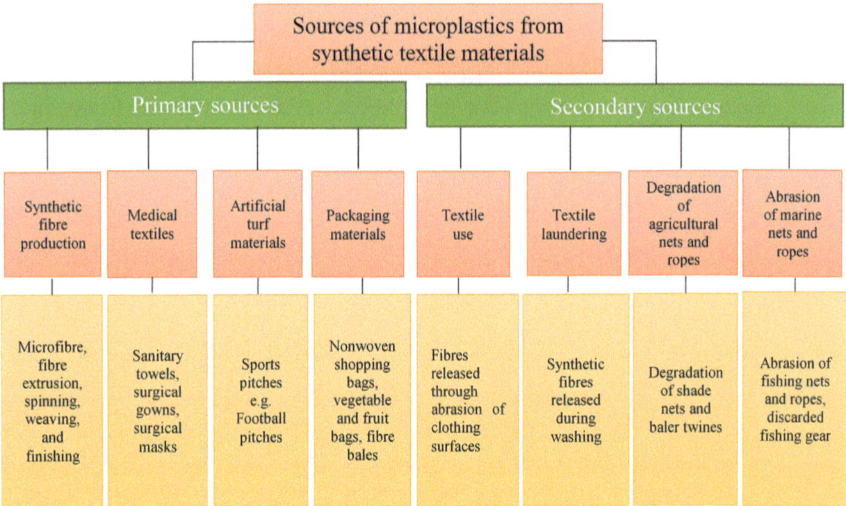

Fig. 7.2 Sources of microplastics originating from synthetic textile materials

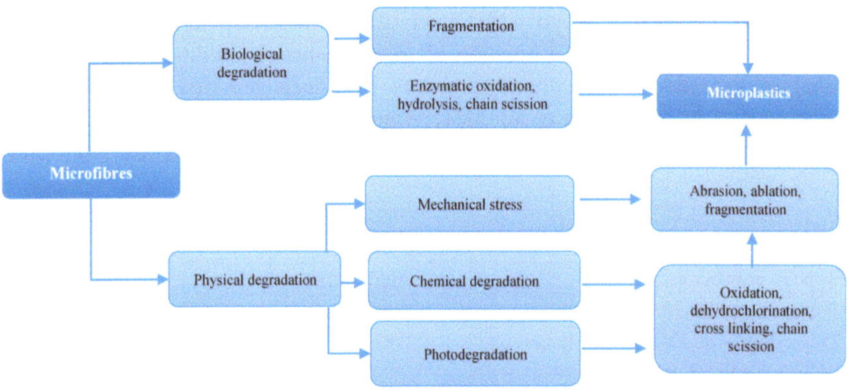

Fig. 7.3 Transformation process of microfibres to microplastics

recovery. Such kind of materials would gradually mount up in cultivated soils and affect the balance of the ecosystem of soils (Henry et al. 2019; Wang et al. 2022). Figure 7.3 shows the transformation process of microfibres to microplastics.

Figure 7.3 shows that the transformation process of microfibres to microplastics begins with degradative processes which could be physical or biological in nature. Fragmentation, oxidative attack, hydrolytic attack, chain scission, abrasion, and ablation actions result in the final generation of microplastics.

Textiles made from synthetic fibres are accountable for microplastic fibre shedding at all stages of their life cycle, that is, production, use, laundry and after disposal. Major release of microfibres from textiles occurs during the wearing and washing of synthetic clothing. Waste water from washing is the predominant pathway for release of these fibres into the environment. (Boucher and Friot 2017). Waste washing effluent is either disposed of into environment directly (in cases of handwashing), or directed towards municipal wastewater treatment plants (WWTPs). About 96% of the microfibres are removed as sludge and the remaining fraction is emitted directly into the environment (Carr et al. 2016). The sludge is sometimes used in agricultural land where it has a high chance of entering soils and waterbodies via runoff. Additionally, microfibres are also emitted during textile manufacturing processes and dispersed in water, air and soil. Moreover, other forms of microplastics can originate from textiles accessories such a prints, coatings, buttons, and zippers. Release of microplastics from textiles is achieved via the shedding of fibres within the garment construction. Studies have shown that shedding is dependent on properties of the textile material such as fibre material, yarn size, fabric construction, fabric density and fabric finishing (Cesa et al. 2017). In addition, manufacturing processes such as knitting, weaving, sewing, and cutting edges have been reported to increase the rate of shedding of textiles (Dalla Fontana et al. 2021). Continuing trends in fast-fashion, fuelled by growing global consumerism, wealth and population means more synthetic fibres would end up in the landfill in a very short space of time. Due to the cheaper cost of synthetic textiles, most average people in developed countries can afford to buy new clothing often and can dispose them of as soon as a new fashion trend emerges. However, in most African countries, buying behaviour is more occasional with more people relying on the second hand clothing market imported from other countries. Most of the second-hand clothing would be already reaching their end life thus they would eventually end up in landfills contributing more to microplastic release. Fishing nets produced using polypropylene and nylon also contribute to the release of microplastic waste as abrasion of these products leads to release of microplastic particles direct into the water. In Africa, agricultural produce such as butternuts, potatoes, onions, and sweet potatoes are packed and sold in sacks made from synthetic fibres. Most of these sacks find a second life application in most households as washing rags and bathing sponges. These activities have abrasive effects to the materials thereby generating micro-particulate fragments which are then released into drainage and sewer systems. Figure 7.4 shows the transportation systems for microplastics originating from synthetic textile materials.

Figure 7.4 shows that the major transport systems are wind, drainage and sewer systems. These deposit the microparticles into the air, freshwater bodies and soil. Microplastics from all these environments finally find their way into the ocean.

7.5 Microplastic from Production Processes of Synthetic Textile Materials

Fibres are exposed to strong mechanical forces during manufacturing and are responsible for the formation of microfibres at the production stage. Production of synthetic fibres begins with polymer granules which are then meltspun into filaments. The filaments can then be used in the filament state converted into short

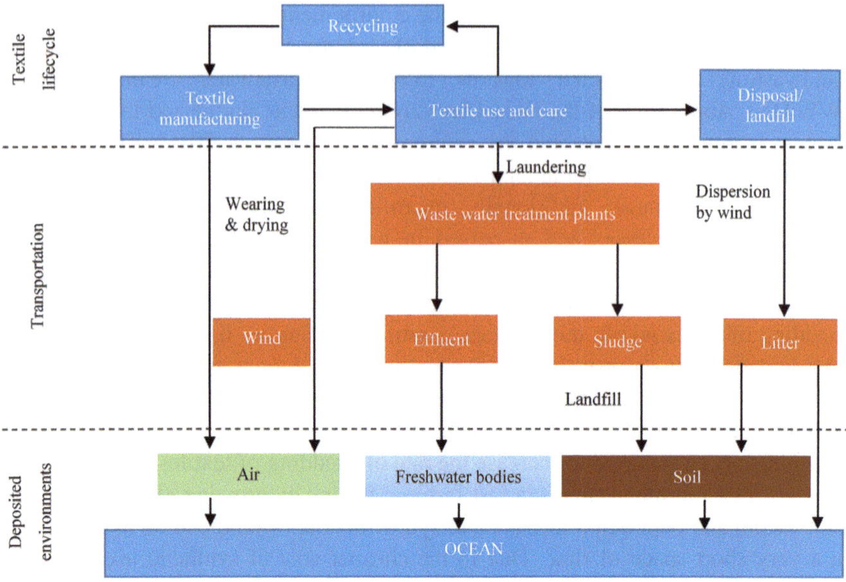

Fig. 7.4 Transportation systems for microplastics originating from synthetic textile materials

staple fibres and texturised. Textured fabrics have been reported to produce five time more the amount of microplastics compared to those with untextured surfaces (Cai et al. 2020). The abreaction friction between yarns and weaving or knitting machines maybe another critical factor for formation of microplastics. An investigation on the presence of microplastic fibres (MPFs) in various intermediary fibre products by Cai et al. (2020) showed that sliver production forms microfibres (Cai et al. 2020). Formation of microfibres during sliver production was observed to be 40 times more in carded sliver than in filament fibre production (Cai et al. 2020). In addition, it was concluded that the number of extracted MPFs from spun yarns was also influenced by the spinning method used. Rotor spinning were found to have more microfibres than ring and airjet spun yarns (Cai et al. 2020; Pinlova et al. 2022). Pinlova et al. (2022) reported that MPFs were found present from initial bale to all types of yarns, confirming their hypothesis that MPF are primarily generated during production (Pinlova et al. 2022). The presence of MPFs in the bales suggested that the MPFs are also formed during staple fibre production. Furthermore, laser-cut textiles were found to have 3 to 31 times less amount of extracted MPFs than scissors-cut textiles (Cai et al. 2020; Jönsson et al. 2018). During textile production, textile dust produced is pneumatically extracted from the machines through air ducts and collected into filter bags which are then disposed of when full. Such dust and impurities may be exposed into the air releasing thousands of microplastic fibres into the environment (Goyal and Nayak 2020, Mellin et al. 2016).

7.6 Release of Microfibres from Textiles During Use and Laundering

Laundering activities on synthetic textile materials especially household textile releases microfibres. An analysis of washing effluent in domestic laundry machines showed that synthetic household textiles shed fibres during laundering process thereby contributing to microplastic pollution (Henry et al. 2019). A study by Browne et al. (2021) indicated that effluent discharged from sewage treatment plants had microfibres of 67%, 17%, and 16%, polyester, acrylic, and polyamide type, respectively (Browne et al. 2011). Several other studies have also shown that laundering releases both natural and synthetic microfibres to the environment (Mishra et al. 2020; Pirc et al. 2016; Rathinamoorthy and Balasaraswathi 2021, Rathinamoorthy and Balasaraswathi 2022). Sillanpää and Sainio (2017) compared the release of microfibres during several machine washings with the wastewater from synthetic textiles. The amount of microfibres released from polyester fabrics in the first wash varied in the range from 2.1×10^5 to 1.3×10^7 per kilogram depending on the fabrics (Sillanpää and Sainio 2017). Several traits of microfibre release from clothes during domestic laundering showed reliance on cloth type, washing cycle, detergent used and garment age (Mishra et al. 2020). Studies have shown that microfibres are released during the first few wash cycles, after which the shedding starts to stabilise (Carney et al. 2018; Hernandez et al. 2017; Lant et al. 2020; Vassilenko et al. 2021). This could be due to the loose fibres and particles embedded on the surface of the materials during manufacturing which are then released during the first few garment washes.

7.7 Influence of Type of Synthetic Textile Material on Release of Microfibres

The type of synthetic fibre used to make the textile material and how the textile material has been constructed have a great influence on the amount of microfibres released from the textile material. In a study of shedding properties of synthetic textiles made up of nylon, acrylic and polyester, polyester was showed the highest degree of shedding due to its high pilling nature attributed to its high tenacity (Rathinamoorthy and Balasaraswathi 2020). High tenacity and breaking strength of the central filament allow the outer filaments to get scratched. Repeated washings lead to formation of fuzz fibres, which develop protrusion of fibres on the surface. The protruded fibres form fibre pills which are then removed by the mechanical activities of the washing process. Furthermore, the shedding property is influenced by friction between fibres, orientation, thinness, toughness and abrasion resistance of the fibres (Zambrano et al. 2019). The structural compactness of the yarns used is also a major influence during washing. Generally, fabrics highly twisted yarns have high resistance to abrasion and hence high resistance to shedding during laundering or other abrasive actions. Carney et al. (2018) investigated the ability of polyester fabric to shed when it is knitted with tight or loose with staple and filament fibres

(Carney et al. 2018). The results showed that the shedding of microfibre is significantly higher in knitted fabric with staple yarn and also in the case with the tightly knitted fabric (Carney et al. 2018). Hernandez et al. (2017) deduced that the amount of stress exerted on yarn during manufacturing and the yarn structure along with other physical and chemical treatments of the fibre significantly influence the shedding of microfibres (Hernandez et al. 2017). A greater compact yarn structure can hold the fibres tight and reduce the amount of microfibres released during the washing process than a low compact yarn structure. Similarly, the garments made from short-staple fibres tend to release higher amount of microfibres than the ones constructed from continuous filaments (Rathinamoorthy and Balasaraswathi 2021). Staple fibres easily slide away when they are wet leading to higher microfibre release (De Falco et al. 2018).

Even though the world has moved toward a closed loop economy through recycling of textile materials, recycled fibres tend to release more microfibres than virgin fibres. The high fibre release in the recycled polyester might be the results of the extensive thermal exposure of those fibres during the remelting process which eventually decreases the crystallinity of the fibre. Such changes in its structure result in decreased tenacity and breaking strength thereby increasing the chances of microfibre shedding in recycled materials than virgin fibres. With aging, the molecular level of the polymer changes resulting in a harder, denser and stiffer fibre which is weak to mechanical stresses. Hartline et al. (2016) assessed the effect of ageing of textiles on fibre shedding and the results showed that there was a 25% microfibre shedding released in the aged garments (Hartline et al. 2016).

7.7.1 Effect of Washing Conditions and Use of Detergents on the Release of Microplastic

Washing conditions and use of detergents have a significant effect on the rate at which microplastics from textiles are released into the environment. Hartline et al. (2016) recovered microfibre weight per garment ranging from roughly 0 to 2 g, or approximately more than 0.3% of the unwashed garment after replicated washings without detergent with front and top load home machines were conducted (Hartline et al. 2016). Pirc et al. (2016) estimated that 0.0012 wt. % of loose microfibres are released into sewer water during every washing of long fibre plush fleece, however, the effect of fabric softener and detergent use is relatively small (Pirc et al. 2016). Similarly, Hernandez et al. (2017) reported that there is no noticeable difference in the quantity of microfibres released with variation with duration of wash cycle, temperature or when surfactants are used (Hernandez et al. 2017). However, microplastics were released four times more when washing with either liquid or powder detergent than when washing with water alone (Hernandez et al. 2017). Carney et al. (2018) also reported that detergents results in significantly higher microfibre release compared to washing in water alone for 3 out of 4 fabric types tested (Carney et al. 2018). Yang et al. (2019) discovered that use of a detergent significantly increased microfibres release, especially for polyester fibres when washed at low temperatures (Yang et al. 2019). De Falco et al. (2018) reported that softener products reduce microfibre discharge by more than 35% (De Falco et al. 2018). In support of this study, Rathinamoorthy and Raja Balasaraswathi (2022) also concluded that the use of detergent and fabric softener during the washing process reduces microfibre shedding significantly, due to the reduction of inter-friction forces between the fibres in the presence of softener (Rathinamoorthy and Raja Balasaraswathi 2022). Higher water volume to fabric ratio was reported as an important driver of microfibre loss, this is caused by increased flow of water through yarns and hence increased fibre extraction (De Falco et al. 2020; Lant et al. 2020; Rathinamoorthy and Raja Balasaraswathi 2022). However, there are some contradicting findings (Pirc et al. 2016; Cesa et al. 2017) that concludes, detergents have no significant effect and reduces the emissions of microfibres from synthetic fibres during washing. Similar observations that the use of detergent pods or fabric softener had no effect on microfibre release were also reported in literature (Lant et al. 2020). The variance in the results presented in literature could be a result of the different methodologies used or the type of machines used during experimentation.

7.7.2 Effects of Drying on the Release of Microplastic

Microfibres are also released during drying of textile materials. Drying can be carried out in the direct sunlight in free air or using machine called a tumble dryer. Drying in the air is a primary source of microplastics as loose microfibres on the surface can be blown away by wind into the atmosphere. On the other hand, household tumble dryers are also a source of microplastic discharge into the ambient atmosphere. Microfibres are released directly through a ventilation pipe connected to the dryer to atmospheric air (Kapp and Miller 2020; O'Brien et al. 2020). Rotation of fabrics in the dryer leads to shedding of fibres especially at higher temperatures thereby releasing microfibres (Tao et al. 2022). An investigated on the influence of tumble drying on release of microfibres from textiles showed that an estimate of 9×10^7 to 12×10^7 microfibres are released from a single dryer in every Canadian household (Tao et al. 2022).

7.8 Mitigation of Microplastics Originating from Paints and Synthetic Textile Materials

There is a need to come up with mitigatory measures to curb or reduce the emission of microplastics from paints and synthetic textile materials. The first mitigatory measure could be at the production phase of paints and synthetic textile materials. A holistic approach to research and development of green chemistry and manufacturing processes in the production of paints and synthetic textiles would reduce emission of microplastics from paints and synthetic textile materials. Green chemistry, green manufacturing processes and green economy initiatives would result in eco-friendly paints and synthetic textile materials. Biopolymers such as polylactic acid (PLA), polyhydroxyalkanoates (PHA) or polyhydroxybutyrate (PHB) could be used in textile materials. Literature reports the use of polylactic acid in the production of biodegradable fibre-reinforced biocomposites (Moyo et al. 2020a; Moyo et al. 2020b). Release of microplastics into the environment can also be mitigated at the point of release. This means that processes can be improved throughout the manufacturing and application of paints and textile materials to reduce the emission of microplastics from paints and synthetic textile materials. Therefore, funding of research and development activities for the development of eco-friendly paints and synthetic textile materials would result in improved technologies of paint and textile materials production and improved paint application methods that would reduce spills. Pinlova et al. (2022) suggested that using ring spinning or air jet spinning in place of rotor spinning in production of synthetic yarns would lower the emission of microfibres from yarn spinning (Pinlova et al. 2022). However, in some cases, yarn production through rotor spinning is inevitable; instead, lower speeds of the opening roller speed should be used to minimize the amount of microplastic in the resultant yarns (Pinlova et al. 2022). To mitigate the release of microplastics during washing, Lant et al. (2020) suggested that using cold water, less washing time and not over filled washing loads can significantly reduce microfibre release. (Lant et al. 2020). Use of laundry machine filters can also present microplastic fibre leaks out into the sewer with the water. Air-drying clothes instead of tumble drying can also reduce the amount of microplastics released since tumble drying has more mechanical action on the materials. Studies show that more than 90% of microplastics can be removed from water by wastewater and drinking water treatment plants (Carr et al. 2016). Formulation of policies that would compel manufacturers of paints and synthetic fibres and shipyards to have wastewater treatment systems would minimise the release of microplastics. Contaminated water would then be treated before being discharged. There must also be legislative actions against DIY painting and maintenance of houses and marine vessels. Such jobs should be done by registered professional whose licences could be revoked if found not adhering to safe handling and use of paints as well as facing legislative actions.

7.9 Conclusion

This chapter showed that there is a significant amount of microplastics originating from paints and synthetic textile materials. In the case of paints, the major sources of microplastics are marine (boats and ship) paints, road markings, building (architectural) paints and spray painting. Major contributing factors in the release of microplastics from paints to the environment are paint production processes, weathering and abrasion, paint spilling, residual paints on paint application tools and storage containers, paint EoL disposal and paint applicator's expertise. In the case of microplastics from synthetic textile materials, major emissions of the microplastics occur during the production process of the synthetic textile materials, during use and laundering, washing conditions and drying. As mitigatory measures, continued research and development is required so as to reduce the influence of the manufacturing processes of paints and synthetic textile materials in the generation of microplastics. Other mitigatory measures include application of green chemistry, green manufacturing processes and green economy initiatives in paints and synthetic textile materials. Lastly, developing policies and legislation on wastewater treatment systems for all responsible stakeholders and legislation on DIY painting activities would also contribute to a reduction in the generation and release of microplastics from paints and synthetic textile materials.

References

(OECD) OfEC-oaD (2009) Emission scenario documents on coating industry (paints, laquers and varnishes), ENV/JM/MONO(2009), 24th edn. OECD, Paris

Bentley J, Turner GPA (1998) Introduction to paint chemistry and principles of paint technology, 4th edn. Chapman & Hall. https://doi.org/10.1007/978-1-4899-3180-1

Boucher J, Friot D (2017) Primary microplastics in the oceans: a global evaluation of sources. IUCN, Gland. https://doi.org/10.2305/IUCN.CH.2017.01.en

Browne MA, Crump P, Niven SJ, Teuten E, Tonkin A, Galloway T, Thompson R (2011) Accumulation of microplastic on shorelines woldwide: sources and sinks. Environ Sci Technol 45:9175–9179

Cai Y, Mitrano DM, Heuberger M, Hufenus R, Nowack B (2020) The origin of microplastic fibre in polyester textiles: the textile production process matters. J Clean Prod 267:121970

Carney AB, Åström L, Roslund S, Petersson H, Johansson M, Persson NK (2018) Quantifying shedding of synthetic fibres from textiles; a source of microplastics released into the environment. Environ Sci Pollut Res 25:1191–1199

Carr SA, Liu J, Tesoro AG (2016) Transport and fate of microplastic particles in wastewater treatment plants. Water Res 91:174–182

Cesa FS, Turra A, Baruque-Ramos J (2017) Synthetic fibres as microplastics in the marine environment: a review from textile perspective with a focus on domestic washings. Sci Total Environ 598:1116–1129

Cunningham EM, Ehlers SM, Dick JTA, Sigwart JD, Linse K, Dick JJ, Kiriakoulakis K (2020) High abundances of microplastic pollution in deep-sea sediments: evidence from Antarctica and the Southern Ocean. Environ Sci Technol 54:13661–13671. https://doi.org/10.1021/acs.est.0c03441

Dalla Fontana G, Mossotti R, Montarsolo A (2021) Influence of sewing on microplastic release from textiles during washing water. Air, Soil Pollut 232:1–9

De Falco F et al (2018) Evaluation of microplastic release caused by textile washing processes of synthetic fabrics. Environ Pollut 236:916–925

De Falco F, Cocca M, Avella M, Thompson RC (2020) Microfibre release to water, via laundering, and to air, via everyday use: a comparison between polyester clothing with differing textile parameters. Environ Sci Technol 54:3288–3296

Eyerer P (2010) Plastics: classification, characterization, and economic data. In: Polymers-opportunities and risks I. Springer, pp 1–17

Faber M, Marinković M, de Valk E, Waaijers-van der Loop SL (2021) Paints and microplastics: exploring recent developments to minimise the use and release of microplastics in the Dutch paint value chain. National Institute for Public Health and the Environment, RIVM, Utrecht. https://doi.org/10.21945/RIVM-2021-0037

Gaylarde C, Neto J, Fonseca E (2021) Paint fragments as polluting microplastics: a brief review. Mar Pollut Bull 162:111847. https://doi.org/10.1016/j.marpolbul.2020.111847

Goyal A, Nayak R (2020) Sustainability in yarn manufacturing. In: Sustainable Technologies for Fashion and Textiles. Woodhead Publishing, pp 33–55

Hartline NL, Bruce NJ, Karba SN, Ruff EO, Sonar SU, Holden PA (2016) Microfibre masses recovered from conventional machine washing of new or aged garments. Environ Sci Technol 50:11532–11538

Harvey E, Sullivan R, Amos N (2017) Microplastic pollution: the causes, consequences and issues for investors. First Sentier MUFG Sustainable Investment Institute, Tokyo

Henry B, Laitala K, Klepp IG (2019) Microfibres from apparel and home textiles: prospects for including microplastics in environmental sustainability assessment. Sci Total Environ 652:483–494

Hernandez E, Nowack B, Mitrano DM (2017) Polyester textiles as a source of microplastics from households: a mechanistic study to understand microfibre release during washing. Environ Sci Technol 51:7036–7046

Imhof H, Laforsch C, Wiesheu A, Schmid J, Anger P, Niessner R, Ivleva N (2016) Pigments and plastic in limnetic ecosystems: a qualitative and quantitative study on microparticles of different size classes. Water Res 98:64. https://doi.org/10.1016/j.watres.2016.03.015

Jönsson C, Levenstam Arturin O, Hanning A-C, Landin R, Holmström E, Roos S (2018) Microplastics shedding from textiles—developing analytical method for measurement of shed material representing release during domestic washing. Sustainability 10:2457

Kapp KJ, Miller RZ (2020) Electric clothes dryers: an underestimated source of microfibre pollution. PLoS One 15:e0239165

Kim S-K, Kim J-S, Lee H, Lee H-J (2021) Abundance and characteristics of microplastics in soils with different agricultural practices: importance of sources with internal origin and environmental fate. J Hazard Mater 403:123997

Lant NJ, Hayward AS, Peththawadu MM, Sheridan KJ, Dean JR (2020) Microfibre release from real soiled consumer laundry and the impact of fabric care products and washing conditions. PLos One 15:e0233332

Lassen C et al (2015) Microplastics: occurrence, effects and sources of releases to the environment in Denmark. The Danish Environmental Protection Agency, Denmark

Mahbub MS, Shams M (2022) Acrylic fabrics as a source of microplastics from portable washer and dryer: impact of washing and drying parameters. Sci Total Environ 834:155429

Mellin P, Jönsson C, Åkermo M, Fernberg P, Nordenberg E, Brodin H, Strondl A (2016) Nano-sized by-products from metal 3D printing, composite manufacturing and fabric production. Journal of cleaner production 139:1224–1233

Mishra S, Singh RP, Rath CC, Das AP (2020) Synthetic microfibres: source, transport and their remediation. J Water Process Eng 38:101612

Moyo M, Falayi T (2022) Generation and management of microplastic waste. In: Shahnawaz M, Sangale MK, Daochen Z, Ade AB (eds) Impact of plastic waste on the marine biota. Springer Nature, Singapore, pp 3–20. https://doi.org/10.1007/978-981-16-5403-9_1

Moyo M, Kanny K, Velmurugan R (2020a) The efficacy of nanoclay loading in the medium velocity impact resistance of kenaf/PLA biocomposites. Appl Nano 11:441–453. https://doi.org/10.1007/s13204-020-01602-9

Moyo M, Kanny K, Velmurugan R (2020b) Performance of kenaf non-woven mat/PLA biocomposites under medium velocity impact. Fibres Polym 21:2642–2651. https://doi.org/10.1007/s12221-020-1130-z

O'Brien S et al (2020) Airborne emissions of microplastic fibres from domestic laundry dryers. Sci Total Environ 747:141175

Paola Paruta MP, Boucher J (2022) Plastic paints the environment. EA – Environmental Action, Lausanne

Pinlova B, Hufenus R, Nowack B (2022) Systematic study of the presence of microplastic fibres during polyester yarn production. J Clean Prod 363:132247

Pirc U, Vidmar M, Mozer A, Kržan A (2016) Emissions of microplastic fibres from microfibre fleece during domestic washing. Environ Sci Pollut Res 23:22206–22211

Poth U (2008) 2. General aspects of coatings. In: Automotive coatings formulation. Vincentz Network, Hannover, pp 17–40. https://doi.org/10.1515/9783748602002-003
Rathinamoorthy R, Balasaraswathi SR (2020) A review of the current status of microfibre pollution research in textiles. Int J Cloth Sci Technol 33:364
Rathinamoorthy R, Raja Balasaraswathi S (2021) Domestic laundry and microfibre shedding of synthetic textiles. In: Microplastic pollution. Springer, pp 127–155
Rathinamoorthy R, Raja Balasaraswathi S (2022) Investigations on the impact of handwash and laundry softener on microfibre shedding from polyester textiles. J Text Inst 113:1428–1437
Roebroek C, Harrigan S, van Emmerik T, Baugh C, Eilander D, Prudhomme C, Pappenberger F (2021) Plastic in global rivers : are floods making it worse? Environ Res Lett. https://doi.org/10.1088/1748-9326/abd5df
Šaravanja A, Pušić T, Dekanić T (2022) Microplastics in wastewater by washing polyester fabrics. Materials 15:2683
Sazali N, Ibrahim H, Jamaludin A, Mohamed M, Salleh W, Abidin M (2020) A short review on polymeric materials concerning degradable polymers. In: IOP conference series: materials science and engineering, vol 1. IOP Publishing, p 012047
Sillanpää M, Sainio P (2017) Release of polyester and cotton fibres from textiles in machine washings. Environ Sci Pollut Res 24:19313–19321
Tao D et al (2022) Microfibres released into the air from a household tumble dryer. Environ Sci Technol Lett 9:120–126
The European Council of the Paint PIaACI (2018) CEPE annual report 2018. CEPE, Brussels
Toben M (2018) Microplastic pollution originating from textiles and paints: environmental impacts and solutions. In: Bund für Umwelt und Naturschutz Deutschland e.V. BUND
Turner A (2021) Paint particles in the marine environment: an overlooked component of microplastics. Water Res X 12:100110. https://doi.org/10.1016/j.wroa.2021.100110
van der Schyff V et al (2020) Impacts of a shallow shipwreck on a coral reef: a case study from St. Brandon's Atoll, Mauritius, Indian Ocean. Mar Environ Res 156:104916. https://doi.org/10.1016/j.marenvres.2020.104916
Van der Vegt A (2006) From polymers to plastics. VSSD, Delft
Vassilenko E, Watkins M, Chastain S, Mertens J, Posacka AM, Patankar S, Ross PS (2021) Domestic laundry and microfibre pollution: exploring fibre shedding from consumer apparel textiles. PLoS One 16:e0250346
Wang J et al (2021) Distinct microplastic distributions in soils of different land-use types: a case study of Chinese farmlands. Environ Pollut 269:116199
Wang K et al (2022) Accumulation of microplastics in greenhouse soil after long-term plastic film mulching in Beijing, China. Sci Total Environ 828:154544
Xu G, Wang Q, Gu Q, Cao Y, Du X, Li F (2006) Contamination characteristics and degradation behavior of low-density polyethylene film residues in typical farmland soils of China. J Environ Sci Health B 41:189–199
Xu Y, Rillig MC, Waldman WR (2022) New separation protocol reveals spray painting as a neglected source of microplastics in soils. Environ Chem Lett 20:3363–3369. https://doi.org/10.1007/s10311-022-01500-2
Yang L, Qiao F, Lei K, Li H, Kang Y, Cui S, An L (2019) Microfibre release from different fabrics during washing. Environ Pollut 249:136–143
Zambrano MC, Pawlak JJ, Daystar J, Ankeny M, Cheng JJ, Venditti RA (2019) Microfibres generated from the laundering of cotton, rayon and polyester based fabrics and their aquatic biodegradation. Mar Pollut Bull 142:394–407

Fishing Gears and Nets as a Source of Microplastic

8

Drishti Sharma, Raunak Dhanker, Bhawna, Ashi Tomar, Sazid Raza, and Anjali Sharma

Abstract

Fishing gears and nets are a crucial part of the fishing industry and play a significant role in the livelihood of coastal communities. These gears and nets are now a significant source of microplastics in the oceans made of synthetic materials including nylon, polypropylene, and polyester due to the tremendous rise in plastic production. Microplastics are minute plastic flecks that are detrimental to the environment and marine life and have a diameter of less than 5 mm. Fishing gears and nets contribute significantly to the accumulation of microplastics in the oceans through their constant use, shedding of microfibers, and accidental loss or abandonment. The primary sources of microplastics from fishing gears and nets are from synthetic materials such as nylon, polypropylene, and polyester, which are widely used in fishing nets, ropes, and lines. The impact of microplastics on marine organisms and ecosystems is a significant concern for marine conservationists and researchers. Microplastics have been found in the stomachs of marine organisms, including fish, birds, and sea turtles, which can lead to physical damage, blockage of digestive tracts, and reduced nutrient absorption. Microplastics can enter the food chain through the ingestion of contaminated seafood, which puts human health at risk. Therefore, reducing the amount of microplastics in our oceans is crucial for protecting the health of marine life and the safety of our food supply. In the present chapter, an attempt was made to overview the collective actions to address the issue of microplastic

D. Sharma · R. Dhanker (✉) · Bhawna · A. Sharma
School of Engineering and Sciences, G D Goenka University, Sohna, Haryana, India

A. Tomar
Banasthali Vidyapith, Jaipur, Rajasthan, India

S. Raza
Jamia Hamdard University, New Delhi, India

M. Shahnawaz et al. (eds.), *Microplastic Pollution*,
https://doi.org/10.1007/978-981-99-8357-5_8

pollution from fishing gears and nets and to work towards a more sustainable and plastic-free future.

Keywords

Microplastics · Fishing gears and nets · Marine ecosystem · Human health · Mitigation measures · sustainable and plastic free future

8.1 Introduction

Fishing gears are the equipment used for fishing and can vary in their design, materials, and methods of operation. Fishermen frequently employ a variety of fishing techniques and gear depending on the type of fish, the terrain, and the surrounding environment. Fishing is the act of capturing fish, and fishing gear refers to any equipment used to harvest or capture fish from bodies of water. Fishing gear can encompass a range of tools, implements, or equipment used for this purpose, such as traps, hooks and lines, gill nets, trawls, seine nets, lift nets, clap nets, spears, cast nets, entangling nets, drift nets, among others. The fishing method refers to how the gear is employed in the fishing process.

Derelict fishing gear, which refers to abandoned, lost, or discarded fishing gear, is a major source of stress for coastal and marine ecosystems. With an estimated 4.6 million marine fishing vessels operating worldwide, the amount of fishing gear that becomes derelict and continues to interact with the marine environment is a significant problem (Hong et al. 2017; Gilman et al. 2021). This can lead to entanglement and ingestion by marine wildlife, as well as damage to habitats and the release of microplastics into the environment. Fishing gear is designed to capture marine organisms, which means that ALDFG can continue to catch and kill marine life for an extended period, leading to substantial fishing mortalities. This poses a significant risk not only to target species but also to bycatch, including vulnerable species, which can have severe ecological and economic consequences (Gilman et al. 2021).

Discarded plastic is a major global issue, with 4.9 billion tonnes of it accumulated so far, and the marine environment is particularly affected (Geyer et al. 2017). The majority of marine debris is composed of synthetic plastics, some of which are easily visible, while others are microscopic (Gilman et al. 2021). It is estimated that approximately ten million tonnes of plastic find their way into the oceans every year, and this number is expected to significantly increase over the next decade (Borrelle et al. 2020). While most marine debris comes from sources on land, at the local level, sources can vary widely. Currents and winds can scatter floating debris across vast distances, causing it to accumulate into large garbage patches in oceanic gyres (Gilman et al. 2021). Debris can move up and down the water column, sinking when biofouling increases its density and rising when fouling decreases (Kooi et al. 2017). Some floating debris may wash up on shore, while most plastic debris sinks to the deep seabed (Choy et al. 2019). Bottom thermohaline currents have the ability to concentrate microplastics in benthic ecosystems, frequently in places where there is great biodiversity. This is similar to how floating trash does it. Plastics can persist in marine environments with little degradation, particularly in areas below the photic zone, such as the deep seabed (Gilman et al. 2021). Approximately 1.15 million tonnes of plastic waste in the seas and oceans annually are made up of fishing nets that are lost or intentionally abandoned. The issue has become more severe due to increased fishing activity and the use of durable synthetic materials. Abandoned fishing gear (ALDFG), can trap marine life in a phenomenon called "ghost fishing" and can remain in the environment for long periods due to their slow biodegradation. ALDFG can either get caught on the ocean floor or float in open water, and their small size can make them difficult to detect (Montarsolo et al. 2018). The presence of ALDFG can obstruct fishing activities, navigation, and increase safety risks at sea. These negative impacts on marine ecosystems and coastal communities reduce the socioeconomic value of coastal and nearshore habitats (Gilman et al. 2021). The main objective of the present chapter is to overview the fishing gear and nets, as well as the problem of abandoned, lost, or discarded fishing gear (ALDFG) contributing to microplastic pollution in the marine environment. Additionally, this chapter will discuss measures to mitigate this issue and its negative impacts.

8.2 Types of Fishing Gears and Nets

8.2.1 Types of Fishing Gear

8.2.1.1 Passive Gear

Passive fishing systems have fixed gears that don't require dragging, pulling, or towing to catch fish. Methods like gill nets, hook and lines, traps, and wires are considered inefficient. Instead, these systems allow the catch to be collected easily after some time by removing the equipment from the water, without the need for energy-intensive dragging, hauling, or pulling.

8.2.1.2 Hooks and Lines

This is the most basic type of fishing equipment that is available, requiring only a line and a baited hook. Hooks are available in various shapes, sizes, points, wire thicknesses, and shank end types. To use this equipment, the line is cast into the water where the fish are expected to be, and once the fish takes the bait, it is reeled in (Walters 2022). In Nigeria, a stone is sometimes wrapped in a string and thrown into the water from the bank or shore to create a sinker. Fishing with hooks and lines is an inexpensive and uncomplicated method, and the catch is often of high quality and lively. The size of the hook and type of bait used will determine the species of fish that can be caught (Eyo and Akpati 1995).

8.2.1.3 Maze Gear

Maze gear refers to a set of tools designed to lead fish into an enclosed environment from which they cannot easily escape or find a way out (Jha 2016). This type of fishing equipment includes various adaptations such as fishing pots, fyke nets, traps, pound nets, and tidal weirs. These tools are used to guide fish into an enclosed area where they can be caught, and they are difficult or impossible for the fish to escape from.

8.2.1.4 Pot Gear

With sizes ranging from roughly half a metre to two metres, small or big cages (Fig. 8.1(1)) or baskets (Fig. 8.1(2)) are utilised to produce pots used for fishing (Petetta et al. 2021). These pots can be made using various materials such as wood, wicker, metal rods, wire netting, plastic, and more. The typical shape of a fishing pot is rectangular or vertical, with a flattish cylinder or a horizontal half-cylinder, and with the flat side resting on the seafloor (Fig. 8.1(3)) (Okwuosa et al. more openings or entrances for fish to enter.

8.2.1.5 Traps Gear

Traps are typically larger than pots, although there are exceptions (Fig. 8.2). They are often complex nets or wooden structures that are fixed to the seabed, especially in shallow or inshore areas. The top of the trap extends above the water surface, and the placement of the trap takes into account the behaviour of the target species, such as local tidal migrations (Cook 2003). These structures are also referred to as barriers, fences, weirs, corrals, or similar names.

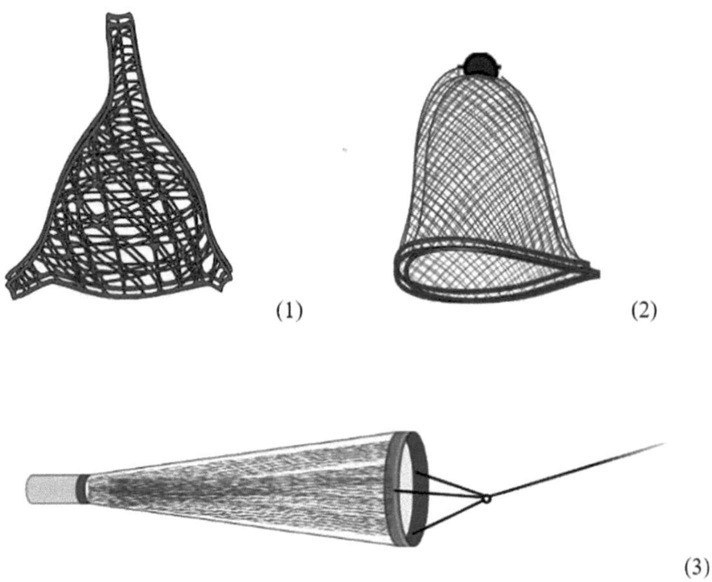

Fig. 8.1 Schematic representation of pot gears: (1) small or big cages, (2) baskets, and (3) fishing pot

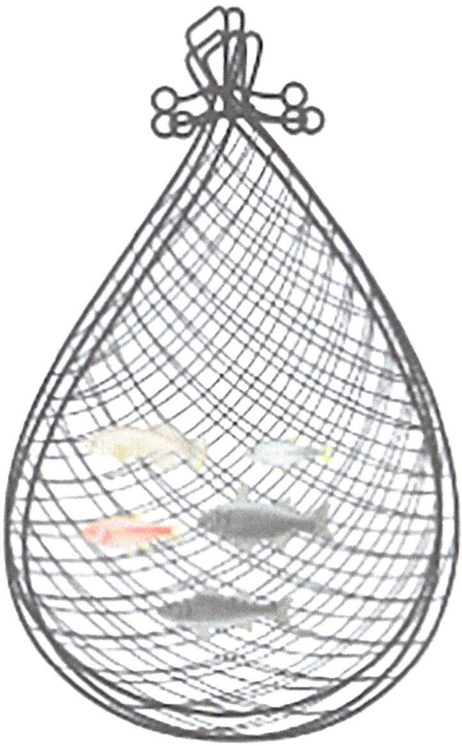

Fig. 8.2 Schematic representation trap gears

8.2.1.6 Fish Fence

The most common method for creating fish fences is to use regular fibres to bind together sticks. In some cases, traditional mats are also used for constructing fish fences (Franco et al. 2022). In Cross River, these barriers are usually built across the main channel, leaving a few meters of space in the centre for canoes and river boats to pass through. Fish fences have the potential to catch almost all fish that migrate upstream since the catching chambers are always oriented downstream (Eyo and Akpati 1995). Additionally, large conical traps with non-returning valves are often attached to these catching chambers, which can be detached when needed.

8.2.2 Types of Nets

8.2.2.1 Gill Net

Gill nets are constructed from fibre netting and form vertical walls underwater (Fig. 8.3). When a fish attempts to swim through a gill net, its mesh captures the fish (Potter and Pawson 2010). If the mesh size is appropriate for the fish being targeted, the fish can fit its head through the opening, but due to its large body size, it becomes trapped as the string slides under its gill cover, preventing escape (Brown 2016).

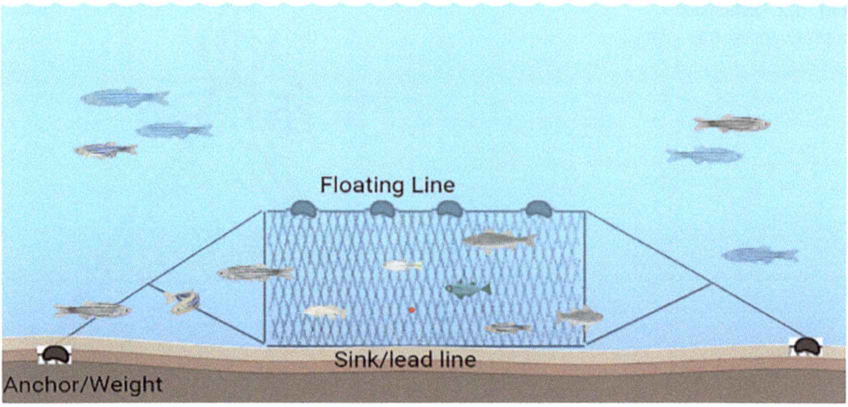

Fig. 8.3 Schematic representation of gill net

8.2.2.2 Entangling Net

Entanglement nets are not vertical walls and are usually positioned parallel to fish migration routes (Franco et al. 2022). Sinkers are utilized to anchor the bottom of the net, while floats keep the top afloat. During gill fishing, a single-walled gill net or a triple-walled net is commonly used.

8.2.2.3 Seine Net

Seine netting is a fishing technique that combines trawling and seining. In Danish seining, the first rope, or warp, is laid out in a semicircle and secured to an anchor with a surface buoy; in Scottish seining, simply a buoy is used. The second warp is then paid out in a second semicircle back to the buoy before connecting it to the anchor in Danish seining.

8.2.2.4 Drift Net

Mobile gill nets called drift nets are typically let to float with the current. To get the net to drop to a certain depth, the number of floats and sinkers can be altered (Franco et al. 2022). A large surface float known as a calabash is attached to one end of the net, while the other end is secured to a canoe when in use. The net and canoe then drift slowly for a considerable distance before the catch is drawn into the canoe along with the net.

8.2.2.5 Trammel Net

Trammel nets consist of three panels of netting that are hung from a single row of floats and connected to a single bottom line (Ern et al. 2022). The netting's interior has a smaller mesh size, while the mesh on the two outer walls is larger than the fish being targeted.

8.2.2.6 Trolling

Trolling lines are commonly used in recreational fishing and commercial fishing to catch a variety of fish species (Eighani et al. 2019). The speed of the boat and the

Fig. 8.4 Schematic representation of (**a**) cast net and (**b**) clap net

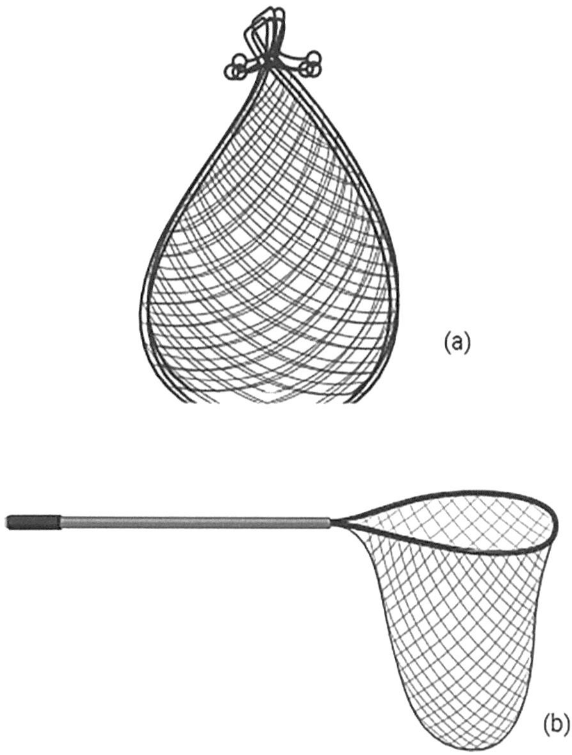

depth at which the hooks are trailed can be adjusted to target specific types of fish. Trolling lines can also be used with lures or artificial baits to attract fish (Eyo and Akpati 1995).

8.2.2.7 Cast Nets

Cast nets are commonly used for fishing in shallow waters or in small streams where other fishing gear may not be suitable. They are typically made of nylon or monofilament and come in various sizes and mesh sizes (Fig. 8.4a) depending on the target species. The technique of using a cast net involves throwing the net with one hand while holding onto the rope with the other. As the net falls into the water, it spreads out in a circular shape and sinks quickly due to the weights attached to the edges. The net is then pulled back to the surface, and any fish caught in the mesh are retrieved (Eyo and Akpati 1995). Cast nets can be very effective for catching baitfish, mullet, and other small to medium-sized fish.

8.2.2.8 Clap Nets

Hausa fishermen frequently employ twin clap nets, particularly in rivers and streams. These nets are designed with two semicircular wooden frames that are covered with netting material (Fig.8.4b). The mesh size of the netting material typically ranges from 10 to 30 mm (Baruah et al. 2018). The fisherman holds one net in each hand and dives into the water to capture the fish by clapping the two nets together. Men often use the larger nets, while women and kids typically use the smaller ones. Twin clap nets are a simple and effective method for catching small to medium-sized fish (Eyo and Akpati 1995).

8.3 Abandoned, Lost, or Discarded Fishing Gear (ALDFG) as Microplastic

Abandoned, lost, or discarded fishing gear (ALDFG) poses significant and far-reaching negative impacts on marine ecosystems and littoral communities. The need for multinational action to address the transboundary negative ecological and socioeconomic goods of abandoned, lost, and discarded fishing gear (ALDFG), also known as derelict fishing gear, has gained transnational attention during the once ten times (Gilman et al. 2021). The fast expansion of fishing trouble and fishing grounds, as well as the switch to synthetic, less-precious, more durable, and more buoyant accoutrements used for fishing gear, have all clearly increased the quantum, distribution, and consequences of ALDFG in recent decades (Kroodsma et al. 2018).

The issue of abandoned, lost, or discarded fishing gear (ALDFG) has been current in the marine terrain since the morning of fishing conditioning and has been linked as a significant problem since the 1980s (Wright et al. 2021). While not all fishing waste becomes abandoned, lost, or discarded fishing gear (ALDFG), the thing is for fishing waste to be disposed of duly through landfilling, incineration, or recycling, analogous to other types of plastic waste (Deshpande et al. 2020). In spite of this, it is estimated that a sizeable amount of fishing equipment is lost at sea, with 5.7% of all fishing nets and 29% of all fishing lines suffering ALDFG. This is a pressing issue, as the quantum of ALDFG has drastically increased since the turn of the century (Ostle et al. 2019; Richardson et al. 2019a, 2019b). In the Northeast Atlantic, fishing gear is prognosticated to regard for 55% of floating microplastic (Ostle et al. 2019), while in the Northeast Pacific, it is at least 46% (Lebreton et al. 2018). On the Mediterranean seafloor, ALDFG can reach consistence of 0.79 particulars per forecourt metre, with fishing lines counting for 68% of the debris, followed by nets (18%), ropes (12%), and pots (0.2%) (Enrichetti et al. 2020).

Abandoned, lost, or discarded fishing gear (ALDFG) poses a serious threat to marine life, particularly to large marine animals that can be harmed by ingesting or getting entangled in fishing gear (Wright et al. 2021). Trawls, fixed and drift nets have been identified as posing the highest risk (Gilman et al. 2021). Recent research conducted in the Ganges river found that large amounts of fishing rope and net increase the risk of entanglement for freshwater turtles such as *Batagur dhongoka, Geoclemys hamiltonii, and Batagur baska*, as well as the endangered Ganges river dolphin (*Platanista gangetica gangetica*) (Nelms et al. 2021). ALDFG can also harm commercial fish species. 1.4% of the 216 gastrointestinal tracts of Atlantic cod (Gadus morhua) that were investigated included entire bait bags used in commercial

Table 8.1 Estimated emission of microplastics from ALDFG

Year	Amount of microplastic per cubic meter (N/m^3)
2010	41.28
2020	112.0
2030	160.9
2050	264.1
2100	520.1

potting or polypropylene thread that most likely came from fishing rope, according to research (Saturno et al. 2020).

The Western English Channel and Celtic Sea coasts in the United Kingdom had 79% more abandoned, lost, or otherwise discarded fishing gear (ALDFG) than any other regional sea, according to data gathered from volunteer beach surveys conducted across the country by the Marine Conservation Society. This is attributed to the high fishing intensity in the region and the widespread use of nets. The north coast of the Southwest Peninsula was analysed in more detail, and it was found that fishing ropes and nets make up 32% of the litter, which is the largest identifiable source of plastic. These polymer ropes can discharge microplastic particles and fibres smaller than 5 mm as they degrade through abrasion and biofouling in the shallow maritime environment (Wright et al. 2021).

ALDFG can also transport and release toxins and microplastics into marine food webs, leading to harmful health effects on marine life and humans who consume them (Table 8.1). The gear can also transport invasive alien species, microalgae that can cause harmful algal blooms, and cause physical damage to marine habitats (Gilman et al. 2021).

8.4 Mitigation Measures

Abandoned, lost, or otherwise discarded equipment fishing (ALDFG), which affects the majority of the world's oceans, has a negative impact on fishermen, seafood enterprises, coastal communities, marine life, and environments (Richardson et al. 2019a, 2019b). When fishing gear is thrown, lost, or abandoned, it can fall to the bottom of the ocean, affecting, dragging, or eroding animals and their habitats (Pham et al. 2014; Du Preez et al. 2020). It can also swim with currents, come into contact with fish and other sea creatures (Lebreton et al. 2018; Stelfox et al. 2020), become a habitat where bacteria can enter, and pose a safety hazard at sea and cruising (Richardson et al. 2021; Hong et al. 2017). Commonly known as "ghost gear," ALDFG can continuously track and detect both targets and non-targets (Richardson et al. 2021; Stelfox et al. 2016). When unused equipment eventually hits the beach, it can have a huge impact on the coastal area, especially tourism and the beauty of the beach (Marcou et al. 2016).

When fishing gear is thrown, lost, or abandoned, it can fall to the bottom of the ocean, affecting, dragging or eroding animals and their habitats (Uhrin 2016; Dağtekin et al. 2019). It can also swim with currents, come into contact with fish and other sea creatures, become a habitat where bacteria can enter, and pose a safety hazard at sea and cruising. Commonly known as "ghost gear," ALDFG can continuously track and detect both targets and non-targets (Gunarathna et al. 2019; Richardson et al. 2019a, 2019b). When unused equipment eventually hits the beach, it can have a huge impact on the coastal area, especially tourism and the beauty of the beach (Yildiz and Karakulak 2016). The chain of events leading to the loss of fishing gear may result from poor management of the farm, such as poor management, management of the area, tags and loss reports, illegal, unreported, and illegal (IUU) fishing activities.

Effective prevention, mitigation, and treatment strategies for ALDFG depend on the type and location of the glass, and the conditions surrounding the lost glass. These tactics work well when the causes and consequences of ALDFG are known. Fishermen can apply prevention, mitigation, and treatment measures, but so can a wide range of other stakeholders, such as farmers, ports, fishery managers, seafood firms, and NGOs (Huntington 2017). ALDFG management and fisheries management systems, the provision of port reception facilities for fishing equipment, and awareness activities are examples of conservation (Wilcox and Hardesty 2016; Huntington 2017). Glass design improvements to reduce overfishing and other ALDFG impacts on marine habitats and species are examples of mitigation (Drinkwin 2017; Huntington 2017; He and Suuronen 2018; FAO 2019).

8.4.1 Prevention Measures for Fishing Gears

A. Gear loss in fishing is an important problem that can lead to economic and environmental losses. Some measures you can take to reduce glass loss are as follows:
B. Regular maintenance and inspection of glass and equipment can help identify problems that could lead to physical damage or an explosion that could cause the glass to fall. Fishermen should properly maintain their equipment and inspect it before and after each use.
C. Equipment suitable for special fishing techniques: Appropriate equipment for specific fishing techniques can help reduce the loss of gear. For example, in areas with high currents or turbulent waves, using larger or stronger nets can help prevent equipment damage.
D. Equipment storage: When equipment is not in use, it should be properly stored to prevent damage. This may require placing the device in the chosen location and protecting it with a suitable cover or carpet.
E. Use of GPS or tracking systems: Some fishing equipment may be equipped with GPS or tracking systems to track its location and retrieve it if lost.
F. Adopt best practices and recommendations: Governments and fisheries organizations can develop best practices and guidelines to help fishermen reduce equipment loss. These recommendations will include recommended equipment, cleaning methods, and appropriate storage procedures.

G. Through training and presentations, fishermen can learn the value of preventing the loss of fishing gear and how to secure and store tackle. This will include outreach events, workshops and training.
H. By implementing these preventive measures, we can reduce tool losses and damage to the environment and economy in fishing.

8.5 Conclusion

In conclusion, abandoned, lost, or discarded fishing gear (ALDFG) is a significant source of microplastic pollution in the marine environment, with fishing gear accounting for a large percentage of floating macroplastic in various regions of the world. ALDFG poses a serious threat to marine life, particularly large marine animals that can be harmed by ingesting or getting entangled in fishing gear. It can also harm commercial fish species and transport toxins and microplastics into marine food webs, leading to harmful health effects on marine life and humans who consume them.

Efforts to address this issue have gained international attention, and measures to mitigate the negative impacts of ALDFG include improved fishing gear disposal practices, recycling initiatives, and the development of biodegradable fishing gear. Additionally, education and awareness campaigns can help to increase understanding of the issue and encourage responsible fishing practices. It is crucial that multilateral action is taken to address this issue, as the consequences of ALDFG extend far beyond the marine environment and can have significant socioeconomic impacts on coastal communities.

Acknowledgments The authors express their gratitude to GD Goenka University for supporting this study.

References

Baruah D, Bhuyan A, Dutta A, Pravin P (2018) Clap net operation in Brahmaputra Valley for capturing Tenualosa ilisha. Research Gate. https://www.researchgate.net/publication/329269222_Clap_net_Operation_in_Brahmaputra_Valley_for_Capturing_Tenualosa_ilisha

Borrelle SB, Ringma J, Law KL, Monnahan CC, Lebreton L, McGivern A, Murphy ES, Jambeck J, Leonard GB, Hilleary MA, Eriksen M, Possingham HP, De Frond H, Gerber LR, Polidoro B, Tahir A, Bernard M, Mallos N, Barnes M, Rochman CM (2020) Predicted growth in plastic waste exceeds efforts to mitigate plastic pollution. Science 369(6510):1515–1518. https://doi.org/10.1126/science.aba3656

Brown E (2016) Fishing gear 101: gillnets – the entanglers - the Safina Center. http://safinacenter.org/2015/03/fishing-gear-101-gillnets-entanglers/. Accessed 20 Aug 2016

Choy CA, Robison BH, Gagne TO, Erwin BE, Firl E, Halden RU, Hamilton JA, Katija K, Lisin S, Rolsky C, Van Houtan KS (2019) The vertical distribution and biological transport of marine microplastics across the epipelagic and mesopelagic water column. Sci Rep 9(1):7843. https://doi.org/10.1038/s41598-019-44117-2

Cook R (2003) The magnitude and impact of by-catch mortality by fishing gear. In: CABI publishing eBooks. CABI Publishing, pp 219–233. https://doi.org/10.1079/9780851996332.0219

Dağtekin M, Özyurt CE, Misir DS, Altuntas C, Cankaya AB, Misir GB, Aydin E (2019) Rate and causes of lost gillnets and entangling nets in the black sea coasts of Turkey. Turk J Fish Aquat Sci 19(8):699. https://doi.org/10.4194/1303-2712-v19_8_08

Deshpande PC, Skaar C, Brattebø H, Fet AM (2020) Multi-criteria decision analysis (MCDA) method for assessing the sustainability of end-of-life alternatives for waste plastics: a case study of Norway. Sci Total Environ 719:137353. https://doi.org/10.1016/j.scitotenv.2020.137353

Drinkwin J (2017) Methods to locate derelict fishing gear in marine waters. A guidance document of the global ghost gear initiative catalyze and replicate solutions working group. Global Ghost Gear Initiative, Seoul. https://doi.org/10.31230/osf.io/f3sq5

Du Preez C, Swan K, Curtis JMR (2020) Cold-water corals and other vulnerable biological structures on a North Pacific seamount after half a century of fishing. Front Mar Sci 7:17. https://doi.org/10.3389/fmars.2020.00017

Eighani M, Paighambari SY, Bayse SM (2019) Comparing handline and trolling fishing methods in the recreational pelagic fishery in the Gulf of Oman. Sci Mar 83(3):215. https://doi.org/10.3989/scimar.04864.10a

Enrichetti F, Dominguez-Carrió C, Toma M, Bavestrello G, Canese S, Bo M (2020) Assessment and distribution of seafloor litter on the deep Ligurian continental shelf and shelf break (NW Mediterranean Sea). Mar Pollut Bull 151:110872. https://doi.org/10.1016/j.marpolbul.2019.110872

Ern R, Molbo K, Jensen TH, Kucheryavskiy S, Møller P, Madsen N (2022) Initial experiments to assess short-term survival of discarded plaice (Pleuronectes platessa) caught in trammel nets during winter season. Fish Res 251:106308. https://doi.org/10.1016/j.fishres.2022.106308

Eyo J, Akpati (1995) Fishing gears and fishing methods. Research Gate. https://www.researchgate.net/publication/233817982_FISHING_GEARS_AND_FISHING_METHODS

FAO (2019) Voluntary Guidelines on the Marking of Fishing Gear. Directives Volontaires sur le Marquage des Engins de peche. In: Directrices Voluntarias Sobre el Marcado de las Artes de Pesca. FAO, Rome

Franco A, Elliott MR, Franzoi P, Nunn AD, Hänfling B, Colclough S, Young M (2022) Appendix A: study methods. In: Fish and fisheries in estuaries: a global perspective. Wiley, Hoboken, pp 874–940. https://doi.org/10.1002/9781119705345.app1

Geyer R, Jambeck J, Law KL (2017) Production, use, and fate of all plastics ever made. Sci Adv 3(7):e1700782. https://doi.org/10.1126/sciadv.1700782

Gilman E, Musyl MK, Suuronen P, Chaloupka M, Gorgin S, Wilson JR, Kuczenski B (2021) Highest risk abandoned, lost and discarded fishing gear. Sci Rep 11(1):7195. https://doi.org/10.1038/s41598-021-86123-3

Gunarathna KTNP, Hallinnage DR, Manikarachchi IU (2019) Assessment of the status of abandoned, lost or otherwise discarded fishing gear (ALDFG) in Southern Coastal waters of Sri Lanka: fisher's perspectives. Proceedings of the National Aquatic Resources Research and Development Agency, Colombo

He P, Suuronen P (2018) Technologies for the marking of fishing gear to identify gear components entangled on marine animals and to reduce abandoned, lost or otherwise discarded fishing gear. Mar Pollut Bull 129(1):253–261. https://doi.org/10.1016/j.marpolbul.2018.02.033

Hong S, Lee J, Lim S (2017) Navigational threats by derelict fishing gear to navy ships in the Korean seas. Mar Pollut Bull 119(2):100–105. https://doi.org/10.1016/j.marpolbul.2017.04.006

Huntington T (2017) Development of a best practice framework for the management of fishing gear. Part 2: best practice framework for the management of fishing gear. Global Ghost Gear Initiative, Hampshire

Jha S (2016) Crafts and gears used for fishing (with diagram). Your Article Library. https://www.yourarticlelibrary.com/fish/applied-fisheries/crafts-and-gears-used-for-fishing-with-diagram/88586

Kooi M, Van Nes EH, Scheffer M, Koelmans AA (2017) Ups and downs in the ocean: effects of biofouling on vertical transport of microplastics. Environ Sci Technol 51(14):7963–7971. https://doi.org/10.1021/acs.est.6b04702
Kroodsma D, Mayorga J, Hochberg T, Miller NH, Boerder K, Ferretti F, Wilson AC, Bergman B, White TJ, Block BA, Woods P, Sullivan BK, Costello C, Worm B (2018) Tracking the global footprint of fisheries. Science 359(6378):904–908. https://doi.org/10.1126/science.aao5646
Lebreton L, Slat B, Ferrari F, Sainte-Rose B, Aitken J, Marthouse R, Hajbane S, Cunsolo S, Schwarz AS, Levivier A, Noble K, Debeljak P, Maral H, Schoeneich-Argent R, Brambini R, Reisser J (2018) Evidence that the great pacific garbage patch is rapidly accumulating plastic. Sci Rep 8(1):4666. https://doi.org/10.1038/s41598-018-22939-w
Marcou M, Marn N, Muñoz MS, Macías FP, Kalachani F, Parenzan T, et al (2016) Med-zero-litter action plan-targeting the marine litter of the tourism industry in the Mediterranean Sea. Massive Open Online Course on Marine Litter Netherlands Open Universiteit. https://bib.irb.hr/datoteka/815695.MOOC_MED.pdf. Accessed 10 Jan 2021
Montarsolo A, Mossotti R, Patrucco A, Caringella R, Zoccola M, Pozzo PD, Tonin C (2018) Study on the microplastics release from fishing nets. Eur Phys J Plus 133(11):494. https://doi.org/10.1140/epjp/i2018-12415-1
Nelms SE, Duncan E, Patel S, Badola R, Bhola S, Chakma S, Chowdhury GW, Godley BJ, Haque AB, Sathyakumar S, Khatoon H, Kumar S, Napper IE, Niloy NH, Akter T, Badola S, Dev A, Rawat S, Santillo D et al (2021) Riverine plastic pollution from fisheries: insights from the Ganges River system. Sci Total Environ 756:143305. https://doi.org/10.1016/j.scitotenv.2020.143305
Ostle C, Thompson RF, Broughton D, Gregory L, Wootton M, Johns DG (2019) The rise in ocean plastics evidenced from a 60-year time series. Nat Commun 10(1):1622. https://doi.org/10.1038/s41467-019-09506-1
Petetta A, Virgili M, Guicciardi S, Lucchetti A (2021) Pots as alternative and sustainable fishing gears in the Mediterranean Sea: an overview. Rev Fish Biol Fish 31(4):773–795. https://doi.org/10.1007/s11160-021-09676-6
Pham CK, Ramirez-Llodra E, Alt CHS, Amaro T, Bergmann M, Canals M, Davies JS, Duineveld GC, Galgani F, Howell KE, Huvenne VA, Isidro E, Jones DB, Lastras G, Morato T, Gomes-Pereira JN, Purser A, Stewart H, Tojeira I et al (2014) Marine litter distribution and density in European seas, from the shelves to deep basins. PLoS One 9(4):e95839. https://doi.org/10.1371/journal.pone.0095839
Potter T, Pawson MG (2010) Gill netting. Research Gate. https://www.researchgate.net/publication/267935131_Gill_Netting
Richardson K, Asmutis-Silvia R, Drinkwin J, Gilardi KVK, Giskes I, Jones G, O'Brien K, Pragnell-Raasch H, Ludwig L, Antonelis K, Barco SG, Henry A, Knowlton AR, Landry S, Mattila DK, MacDonald K, Moore M, Morgan JP, Robbins J et al (2019a) Building evidence around ghost gear: global trends and analysis for sustainable solutions at scale. Mar Pollut Bull 138:222–229. https://doi.org/10.1016/j.marpolbul.2018.11.031
Richardson K, Hardesty BD, Vince JZ, Wilcox C (2021) Global causes, drivers, and prevention measures for lost fishing gear. Front Mar Sci 8:690447
Richardson K, Hardesty BD, Wilcox C (2019b) Estimates of fishing gear loss rates at a global scale: a literature review and meta-analysis. Fish Fish 20(6):1218–1231. https://doi.org/10.1111/faf.12407
Saturno J, Liboiron M, Ammendolia J, Healey N, Earles E, Duman N, Schoot I, Morris T, Favaro B (2020) Occurrence of plastics ingested by Atlantic cod (Gadus morhua) destined for human consumption (Fogo Island, Newfoundland and Labrador). Mar Pollut Bull 153:110993. https://doi.org/10.1016/j.marpolbul.2020.110993
Stelfox M, Hudgins JA, Sweet M (2016) A review of ghost gear entanglement amongst marine mammals, reptiles and elasmobranchs. Mar Pollut Bull 111(1-2):6–17. https://doi.org/10.1016/j.marpolbul.2016.06.034

Stelfox M, Lett C, Reid G, Souch G, Sweet M (2020) Minimum drift times infer trajectories of ghost nets found in the Maldives. Mar Pollut Bull 154:111037. https://doi.org/10.1016/j.marpolbul.2020.111037

Uhrin AV (2016) Tropical cyclones, derelict traps, and the future of the Florida Keys commercial spiny lobster fishery. Mar Policy 69:84–91. https://doi.org/10.1016/j.marpol.2016.04.009

Walters B (2022) 27 Most common types of fishing hooks. Norrik. https://norrik.com/27-most-common-types-of-fishing-hooks/

Wilcox C, Hardesty BD (2016) Biodegradable nets are not a panacea, but can contribute to addressing the ghost fishing problem. Anim Conserv 19(4):322–323. https://doi.org/10.1111/acv.12300

Wright L, Napper IE, Thompson RF (2021) Potential microplastic release from beached fishing gear in Great Britain's region of highest fishing litter density. Mar Pollut Bull 173:113115. https://doi.org/10.1016/j.marpolbul.2021.113115

Yildiz T, Karakulak F (2016) Types and extent of fishing gear losses and their causes in the artisanal fisheries of Istanbul, Turkey. J Appl Ichthyol 32(3):432–438. https://doi.org/10.1111/jai.13046

Tyres, Bitumen Wear and Plastic Bottles, Other Single Used Plastic as Major Sources of Microplastic

9

Babatunde Oluwafemi Adetuyi, John Tsado Mathew, Abel Inobeme, Yetude Olaitan Falana, Charles Oluwaseun Adetunji, Mohd Shahnawaz, Oluwafemi Adebayo Oyewole, K. I. T. Eniola, Mohammed Bello Yerima, and Oluwabukola Atinuke Popoola

Abstract

Microplastic pollution has emerged as a global environmental concern due to its potential adverse impacts on ecosystems and human health. Among the various sources of microplastics, tyres, bitumen wear, and plastic bottles are recognized as significant contributors to the contamination of terrestrial and aquatic environments. The abrasion of tyre treads and degradation of tyre walls release substantial amounts of microplastic particles into the environment. Over time,

B. O. Adetuyi
Department of Natural Sciences, Biochemistry unit, Faculty of Pure and Applied Sciences, Precious Cornerstone University, Ibadan, Nigeria

J. T. Mathew
Department of Chemistry, Ibrahim Badamasi Babangida University, Lapai, Niger State, Nigeria

A. Inobeme
Department of Chemistry, Edo University Iyamho, Auchi, Edo State, Nigeria

Y. O. Falana
Plant Improvement Unit, Agricultural Biotechnology Department, National Biotechnology Development Agency, Abuja, Nigeria

C. O. Adetunji (✉)
Applied Microbiology, Biotechnology and Nanotechnology Laboratory, Department of Microbiology, Edo State University Iyamho, Auchi, Edo State, Nigeria
e-mail: adetunji.charles@edouniversity.edu.ng

M. Shahnawaz
Department of Botany, University of Ladakh, Kargil Campus, Khumbathang, Kargil, Ladakh UT, India

Present Address: Department of Botany, Shri Pandit Prem Nath Dogra Government Degree College Samba, Samba, Jammu and Kashmir UT, India

M. Shahnawaz et al. (eds.), *Microplastic Pollution*,
https://doi.org/10.1007/978-981-99-8357-5_9

traffic and environmental factors lead to the wear and degradation of bitumen surfaces, resulting in the generation of microplastics. The presence of plastic polymers in road bitumen, such as microbeads or polymer-modified bitumen, further exacerbates the release of microplastics into the environment. The mitigation of microplastic pollution requires concerted efforts at multiple levels. Strategies such as promoting sustainable tyre manufacturing practices, developing alternative road construction materials with reduced microplastic release, and enhancing waste management systems to minimize plastic bottle pollution are crucial. Furthermore, public awareness, policy interventions, and international collaborations are essential to address these major sources of microplastics and mitigate their detrimental impacts on the environment.

Keywords

Tyres · Bitumen · Environment · Microplastics Pollution

9.1 Introduction

Microplastics, tiny particles of plastic less than 5 mm in size, have become a significant environmental concern in recent years. They are pervasive in our oceans, rivers, and even the air we breathe. While there are various sources of microplastics, this introduction will focus on two major contributors: tyres and plastic waste, including bitumen wear and single-use plastic bottles (Zolotova et al. 2022).

Tyres are essential for transportation, but they also play a significant role in the release of microplastics into the environment. As vehicles traverse roads, friction between the tyres and the road surface causes wear and tear. This wear generates small rubber particles known as tyre dust, which contains microplastics. Rainwater and winds can transport these particles to nearby water bodies, where they become a major source of pollution. Furthermore, tyre dust can also enter the atmosphere and settle in various ecosystems, affecting soil quality and potentially harming organisms (Tamis et al. 2021). Bitumen is a black, viscous material used in road construction and maintenance. Over time, bitumen wears off due to vehicle traffic, weathering, and other factors. As bitumen deteriorates, it releases microplastics into

O. A. Oyewole
Department of Microbiology, Federal University of Technology, Minna, Nigeria

K. I. T. Eniola
Department of Biological Sciences, Joseph Ayo Babalola University, Ikeji Arakeji, Osun State, Nigeria

M. B. Yerima
Department of Microbiology, Sokoto State University, Sokoto, Sokoto State, Nigeria

O. A. Popoola
Genetics, Genomics and Bioinformatics Department, National Biotechnology Development Agency, Abuja, FCT, Nigeria

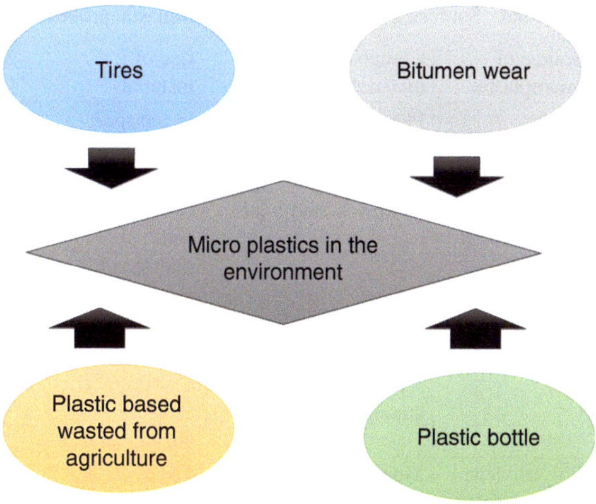

Fig. 9.1 Generation of microplastic in the environment from various sources

the environment. The microplastics in bitumen wear primarily originate from additives and fillers incorporated into the material during manufacturing. Rainwater runoff carries these particles into water bodies, contributing to the microplastic pollution problem (Järlskog et al. 2020). Plastic bottles and other single-use plastics are major contributors to the accumulation of microplastics in the environment. These items, often used for a short period before being discarded, break down over time into smaller fragments. Through various means, such as improper waste management, wind dispersal, and water transport, these microplastics find their way into ecosystems, including marine environments. Marine animals, mistakenly perceiving them as food, can ingest these particles, leading to potential harm to their health and the overall ecosystem (Kumar et al. 2021).

The widespread distribution of microplastics poses risks to both the environment and human health. The microplastics are generated from various sources (Fig. 9.1). These tiny particles can accumulate in aquatic organisms, disrupting their reproductive systems, and find their way into the food chain, ultimately reaching humans. Additionally, microplastics have been found in drinking water, indicating their presence in various water sources. Addressing the issue of microplastic pollution requires a multifaceted approach, including reducing plastic waste, promoting sustainable alternatives, improving waste management systems, and developing innovative technologies to capture and mitigate microplastic pollution. By raising awareness about the significant sources of microplastics, such as tyres, bitumen wear, and plastic bottles, we can work towards implementing effective solutions to protect our environment and preserve the health of ecosystems for future generations (Inobeme et al. 2023a; Li et al. 2023). There are various sources and mechanisms in the formation of microplastics (Table 9.1).

Table 9.1 Sources and mechanism of microplastic production

Source of microplastic	Composition of material	Use of materials	Mechanism of formation	Reference
Tyers	Plastic-based polymers	In transport industry	Abrasion due to friction	Zolotova et al. (2022)
Bitumen	Additives, and fillers	Construction and maintenance of road	Deterioration of bitumen over time	Järlskog et al. (2020)
Plastic bottles	Polymers from monomers such as styrene, ethylene amongst others	Storage of water, beverages and various chemicals	Disintegration and abrasion of larger plastic materials	Kumar et al. (2021)
Polyethylene terephthalate fiber	Monomer units such as terephthalate	Fibers for the production of textiles	Shedding of materials during wash and other production finishing	Gong et al. (2018)

9.2 Microplastic from Tyres

There is a global concern against the increasing amount of plastic waste in the environment. By the year 2030, plastic emissions has been projected to reach 53 million metric tons/year due to the increase in plastic production and usage. MP has been estimated to constitute up to 42% of the total microplastics (MPs) transported to the ocean, including vehicle tyres, road marking paints, and bitumen formulations. Particulate matter found in the road are as a result of the continuous passage of vehicles, road dust consisting of combination of organic and inorganic particles, among which MPs were found to contribute up to 0.06% of dry mass within an urban city environment.

The principal point of contact between a car and the road, the tyres are an integral part of an automobile and give it the traction and grip it needs to efficiently accelerate, brake, and steer. When cars are driven on roadways, the friction between the tyres and the road surface can cause the tyres to wear out and shed rubber particles that are released into the environment (Österlund et al. 2023).

They are the primary MP since they are knowingly produced when making tyres. Tyre wear particles (TWP), often known as the debris that comes off tyres, have been discovered in oceans and on highways (Rosso et al. 2023). TWPs is an important part of non-exhaust emissions, and are present at over six million tonnes annually across the world (Zhang et al. 2023). When TWPs are emitted, they are released into the air or settle on the road surface by the wind or interface of air-water created through bubble bursting (Shiu et al. 2022).

Vehicles used for sport that are fast driven contribute more of TWP to the environment (Mengistu et al. 2023). They are 24% heavier than their gas counterparts, leading to a 20% increase in tyre wear and a 30% increase in tyre wear $PM_{2.5}$ as estimated (Timmers and Achten 2016).

9.3 Microplastic from Plastic Bottles

In the natural surroundings, plastic waste undergoes biotic degradation, which involves the breakdown of the material by microorganisms, as well as abiotic degradation, such as photodegradation from exposure to ultraviolet radiation and weathering degradation caused by wave and wind action (Andrady 2011; Magalhães et al. 2020). These processes result in the production of reduced-sized plastic pieces, which are commonly known as microplastics (MPs).

The plastic particles exhibit considerable resistance to decomposition and tend to accumulate progressively in our biological systems, a phenomenon known as bioaccumulation. The chemicals utilized in the production of plastics are of significant concern, particularly in light of their association with various severe illnesses, as previously identified in research conducted by Arthur et al. (2009). The detection of microplastics in human fecal matter implies that individuals are regularly subjected to microplastics in their routine activities. The organisms in question have successfully infiltrated the food chain, thereby evoking apprehension regarding the safety of the food supply. The ubiquitous nature of microplastics is such that they are found in bottled water on a global scale, as reported by Koelmans et al. (2019). Numerous studies have conducted investigations regarding microplastics present in bottled water. These investigations have revealed that particle sizes of less than 1 mm are prevalent in most samples. Such particles have been determined to originate from the bottle material itself, the bottleneck, and the cap. The hue of the particles emanating from the container substrate is characterized by its transparent nature, whereas those that originate from the caps exhibit shades of blue or green. Polyethylene terephthalate (PET) has been identified as the highest polymer of plastics, which serves as a raw material for both the bottle and cap production (Gigault et al. 2018).

It has been established that the chemical substances released from the packaging materials of bottled water are classified as emerging pollutants and endocrine-disrupting chemicals (EDCs) that have the potential to give rise to a myriad of severe health issues such as cancer and developmental abnormalities (Cosmos magazine 2023).

A study was conducted to investigate the presence of microplastic contaminants in bottled water sourced from eleven global brands. The water samples were procured from 19 diverse locations across nine distinct countries. Nile Red tagging was employed as the analytical technique for the aforementioned testing. Out of the processed 259 bottles, a vast majority of 93% demonstrated indications of contamination by microplastics. Upon considering the potential influence of extraneous variables, specifically laboratory contamination, a mean concentration of 10.4

micrometer-sized microplastic particles greater than 100 um per liter of bottled water subjected to processing was discovered by the researchers Mason et al. (2018).

The bottles utilized for water storage are generally composed of polyethylene terephthalate (PET), while the closure mechanism or caps are ordinarily fashioned from polypropylene (PP), as noted in a study conducted by Schymanski et al. in 2018.

Polyethylene terephthalate (PET) fiber, commonly known as polyester within the textile industry, represents the most extensively utilized synthetic fiber globally, accounting for the majority of the PET production. In light of its impressive physical and chemical characteristics, Polyethylene terephthalate (PET) has found extensive application in various domains including as packaging material, functional material, and beverage bottles. Notwithstanding, the substantial quantities of polyethylene terephthalate (PET) that ingress and amass in the biosphere present a formidable ecological predicament (Gong et al. 2018).

Polyethylene terephthalate (PET) is subject to hydrolytic degradation in aqueous environments through the cleavage of ester linkages, particularly when exposed to temperatures above its glass transition temperature (Ballara and Verdu 1989; Allen et al. 1991). The aforementioned procedure culminates in the synthesis of carboxylic acid and alcohol moieties, with its conversion rate being escalated by acidic and alkaline environments. Accordingly, the formation of carboxylic end groups induces an autocatalytic process in the hydrolysis of polyethylene terephthalate (PET), as suggested by Gewert and colleagues in 2015.

Positron emission tomography is widely utilized in various industries owing to its numerous applications in everyday life. The utilization of the aforementioned substance is prevalent within the sectors of plastic film and fiber manufacturing. The underlying cause may be associated with the substance's elevated melting point and outstanding mechanical characteristics (Ma et al. 2018). A composite fiber fabricated from polyethylene terephthalate (PET) and desiccated aerogel particles is employed in the domains of textile manufacturing and transportation engineering. Furthermore, Christiansen et al. state that composite fibre materials are employed in both water filtration and air purification systems. The multifaceted characteristics of this particular polymer, including its lightweight nature, robust capacity for gas containment, superior tensile strength, and transparent composition, render it an optimal choice for the production of thin-walled containers. The employment of this material as a packaging medium in the food and beverage sectors is extensive. Additionally, it is utilized as a supplementary component within the composite of asphalt (Zair et al. 2021). The utilization of plastic bottle wastes as an additional component in road paving initiatives has been observed to augment the properties of stone mastic asphalt blends. According to Ahmadinia et al. (2012), there is evidence suggesting a beneficial impact on the characteristics of stone mastic asphalt, including enhanced mixture stiffness and a heightened ability to resist rutting. The method of shaping diverse plastic bottles, jars, and containers at varying blowing rates and temperatures entails the employment of a deformation technique termed as injection stretch blow moulding (ISBM).

The use of PET (polyethylene terephthalate) bottles is preferred over glass bottles due to the inherent fragility, weight, and propensity to break of the latter. The industrial sector exhibits a predilection for employing PET bottles for carbonated products owing to their transparency, robustness, and longevity, as espoused by Alvarado et al. (2020). Moreover, it is extensively employed in the manufacturing of textile products, as evidenced by Sevigné-Itoiz et al. (2015). In 1969, the Coca-Cola Company conducted the initial Life Cycle Assessment (LCA) study involving Polyethylene Terephthalate (PET), utilizing an academic writing style. According to the investigation conducted by Saleh (2016), it was evidenced that polyethylene terephthalate (PET) bottles portrayed a lower ecological impact in comparison to glass bottles. The deployment of this particular storage vessel is not limited to a singular market and extends to the preservation of fruit juices, sports beverages, salad oils, dressings, and sauces within varied industries. Over the course of the past two decades, a significant increase in the resin market for packaging purposes has been observed. According to PET bottles with a capacity of 1.5 L exhibit remarkable strength as they are nearly indestructible. Specifically, PET bottles weigh approximately one-tenth of glass bottles, with a weight range of 40–45 g. The PET plastic film is additionally employed as a barrier to oxygen in the context of food packaging, as indicated within the study conducted by Gaikwad et al. (cite). In the year of 2018, the following events occurred. PET trays are commonly utilized for the packaging of fresh meat and produce. The majority of polymeric films that possess antifogging properties are comprised of PET, a material commonly utilized in various food packaging applications (Gaikwad et al. 2020). The necessity for PET within the food industry will continue unabated until a viable substitute can be identified. Therefore, the comprehension and reduction of PET are integral.

The plastic bottle cap, which is comprised of polypropylene (PP), constitutes a plastic product that falls under the category of low modulus synthesis fibers. These fibers are attributed with a significant degree of effectiveness in mitigating shrinkage cracking, as substantiated by Foti's (2011) research. According to Siddique et al. (2008), the polypropylene cap of a plastic bottle is characterized as a rigid yet pliable plastic material. Meanwhile, the body of the plastic bottle is primarily composed of PET, which is a durable and transparent plastic substance. The polypropylene cap, due to its elevated melting point, is typically disregarded during the recycling of bottles, in contrast to PET. The discrepancy in the melting points between polypropylene caps and polyethylene terephthalate bottles is around 100 degrees Celsius. As a result, the processing expenses for polypropylene caps are significantly higher. The polypropylene cap was excluded from consideration for recycling as it was found to hold significant monetary value within the market, as noted by Mason et al. (2018).

Secondary microplastics can be produced through indirect means, such as the fragmentation of larger plastic residuals (Sobhani et al. 2020), or through the utilization of plastic materials that have already been subjected to previous use.

9.4 Generation of Microplastic Using Plastic-Based Agricultural Supplies

The production of raw materials in the agricultural sector is heavily dependent on the usage of plastics. As per the findings reported by Miles et al. (2017), plastic-based materials were found to be utilized within agricultural frameworks, specifically in relation to the cultivation of crops and livestock production. Plastic structures, in the form of greenhouses and small tunnels, coupled with mulching films, protective nets, and irrigation pipes, constitute a varied range of materials employed in agricultural practices to stimulate crop production. Similarly, the integration of plastics into animal production systems is facilitated by the utilization of fodder. Plastic materials are frequently employed in the manufacturing of twine, bale nets, and stretch films that are employed for the preservation of straw (Plastic Europe). Plastics remain a matter of significant concern, as highlighted in the Plastics - The Facts 2020 report. Whilst the quantity may not be of significant magnitude, it is worthy of attention that these plastics have the potential to intermingle with both the surrounding soil and water. In the present scenario, the polymers that were predominantly employed were polypropylene (PP) and low-density polyethylene (LDPE). These polymers are mainly utilized in the domain of plastic films that are designated for various uses such as greenhouses, small tunnels, mulching, and stretch applications. Following this, polyvinylchloride (PVC) became a prominent component utilized within irrigation pipelines and emitters. Polystyrene, denoted as PS, plays a crucial role in the manufacturing of twine and nets, and is encompassed in the assemblage of materials (Lwanga et al. 2022; Briassoulis et al. 2013).

9.5 Generation of Microplastic Waste During Plastic Production

The process of producing plastic materials through various industrial methods, typically involving the melting and shaping of polymers, is referred to as plastic manufacturing. The production process of plastic goods entails the utilization of granules and tiny resin pellets as the primary form of raw material. Raw materials have the propensity to infiltrate aquatic ecosystems as a result of inadvertent leakage that occurs during land or sea transport, unsuitable deployment as packaging materials, and unmediated discharge from processing plants. When assessing Swedish aquatic environments utilizing an 80 μm mesh, KIMO Sweden discovered that ordinary microplastic densities ranged from 150 to 2400 microplastics per cubic meter. Having examined the waters within a harbor situated in close proximity to a plastic production facility, the concentration of microplastics was notably elevated, with a recorded value of 102,000 microplastics per cubic meter. According to Cole et al. (2011), the following information was provided.

Numerous industrial sites, which frequently utilize easily accessible raw plastics, are situated in proximity to water bodies. If these materials are accidentally released in the course of production, they have the potential to contaminate the immediate surrounding environment, thus causing waterways to be polluted (Stundt et al. 2015).

9.5.1 Generation of Microplastic Using Baby Bottles

Infant feeding bottles, also known as baby bottles, are essential tools for providing nourishment to infants and young children during their early developmental stages. These instruments come in a wide range of shapes, sizes, and materials to cater to the diverse and often unique needs of the users and caregivers. The selection of the appropriate baby bottle is influenced by various factors including the age, feeding capacity, motor skills, allergies, personal preferences, health conditions, and cultural traditions of the infants and their families. Infants are particularly vulnerable to infections, and as such, the proper cleaning, sterilization and maintenance of baby bottles is crucial to minimize the risk of contamination and ensure optimal hygienic conditions.

In 2020, scholars reported that the employment of contemporary preparation procedures in tandem with polypropylene infant feeding bottles resulted in the exposure of infants to microplastics. The aforementioned exposure ranged from 14,600 to 4,550,000 particles per capita per day and was observed across 48 regions. The release of microplastics is observed to escalate in the presence of warmer liquids, while akin levels of release are exhibited by various other polypropylene commodities, including lunchboxes. A recent research study in 2021 has revealed the surprising phenomenon of progressive degradation of silicone rubber baby bottle nipples triggered by recurrent sterilization procedures utilizing steam. Consequently, these sterilization procedures induce a shedding of minute silicone rubber particles that encompass both micro and nano scales. According to Zangmeister et al. (2022), an assessment suggests that if heat-degraded nipples are utilized for a year, the infant is predicted to consume over 660,000 particles.

9.5.2 Production of Microplastic Using Single-Use Plastic Commodities

Everyday disposable plastic items, such as paper cups used for coffee, which are coated with a delicate plastic layer on the interior, discharge trillions of microplastic nanoparticles per liter of water during typical usage. The infiltration of single-use plastic products into aquatic ecosystems is a matter of concern. It has been pointed out that effective legislative measures to counter plastic pollution can be implemented at the local and statewide level, as emphasized by Li et al. (2020).

9.5.3 Production of Microplastic Using Mulching Film

Mulching films are thin, plastic sheets that are primarily utilized to cover the soil in agricultural fields. These films provide a range of benefits, such as conserving moisture, controlling soil temperature, suppressing weed growth, preserving soil nutrients, and protecting crops from environmental stressors. Mulching films can also improve crop yields and reduce the need for herbicides and pesticides, making them a valuable tool for sustainable agriculture. Furthermore, the use of mulching films can mitigate soil erosion and promote soil structure, resulting in improved soil health and fertility. Thus, mulching films play a crucial role in modern farming practices, contributing to the success and sustainability of agricultural operations worldwide.

Plastic mulches are employed by farmers with the aim of augmenting soil temperature and optimizing water utilization, while concurrently mitigating weed growth (Espi et al. 2006). One potential drawback of utilizing mulch in agriculture is the inability of farmers to entirely remove the plastic material following the conclusion of the growth cycle. This limitation results in residual plastic fragments which pose a hazard to the natural environment. The utilization of plastic mulches by agricultural practitioners results in the escalation of plastic waste in soil, thereby elevating the likelihood of its transference and propagation into other ecological systems, as posited by.

In recent years, scholars have expressed concerns regarding the employment of plastic mulches in agriculture, citing their propensity to heighten the prevalence of microplastics within soil compositions (Wang et al. 2019). The preponderance of evidence originates from China, which stands as one of the nations that implement plastic mulching to a significant degree. In the year 2015, a total of over 1.5 million tonnes of plastic mulch was utilized to cover an expanse of 20 million hectares of arable land in China.

9.6 The Utilization of Plastics as a Method of Safeguarding Plants

The usage of plastic films and plastic-derived non-woven fabrics, in the context of their application in greenhouses, polytunnels, shade nets, and wind barriers, has been identified as a contributing factor in the prevalent occurrence of microplastics within soil samples. There have been various investigations conducted on the ascertainable impacts of plant safeguarding plastics on the concentrations of microplastics detected within the natural ecosystem. Several studies have been conducted to quantify the levels of microplastics likely to be present in soils situated in the proximity of productive systems utilizing such plastic materials.

Limited research has been conducted with a specific focus on the release of microplastics from plastics used in plant protection. Li et al. (2021) demonstrates that, in comparison to farms with greenhouses, farms without greenhouses exhibited lower levels of microplastics in soil samples obtained from equivalent productive areas. The scholars emphasize that a direct correlation exists between the extent of plastic usage and the quantities of microplastics evident in specimens of soil. Dahl et al. utilized a comparable methodology. In 2021, a chronological sequence of plastic materials within sediments obtained from a bay located in Spain was established. The researchers also established a correlation between the rising levels of microplastics detected within the sediments and the corresponding increase of land surface covered by greenhouses in the region.

The investigation into microplastic pollution has not frequently incorporated other forms of plant-protecting plastics.

The utilization of silage and bale nets for agricultural purposes is also of interest.

The scholarly work authored by Piehl et al. (2018) discussed the inaugural investigation on the matter of microplastics in soils at a farm scale and asserts that the employment of plastics to cover grass bales and silage presents a conceivable source of contamination of the feedstuff with microplastics. Said microplastics can proceed to permeate the soil through the process of excretion subsequent to ingestion. Beriot et al. (2021) discovered that plastic mulch fragments, when consumed by sheep, can lead to the excretion of microplastics.

Twine is a digital tool commonly used for creating interactive narratives. It allows users to connect different passages of text and media, creating a branching storyline. This tool has gained popularity in the field of game design and interactive storytelling due to its accessibility and ease of use. Twine has also become a platform for exploring a wide range of topics, from personal narratives to social issues. Its versatility has made it a valuable tool for educators, writers, and game developers alike.

In 2019, the European Union witnessed a demand of 11% for plastic twine in the agricultural sector. Polypropylene, commonly abbreviated as PP, is predominantly utilized in the manufacturing of plastic twine, which finds significant application in agriculture (Anonymous 2023). The utilization of twine as a fastening mechanism for the anchorage of plants, notably in agricultural pursuits surrounding the cultivation of paramount produce varieties such as tomatoes, sweet pepper, and hops, has been documented. In the cultivation of bananas, the application of binding agents to interconnect individual plants represents an effective measure for ensuring that the plants remain upright and thus avoid instances of falling over (Hernandez and Witter 1996). Twine is utilized for the purpose of enveloping hay bales. During the harvesting process, the twine is frequently severed and consequently abandoned in the agricultural fields in a careless manner. This neglectful practice poses a potential threat, as the residual twine can degrade into microscopic plastic particles that could accumulate within the soil or compost. Guerrini et al. (2017) have reported the existence of programs designed to promote the adoption of compostable twine, which is capable of being conveniently gathered in tandem with plant residues for composting purposes.

The utilization of coated fertilizers and pesticides has become a common practice in modern agriculture. The utilization of plastic polymers as coating agents for slow-release fertilizers and pesticides is a common practice in the fertilizer industry. Upon successful delivery, the plastic coating of the fertilizer pellet persists within the soil.

According to Katsumi et al. (2020), the issue at hand constitutes a plausible origin of microplastics within agricultural terrain.

Polymeric pipes have become increasingly popular in recent years and are now a common feature of various industries. These pipes, commonly referred to as plastic pipes, are widely used due to their low cost, light weight, corrosion resistance, and ease of installation. Polymeric materials such as PVC, PE, PEX, and PP are commonly used to manufacture these pipes. Plastic pipes are commonly used in applications where the transportation of water, gases, and other fluids is required. Their unique properties make them suitable for a variety of applications, including irrigation, potable water supply, sewage disposal, and industrial piping systems. In addition, plastic pipes have proven to be a sustainable alternative to traditional metal pipes, reducing cost, environmental impact, and maintenance requirements.

According to Beieler (2013), the distribution of water across a field is facilitated through the utilization of plastic pipes in drip and sprinkler irrigation. It is imperative for farmers to carry out a systematic replacement of the pipes. Subsequent to the aforementioned action, it remains ambiguous how farmers proceed. The course of action for farmers includes either opting for the retention of aged pipes which are currently submerged and simultaneously situating a new pipeline beside it, or implementing a method of careful extraction of the previous plumbing so as to prevent fragmentation and the dispersion of plastic debris.

The inappropriate management of goods in storage facilities, oftentimes resulting in suboptimal conditions or unsuitable environments for the stored items, is a significant issue in logistics and supply chain management. This problem may lead to damage, theft, spoilage, or other adverse consequences, further exacerbating the associated economic costs and operational inefficiencies. As such, the proper storage of materials and products is a critical requirement for successful business operations and sustainable development.

As elucidated in the preceding sections, the presence of plastic on a farm is often observed in significant quantities. The requisite acquisition and appropriate preservation of utilized plastics entails supplementary efforts from farmers who occasionally lack the resources and expertise to effectively sanitize and sustainably warehouse these materials. According to Anonymous (2021), proper storage of plastic waste entails locating it in a dry environment that is shielded from wind currents. This measure is aimed at ensuring the cleanliness of the plastic waste and mitigating the risk of its dispersal through wind-induced displacement. The improper handling and storage of plastic wastes within agricultural zones has emerged as a significant problem that calls for urgent attention and intervention. The mismanagement of plastic waste results in the ingestion of plastic particles by various vertebrates, including sheep, cows, dogs, and vultures.

The subject matter that is being referred to pertains to the residual by-products of wastewater treatment, commonly known as sewage sludge and biosolids.

The implementation of wastewater treatment plants has been found to be effective in mitigating the contamination of surface water caused by microplastics present in urban wastewater (Ziajahromi et al. 2017). Plants have the ability to capture micropollutants (MPs) present in incoming water through a high-density organic matrix referred to as sludge. According to Du et al. (2020), microplastics of urban origin, such as those from industrial processes, domestic grey water, and traffic, accumulate in sludge. In numerous nations, the utilization of sludge as a source of crop fertilization has been observed, which raises concerns regarding the transportation of microplastics to agricultural fields.

Composting is the process of breaking down organic waste such as food scraps, paper, and yard trimmings into a rich soil amendment known as compost. This practice provides numerous benefits such as reducing landfill waste and greenhouse gas emissions, improving soil quality, and promoting sustainable agriculture. Composting is a fundamental component of the circular economy, as it facilitates the transformation of waste into a valuable resource. Proper management of composting systems is essential to ensure air quality, prevent odors and pest problems, and maintain optimal compost conditions.

It has been widely postulated among scientists that organic fertilizers serve as a potential origin for the existence of microplastics in soil. Despite its perceived plausibility, empirical investigations regarding the plastic content in compost have remained relatively scarce. Weithmann et al. (surname of lead author followed by "et al." to denote other authors) In 2018, a study found the presence of 24 microplastic particles per kilogram, which were in the size range of 1 mm to 5 mm, in German compost derived from the amalgamation of municipal organic waste and green clippings. With regards to the preponderance of polymers, Gui et al. The findings of the study conducted in 2021 revealed that PES, PP, and PE were responsible for covering approximately 70 to 80% of the entire quantity of microplastics that were detected in compost. The documented data aligns with the current demand for plastics within the European continent, as stated by Plastic Europe. According to the report titled "Plastics - The Facts 2020," ... Limited available evidence suggests that the utilization of compost may facilitate the conveyance of microplastics to agricultural fields.

The producers of plastic pellets. Pellets constitute the fundamental state of various plastics and generally possess a diameter of approximately 2.55 mm. The aforementioned pellets are utilized in the production of plastic goods. According to Fabbri et al. (2000), the spherules commonly known as pellets constitute 79.4% of the overall plastic debris present in the water of river Danube. Dhodapkar and colleagues have demonstrated in their research that... observed that the manufacturing process of plastic not only produces pellets but also results in the accumulation of plastic dust. Furthermore, relocation and transportation activities of plastic materials also generate such dust. This phenomenon was previously noted in 2009. Pellets are found to contain hazardous additives, including plasticizers and flame retardants, that exacerbate ecological toxicity. The aforementioned additives are either blended prior to the production stage or introduced during the process of conversion. These particular small cylindrical objects are commonly known as nibs and nurdles, according to Lassen et al. (2015).

Weathering and abrasion are physical processes that occur in the natural environment. These phenomena are characterized by the breaking down and wearing away of rocks, soils, and other geological materials over time. Such changes can result from exposure to various weather conditions, such as wind, rain, and extreme temperatures, as well as the constant rubbing and scraping of materials against one another. Weathering and abrasion play a significant role in shaping the surface of the Earth's landscapes and can have important implications for geological history and soil fertility.

The urban environment is frequently confronted with the ubiquitous presence of particulate matter, colloquially referred to as city dust. The process of abrasion stemming from various sources such as the use of synthetic cooking utensils, wear and tear of footwear soles, deterioration of infrastructure, and employment of blasting abrasives, collectively contribute to the formation of particulate matter commonly known as urban dust. Collectively, the impact of these factors on the country's losses is considerable, whereas their individual significance is inconsequential. In order to facilitate the progress of roadway infrastructure, road markings are implemented. Thermoplastic, paint, and polymer tapes have been shown to be the preferred materials for utilization in this particular process. The phenomenon of microplastics loss induced by vehicular weathering entails the detachment of said particles which are then dispersed through the action of natural forces such as wind or precipitation, and eventually deposited into marine ecosystems. The application of anticorrosive paints on boat coatings is executed through diverse materials such as polyurethane, lacquers, and vinyl (OECD 2009).

The prioritization of packaging materials as a primary contributor to microplastics is warranted. This phenomenon accounts for approximately 62% of the total plastic amassed. Generally, the issue incorporates secondary microplastics. The accumulation of plastic bags and soft drink bottles has been pinpointed as a significant issue (Verschoor et al. 2014) within the context of environmental degradation. Packaging materials represent a substantial component of litter. According to Raynaud (2014), articles such as toys, cutlery, shoes, and clothing constitute additional manifestations of litter. It has been observed that agricultural plastics litter is characterized by their non-biodegradable nature. Nevertheless, the prevalence of biodegradable plastics cannot be disregarded in contemporary times. The inclusion of preservatives within plastics has been found to decrease their biodegradability, leading to the subsequent formation of increasingly smaller fragments. This process ultimately results in the introduction of microplastics into surface water via nearby streams (Verschoor et al. 2014).

The aforementioned commodities, comprising household articles, consumable productions, and playthings, are under consideration.

As per the statement made by the United Nations Environment Programme (UNEP) in 2014 (Raynaud 2014), it has been observed that domestic items are a significant contributor to marine pollution. Numerous articles such as plastic cutlery, cups, straws, and bottles are found in great quantities within the ocean environment. Foodstuffs and snacks are also a significant source of microplastics. The composition of chewing gum comprises of microplastics fillers. The investigation conducted along the Dutch coastline resulted in the identification of an average of 105 microparticles of plastic per gram in mussels, while oysters exhibited a concentration of up to 87 microparticles per gram (Leslie et al. 2013). Party paraphernalia, such as balloons, confetti, firework remnants, and toy fragments, collectively contribute towards the accumulation of waterborne debris. Loom bracelets are comprised of microplastics that have the potential to contaminate the surrounding environment through the mechanism of surface runoff, as noted by Siegle (2014).

The allocation and distribution of medical resources is a critical aspect in the healthcare industry. The availability of medical resources greatly influences the delivery and quality of healthcare services provided to patients. Adequate and efficient distribution of these resources can ultimately lead to improved health outcomes for patients. Therefore, healthcare organizations and policymakers must make informed decisions regarding the allocation and distribution of medical resources in order to ensure optimal healthcare delivery.

Microplastics have been extensively employed in the medical realm (Essel et al. 2015). The treatment of the reflux of gastric juices is facilitated through the deployment of microplastics. Compounds coated with aluminum are known to have detrimental effects on the health of individuals. Microplastics have been identified as the replacements for them. The items commonly referred to as nappies, sanitary towels, and plasters are considered to be forms of litter. Capsules utilized in the domain of medicine contain plastic material. According to Verschoor et al. (2014), spectacles and contact lenses are significant contributors to the occurrence of microplastics.

The fecal and urinary excreta of animals, often used as an organic fertilizer to enrich soil nutrient content, is commonly referred to as manure (Krause et al. 2021). Nevertheless, it has been reported by the authors that a considerably greater proportion of microplastics were identified in soil specimens obtained from agricultural lands treated with pig manure (Zhang et al. 2022).

Irrigation is a critical practice that has been employed for centuries to enhance agricultural productivity. The technique involves the artificial provision of water to crops, leading to increased yield and crop quality. In modern times, the use of irrigation has been further enhanced through the development of advanced technologies and methodologies, such as drip irrigation and sprinkler irrigation. These innovations have contributed immensely to the sustainable utilization of water resources and the conservation of the environment. Therefore, the role of irrigation in the agricultural sector cannot be underestimated (Koech and Langat 2018).

Zhou and colleagues did a comprehensive assessment of agricultural soils in Hangzhou Bay (China) in 2020 revealed that microplastics are introduced to farm soils through irrigation water. It can be postulated that in cases where superficial bodies of water are contaminated by microplastics, it is highly probable that such plastic pollutants will permeate agricultural soil if farmers employ irrigation systems utilizing water from these polluted sources. This proposal is congruent with the global plastic cycle, a novel conceptual framework that affords an enhanced comprehension of the issues engendered by plastic contamination.

9.7 Conclusion and Future Prospects

In conclusion, this chapter highlights the diverse sources of microplastic pollution, including tyres, plastic bottles, plastic-based agricultural supplies, and plastic production waste. These findings underscore the urgent need for comprehensive strategies to tackle microplastic pollution and promote sustainable alternatives in safeguarding plants without further harming our environment (Inobeme et al. 2023b; Adetunji et al. 2021a, b, c; Adetunji et al. 2020; Adetunji 2019; Adetunji et al. 2019; Adetunji and Anani 2021, Anani et al. 2020; Anani et al. 2022; Adetunji et al. 2022a, b). Future prospects include advancing microplastic research, promoting sustainable alternatives, enforcing policies and regulations, increasing consumer awareness, adopting circular economy practices, fostering international collaboration, investing in green technology, and implementing restoration and remediation efforts to mitigate microplastic pollution and safeguard the environment.

References

Adetunji CO (2019) Environmental Impact and Ecotoxicological Influence of Biofabricated and Inorganic Nanoparticle on Soil Activity. In: Panpatte D, Jhala Y (eds) Nanotechnology for Agriculture. Springer, Singapore. https://doi.org/10.1007/978-981-32-9370-0_12

Adetunji CO, Panpatte DG, Bello OM, Adekoya MA (2019) Application of Nanoengineered Metabolites from Beneficial and Eco-friendly Microorganisms as a Biological Control Agents for Plant Pests and Pathogens. In: Panpatte D, Jhala Y (eds) Nanotechnology for Agriculture: Crop Production & Protection. Springer, Singapore. https://doi.org/10.1007/978-981-32-9374-8_13

Adetunji CO, Oloke JK, Prasad G (2020) Effect of carbon-to-nitrogen ratio on eco-friendly mycoherbicide activity from *Lasiodiplodia pseudotheobromae* C1136 for sustainable weeds management in organic agriculture. *Environ Dev Sustain* 22:1977–1990. https://doi.org/10.1007/s10668-018-0273-1

Adetunji CO, Anani OA (2021) Plastic-Eating Microorganisms: Recent Biotechnological Techniques for Recycling of Plastic. In: Panpatte DG, Jhala YK (eds) Microbial Rejuvenation of Polluted Environment. Microorganisms for Sustainability, vol 25. Springer, Singapore. https://doi.org/10.1007/978-981-15-7447-4_14

Adetunji CO, Akram M, Olaniyan OT, Ajayi OO, Inobeme A, Olaniyan S, Hameed L, Adetunji JB (2021a) Targeting SARS-CoV-2 Novel Corona (COVID-19) Virus Infection Using Medicinal Plants. In: Dua K, Nammi S, Chang D, Chellappan DK, Gupta G, Collet T (eds) Medicinal Plants for Lung Diseases. Springer, Singapore. https://doi.org/10.1007/978-981-33-6850-7_21

Adetunji CO, Ajayi OO, Akram M, Olaniyan OT, Chishti MA, Inobeme A, Olaniyan S, Adetunji JB, Olaniyan M, Awotunde SO (2021b) Medicinal Plants Used in the Treatment of Influenza A Virus Infections. In: Dua K, Nammi S, Chang D, Chellappan DK, Gupta G, Collet T (eds) Medicinal Plants for Lung Diseases. Springer, Singapore. https://doi.org/10.1007/978-981-33-6850-7_19

Adetunji CO, Inobeme A, Olaniyan OT, Ajayi OO, Olaniyan S, Adetunji JB (2021c) Application of Nanodrugs Derived from Active Metabolites of Medicinal Plants for the Treatment of Inflammatory and Lung Diseases: Recent Advances. In: Dua K, Nammi S, Chang D, Chellappan DK, Gupta G, Collet T (eds) Medicinal Plants for Lung Diseases. Springer, Singapore. https://doi.org/10.1007/978-981-33-6850-7_26

Adetunji CO, Inobeme A, Tadso J, Olaniyan OT, Abimbola OF, Shahnawaz M, Anani O (2022a) Potential of Plastic Waste in Enhancing the level of Pathogenicity of diverse Pathogens in the

Marine Biota. In: Shahnawaz M, Sangale MK, Daochen Z, Ade AB (eds) Impact of Plastic Waste on the Marine Biota. Springer, Singapore. https://doi.org/10.1007/978-981-16-5403-9_16

Adetunji CO, Oloke JK, Osemwegie OOC, Ehis-Eriakha B (2022b) Use of agro-wastes for *Lasiodiplodia pseudotheobromae* (C1136) production with sustainable bioefficacy. *Environ Dev Sustain* 24:7794–7809. https://doi.org/10.1007/s10668-021-01758-6

Ahmadinia E, Zargar M, Karim MR, Abdelaziz M, Ahmadinia E (2012) Performance evaluation of utilization of waste Polyethylene Terephthalate (PET) in stone mastic asphalt. Constr Build Mater 36:984–989

Allen NS, Edge M, Mohammadian M, Jones K (1991) Hydrolytic degradation of poly (ethylene terephthalate): importance of chain scission versus crystallinity. Eur Polym J 27(12):1373–1378

Alvarado CF, Brouwer MT, Thoden van Velzen EU (2020) Effect of recycled content and rPET quality on the properties of PET bottles, part I: optical and mechanical properties. Packag Technol Sci 33:347–357

Anani OA, Mishra RR, Mishra P, Olomukoro JO, Imoobe TOT, Adetunji CO (2020) Influence of Heavy Metal on Food Security: Recent Advances. In: Mishra P, Mishra RR, Adetunji CO (eds) Innovations in Food Technology. Springer, Singapore. https://doi.org/10.1007/978-981-15-6121-4_18

Anani OA, Adetunji CO, Anani GA, Olomukoro JO, Imoobe TOT, Enenuku AA, Tongo I (2022) Effect of Meso-, Micro-, and Nano-Plastic Waste on the Benthos. In: Shahnawaz M, Sangale MK, Daochen Z, Ade AB (eds) Impact of Plastic Waste on the Marine Biota. Springer, Singapore. https://doi.org/10.1007/978-981-16-5403-9_12

Andrady AL (2011) Microplastics in the marine environment. Mar Pollut Bull 62(8):1596–1605

Anonymous (2021) Step-by-step recycling of plastic in the agricultural sector. http://svepretur.se/en/our-services/proceed-as-follows/. Accessed 15 June 2023

Anonymous (2023) Statistics: plasticulture in Europe. https://apeeurope.eu/statistics/#:~:text=The%20volume%20of%20plasticulture%20marketed,is%20considered%20relatively%20stable%20overall. Accessed 18 May 2023

Arthur C, Baker J, Bamford H (2009) Proceedings of the international research workshop on the occurrence, effects and fate of microplastic marine debris. National Oceanic and Atmospheric Administration (NOAA); Tacoma, WA, USA: 2009. NOAA Technical Memorandum NOS-OR&R-30

Ballara A, Verdu J (1989) Physical aspects of the hydrolysis of polyethylene terephthalate. Polym Degrad Stab 26(4):361–374

Beieler RW (2013) Pipelines for water conveyance and drainage. American Society of Civil Engineers Virginia, American Society of Civil Engineers, Reston

Beriot N, Peek J, Zornoza R, Geissen V, Lwanga EH (2021) Low density-microplastics detected in sheep faeces and soil: a case study from the intensive vegetable farming in Southeast Spain. Sci Total Environ 755:142653. https://doi.org/10.1016/j.scitotenv.2020.142653

Briassoulis D, Hiskakis M, Babou E (2013) Technical specifications for mechanical recycling of agricultural plastic waste. Waste Manag 33(6):1516–1530. https://doi.org/10.1016/j.wasman.2013.03.004

Cole M, Lindeque P, Halsband C, Galloway TS (2011) Microplastics as contaminants in the marine environment: a review. Mar Pollut Bull 62(12):2588–2597

Cosmos Magazine (2023). https://cosmosmagazine.com/news/microplastics-bottled-water/. Accessed 19 June 2023

Dahl M, Bergman S, Björk M, Diaz-Almela E, Granberg M, Gullström M, Leiva-Dueñas C, Magnusson K, Marco-Méndez C, Piñeiro-Juncal N, Mateo MÁ (2021) A temporal record of microplastic pollution in Mediterranean seagrass soils. Environ Pollut 273:116451

Dhodapkar S, Trottier R, Smith B (2009) Measuring dust and fines in polymer pellets. Chem Eng 116(9):24

Du F, Cai H, Zhang Q, Chen Q, Shi H (2020) Microplastics in take-out food containers. J Hazard Mater 399:122969

Espi E, Salmerón A, Fontecha A, García Y, Real AI (2006) Plastic films for agricultural applications. J Plast Film Sheeting 22(2):85–102. https://doi.org/10.1177/8756087906064220
Essel R, Engel L, Carus M, Ahrens RH (2015) Sources of microplastics relevant to marine protection in Germany. Texte 64(2015):1219–1226
Fabbri D, Tartari D, Trombini C (2000) Analysis of poly (vinyl chloride) and other polymers in sediments and suspended matter of a coastal lagoon by pyrolysis-gas chromatography-mass spectrometry. Anal Chim Acta 413(1-2):3–11
Foti D (2011) Preliminary analysis of concrete reinforced with waste bottles PET fibers. Constr Build Mater 25(4):1906–1915
Gaikwad KK, Singh S, Lee YS (2018) Oxygen scavenging films in food packaging. Environ Chem Lett 16:523–538
Gaikwad KK, Singh S, Negi YS (2020) Ethylene scavengers for active packaging of fresh food produce. Environ Chem Lett 18:269–284
Gewert B, Plassmann MM, MacLeod M (2015) Pathways for degradation of plastic polymers floating in the marine environment. Environ Sci: Processes Impacts 17(9):1513–1521
Gigault J, Ter Halle A, Baudrimont M, Pascal PY, Gauffre F, Phi TL, El Hadri H, Grassl B, Reynaud S (2018) Current opinion: what is a nanoplastic? Environ Pollut 235:1030–1034
Gong J, Kong T, Li Y, Li Q, Li Z, Zhang J (2018) Biodegradation of microplastic derived from poly (ethylene terephthalate) with bacterial whole-cell biocatalysts. Polymers (Basel) 10(12):1326
Guerrini S, Borreani G, Voojis H (2017) Biodegradable materials in agriculture: case histories and perspectives. In: Malinconico M (ed) Soil degradable bioplastics for a sustainable modern agriculture. Springer, Berlin, pp 35–65
Gui J, Sun Y, Wang J, Chen X, Zhang S, Wu D (2021) Microplastics in composting of rural domestic waste: abundance, characteristics, and release from the surface of macroplastics. Environ Pollut 274:116553
Hernandez CE, Witter S (1996) Evaluating and managing the environmental impact of banana production in Costa Rica: a systems approach. Ambio 25(3):171–178
Inobeme A, Ikechukwu Ajai A, Inobeme J, Adetunji CO, Obar A, Mathew JT, Jacob JO, Nwakife N (2023a) Superabsorbent polymers for the development of nanofiltration. In: Properties and applications of superabsorbent polymers: smart applications with smart polymers. Springer Nature, Singapore, pp 157–170
Inobeme A, Mathew JT, Adetunji CO, Ajai AI, Inobeme J, Maliki M, Okonkwo S, Adekoya MA, Bamigboye MO, Jacob JO, Eziukwu CA (2023b) Recent advances in nanotechnology for remediation of heavy metals. *Environ Monit Assess* 195:111. https://doi.org/10.1007/s10661-022-10614-7
Järlskog I, Strömvall AM, Magnusson K, Gustafsson M, Polukarova M, Galfi H, Aronsson M, Andersson-Sköld Y (2020) Occurrence of tire and bitumen wear microplastics on urban streets and in sweepsand and washwater. Sci Total Environ 729:138950. https://doi.org/10.1016/j.scitotenv.2020.138950
Katsumi N, Kusube T, Nagao S, Okochi H (2020) The role of coated fertilizer used in paddy fields as a source of microplastics in the marine environment. Mar Pollut Bull 161:111727
Koech R, Langat P (2018) Improving irrigation water use efficiency: a review of advances, challenges and opportunities in the Australian context. Water 10(12):1771
Koelmans AA, Nor NH, Hermsen E, Kooi M, Mintenig SM, De France J (2019) Microplastics in freshwaters and drinking water: critical review and assessment of data quality. Water Res 155:410–422
Krause A, Häfner F, Augustin F, Udert KM (2021) Qualitative risk analysis for contents of dry toilets used to produce novel recycling fertilizers. Circ Econ Sustain 1:1107–1146
Kumar R, Verma A, Shome A, Sinha R, Sinha S, Jha PK, Kumar R, Kumar P, Shubham DS, Sharma P (2021) Impacts of plastic pollution on ecosystem services, sustainable development goals, and need to focus on circular economy and policy interventions. Sustainability 13(17):9963. https://doi.org/10.3390/su13179963

Lassen C, Hansen SF, Magnusson K, Norén F, Hartmann NI, Jensen PR, Nielsen TG, Brinch A (2015) Microplastics—occurrence, effects and sources of releases to the environment in Denmark, Environmental Project No. 1973. Danish Ministry of the Environment—Environmental Protection Agency, Denmark, p 204

Leslie HA, Van Velzen MJ, Vethaak AD (2013) Microplastic survey of the Dutch environment. Novel data set of microplastics in North Sea sediments, treated wastewater effluents and marine biota. The Netherlands, pp 1–30

Li C, Busquets R, Campos LC (2020) Assessment of microplastics in freshwater systems: a review. Sci Total Environ 707:135578

Li Q, Zeng A, Jiang X, Gu X (2021) Are microplastics correlated to phthalates in facility agriculture soil? J Hazard Mater 412:125164. https://doi.org/10.1016/j.jhazmat.2021.125164

Li X, Bao L, Wei Y, Zhao W, Wang F, Liu X, Su H, Zhang R (2023) Occurrence, bioaccumulation, and risk assessment of microplastics in the aquatic environment: a review. Water 15(9):1768. https://doi.org/10.3390/w15091768

Lwanga EH, Beriot N, Corradini F, Silva V, Yang X, Baartman J, Rezaei M, van Schaik L, Riksen M, Geissen V (2022) Review of microplastic sources, transport pathways and correlations with other soil stressors: a journey from agricultural sites into the environment. Chem Biol Technol Agric 9(1):1–20. https://doi.org/10.1186/s40538-021-00278-9

Ma J, Yu L, Chen S, Chen W, Wang Y, Guang S, Zhang X, Lu W, Wang Y, Bao J (2018) Structure–property evolution of poly (ethylene terephthalate) fibers in industrialized process under complex coupling of stress and temperature field. Macromolecules 52(2):565–574

Magalhães S, Alves L, Medronho B, Romano A, da GraçaRasteiro M (2020) Microplastics in ecosystems: from current trends to bio-based removal strategies. Molecules 25:3954

Mason SA, Welch VG, Neratko J (2018) Synthetic polymer contamination in bottled water. Front Chem 6:407. https://doi.org/10.3389/fchem.2018.00407

Mengistu D, Nilsen V, Coutris C, Amdal HM, Heistad A (2023) Microplastics concentrations in soil along a racetrack. Water Air Soil Pollut 234(1):9

Miles C, DeVetter L, Ghimire S, Hayes DG (2017) Suitability of biodegradable plastic mulches for organic and sustainable agricultural production systems. HortScience 52(1):10–15

OECD (2009) Series on emissions documents. Emission scenario documents on coating industry (Paints, Lacquers and Varnishes)

Österlund H, Blecken G, Lange K, Marsalek J, Gopinath K, Viklander M (2023) Microplastics in urban catchments: review of sources, pathways, and entry into stormwater. Sci Total Environ 858:159781

Piehl S, Leibner A, Löder MG, Dris R, Bogner C, Laforsch C (2018) Identification and quantification of macro-and microplastics on an agricultural farmland. Sci Rep 8(1):17950. https://doi.org/10.1038/s41598-018-36172-y

Raynaud J (2014) Valuing plastics: the business case for measuring, managing and disclosing plastic use in the consumer goods industry. United Nations Environment Programme (UNEP). https://wedocs.unep.org/bitstream/handle/20.500.11822/25302/Valuing_Plastic_ES.pdf. Accessed 18 May 2023

Rosso B, Gregoris E, Litti L, Zorzi F, Fiorini M, Bravo B, Barbante C, Gambaro A, Corami F (2023) Identification and quantification of tire wear particles by employing different cross-validation techniques: FTIR-ATR Micro-FTIR, Pyr-GC/MS, and SEM. Environ Pollut 326:121511

Saleh Y (2016) Comparative life cycle assessment of beverages packages in Palestine. J Clean Prod 131:28–42

Schymanski D, Goldbeck C, Humpf HU, Fürst P (2018) Analysis of microplastics in water by micro-Raman spectroscopy: release of plastic particles from different packaging into mineral water. Water Res 129:154–162

Sevigné-Itoiz E, Gasol CM, Rieradevall J, Gabarrell X (2015) Contribution of plastic waste recovery to greenhouse gas (GHG) savings in Spain. Waste Manag 46:557–567

Shiu RF, Chen LY, Lee HJ, Gong GC, Lee C (2022) New insights into the role of marine plastic-gels in microplastic transfer from water to the atmosphere via bubble bursting. Water Res 222: 118856

Siddique R, Khatib J, Kaur I (2008) Use of recycled plastic in concrete: a review. Waste Manag 28(10):1835–1852

Siegle L (2014) Are loom bands the next environmental disaster? https://www.theguardian.com/environment/2014/aug/10/are-loom-bands-next-environmental-disaster. Accessed 19 June 2023

Sobhani Z, Lei Y, Tang Y, Wu L, Zhang X, Naidu R, Megharaj M, Fang C (2020) Microplastics generated when opening plastic packaging. Sci Rep 10(1):4841

Stundt P, Schulze P, Syversen F (2015) Sources of microplastics-pollution to the marine environment. Mepex for the Norwegian Environment Agency Report no M-321

Tamis JE, Koelmans AA, Dröge R (2021) Environmental risks of car tire microplastic particles and other road runoff pollutants. Microplast Nanoplast 1(1):1–10. https://doi.org/10.1186/s43591-021-00008-w

Timmers VR, Achten PA (2016) Non-exhaust PM emissions from electric vehicles. Atmos Environ 134:10–17

Verschoor A, De Poorter L, Roex E, Bellert B (2014) Quick scan and prioritization of microplastic sources and emissions. RIVM Lett Rep 156:1–41

Wang J, Liu X, Li Y, Powell T, Wang X, Wang G, Zhang P (2019) Microplastics as contaminants in the soil environment: a mini-review. Sci Total Environ 691:848–857. https://doi.org/10.1016/j.scitotenv.2019.07.209

Weithmann N, Möller JN, Löder MG, Piehl S, Laforsch C, Freitag R (2018) Organic fertilizer as a vehicle for the entry of microplastic into the environment. Sci Adv 4(4):eaap8060

Zair MMB, Jakarni FM, Muniandy R, Hassim S (2021) A brief review: application of recycled polyethylene terephthalate in asphalt pavement reinforcement. Sustainability 13(3):1303

Zangmeister CD, Radney JG, Benkstein KD, Kalanyan B (2022) Common single-use consumer plastic products release trillions of sub-100 nm nanoparticles per liter into water during normal use. Environ Sci Technol 56(9):5448–5455

Zhang M, Yin H, Tan J, Wang X, Yang Z, Hao L, Du T, Niu Z, Ge Y (2023) A comprehensive review of tyre wear particles: formation, measurements, properties, and influencing factors. Atmos Environ 19:11959

Zhang S, Li Y, Chen X, Jiang X, Li J, Yang L, Yin X, Zhang X (2022) Occurrence and distribution of microplastics in organic fertilizers in China. Sci Total Environ 844:157061

Zhou B, Wang J, Zhang H, Shi H, Fei Y, Huang S, Tong Y, Wen D, Luo Y, Barceló D (2020) Microplastics in agricultural soils on the coastal plain of Hangzhou Bay, east China: multiple sources other than plastic mulching film. J Hazard Mater 388:121814. https://doi.org/10.1016/j.jhazmat.2019.121814

Ziajahromi S, Neale PA, Rintoul L, Leusch FD (2017) Wastewater treatment plants as a pathway for microplastics: development of a new approach to sample wastewater-based microplastics. Water Res 112:93–99

Zolotova N, Kosyreva A, Dzhalilova D, Fokichev N, Makarova O (2022) Harmful effects of the microplastic pollution on animal health: a literature review. PeerJ 10:e13503. https://doi.org/10.7717/peerj.13503

Part III

Reports on Collection, Identification, Estimation and Removal of Microplastics from the Contaminated Sites

Reports on Collection, Identification and Estimation of Microplastic in Freshwater Bodies

10

Thabo Falayi, Harmony Musiyarira, Thandiwe Sithole, and Mufaro Moyo

Abstract

Microplastics (MP) are plastic materials whose particle size is less than 5 μm. MP are pervasive and are an emerging threat to fauna and flora. MPs are now being found in fresh water sources and this has a threat on the wellbeing of human beings. To properly manage the pollution problem of MP, the collection, identification, and estimation of MPs in freshwater is important. The methods described should be robust with very low estimates of uncertainty. This chapter, therefore, takes a systematic review of the current and advances in the identification, collection and estimation of MPs in freshwater. The use of non-validated marine methods for freshwater analysis is problematic. The expensive nature of some of the MPs detection method render the monitoring of MPs in resource challenged countries a problem.

T. Falayi
Namibia University of Science and Technology, Department of Mining and Process Engineering, Windhoek, Namibia

University of Johannesburg, Chemical Engineering Technology Department, Johannesburg, South Africa

H. Musiyarira
Namibia University of Science and Technology, Department of Mining and Process Engineering, Windhoek, Namibia

T. Sithole
University of Johannesburg, Chemical Engineering Technology Department, Johannesburg, South Africa

M. Moyo (✉)
Malawi University of Science and Technology, Engineering Department, Limbe, Malawi

National University of Science and Technology, Department of Fibre and Polymer Materials Engineering, Bulawayo, Zimbabwe

M. Shahnawaz et al. (eds.), *Microplastic Pollution*,
https://doi.org/10.1007/978-981-99-8357-5_10

Keywords

Microplastics · Raman microscopy · Fourier transform infra-red spectroscopy · Pyrolysis-gas chromatography · Detection · Sampling techniques

10.1 Introduction

Microplastics (MPs) are plastic materials whose particle size ranges between 1 μm and 5 mm (Lu et al. 2021). MPs get into the aquatic ecosystem as large chunks which are then degraded through a variety of processes leading to small-sized particles of which the less than 5 mm are deemed MPs. These processes include photolytic, mechanical, and biological processes (Onoja et al. 2022). Due to their durability which results in high persistence in the environment, they have become a pollutant of concern more so that they are now being detected or reported in freshwater bodies. The problem of MPs is most likely to be increased in the future, given that the global estimates are that over 12 billion tons of plastic will be produced by the year 2050 with 80% of that production will not be recycled (Zhao et al. 2022). There are many routes by which MPs are found in freshwater systems. These include continued breakdown of improperly discarded plastics, washing of clothes, from personal hygiene products like toothpaste and facial scrubs, wastewater sludge and breaking of tyres during car motion (Koutnik et al. 2021).

The bulk of the research in MPs (54%) is from marine environments (Lu et al. 2021), however, fresh water bodies have become avenues through which MPs are transported to the marine environments (Yao et al. 2019). 11% of the plastic waste generated in the world has found the aquatic system as a sink (Onoja et al. 2022). MPs are known to affect aquatic life through interactions with their additives which include metals, polyaromatic hydrocarbons and polychlorinated bisphenols(Lu et al. 2021). The movement and fate of MPs in water bodies is regulated by hydrological processes like turbulent transport, settling, aggregation, biofouling, resuspension, and burial (Phuong et al. 2022). Aggregation and biofouling increases the chances of MPs settling in freshwater systems. (Phuong et al. 2022). The most commonly reported plastics are polyethylene (PE), polypropylene (PP), polystyrene (PS), polyvinylchloride (PVC), polyethylene terephthalate (PET), and polyester, polyamide, and acrylic (PP&A).

It has also been shown that the reporting of MPs in aquatic bodies is not uniform due to varying sample collection and assaying methods (Uurasjärvi et al. 2021). This chapter, therefore, focuses on the various sampling and assaying methods and their influences on reported data. The chapter also focuses on selected reports of MPs in freshwater. It is important to note that the reports are not an exhaustive list since this is a fluid area of research.

A systematic scientometric and manual analysis was carried out. The limitation of this method was that only the output from the Scopus site was used. The key words included MPs, MPs analysis, MPs identification, MPs in freshwater. The search period was from 2000 to 2022. There were over 20,000 articles that were returned the articles were manually eliminated through key words and abstract relevance. Finally, 95 relevant articles were used for this exercise.

10.2 Sampling and Sample Preparation Methods

There are basically two sampling methods for MPs in fresh water. These are bulk and volume reduced sampling (Lu et al. 2021). Bulk sampling involves collection of the whole sample whether liquid or solid with further sample handling happening at the laboratory. Volume reduction is when there is a sample size reduction at the source either by coning and quartering for solid samples or by filtration for liquid samples. Filtration at the source is time consuming hence becomes a laborious process. Volume reduction has been reported to be done using 300 mesh stainless steel sieve with storage of sieved samples in sterile glass bottles, however, the collection of samples was done using a plastic container (Li et al. 2022a). It also recommended to avoid use of plastic containers for sampling to avoid cross contamination. Other researchers do specify the steel sieve sizes to 48 μm whilst the sampling is done at 20–50 cm below the surface for water samples (Wang et al. 2022). Sediment samples are usually collected using shovels at 2–3 m sampling length (Wang et al. 2022). Bulk sampling is usually achieved through the use of trawling technique to collect a large sample over large areas (Bashir and Hashmi 2022). Therefore, the speed with which trawling is done and the duration of the trawling will therefore affect the integrity of the sample collected. This method is expensive in that there is a need to have a boat and there is possible contamination from the boat tow lines and nets. Various types of methods have been used for sampling sediment and these include Van veen/grab sampling, box coring, shovelling and metal scooping (Baldwin et al. 2020; Ramírez-Álvarez et al. 2020).

Sample handling usually includes H_2O_2 digestion of organic matter before further filtration of the sample using a Millipore filter membrane of size 0.45 μm although other researchers have used of KOH for the removal of organic matter from samples (Zarfl 2019). Peroxide treatment of samples have been shown to affect the integrity of some MPs, so care should be taken into account when reporting the results (Koelmans et al. 2019). The use of filters can present its own challenges as traditional filter papers may result in contamination of cellulose polymers and increases the risk of identifying microfibres from the filter paper as MPs (Bashir and Hashmi 2022). Temperatures at which the digestion and drying are done are also critical as they may affect the MPs integrity. One study recommended keeping the temperature below 50 °C (Koelmans et al. 2019). Enzymatic digestion has been reported to be less damaging to MPs although the method is time consuming (Razeghi et al. 2022). Protease, amylase, and lipase have been used in conjunction with hydrogen peroxide in some studies.

Density separation for using sodium chloride has also been reported to be a sample handling technique (Bashir and Hashmi 2022) whilst other researchers also use zinc chloride as an agent for density separation (Ajay et al. 2021). A number of other less common chemicals have been used as density separators for sediment

Table 10.1 Density separator efficiency

Density separator agent	Efficiency %	Reference
NaCl	90–100	Duong et al. (2022)
$ZnCl_2$	60–100	Duong et al. (2022)
NaI	50–98	Duong et al. (2022)
$CaCl_2$	30–100	Crichton et al. (2017), Duong et al. (2022)
Water	<70	
NaI	50–100	Crichton et al. (2017)
Canola oil	90–100	Crichton et al. (2017)

samples and these include lithium metatungstate, sodium polytungstate, potassium formate and NaI + NaCl mixture (Corcoran et al. 2015; Ding et al. 2017; Eo et al. 2018; Zhang et al. 2018). MPs have a low density making it easy to achieve their separation from the rest of the sample using density separation. Density separation usually follows 4 steps which are addition of a solution of density separation agent, shaking of the slurry followed by gravity separation. Elutriation can also be used to separate MPs from sediment. The separation takes advantage of the MPs terminal falling velocities (Kedzierski et al. 2016). Crichton et al. 2017 has reported the use of an oil elution method using canola oil for MPs separation (Crichton et al. 2017). NaCl density separation has been shown not to be capable of separating high density MPs like polyvinyl chloride, and polyethylene terephthalate (Zarfl 2019). The recoveries of different methods are given Table 10.1.

It can be seen from Table 10.1 that NaCl and Canola oil can achieve a removal efficiency of 90–100%. MPs recovery. A study by Monteiro et al. 2022 showed that centrifuging freshwater samples at 3500 rpm for 5 min preserved the integrity of MPs with recoveries as high as 93% (Monteiro et al. 2022). The study further showed that the use of 7% NaClO at 50 °C for 6 h results in 92% recovery rates of MPs in organic-rich freshwater samples. Most papers, however, use NaCl because of availability and low cost.

Vacuum filtration using a cellulose acetate filter followed by a sodium iodine solution washing can also been used for density separation (Bertoldi et al. 2021). Fe (II) can also be used as a catalyst in the digestion of organic matter. Other researchers also use nylon filters (Nan et al. 2020). Filtration using polymer-based filters may also introduce contamination, which can be avoided using blank samples and observing the filter afterwards. Grab sampling or random sampling can also be used (Ajay et al. 2021; Nan et al. 2020). Another filtration system suggested for storm water was the use of a cascade filtration setup (Herath et al. 2022).

10.3 MPs Identification and Quantification

The counting of MPS can be done using stereomicroscope. A digital camera can enhance the use of the stereomicroscope technique (Bertoldi et al. 2021) although black coloured MPs were shown to be problematic to count using this technique.

Visual inspection of the images is subjective and can result in wide variability of results. It has been reported that visual identification only is inadequate and hence the use of spectroscopic techniques (Wang et al. 2022). Visual inspection and sorting can be used with some degree of accuracy when the MPs size>300 μm (Koelmans et al. 2019). Visual inspection is prone to underreporting or overreporting of MPs. Visual methods can be improved by Nile red staining for the quantification of MPs (Zarfl 2019). Florescence staining has been reported as a fast and easy method for MP identification (Liu et al. 2022). This technique can be used both in the laboratory and in the field, however, dyed organic matter can interfere with the analysis. 1-pyrenebutyric acid N-hydroxysuccinimidyl ester (PBN) has been used to stain MPs, particularly for polyethylene (PE), poly(ethylene terephthalate) (PET), polyvinyl chloride (PVC), polyamide-6 (PA6), polypropylene (PP), polystyrene (PS), polycarbonate (PC), polyvinylidenechloride (PVDC), and polyurethane (PU) (Lee et al. 2021).

A dissecting microscope with image view microscope has been used to characterise and quantify MPs (Li et al. 2022a) whilst the type of polymer was investigated using Micro-Fourier transformed infrared spectroscopy. The identification of MPs is then done through established sample libraries using comparative analysis. Attenuated Total Reflection Fourier Transform Infrared (ATR-FTIR) and Raman spectroscopy can also be used for sample identification (Ajay et al. 2021).

The estimation of MPs can be achieved using scanning electron, fluorescence, compound and ultra violet microscopy techniques (Lu et al. 2021). Scanning electron microscopy (SEM) has been reported to give very clear pictures of MPS and when coupled to an energy dispersive spectroscopy (EDS) below 100 μm (Käppler et al. 2018). The FTIR technique can also be improved by the use of a silicon filter, this also allows for differentiation of MPs with similar chemical structure (Käppler et al. 2015). Pyrolysis-gas chromatography together with mass spectrometry (Py-GC-MS) can be used for the identification of the type of polymer e together with additives but suffers from sample destruction (Zarfl 2019). This works for the type of quantification where the number of particles is to be reported. When mass is the quantification unit, then any pyrolysis method can be used. The methods can be grouped according to precision and time required for sample preparation. The highest accuracy can be achieved using a Raman spectroscopy has the highest precision and whilst the downside is the sample preparation time, PY-GC-MS have a relatively high precision with the shortest preparation time (Zarfl 2019) in terms of microscopy. Although the EDS has a higher accuracy than electron microscopy it suffers from a long sample preparation time. A combination pressurized liquid extraction (PLE) and pyrolysis GC-MS has been used to improve detection limits of MPs (Dierkes et al. 2019). This method managed to achieve quantification limits as low as 0.007 mg/g for polyethylene (PE), polypropylene (PP), and polystyrene (PS). The PLE was achieved using tetrahydrofuran. Peez et al. 2019 investigated the use of ^{1}H-NMR spectroscopy for the quantification of MPs (Peez et al. 2019). The method achieved high recoveries of above 94% for freshwater samples for polyethylene terephthalate (PET) fibers.

Due to the laborious nature of sample preparation and analysis, automation has been developed. In a study by Dong et al. 2022, they compared three methods namely laser direct infrared (LDIR)–based particle analysis, Raman-based particle analysis, and focal plane array-Fourier transform infrared (FPA-FTIR) imaging (Dong et al. 2022). They study claims that LDIR-method had the shortest analysis time, but under reported particles less than 60 μm as compared to the Raman-based method. The Raman-based method suffered from fluorescence interference. FPA-FTIR was found to be accurate for samples that are bigger than 20 μm but MPs shapes could not be ascertained with certainty. In order to gain a full picture of MPs analysis, Li et al. (2022b) proposed a multiple technique for detection of MPs. The technique combined microscope-FTIR (μ-FTIR) with pyrolysis-GC/MS (Py-GC/MS). The basis of this method was to identify most of the characteristic tests for MPs (Li et al. 2022b). However, the expensive nature of the equipment involved would be a deterrent for continuous environmental MPs monitoring. Microfluidic approach has been proposed where MPs develop colour through the use of Nile red and dispersed in microchannel with a suitable dispersant. Detection is done using a fluorescence microscope. The matching rate for the technique was 0.91, 0.95, and 0.79 for PP, PE, and PS, respectively, (Zhang et al. 2022). Hyperspectral imaging (HSI) technology has been developed for the simple and rapid detection of MPs (Baruah et al. 2022). The technique can be improved by utilisation of Support Vector Machine (SVM) models.

A relatively new method using sensors has been developed for the detection of MPs (Baruah et al. 2022). These sensors can be biological or chemical and they include optical sensors capable of detecting translucent MPs.

10.4 Selected Reports on MPs in Freshwater

There is a wide variability in data regarding the presence of microplastic in fresh water which may be due to different locations (Table 10.2). The data also shows evidence of pollution in a variety of freshwater systems which then calls for increased monitoring to allow for better management practises. Fresh water systems close to densely populated areas are also associated with higher concentration of MPs indirectly indicating the role of humans in MPs pollution routes. The prohibitive costs of the instruments of MPs detection limit the ability of poor resourced countries to monitor MPs. This would explain why in a recent review article of MPs reporting in Africa; the report only had a handful of countries like South Africa, Tunisia, Nigeria, Ghana, and Ethiopia with South Africa dominating at 58% of the reports (Aragaw 2021).

Table 10.2 Selected reports on MPs in freshwater

Location	Type of sample	Sample treatment	MP identification	MP type	Abundance	Size range	Reference
Rawal Lake, Pakistan	Water	Wet sieving and peroxide digestion	Fourier transform infrared spectrometer	Polyethylene (PE) and polypropylene (PP)	6.4–8.8 particles/m^3	0.3–1 mm	Bashir and Hashmi (2022)
Lesser Himalayas	Water and sediment	Water-fibre glass filtration Sediment-wet peroxide oxidation and NaCl density separation	Attenuated Total reflection Fourier transform infrared (ATR-FTIR) and Raman spectroscopy	Polyethylene and polystyrene	Water-02–64 particles/L Sediment- 15–632 particles/kg dw	N/A	Ajay et al. (2021)
Saint-Lawrence, Canada	Water and bivalves	Filtration	Infrared spectroscopy	RCel: Regenerated cellulose; PET: Polyester; PS: Polystyrene; PE: Polyethylene; PVA: Polyvinyl alcohol; PP: Polypropylene.	0.0319 ± 0.0147 items/L and 0.0007 ± 0.0006 items/L	20–3200 μm	Rowenczyk et al. (2022)
Xijin Wetland Park, Nanning, South China	Water and sediment	Filtration and density separation	FTIR spectroscopy Stereomicroscope	Rayon, polyester, and polyethylene	Water 150 to 1067 n/m^3 Sediment 4 to 148 n/kg dry weight (dw)	<5 mm	Wang et al. (2022)
Northwest Himalaya, India	Sediment	Wet peroxide oxidation and NaCl density separation	Attenuated Total reflectance Fourier transform infrared (ATR-FTIR) spectroscopy	Polyamide, polyethylene terephthalate, polystyrene polyvinyl chloride and polypropylene	606 ± 360 n/kg of sample	<5 mm	Neelavannan et al. (2022)
	Water			N/A			

(continued)

Table 10.2 (continued)

Location	Type of sample	Sample treatment	MP identification	MP type	Abundance	Size range	Reference
Biobío river, located in Chile		Filtration and density separation	Micro-FTIR and infrared spectroscopy		22–46 particles/m^3	847.1 ± 134.3 (μm) and 1482.6 ± 494.2	Correa-Araneda et al. (2022)
Bangkok, Thailand	Water	Wet peroxide oxidation and NaCl density separation	μ-Raman spectroscopy	Polyethylene	0.40–2.40 particles/L	<300 μm	Chanpiwat and Damrongsiri (2021)
Barcelona metropolitan area	Water	Filtration	Stereomicroscope, y μ-FT-IR microscope and	Polyester (PES) and polypropylene (PP)	0.96 ± 0.46 microplastics/L 0.06 ± 0.04 microplastics/L	*0.3 to 4.5 mm*	Dalmau-Soler et al. (2021)
Rietvlei Dam, South Africa	Water	Filtration, wet peroxide oxidation and NaCl density separation	Pyrolysis gas chromatography–time of flight–mass spectrometry	PE, PS, PA, PVC, and PET	N/A	N/A	Vilakati et al. (2022)

10.5 Conclusion

The varied non-standardised analysis and reporting of MPs calls for standardisation of the analysis method. This also calls for the validation of the current methods for freshwater MPs which have been adapted from marine analytical methods. There is also a need for innovative detection methods or preliminary methods to allow resource challenged countries to be able to start monitoring freshwater MPs as some freshwater rivers are transboundary resources which would affect many countries at the same time.

References

Ajay K, Behera D, Bhattacharya S, Mishra PK, Ankit Y, Anoop A (2021) Distribution and characteristics of microplastics and phthalate esters from a freshwater lake system in Lesser Himalayas. Chemosphere 283:131132. https://doi.org/10.1016/j.chemosphere.2021.131132

Aragaw TA (2021) Microplastic pollution in African countries' water systems: a review on findings, applied methods, characteristics, impacts, and managements. SN Appl Sci 3(6):629. https://doi.org/10.1007/s42452-021-04619-z

Baldwin AK, Spanjer AR, Rosen MR, Thom T (2020) Microplastics in Lake Mead national recreation area, USA: occurrence and biological uptake. PLoS One 15(5):1–20. https://doi.org/10.1371/journal.pone.0228896

Baruah A, Sharma A, Sharma S, Nagraik R (2022) An insight into different microplastic detection methods. Int J Environ Sci Technol 19(6):5721–5730. https://doi.org/10.1007/s13762-021-03384-1

Bashir A, Hashmi I (2022) Detection in influx sources and estimation of microplastics abundance in surface waters of Rawal Lake, Pakistan. Heliyon 8(3):e09166. https://doi.org/10.1016/j.heliyon.2022.e09166

Bertoldi C, Lara LZ, Mizushima FA, de L. Mizushima FA, Martins FCG, Battisti MA, Hinrichs R, Fernandes AN (2021) First evidence of microplastic contamination in the freshwater of Lake Guaíba, Porto Alegre, Brazil. Sci Total Environ 759:143503. https://doi.org/10.1016/j.scitotenv.2020.143503

Chanpiwat P, Damrongsiri S (2021) Abundance and characteristics of microplastics in freshwater and treated tap water in Bangkok, Thailand. Environ Monit Assess 193(5):1–15. https://doi.org/10.1007/s10661-021-09012-2

Corcoran PL, Norris T, Ceccanese T et al (2015) Hidden plastics of Lake Ontario, Canada and their potential preservation in the sediment record. Environ Pollut 204:17–25

Correa-Araneda F, Pérez J, Tonin AM, Esse C, Boyero L, Díaz ME, Figueroa R, Santander-Massa R, Cornejo A, Link O, Jorquera E, Urbina MA (2022) Microplastic concentration, distribution and dynamics along one of the largest Mediterranean-climate rivers: a whole watershed approach. Environ Res 209:112808. https://doi.org/10.1016/j.envres.2022.112808

Crichton EM, Noël M, Gies EA, Ross PS (2017) A novel, density-independent and FTIR-compatible approach for the rapid extraction of microplastics from aquatic sediments. Anal Methods 9(9):1419–1428. https://doi.org/10.1039/c6ay02733d

Dalmau-Soler J, Ballesteros-Cano R, Boleda MR, Paraira M, Ferrer N, Lacorte S (2021) Microplastics from headwaters to tap water: occurrence and removal in a drinking water treatment plant in Barcelona Metropolitan area (Catalonia, NE Spain). Environ Sci Pollut Res 28(42):59462–59472. https://doi.org/10.1007/s11356-021-13220-1

Dierkes G, Lauschke T, Becher S, Schumacher H, Földi C, Ternes T (2019) Quantification of microplastics in environmental samples via pressurized liquid extraction and pyrolysis-gas

chromatography. Anal Bioanal Chem 411(26):6959–6968. https://doi.org/10.1007/s00216-019-02066-9
Ding J, Zhang S, Zou H et al (2017) Occurrence, source and ecotoxicological effect of microplastics in freshwater environment. Ecol Environ Sci 26(9):1619–1626
Dong M, She Z, Xiong X, Ouyang G, Luo Z (2022) Automated analysis of microplastics based on vibrational spectroscopy: are we measuring the same metrics? Anal Bioanal Chem 414:3359–3372. https://doi.org/10.1007/s00216-022-03951-6
Duong TT, Le PT, Nguyen TNH, Hoang TQ, Ngo HM, Doan TO, Le TPQ, Bui HT, Bui MH, Trinh VT, Nguyen TL, Da Le N, Vu TM, Tran TKC, Ho TC, Phuong NN, Strady E (2022) Selection of a density separation solution to study microplastics in tropical riverine sediment. Environ Monit Assess 194(2):65. https://doi.org/10.1007/s10661-021-09664-0
Eo S, Hong S, Song Y et al (2018) Abundance, composition, and distribution of microplastics larger than 20 mu m in sand beaches of South Korea. Environ Pollut 238:894–902
Herath S, Hagare D, Siddiqui Z, Maheshwari B (2022) Microplastics in urban stormwater—developing a methodology for its monitoring. Environ Monit Assess 194(3):173. https://doi.org/10.1007/s10661-022-09849-1
Käppler A, Fischer M, Scholz-Böttcher BM, Oberbeckamann S, Labrenz M, Fischer D, Eichhorn K-J, Voit B (2018) Käppler2018_Article_ComparisonOfM-ATR-FTIRSpectros.pdf. Anal Bioanal Chem:5313–5327
Käppler A, Windrich F, Löder MGJ, Malanin M, Fischer D, Labrenz M, Eichhorn KJ, Voit B (2015) Identification of microplastics by FTIR and Raman microscopy: a novel silicon filter substrate opens the important spectral range below 1300 cm^{-1} for FTIR transmission measurements. Anal Bioanal Chem 407(22):6791–6801. https://doi.org/10.1007/s00216-015-8850-8
Kedzierski M, Le Tilly V, Bourseau P, Bellegou H, César G, Sire O, Bruzaud S (2016) Microplastics elutriation from sandy sediments: a granulometric approach. Mar Pollut Bull 107(1):315–323. https://doi.org/10.1016/j.marpolbul.2016.03.041
Koelmans AA, Mohamed Nor NH, Hermsen E, Kooi M, Mintenig SM, De France J (2019) Microplastics in freshwaters and drinking water: critical review and assessment of data quality. Water Res 155:410–422. https://doi.org/10.1016/j.watres.2019.02.054
Koutnik VS, Leonard J, Alkidim S, DePrima FJ, Ravi S, Hoek EMV, Mohanty SK (2021) Distribution of microplastics in soil and freshwater environments: global analysis and framework for transport modeling. Environ Pollut 274:116552. https://doi.org/10.1016/j.envpol.2021.116552
Lee EH, Lee S, Chang Y, Lee SW (2021) Simple screening of microplastics in bottled waters and environmental freshwaters using a novel fluorophore. Chemosphere 285:131406. https://doi.org/10.1016/j.chemosphere.2021.131406
Li H, Wu Q, Ng J, Yu D, Harn S, Angela C (2022b) Identification and quantification of common microplastics in table salts by a multi - technique - based analytical method. Anal Bioanal Chem 414(22):6647. https://doi.org/10.1007/s00216-022-04226-w
Li W, Chen X, Cai Z, Li M, Liu Z, Gong H, Yan M (2022a) Characteristics of microplastic pollution and analysis of colonized-microbiota in a freshwater aquaculture system. Environ Pollut 306:119385. https://doi.org/10.1016/j.envpol.2022.119385
Liu S, Shang E, Liu J, Wang Y, Bolan N, Kirkham MB, Li Y (2022) What have we known so far for fluorescence staining and quantification of microplastics: a tutorial review. Front Environ Sci Eng 16(1):1–14. https://doi.org/10.1007/s11783-021-1442-2
Lu HC, Ziajahromi S, Neale PA, Leusch FDL (2021) A systematic review of freshwater microplastics in water and sediments: recommendations for harmonisation to enhance future study comparisons. Sci Total Environ 781:146693. https://doi.org/10.1016/j.scitotenv.2021.146693
Monteiro SS, Rocha-Santos T, Prata JC, Duarte AC, Girão AV, Lopes P, Cristovão T, da Costa JP (2022) A straightforward method for microplastic extraction from organic-rich freshwater samples. Sci Total Environ 815:152941. https://doi.org/10.1016/j.scitotenv.2022.152941

Nan B, Su L, Kellar C, Craig NJ, Keough MJ, Pettigrove V (2020) Identification of microplastics in surface water and Australian freshwater shrimp Paratya australiensis in Victoria, Australia. Environ Pollut 259:113865. https://doi.org/10.1016/j.envpol.2019.113865

Neelavannan K, Sen IS, Lone AM, Gopinath K (2022) Microplastics in the high-altitude Himalayas: assessment of microplastic contamination in freshwater lake sediments, Northwest Himalaya (India). Chemosphere 290:133354. https://doi.org/10.1016/j.chemosphere.2021.133354

Onoja S, Nel HA, Abdallah MAE, Harrad S (2022) Microplastics in freshwater sediments: analytical methods, temporal trends, and risk of associated organophosphate esters as exemplar plastics additives. Environ Res 203:111830. https://doi.org/10.1016/j.envres.2021.111830

Peez N, Becker J, Ehlers SM, Fritz M, Fischer CB, Koop JHE, Winkelmann C, Imhof W (2019) Correction to: Quantitative analysis of PET microplastics in environmental model samples using quantitative 1H-NMR spectroscopy: validation of an optimized and consistent sample clean-up method. Anal Bioanal Chem 411(28):7603. https://doi.org/10.1007/s00216-019-02160-y

Phuong NN, Duong TT, Le TPQ, Hoang TK, Ngo HM, Phuong NA, Pham QT, Doan TO, Ho TC, Da Le N, Nguyen TAH, Strady E, Fauvelle V, Ourgaud M, Schmidt N, Sempere R (2022) Microplastics in Asian freshwater ecosystems: current knowledge and perspectives. Sci Total Environ 808:151989. https://doi.org/10.1016/j.scitotenv.2021.151989

Ramírez-Álvarez N, Rios Mendoza LM, Macías-Zamora JV, Oregel-Vázquez L, Alvarez-Aguilar-A, Hernández-Guzmán FA, Sánchez-Osorio JL, Moore CJ, Silva-Jiménez H, Navarro-Olache LF (2020) Microplastics: sources and distribution in surface waters and sediments of Todos Santos Bay, Mexico. Sci Total Environ 703:134838. https://doi.org/10.1016/j.scitotenv.2019.134838

Razeghi N, Hamidian AH, Mirzajani A, Abbasi S, Wu C, Zhang Y, Yang M (2022) Sample preparation methods for the analysis of microplastics in freshwater ecosystems: a review. Environ Chem Lett 20(1):417–443. https://doi.org/10.1007/s10311-021-01341-5

Rowenczyk L, Cai H, Nguyen B, Sirois M, Côté-Laurin MC, Toupoint N, Ismail A, Tufenkji N (2022) From freshwaters to bivalves: microplastic distribution along the Saint-Lawrence river-to-sea continuum. J Hazard Mater 435:128977. https://doi.org/10.1016/j.jhazmat.2022.128977

Uurasjärvi E, Sainio E, Setälä O, Lehtiniemi M, Koistinen A (2021) Validation of an imaging FTIR spectroscopic method for analyzing microplastics ingestion by Finnish lake fish (Perca fluviatilis and Coregonus albula). Environ Pollut 288:117780. https://doi.org/10.1016/j.envpol.2021.117780

Vilakati B, Venkataraman S, Nyoni H, Mamba BB, Omine K, Msagati TAM (2022) Qualitative characterisation and identification of microplastics in a freshwater dam at Gauteng Province, South Africa, using pyrolysis–gas chromatography–time of flight–mass spectrometry (Py–GC–ToF–MS). Environ Sci Pollut Res 29(55):83452. https://doi.org/10.1007/s11356-022-21510-5

Wang Q, Huang K, Li Y, Zhang Y, Yan L, Xu K, Huang S, Junaid M, Wang J (2022) Microplastics abundance, distribution, and composition in freshwater and sediments from the largest Xijin Wetland Park, Nanning, South China. Gondwana Res 108:13–21. https://doi.org/10.1016/j.gr.2021.07.009

Yao P, Zhou B, Lu YH, Yin Y, Zong YQ, Chen MT, O'Donnell Z (2019) A review of microplastics in sediments: spatial and temporal occurrences, biological effects, and analytic methods. Quat Int 519:274–281

Zarfl C (2019) Promising techniques and open challenges for microplastic identification and quantification in environmental matrices. Anal Bioanal Chem 411:3743–3756. https://doi.org/10.1007/s00216-019-01763-9

Zhang K, Shi H, Peng J et al (2018) Microplastic pollution in China's inland water sys_tems: a review of findings, methods, characteristics, effects, and management. Sci Total Environ 630:1641–1653

Zhang Y, Zhang M, Fan Y (2022) Assessment of microplastics using microfluidic approach. Environ Geochem Health 45:1045. https://doi.org/10.1007/s10653-022-01262-4

Zhao M, Cao Y, Chen T, Li H, Tong Y, Fan W, Xie Y, Tao Y, Zhou J (2022) Characteristics and source-pathway of microplastics in freshwater system of China: a review. Chemosphere 297: 134192. https://doi.org/10.1016/j.chemosphere.2022.134192

Collection, Detection, Identification, and Estimation of Microplastic from the Marine Environment: An Overview

11

Manisha K. Sangale, Rahul Kapoore, and Anthony Gravell

Abstract

Microplastic (≤5 mm) is one of the potential pollutants in the marine environment. It leads to various health issues in marine biota. The microplastic pollution is increasing exponentially in all the zones of the biosphere. Microplastic is reported in seawater and sea sediments. Microplastic is reported to enter food chain, and consumption of the seafood is becoming risky day by day. Microplastic has not only spread throughout the aquatic environment, but also reported in sea ice. In the present chapter, an attempt was made to overview the methods currently used for collecting, detecting, identifying, and estimating microplastics from the marine environment under the following lines: (1) to discuss the techniques used to collect, detect, identify, and estimate the microplastic, (2) to enlist the methods used to detect the microplastic pollution in the marine environment, (3) to highlight techniques used to identify and estimate different types of the microplastic, and (4) to discuss some of the important reports of microplastic pollution in the marine environment.

Keywords

Microplastic · Deep sea · Seawater · Sediment · Detection · SEM · FTIR

M. K. Sangale (✉) · A. Gravell
Natural Resources Wales, Swansea, UK
e-mail: manisha.sangale@cyfoethnaturiolcymru.gov.uk

R. Kapoore
Natural Resources Wales, Swansea, UK

Beacon Development, King Abdullah University of Science and Technology (KAUST), Thuwal, Saudi Arabia

M. Shahnawaz et al. (eds.), *Microplastic Pollution*,
https://doi.org/10.1007/978-981-99-8357-5_11

11.1 Introduction

Plastic particles with particle size below 5 mm are referred to as microplastic and are reported as an emerging environmental pollutant and poses serious threats to the marine biota and humans (Eerkes-Medrano et al. 2015; Akdogan and Guven 2019). They are categorized into two main groups based on their mode of production, viz. primary and secondary microplastics. The first type of microplastics is produced intentionally like plastic beads and has various applications in pharmaceutical, toiletry, and cosmetics industries, whereas second type of microplastics gets generated due to fragmentation of large plastic with various environmental forces, viz. degradation by sunlight, abrasion, and finally fragmentation of large plastic into microplastic particles (Munari et al. 2017; Osman et al. 2023). In the maritime environment, microplastics are pervasive and persist for long periods of time, harming both marine biota and seafood consumers including human (Alimba and Faggio 2019; Elizalde-Velázquez and Gómez-Oliván 2021).

The health and function of the marine ecosystem and marine creatures are threatened by the presence of microplastics in the environment (Al-Thawadi 2020). Microplastics have been discovered in a variety of marine creatures recently, including fish, crabs, and mollusks. This has led to ingestion, entanglement, and habitat modification in these organisms (Franzellitti et al. 2019; Anastasopoulou and Fortibuoni 2019; Freitas et al. 2023). In the marine environment, microplastics also absorb and gather contaminants, potentially worsening their effects on the ecosystem and human health (Huang et al. 2021).

Collecting, detecting, identifying, and estimating microplastics from the marine environment are crucial to mitigate their impact and understand their distribution and abundance (Eerkes-Medrano et al. 2015; Chen et al. 2020a, b). A variety of techniques have been developed for these tasks, with advantages and limitations that must be considered (Wang and Wang 2018; Caputo et al. 2021; Guo et al. 2022). Understanding the challenges and limitations of these techniques is crucial to produce reliable and accurate data, which can inform appropriate mitigation measures (Fu et al. 2020; Malafaia and Barceló 2023).

So, the main objective of the present chapter was to enlist the methods/techniques currently used for collecting samples from seawater and sediments to isolate, detect, and quantify different types of microplastic followed by discussing some important case studies of microplastic collection, detection, identification, and quantification in different oceans and seas at global level.

11.2 Sampling/Collection in the Marine Environment for Isolation of Microplastic

Understanding the prevalence and effects of plastic contamination on aquatic ecosystems requires the collecting of microplastics from the marine environment (Su et al. 2018). Microplastics are collected using a variety of techniques, each with their own benefits and drawbacks (Wang and Wang 2018; Li et al. 2018; Prata et al.

2019). Surface trawls, which drag a net along the water's surface to catch floating trash, are considered one of the commonly employed techniques for gathering microplastics (Lindeque et al. 2020; Ryan et al. 2020). Another method is sediment sampling, which involves collecting sediment from the ocean floor and analyzing it for microplastics (Gomiero et al. 2019; Barrett et al. 2020). Neuston nets and manta trawls can also be used to collect samples of the water column for isolation of microplastics (Eriksen et al. 2018). New techniques for collecting microplastic have been developed as a result of recent technological breakthroughs, such as using drones with cameras to find floating plastics (Martin et al. 2018; Faltynkova et al. 2021) or sonar technology to locate microplastic hotspots in the water column (Stolte et al. 2022). These methods have the advantage of being less invasive and less harmful to marine life than traditional sampling methods (De Taxis and Poët 2018; Mukonza and Chiang 2022). Once collected, microplastics can be analyzed using various techniques, such as visual inspection, Fourier transform infrared (FTIR) spectroscopy, and Raman spectroscopy, to identify the type and size of plastic particles present (Löder et al. 2015). This information is important for understanding the sources and distribution of microplastics in the marine environment and for assessing the potential impacts on marine organisms (Horton et al. 2017a, b; Akinyede, 2023).

11.2.1 Sampling Strategies for Collecting Microplastics

Depending on the kind of matrix being gathered, several sampling techniques can be used to detect microplastics; e.g., various nets and filtering systems can be used to gather samples of water (Campanale et al. 2020). Some common sampling strategies include surface trawls, manta trawls, Neuston nets, and plankton nets. Surface trawls are typically used to collect larger debris and microplastics floating on the surface of the water, while manta trawls are used to collect smaller microplastics near the surface (Eriksen et al. 2018). Plankton nets are used to catch microplastics and other particles in the water column, whereas debris and microplastics at the air-water interface are collected with Neuston nets (Kooi et al. 2016; Fox et al. 2022).

In sediments, microplastic samples can be collected using corers, grabs, and dredges (Pagter et al. 2018). Corers are typically used to collect surface sediments, while grabs and dredges are used to collect deeper sediment layers (Näkki et al. 2017; Kutralam-Muniasamy et al. 2021a, b). The use of sediment traps is another method for collecting microplastics from the water column. Sediment traps are deployed vertically in the water column and collect particles settling out of the water over time (Galgani et al. 2022).

Biota sampling can also be used to collect microplastics. Biota samples can be collected by dissecting marine organisms or by analyzing fecal material. These techniques make it possible to detect microplastics in digestive systems of marine animals (Lusher et al. 2017; Karlsson et al. 2017; Maxwell et al. 2020).

11.2.2 Strategies Used to Isolate Microplastic from the Collected Samples

After sampling, microplastics must be separated from the matrix for further analysis. The most common methods for separating microplastics from marine matrices are discussed as follows.

11.2.2.1 Filtration

Filtration is a widely used method for separating microplastics from water samples (Cutroneo et al. 2020). Water samples are passed through a filter, and the filter is then rinsed with a solvent to remove any organic material. Microplastics are left behind on the filter, which can then be analyzed (Möller et al. 2022). One of the disadvantages of filtration is that it can be time-consuming and may not capture all the available microplastics particles in the collected samples (Lv et al. 2021).

11.2.2.2 Density Separation

Microplastics can also be removed from marine matrices using a density-based technique (He et al. 2021). A high-density solution, like sodium chloride, is added to the sample using this technique. Microplastics will float to the surface, while other particles will sink. The microplastics can then be skimmed off the surface and analyzed. Microplastics can be removed from sediments and biota via density separation, which is an efficient technique (Hidalgo-Ruz et al. 2012; Campanale et al. 2020).

11.2.2.3 Visual Sorting

Visual sorting entails manually removing microplastics from marine matrices (Enders et al. 2020). When working with minute microplastics or complicated matrices like sediments, this process takes a lot of time and can be challenging (Mai et al. 2018). However, visual sorting is still used in certain circumstances, such as when analyzing microplastics in biota samples (Löder and Gerdts 2015).

11.2.3 Factors Affecting Collection Efficiency

Effective microplastic collection from the marine environment necessitates careful consideration of a range of variables that may influence the sampling procedure (Malankowska et al. 2021). Some of the factors that can impact the collection efficiency are as follows.

11.2.3.1 Sampling Location

The kind and amount of microplastics found in the samples that were collected are greatly influenced by the sampling site (Yonkos et al. 2014; Natesan et al. 2021). In accordance with the kind of microplastics being investigated, it is crucial to choose suitable sampling sites, the type of marine environment, and the prevailing environmental conditions (Miller et al. 2021).

11.2.3.2 Water Depth

The effectiveness of collecting microplastics can also be impacted by the water column's depth. Due to variations in how microplastics are distributed throughout the water column, sampling at various depths of water may result in varied quantities of the particles (Lenaker et al. 2019; Lofty et al. 2023).

11.2.3.3 Water Flow

Microplastic dispersion and concentration in the water column can be impacted by the speed and direction of water flow (Dai et al. 2018; Lenaker et al. 2019). Sampling in areas with high water flow can increase the chances of collecting a greater number of microplastics (Lusher et al. 2015).

11.2.3.4 Sampling Device

The effectiveness of collecting microplastics can also be impacted by the type of sampling tool (Shi et al. 2023). Plankton nets, manta trawls, and sediment grabs are a few tools that can be used to gather microplastics from various marine matrices (Razeghi et al. 2021). The study goals and the kind of microplastics being researched should be taken into consideration when choosing a sampling instrument (Al-Azzawi et al. 2020).

11.2.3.5 Sampling Duration

The amount of time spent in sampling can also impact how effectively microplastics are collected (Barrows et al. 2017). However, the length of sampling should be weighed against practical factors including the availability of sampling resources and the time needed for sample processing and analysis (Thomas et al. 2020). Longer sampling periods may reveal higher levels of microplastics.

11.2.3.6 Particle Size

The collecting efficiency of the microplastics under study may also be impacted by their size (Alimi et al. 2018). In comparison to bigger microplastics, smaller microplastics might need different sampling tools and methods (Lv et al. 2021). Based on the particle size distribution of the microplastics under study, suitable sample tools and techniques should be chosen.

11.2.3.7 Interference from Other Particles

The accumulation of microplastics can be hampered by other marine environment particles such as silt, plankton, and organic materials (Galloway et al. 2017). To ensure high collection efficiency, techniques for isolating microplastics from other particles should be carefully chosen (He et al. 2021). The variables influencing the effectiveness of collecting microplastics from the marine environment have been assessed by several researches. For instance, Horton et al.'s (2017a, b) study looked into how water movement affected the amount and distribution of microplastics in the water column. According to the study, locations with higher water flow rates had higher microplastic concentrations. The effectiveness of several sample tools for removing microplastics from the water's surface was assessed in a different study by

Lusher et al. (2013). According to the study, Neuston nets were the best sample tool for removing microplastics from the water's surface. Studies have also looked at how sampling duration affects how effectively microplastics are collected. For instance, Koelmans et al.'s (2017) study looked at the ideal sample period for removing microplastics from the water column. According to the study, microplastic concentrations in the samples that were obtained increased with longer sampling intervals.

11.3 Detection of Microplastics from the Marine Environment

Due to their small size and varying shapes, colors, and densities, microplastics are difficult to detect in the marine environment (Hidalgo-Ruz et al. 2012; Vivekanand et al. 2021). Visual identification, microscopy, and spectroscopy are only a few of the approaches that have been developed (Wirnkor et al. 2019; Chen et al. 2020a, b). Visual identification is not accurate enough to identify small or translucent particles when it comes to microplastics (Lenz et al. 2015; Shim et al. 2017). On the other hand, microscopy offers excellent resolution and magnification, making it possible to identify tiny, translucent particles (von der Esch et al. 2020). Microplastics can be confused with other microparticles found in the marine environment, and microscopy has limits in detecting particles smaller than 20 μm (Ahmed et al. 2021). The identification of microplastics has recently benefited from the development of spectroscopic methods including Raman spectroscopy, FTIR, and X-ray fluorescence (XRF) (Michel et al. 2020; Yadav et al. 2023). In order to identify even tiny particles that would have evaded detection using other techniques, spectroscopy offers a fingerprint of the chemical makeup of the microplastic particle (Lenz et al. 2015), but spectroscopy calls for specialized tools and knowledge, which can be costly and time-consuming (Zhu and Wang 2020).

11.4 Identification of Microplastics in Collected Samples

In order to comprehend the magnitude of plastic pollution in the oceans, it is crucial to identify microplastics from the marine environment (Lusher and Primpke 2023). Based on their physical, chemical, and biological characteristics, numerous approaches have been created to identify microplastics (Xu et al. 2019). The methods used to identify microplastics are varied and include everything from visual inspection to cutting-edge spectroscopic methods (Zarfl 2019; Wirnkor et al. 2019; Ye et al. 2022; Kutralam-Muniasamy et al. 2023).

11.4.1 Techniques for Identifying Microplastics

11.4.1.1 Visual Identification

Visual detection of microplastics is the most straightforward and widely utilized method (Lv et al. 2021). To find and count the microplastics, the sample must be visually examined under a microscope. The method depends on the microplastics' morphological properties, including their size, shape, and color. This method is rapid and economical, but it is unable to distinguish between microplastics and other particulate matter, such as organic matter, silt, and natural particles, which are smaller than 20 μm (Mariano et al. 2021).

11.4.1.2 Spectroscopic Techniques

Microplastics in the marine environment are typically identified using spectroscopic methods like Fourier transform infrared (FTIR) spectroscopy and Raman spectroscopy (Cabernard et al. 2018). In order to measure the absorbance spectrum, FTIR spectroscopy exposes the sample to infrared radiation. Through the examination of the sample's molecular vibrations, the approach can be used to determine the chemical makeup of microplastics (Käppler et al. 2016). The sample is excited by laser light in Raman spectroscopy, in contrast, and the light that is scattered from the sample is examined to determine the molecular vibrations. These methods are helpful in that they can give specific details about the structure and chemical makeup of the microplastics. Faltynkova et al. (2021) noted that they are expensive and operate with specialized technical knowledge.

11.4.1.3 Microscopy

Microscopy techniques, such as transmission electron microscopy (TEM) and scanning electron microscopy (SEM), are sophisticated methods for locating microplastics in the marine environment (Chou et al. 2022). SEM allows extensive morphological study through the provision of high-resolution pictures of the microplastics' surface. TEM, on the other hand, provides images of the internal structure of the microplastics (Girão 2020). These techniques are advantageous in providing detailed information on the morphology and structure of the microplastics. However, they require costly equipment and specialized technical expertise to operate (Kalaronis et al. 2022).

11.5 Estimation of Microplastics Isolated from the Collected Samples

To comprehend their distribution, abundance, and potential effects on the ecosystem, it is essential to estimate the amount of microplastics in the marine environment. Chemical analysis, microscopy, and spectroscopy are a few of the methods that can be used to quantify microplastics.

11.5.1 Quantification Techniques for Estimating Microplastics

11.5.1.1 Chemical Analysis

Chemical analysis involves the digestion of the organic material in seawater or sediment, leaving only the inorganic particles, including microplastics, which can then be quantified (Lusher et al. 2017). The most commonly used chemical analysis method is Fourier transform infrared spectroscopy (FTIR) (Chen et al. 2020a, b; Veerasingam et al. 2021 In order to determine the type of plastic, FTIR identifies the particular infrared spectrum of the microplastic's polymer material and compares it to a database of other polymer spectra. This technique can recognize a variety of microplastic sizes and forms and is comparatively quick. However, it is limited in its ability to detect microscopic microplastics, and its analysis may be hampered by the presence of other inorganic particles, leading to false positives or negatives (Woo et al. 2021).

11.5.1.2 Microscopy

Microscopy techniques include light microscopy, scanning electron microscopy (SEM), and transmission electron microscopy (TEM). Light microscopy is commonly used for analyzing microplastics larger than 20 μm, while SEM and TEM can detect smaller particles (Jakubowicz et al. 2021). The capacity to analyze the size, shape, and color of microplastics is microscopy's main benefit (Huppertsberg and Knepper 2018). However, it can be time-consuming, and the technique requires skilled operators. In addition, the presence of organic materials can make it challenging to distinguish between microplastics and other particles (Shim et al. 2017).

11.5.1.3 Spectroscopy

Spectroscopy techniques include Raman spectroscopy and X-ray fluorescence (XRF) spectroscopy. Raman spectroscopy is a nondestructive method that can identify the chemical composition of microplastics based on the spectrum of scattered light (Dąbrowska 2021). It is a fast and efficient method and can identify small microplastics. XRF spectroscopy uses X-rays to excite the elements present in the sample, and the emitted radiation is analyzed to determine the elements' identity and concentration (Kachhawa et al. 2022). The technique is nondestructive and can identify heavy metals present in microplastics. However, it cannot identify the type of polymer present in the sample (Kutralam-Muniasamy et al. 2021a, b).

11.6 Reports of Microplastic Pollution from the Marine Environment

Various people have reported microplastic in different regions of the oceans, including Northwestern Mediterranean Sea, South Pacific region, Arctic Sea, South China Sea, and Northern South China Sea (Table 11.1). Microplastic was reported in both seawater and sea sediments. Various types of microplastics were found abundantly in the marine environment viz. Beached microplastic (fiber), cellulose acetate,

Table 11.1 Reports of microplastic pollution in different regions of the marine environment at global level

Sr. No.	Source of microplastic	Sampling sites	Types of microplastic	Collection/ sampling method	Microplastic detection and identification method	Overall conclusion	Reference
1	Two beaches (northern and southern site) of the western Gulf of Lion, Northwestern Mediterranean Sea	Sediments	Beached microplastic (fiber)	Sampling was performed along the three deposit lines named from here onward "lower," "mid," and "upper" at the crossing of four cross-beach transects termed "north," "mid-north," "mid-south," and "south	FTIR spectroscopy	They have reported varied concentration of microplastic particles in both the beaches (from 33 to 798 and from 12 to 187 microplastics per kg of dry sediment)	Constant et al. (2019)
2	Vanuatu (Vila Bay and Mele Bay) and Solomon Islands, South Pacific region	Surface, sediments, and biota	Mesoplastic and microplastic (polyethylene polystyrene and polypropylene)	Surface samples were trawled with a manta trawl. Sediment samples were collected with a small Van Veen grab. For biota-based samples, microplastic was extracted from the stomach of the collected	Nile red screening method and attenuated total reflection Fourier transform infrared spectroscopy (ATR-FT-IR)	Widespread occurrence of microplastics was reported in all surface water and sediment samples for both Vanuatu and the Solomon Islands	Bakir et al. (2020)

(continued)

Table 11.1 (continued)

Sr. No.	Source of microplastic	Sampling sites	Types of microplastic	Collection/ sampling method	Microplastic detection and identification method	Overall conclusion	Reference
				fishes from the Vanuatu and some edible fishes from the market			
3	Polar front: Southern Ocean, Porcupine Abyssal Plain: North Atlantic Ocean, distal lobe of Congo: Canyon Gulf of Guinea, Nile deep-sea fan: Mediterranean Sea	Eleven deep-sea sediment samples collected from several locations ranging in depth from 1176 to 4844 m	Low-density polyethylene and polypropylene	A surface area of 25 cm^2 was sampled for sediments at each place where they were collected. After recovery, the cores were immediately extruded into rings, and the sediment was cut into horizontal slices using a metal plate. The cores were cut into slices that were 1 cm thick. Only the top centimeter of the sediment cores	Micro-Raman spectroscopy	They reported the presence of microscopic particles in all regions of marine environment including the deep-sea region	Van Cauwenberghe et al. (2013)

				were analyzed for the presence of microplastics. Microplastic was extracted using the flotation process			
4	Sea food (crabs and fish)	Muscles of crabs and fish	Fifteen types of microplastic (cellulose acetate; expanded polystyrene; high-density polyethylene; low-density polyethylene; polyamide; polylauryllactam; polycaprolactam; polycarbonate; polyethylene; polyethylene terephthalate; poly (methyl-methacrylate); polypropylene; polystyrene; polytetrafluoroethylene; polyurethane; crosslinked polystyrene; and unplasticized polyvinyl chloride)	Fish, crab, and mussel tissues were chemically degraded before being filtered via glass fiber filters, but the microplastic was not broken down	Raman analysis and Pyr-GC/MS analysis	After testing various protocols, the method employing KOH 10% solution to digest the biological tissues at 60 °C for 24 h was reported as the most efficient method to extract microplastic	Dehaut et al. (2016)
5	Marine sediments,	Sediments	Polypropylene, polypropylene-	Grab sampling was used to		Overall, the South China Sea's	Chen et al. (2023)

(continued)

Table 11.1 (continued)

Sr. No.	Source of microplastic	Sampling sites	Types of microplastic	Collection/ sampling method	Microplastic detection and identification method	Overall conclusion	Reference
	Eastern Guangdong, South China Sea		polyethylene copolymer, polyethylene terephthalate, and polyethylene	collect the top 10 cm sediments from 26 different location during three seasons (spring, summer, and winter) of the year 2021	Dissecting stereomicroscope (Mshot MZ62)	Eastern Guangdong multi-use zones have microplastic pollution levels that are equivalent to or even somewhat lower than many other marine sediments from heavily populated areas	
6	Arctic Ocean	Sea ice and seawater beneath ice floes	Nine different types of microplastic were reported from sea ice, and three types were reported from seawater beneath the ice floes. In both the locations, the majority of microplastics were comprised of fibrous polyesters (57% sea ice cores and 70% surface waters) and polyamides (19% sea ice cores and 23% surface waters)	Narcotic ice drill was used to collect ice cores. A hole was also made in the ice to collect water beneath	Fourier transform infrared (FT-IR) spectroscopy	Arctic sea ice was reported as an intermediate sink for microplastics	Kanhai et al. (2020)

7	Haikou Bay, Northern South China Sea	Seawater and sediment	The predominant microplastic in seawater samples and sediment samples was reported as fiber and rayon PE polyolefin	During sunny days, boat was used for collecting samples from seawater and sediment. The method Ryan et al. (2009) was used to determine the abundance of the microplastic accurately	ImageJ, SEM, and ATR-FTIR	They have reported microplastic in both coastal seawater (most abundant) and sediment	Qi et al. (2020)

crosslinked polystyrene, expanded polystyrene, high-density polyethylene, low-density polyethylene, poly(methyl-methacrylate), polyamide, polycaprolactam, polycarbonate, polyethylene, polyethylene polystyrene, polyethylene terephthalate, polycaprolactam, polypropylene, polypropylene-polyethylene copolymer, polystyrene, polytetrafluoroethylene, polyurethane, and unplasticized polyvinyl chloride (Van Cauwenberghe et al. 2013; Dehaut et al. 2016; Constant et al. 2019; Bakir et al. 2020; Kanhai et al. 2020; Qi et al. 2020; Chen et al. 2023). Attenuated total reflection Fourier transform infrared spectroscopy (ATR-FTIR), micro-Raman spectroscopy, Nile red screening method, Pyr-GC/MS analysis, ImageJ, scanning electron microscopy (SEM), and dissecting stereomicroscope are some of the common instruments and methods used to detect the abundance and characterization of the microplastic. The microplastic particles in the marine environment are increasing exponentially, and if the entry of the plastic from the terrestrial environment to marine environment is not stopped/minimized, it will prove more deadly to the marine biota and humans (Shahnawaz et al. 2022). It is also anticipated that by 2050, there will be more microplastic particles in the ocean than fish. (Jovanović 2017.

11.7 Conclusion

Measurements are becoming increasingly accurate and precise as the methods for locating and classifying microplastics continue to advance. To overcome the many difficulties related to microplastic detection, collection, identification, and quantification, more research and the creation of fresh procedures are required. The development of technologies such as machine learning and artificial intelligence presents exciting potential for more effective and precise microplastic detection, collection, and quantification.

References

Ahmed MB, Rahman MS, Alom J, Hasan MS, Johir MAH, Mondal MIH, Yoon MH (2021) Microplastic particles in the aquatic environment: a systematic review. Sci Total Environ 775: 145793

Akdogan Z, Guven B (2019) Microplastics in the environment: a critical review of current understanding and identification of future research needs. Environ Pollut 254:113011

Akinyede KA (2023) Microplastic sources, formation, toxicity and remediation: a review. Environ Chem Lett, 1–41

Al-Azzawi MS, Kefer S, Weißer J, Reichel J, Schwaller C, Glas K et al (2020) Validation of sample preparation methods for microplastic analysis in wastewater matrices—reproducibility and standardization. Water 12(9):2445

Alimba CG, Faggio C (2019) Microplastics in the marine environment: current trends in environmental pollution and mechanisms of toxicological profile. Environ Toxicol Pharmacol 68:61–74

Alimi OS, Farner Budarz J, Hernandez LM, Tufenkji N (2018) Microplastics and nanoplastics in aquatic environments: aggregation, deposition, and enhanced contaminant transport. Environ Sci Technol 52(4):1704–1724

Al-Thawadi S (2020) Microplastics and nanoplastics in aquatic environments: challenges and threats to aquatic organisms. Arab J Sci Eng 45(6):4419–4440

Anastasopoulou A, Fortibuoni T (2019) Impact of plastic pollution on marine life in the Mediterranean sea. In: Plastics in the aquatic environment—part I: current status and challenges. Springer International Publishing, Cham, pp 135–196

Bakir A, Desender M, Wilkinson T, Van Hoytema N, Amos R, Airahui S, Maes T (2020) Occurrence and abundance of meso and microplastics in sediment, surface waters, and marine biota from the South Pacific region. Mar Pollut Bull 160:111572

Barrett J, Chase Z, Zhang J, Holl MMB, Willis K, Williams A, . . ., Wilcox C (2020) Microplastic pollution in deep-sea sediments from the great Australian bight. Front Mar Sci, 808

Barrows AP, Neumann CA, Berger ML, Shaw SD (2017) Grab vs. neuston tow net: a microplastic sampling performance comparison and possible advances in the field. Anal Methods 9(9): 1446–1453

Cabernard L, Roscher L, Lorenz C, Gerdts G, Primpke S (2018) Comparison of Raman and Fourier transform infrared spectroscopy for the quantification of microplastics in the aquatic environment. Environ Sci Technol 52(22):13279–13288

Campanale C, Savino I, Pojar I, Massarelli C, Uricchio VF (2020) A practical overview of methodologies for sampling and analysis of microplastics in riverine environments. Sustainability 12(17):6755

Caputo F, Vogel R, Savage J, Vella G, Law A, Della Camera G, Calzolai L (2021) Measuring particle size distribution and mass concentration of nanoplastics and microplastics: addressing some analytical challenges in the sub-micron size range. J Colloid Interface Sci 588:401–417

Chen G, Fu Z, Yang H, Wang J (2020a) An overview of analytical methods for detecting microplastics in the atmosphere. TrAC Trends Anal Chem 130:115981

Chen Y, Wen D, Pei J, Fei Y, Ouyang D, Zhang H, Luo Y (2020b) Identification and quantification of microplastics using Fourier-transform infrared spectroscopy: current status and future prospects. Curr Opin Environ Sci Health 18:14–19

Chen X, Zhao P, Wang D, Wang L, Zhao H, Wang X et al (2023) Microplastics in marine sediments in eastern Guangdong in the South China Sea: factors influencing the seasonal and spatial variations. Water 15(6):1160

Chou SH, Chuang YK, Lee CM, Chang YS, Jhang YJ, Yeh CW et al (2022) Visualization and (semi-) quantification of submicrometer plastics through scanning electron microscopy and time-of-flight secondary ion mass spectrometry. Environ Pollut 300:118964

Constant M, Kerhervé P, Mino-Vercellio-Verollet M, Dumontier M, Vidal AS, Canals M, Heussner S (2019) Beached microplastics in the northwestern Mediterranean Sea. Mar Pollut Bull 142: 263–273

Cutroneo L, Reboa A, Besio G, Borgogno F, Canesi L, Canuto S, Capello M (2020) Microplastics in seawater: sampling strategies, laboratory methodologies, and identification techniques applied to port environment. Environ Sci Pollut Res 27:8938–8952

Dąbrowska A (2021) Raman spectroscopy of marine microplastics—A short comprehensive compendium for the environmental scientists. Mar Environ Res 168:105313

Dai Z, Zhang H, Zhou Q, Tian Y, Chen T, Tu C et al (2018) Occurrence of microplastics in the water column and sediment in an inland sea affected by intensive anthropogenic activities. Environ Pollut 242:1557–1565

De Taxis D, Poët F (2018). Marine litter: how to monitor, reduce and prevent ocean debris. Focus on plastics and microplastics. Vrie Universiteit, Amsterdam, 52pp. https://doi.org/10.25607/OBP-871

Dehaut A, Cassone AL, Frère L, Hermabessiere L, Himber C, Rinnert E, Paul-Pont I (2016) Microplastics in seafood: benchmark protocol for their extraction and characterization. Environ Pollut 215:223–233

Eerkes-Medrano D, Thompson RC, Aldridge DC (2015) Microplastics in freshwater systems: a review of the emerging threats, identification of knowledge gaps and prioritisation of research needs. Water Res 75:63–82

Elizalde-Velázquez GA, Gómez-Oliván LM (2021) Microplastics in aquatic environments: a review on occurrence, distribution, toxic effects, and implications for human health. Sci Total Environ 780:146551

Enders K, Lenz R, do Sul JAI, Tagg AS, Labrenz M (2020) When every particle matters: QuEChERS approach to extract microplastics from environmental samples. MethodsX 7: 100784

Eriksen M, Liboiron M, Kiessling T, Charron L, Alling A, Lebreton L, Thiel M (2018) Microplastic sampling with the AVANI trawl compared to two neuston trawls in the Bay of Bengal and South Pacific. Environ Pollut 232:430–439

Faltynkova A, Johnsen G, Wagner M (2021) Hyperspectral imaging as an emerging tool to analyze microplastics: a systematic review and recommendations for future development. Microplastics Nanoplastics 1(1):1–19

Fox JM, Schwoerer GD, Schreiner KM, Minor EC, Maurer-Jones MA (2022) Microplastics in the water column of Western lake superior. ACS ES&T Water 2(10):1659–1666

Franzellitti S, Canesi L, Auguste M, Wathsala RH, Fabbri E (2019) Microplastic exposure and effects in aquatic organisms: a physiological perspective. Environ Toxicol Pharmacol 68:37–51

Freitas TB, Leite TS, de Ramos B, Proietti MC (2023) Understanding the interactions between cephalopods and marine litter: a research evaluation with identification of gaps and future perspectives. Mar Pollut Bull 190:114814

Fu W, Min J, Jiang W, Li Y, Zhang W (2020) Separation, characterization and identification of microplastics and nanoplastics in the environment. Sci Total Environ 721:137561

Galgani L, Goßmann I, Scholz-Böttcher B, Jiang X, Liu Z, Scheidemann L, Engel A (2022) Hitchhiking into the deep: how microplastic particles are exported through the biological carbon pump in the North Atlantic Ocean. Environ Sci Technol 56(22):15638–15649

Galloway TS, Cole M, Lewis C (2017) Interactions of microplastic debris throughout the marine ecosystem. Nat Ecol Evol 1(5):0116

Girão AV (2020) SEM/EDS and optical microscopy analysis of microplastics. In: Handbook of microplastics in the environment, pp 1–22

Gomiero A, Øysæd KB, Agustsson T, van Hoytema N, van Thiel T, Grati F (2019) First record of characterization, concentration and distribution of microplastics in coastal sediments of an urban fjord in south west Norway using a thermal degradation method. Chemosphere 227:705–714

Guo X, Lin H, Xu S, He L (2022) Recent advances in spectroscopic techniques for the analysis of microplastics in food. J Agric Food Chem 70(5):1410–1422

He D, Zhang X, Hu J (2021) Methods for separating microplastics from complex solid matrices: comparative analysis. J Hazard Mater 409:124640

Hidalgo-Ruz V, Gutow L, Thompson RC, Thiel M (2012) Microplastics in the marine environment: a review of the methods used for identification and quantification. Environ Sci Technol 46(6): 3060–3075

Horton AA, Svendsen C, Williams RJ, Spurgeon DJ, Lahive E (2017a) Large microplastic particles in sediments of tributaries of the river Thames, UK—abundance, sources and methods for effective quantification. Mar Pollut Bull 114(1):218–226

Horton AA, Walton A, Spurgeon DJ, Lahive E, Svendsen C (2017b) Microplastics in freshwater and terrestrial environments: evaluating the current understanding to identify the knowledge gaps and future research priorities. Sci Total Environ 586:127–141

Huang W, Song B, Liang J, Niu Q, Zeng G, Shen M, Zhang Y (2021) Microplastics and associated contaminants in the aquatic environment: a review on their ecotoxicological effects, trophic transfer, and potential impacts to human health. J Hazard Mater 405:124187

Huppertsberg S, Knepper TP (2018) Instrumental analysis of microplastics—benefits and challenges. Anal Bioanal Chem 410:6343–6352

Jakubowicz I, Enebro J, Yarahmadi N (2021) Challenges in the search for nanoplastics in the environment—A critical review from the polymer science perspective. Polym Test 93:106953

Jovanović B (2017) Ingestion of microplastics by fish and its potential consequences from a physical perspective. Integr Environ Assess Manag 13(3):510–515

Kachhawa KS, Singh TP, Awasthi YK (2022) Methods of detection of micro-plastic in environment with effect. JIMS8I Int J Inform Commun Comput Technol 10(2):620–627
Kalaronis D, Ainali NM, Evgenidou E, Kyzas GZ, Yang X, Bikiaris DN, Lambropoulou DA (2022) Microscopic techniques as tools for the determination of microplastics and nanoplastics in the aquatic environment: a short review. Green Anal Chem 3:100036
Kanhai LDK, Gardfeldt K, Krumpen T, Thompson RC, O'Connor I (2020) Microplastics in sea ice and seawater beneath ice floes from the Arctic ocean. Sci Rep 10(1):5004
Käppler A, Fischer D, Oberbeckmann S, Schernewski G, Labrenz M, Eichhorn KJ, Voit B (2016) Analysis of environmental microplastics by vibrational microspectroscopy: FTIR, Raman or both. Anal Bioanal Chem 408:8377–8391
Karlsson TM, Vethaak AD, Almroth BC, Ariese F, van Velzen M, Hassellöv M, Leslie HA (2017) Screening for microplastics in sediment, water, marine invertebrates and fish: method development and microplastic accumulation. Mar Pollut Bull 122(1–2):403–408
Koelmans AA, Kooi M, Law KL, Van Sebille E (2017) All is not lost: deriving a top-down mass budget of plastic at sea. Environ Res Lett 12(11):114028
Kooi M, Reisser J, Slat B, Ferrari FF, Schmid MS, Cunsolo S, Koelmans AA (2016) The effect of particle properties on the depth profile of buoyant plastics in the ocean. Sci Rep 6(1):1–10
Kutralam-Muniasamy G, Pérez-Guevara F, Elizalde-Martínez I, Shruti VC (2021a) How well-protected are protected areas from anthropogenic microplastic contamination? Review of analytical methods, current trends, and prospects. Trends Environ Anal Chem 32:e00147
Kutralam-Muniasamy G, Pérez-Guevara F, Martínez IE, Shruti VC (2021b) Overview of microplastics pollution with heavy metals: analytical methods, occurrence, transfer risks and call for standardization. J Hazard Mater 415:125755
Kutralam-Muniasamy G, Shruti VC, Pérez-Guevara F, Roy PD (2023) Microplastic diagnostics in humans: "the 3Ps" progress, problems, and prospects. Sci Total Environ 856:159164
Lenaker PL, Baldwin AK, Corsi SR, Mason SA, Reneau PC, Scott JW (2019) Vertical distribution of microplastics in the water column and surficial sediment from the Milwaukee River basin to Lake Michigan. Environ Sci Technol 53(21):12227–12237
Lenz R, Enders K, Stedmon CA, Mackenzie DM, Nielsen TG (2015) A critical assessment of visual identification of marine microplastic using Raman spectroscopy for analysis improvement. Mar Pollut Bull 100(1):82–91
Li J, Liu H, Chen JP (2018) Microplastics in freshwater systems: a review on occurrence, environmental effects, and methods for microplastics detection. Water Res 137:362–374
Lindeque PK, Cole M, Coppock RL, Lewis CN, Miller RZ, Watts AJ, Galloway TS (2020) Are we underestimating microplastic abundance in the marine environment? A comparison of microplastic capture with nets of different mesh-size. Environ Pollut 265:114721
Löder MG, Gerdts G (2015) Methodology used for the detection and identification of microplastics—a critical appraisal. In: Marine anthropogenic litter, pp 201–227
Löder MGJ, Kuczera M, Mintenig S, Lorenz C, Gerdts G (2015) Focal plane array detector-based micro-Fourier-transform infrared imaging for the analysis of microplastics in environmental samples. Environ Chem 12(5):563–581
Lofty J, Ouro P, Wilson CAME (2023) Microplastics in the riverine environment: meta-analysis and quality criteria for developing robust field sampling procedures. Sci Total Environ 863: 160893
Lusher AL, Primpke S (2023) Finding the balance between research and monitoring: when are methods good enough to understand plastic pollution? Environ Sci Technol 57(15):6033–6039
Lusher AL, Mchugh M, Thompson RC (2013). Occurrence of microplastics in the gastrointestinal tract of pelagicand demersal fish from the English Channel. Marine pollution Bulletin, 67(1-2): 94-99.
Lusher AL, Tirelli V, O'Connor I, Officer R (2015) Microplastics in Arctic polar waters: the first reported values of particles in surface and sub-surface samples. Sci Rep 5(1):14947
Lusher AL, Welden NA, Sobral P, Cole M (2017) Sampling, isolating and identifying microplastics ingested by fish and invertebrates. Anal Methods 9(9):1346–1360

Lv L, Yan X, Feng L, Jiang S, Lu Z, Xie H, Li C (2021) Challenge for the detection of microplastics in the environment. Water Environ Res 93(1):5–15

Mai L, Bao LJ, Shi L, Wong CS, Zeng EY (2018) A review of methods for measuring microplastics in aquatic environments. Environ Sci Pollut Res 25:11319–11332

Malafaia G, Barceló D (2023) Microplastics in human samples: recent advances, hot-spots, and analytical challenges. TrAC Trends Anal Chem 161:117016

Malankowska M, Echaide-Gorriz C, Coronas J (2021) Microplastics in marine environment: a review on sources, classification, and potential remediation by membrane technology. Environ Sci Water Res Technol 7(2):243–258

Mariano S, Tacconi S, Fidaleo M, Rossi M, Dini L (2021) Micro and nanoplastics identification: classic methods and innovative detection techniques. Front Toxicol 3:636640

Martin C, Parkes S, Zhang Q, Zhang X, McCabe MF, Duarte CM (2018) Use of unmanned aerial vehicles for efficient beach litter monitoring. Mar Pollut Bull 131:662–673

Maxwell SH, Melinda KF, Matthew G (2020) Counterstaining to separate Nile red-stained microplastic particles from terrestrial invertebrate biomass. Environ Sci Technol 54(9): 5580–5588

Michel AP, Morrison AE, Preston VL, Marx CT, Colson BC, White HK (2020) Rapid identification of marine plastic debris via spectroscopic techniques and machine learning classifiers. Environ Sci Technol 54(17):10630–10637

Miller E, Sedlak M, Lin D, Box C, Holleman C, Rochman CM, Sutton R (2021) Recommended best practices for collecting, analyzing, and reporting microplastics in environmental media: lessons learned from comprehensive monitoring of San Francisco Bay. J Hazard Mater 409: 124770

Möller, J. N., Heisel, I., Satzger, A., Vizsolyi, E. C., Oster, S. J., Agarwal, S., . & Löder, M. G. (2022). Tackling the challenge of extracting microplastics from soils: a protocol to purify soil samples for spectroscopic analysis. Environ Toxicol Chem, 41(4), 844–857

Mukonza SS, Chiang JL (2022) Satellite sensors as an emerging technique for monitoring macro- and microplastics in aquatic ecosystems. Water Emerg Contam Nanoplast 1(4):17

Munari C, Scoponi M, Mistri M (2017) Plastic debris in the Mediterranean sea: types, occurrence and distribution along Adriatic shorelines. Waste Manag 67:385–391

Näkki P, Setälä O, Lehtiniemi M (2017) Bioturbation transports secondary microplastics to deeper layers in soft marine sediments of the northern Baltic sea. Mar Pollut Bull 119(1):255–261

Natesan U, Vaikunth R, Kumar P, Ruthra R, Srinivasalu S (2021) Spatial distribution of microplastic concentration around landfill sites and its potential risk on groundwater. Chemosphere 277:130263

Osman AI, Hosny M, Eltaweil AS, Omar S, Elgarahy AM, Farghali M, Yap PS, Wu YS, Nagandran S, Batumalaie K, Gopinath SC (2023). Microplastic sources, formation, toxicity and remediation: a review. Environmental Chemistry Letters, 21(4):2129-2169.

Pagter E, Frias J, Nash R (2018). Microplastics in Galway Bay: A comparison of sampling and separation methods. Marine Pollution Bulletin, 135:932–940.

Prata JC, da Costa JP, Duarte AC, Rocha-Santos T (2019) Methods for sampling and detection of microplastics in water and sediment: a critical review. TrAC Trends Anal Chem 110:150–159

Qi H, Fu D, Wang Z, Gao M, Peng L (2020) Microplastics occurrence and spatial distribution in seawater and sediment of Haikou Bay in the northern South China Sea. Estuar Coast Shelf Sci 239:106757

Razeghi N, Hamidian AH, Wu C, Zhang Y, Yang M (2021) Microplastic sampling techniques in freshwaters and sediments: a review. Environ Chem Lett 19(6):4225–4252

Ryan PG, Moore CJ, Van Franeker JA, Moloney CL (2009). Monitoring the abundance of plastic debris in the marine environment. Philosophical Transactions of the Royal Society B: Biological Sciences, 364(1526):1999–2012.

Ryan PG, Suaria G, Perold V, Pierucci A, Bornman TG, Aliani S (2020) Sampling microfibres at the sea surface: the effects of mesh size, sample volume and water depth. Environ Pollut 258: 113413

Shahnawaz M, Sangale MK, Daochen Z, Ade AB (eds) (2022) Impact of plastic waste on the marine biota. Springer, Singapore
Shi H, Wang X, Zhu L, Li D (2023) Comprehensive comparison of various microplastic sampling methods in sea water: implications for data compilation. Water 15(6):1035
Shim WJ, Hong SH, Eo SE (2017) Identification methods in microplastic analysis: a review. Anal Methods 9(9):1384–1391
Stolte A, Dederer G, Lamp J, Fenn C, Lee M, Frank W, Werner S (2022) The quest for ghost gear in the German Baltic Sea: a team effort between WWF, divers, fisherfolk, and public authorities. Front Mar Sci 9:981840
Su L, Cai H, Kolandhasamy P, Wu C, Rochman CM, Shi H (2018) Using the Asian clam as an indicator of microplastic pollution in freshwater ecosystems. Environ Pollut 234:347–355
Thomas D, Schütze B, Heinze WM, Steinmetz Z (2020) Sample preparation techniques for the analysis of microplastics in soil—a review. Sustainability 12(21):9074
Van Cauwenberghe L, Vanreusel A, Mees J, Janssen CR (2013) Microplastic pollution in deep-sea sediments. Environ Pollut 182:495–499
Veerasingam S, Ranjani M, Venkatachalapathy R, Bagaev A, Mukhanov V, Litvinyuk D et al (2021) Contributions of Fourier transform infrared spectroscopy in microplastic pollution research: a review. Crit Rev Environ Sci Technol 51(22):2681–2743
Vivekanand AC, Mohapatra S, Tyagi VK (2021) Microplastics in aquatic environment: challenges and perspectives. Chemosphere 282:131151
von der Esch E, Kohles AJ, Anger PM, Hoppe R, Niessner R, Elsner M, Ivleva NP (2020) TUM-ParticleTyper: a detection and quantification tool for automated analysis of (microplastic) particles and fibers. PLoS One 15(6):e0234766
Wang W, Wang J (2018) Investigation of microplastics in aquatic environments: an overview of the methods used, from field sampling to laboratory analysis. TrAC Trends Anal Chem 108:195–202
Wirnkor VA, Ebere EC, Ngozi VE (2019) Microplastics, an emerging concern: a review of analytical techniques for detecting and quantifying microplatics. Anal Methods Environ Chem J 2:13–30
Woo H, Seo K, Choi Y, Kim J, Tanaka M, Lee K, Choi J (2021) Methods of analyzing microsized plastics in the environment. Appl Sci 11(22):10640
Xu JL, Thomas KV, Luo Z, Gowen AA (2019) FTIR and Raman imaging for microplastics analysis: state of the art, challenges and prospects. TrAC Trends Anal Chem 119:115629
Yadav S, Kataria N, Khyalia P, Rose PK, Mukherjee S, Sabherwal H, Khoo KS (2023) Recent analytical techniques, and potential eco-toxicological impacts of textile fibrous microplastics (FMPs) and its associated contaminates: a review. Chemosphere 326:138495
Ye Y, Yu K, Zhao Y (2022) The development and application of advanced analytical methods in microplastics contamination detection: a critical review. Sci Total Environ 818:151851
Yonkos LT, Friedel EA, Perez-Reyes AC, Ghosal S, Arthur CD (2014) Microplastics in four estuarine rivers in the Chesapeake Bay, USA. Environ Sci Technol 48(24):14195–14202
Zarfl C (2019) Promising techniques and open challenges for microplastic identification and quantification in environmental matrices. Anal Bioanal Chem 411:3743–3756
Zhu J, Wang C (2020) Recent advances in the analysis methodologies for microplastics in aquatic organisms: current knowledge and research challenges. Anal Methods 12(23):2944–2957

Detection of Microplastic in Wastewater Treatment Plants

12

Enespa and Prem Chandra

Abstract

There is rising concern over the threat that microplastics pose to aquatic life. In addition to their capability to soak up tiny pollutants, they not merely assist in the accumulation of plastics in the natural world but can also assist in their distribution. Humans absorb microplastics at the tertiary stage of the dietary network, which follows aquatic organisms including fish and numerous crustaceans. Treatment facilities for wastewater (WWTP) have been found to be a substantial single origin of microplastics pollutants, especially in regard to fibres. The sources of plastic trash that enter WWTP include make-up and skin care products, wearable material such as garments, car tyres, and roadway paint. Microplastics may be discharged into reservoirs or combined with filth after entering the WWTP via household wastewater or drainage systems. In complex environmental matrices, microplastics can be examined, removed, filtered, determined, and characterized using a wide range of procedures. Microplastics can be identified by examining the chemical and physical characteristics of plastic particles that have been separated and rinsed out of a combination of inorganic and organic particles. There is a good chance that this identification will provide results that are false-positive or false while studying tiny microplastics. Today's most common approach integrates thermal, chemical, and physiological assessments, such as microscopy and spectroscopic way. This chapter carefully examines the most recent data regarding the finding and prevalence of microplastics in WWTPs.

Enespa (✉)
Department of Plant Pathology, School of Agriculture, MPDC, University of Lucknow, Lucknow, Uttar Pradesh, India

P. Chandra
School of Agriculture, Indira Gandhi National Open University, New Delhi, India

M. Shahnawaz et al. (eds.), *Microplastic Pollution*,
https://doi.org/10.1007/978-981-99-8357-5_12

These investigations provide a thorough analysis of all methods for locating and eliminating the microplastics.

Keywords

Water pollution · Analytical chemistry · Microplastic pollutants · Wastewater treatment plants (WWTPs)

12.1 Introduction

Microplastics, which are frequently characterized as plastic fragments shorter than 5 mm, represent a risk for both aquatic organisms and humanity (Smith et al. 2018). Microplastics are frequently observed in aquatic bodies of water, such as lakes, estuaries, rivers, coastlines, and marine environments because of the widespread consumption of plastic items and the careless management of the disposal of plastic waste (Duis and Coors 2016; Kumar et al. 2021). The amount of plastic garbage that now adversely affects the underwater environment on a yearly basis is predicted to rise by a number of four, or to approximately eight million metric tonnes, by 2050 (Chandra et al. 2020). Most wrapping for soapy products, exfoliation, moisturizers, and other regular products is made of polyethylene, among the most prevalent kinds of plastic (PE), and polypropylene, the majority prevalent type of plastic (PP) (Hernández Fernández et al. 2022).

Polystyrene (PS) materials are frequently utilized to create things that are disposable including insulators and packaging for food, while PET plastic is frequently utilized for making receptacles as well as various types of wrapping (Andrady and Neal 2009). Every single one of these items has the capacity to become microplastics. Many beauty and beauty products contain microplastics, also known as the primary microplastics, which can be manufactured immediately (Murphy et al. 2016; Atugoda et al. 2021). They can develop when significant plastic waste becomes decomposed by exposed to environmental stresses such as water, wind, and sunshine. They are sometimes referred to as secondary microplastics. According to a recent study, wastewater treatment plants (WWTPs) may have had a substantial impact on the release of microplastics throughout the natural environment (Sun et al. 2019). Due to the actions of humans, tiny particles found in toothbrushes and cleanser for the face could wind up in sewage. Additionally, when synthetic materials like polyester (PES) and nylon are washed, several hundred threads may be spilled into the wastewater (Cristaldi et al. 2020). Depending on the disinfection equipment used, some microplastics may be removed by the WWTP. However, it has been demonstrated that microplastics can evade the WWTP, infiltrate aquatic waterways, and eventually build up in the natural world (Liu et al. 2021; Tang and Hadibarata 2021).

Using spectroscopic techniques, such as FT-IR, Raman, or SEM-EDS, which enable more precise detection and categorization of microplastics because they offer numbers and sizes of particles and, consequently, information on the composition of the material, as well as using thermo-analytical products, such as DSC or Pyr-GC/MS, which enable reliable detection and categorization of microplastics (Tagg et al. 2015; Hernández Fernández et al. 2022). To identify, measure, and determine the deterioration of the microplastics contained within a sample using a single thermal experiment, many researchers have developed techniques for estimating the mass amount of microplastics using DSC (Woo et al. 2021). Amorphous polymers plastics cannot be used with this technology, only semi-crystalline polymers may. Following their filtration onto membrane filters, these contaminants have been detected in wastewater and seawater using FPA-based micro-Fourier transform infrared spectroscopic (FPA-based micro-FTIR) imaging. Both the transmission mode and the reflectance mode of infrared imaging have been successfully used to locate microplastics as small as 20–25 mm (Krishnan et al. 2022; Bitter and Lackner 2021).

These BOM-rich samples underwent Fenton's reagent pretreatment to facilitate quick extraction of microplastics without jeopardizing the dimension of the microplastics or the presence and locations of the crucial FTIR absorbency bands for plastic confirmation (Kedzierski et al. 2021; Cunsolo et al. 2021).

We show that the entire membrane filters can be photographed without the need for IR transparent transmission windows or an optical preselection procedure, making it possible to use luminescence micro-FTIR as an efficient and accurate method to find microplastics throughout all stages of treatment (even in particularly difficult sample types like both biologically triggered wastewater) (Harrison et al. 2012). During the treatment process, a percentage of the substance that penetrates WWTPs, including microplastics, is routinely eliminated. Despite the likelihood of a considerable microplastic discharge being possible due to the high disposal efficiency, which in some situations may be more than 90%, steady discharge volumes of the treated wastewater nonetheless allow it (Kwon et al. 2022). The primary tools for locating microplastics in sewage treatment plants are inputs and detection techniques are discussed in this chapter given below in Table 12.1.

12.2 Occurrence of Microplastics in WWTPs

12.2.1 Microplastics Transportation Phenomena

Plastic particles can enter the ecosystem in a number of techniques, including water from storms, sewage treatment plant emissions, household wastewater outflow, and wind (Murphy et al. 2016). Another primary source and environmental channel for MPs, which incorporates numerous synthetic fibres and micro-exfoliates, are sewage outflow from various domestic operations. Domestic wastewater from sources is processed in wastewater treatment plants in towns and cities, while in rural regions, it is either released into the environment unprocessed or with little to no treatment (Ngo et al. 2019). Although these facilities for treatment may remove a sizable number of MPs, a sizable amount of MPs are still discharged into the aquatic environment since they miss multiple plant processing steps (Cheng et al. 2021).

Table 12.1 Summary of technically analysing results

Identification method	Advantages	Drawbacks	References
Microscopy	A quick, low-cost, and easy method for identifying MPs with particles between 1 and 5 mm in size	It is unable to be used on its own to identify MPs. Scarcity of information concerning MPs' chemical make-up. Outcomes are arbitrary and dependent on the assessor	Renner et al. (2018); Gies et al. (2018)
Emerging technology	Improvement aimed at addressing the drawbacks of current technology	Fewer points of information collected than with modern technology	Erni-Cassola et al. (2017)
Spectroscopy FTIR (FPA-FTIR)	Nondestructive. There is no need to prepare the sample in advance, and the sample amount is minimal and recognizes particulates larger than 10–20 mm	Incredibly pricey, tough, and time-consuming. Signal analysis challenges in dealing with the presence of pollutants	Dyachenko et al. (2020); Löder et al. (2015)
Raman spectrometry	Capable to find MPs with fragments larger than 1 mm	Work that is costly, lengthy, and tedious. Possible interference caused by additional bands as well as dye and pigment illumination.	Lenz et al. (2015)
Thermal analysis (Pyro-GC-MS)	Fully automated system Preparation of the sample is not needed. Ability to identify MPs of all sizes and forms	Interpreting information challenges for complicated samples. Inorganic additions are undetectable. Detrimental	Fries et al. (2013); Dekiff et al. (2014)

MPs may be accessible to the soil's elements when settled sludge utilized for landfilling contains a significant amount of plastic fragments from the water processing facilities. Macroplastic and microplastic particles can also reach the aquatic environment by stormwater discharge (Bellasi et al. 2020). Plastic wastes are pervasive and accumulate during rainfall in the agrarian system, especially productive soil, roadways, and garbage dumps as a result of insufficient disposal of waste. 42% of the plastic particles that transfer from streams to the marine ecosystems originate from road water flow, which can cause MPs from tyres and highway wear components to travel to the outside world (Kehinde et al. 2020; Ferronato and Torretta 2019).

Macroparticles and MP particles can be transported to lakes and rivers by the meteorological flow due to their tiny size. The wind has the potential to carry them too far from their source (Lwanga et al. 2022).

The comportment of MPs while commuting is still a mystery. Notwithstanding the fact that terrestrial and freshwater ecosystems are considered to be the primary sources of MPs, oceans are predicted to be the main sinks for plastic particles (Li et al. 2023). MPs might additionally break down during migratory and ultimately end up in aquatic ecosystems and sediments that are close to the shore. MPs' propensity for sinking is correlated with their size, volume, and structure—all of which make them susceptible to microbiological filth and deterioration. The water flow velocity also contributes to the concurrent accumulation of plastics in deposits (Lagarde et al. 2016; Thompson et al. 2009). It is common knowledge that microbiological filth is influenced by sunshine, amount of nutrients, temperature of the water, and the hydraulic system. Once MPs come into contact with the aquatic environment, they are susceptible to a variety of mechanical and physiological problems (Ateia et al. 2022). The photo-oxidation of plastic particles by UV and natural light can cause breakage when floating. However, the presence of biological pollutants and murky water can halt this procedure completely. Degrading and weak MPs may also collapse under mechanical stress (Yousif and Haddad 2013; Oliveira et al. 2020).

A number of variables, including the type of polymers, the type of organisms, the molecular weight of the plastic and rubber, the environmental factors, and further microbial degradation, may take place. This deteriorating mechanism is accelerated by ionization, the process of hydrolysis, and the dissociation of plastic particles (Mohanan et al. 2020). Roughness of the surface alterations that result in flowering could speed up the slow process of plastic waste dispersion. According to Urbanek et al. (2018), this change may also help bacteria to attach to and break up MP rapidly. Studies have shown that aquatic hazards may nourish MPs.

Despite the fact that exposure to sunlight speeds up the disintegration of plastic fragments, it is plausible to believe that surface water waves are the primary mechanism for the transport of MPs (Wojnowska-Baryła et al. 2022). In order to guarantee total MP segregation in the wastewater treatment structure, it is crucial. Slurry must be handled with greater attention before being disposed of in a landfill so as to stop plastic particles from leaking into the aqueous column. Additionally, these drifting plastic particles fall as a result of marine life-consuming MPs (Tursi et al. 2022). Several investigations have demonstrated that once MPs reach an aquatic environment, they may move, bioaccumulate, and get magnified (Miller et al. 2020). The conclusion follows that a significant proportion of plastic particles may be digested by aquatic creatures. Additionally, further dissection has already taken place in their gastrointestinal tracts due to peristalsis, a number of inside microbes, and intestinal excretion (Duis and Coors 2016).

12.2.2 Microplastics Concentrations in Influent and Effluent of WWTPs

Due to their common displays, plastic particles were easily found in the influent and effluent of WWTPs. The observed particle concentrations ranged from 1 to 10,044 particles/L, and only a small percentage of WWTPs found microplastics in the wastewater influent (Sun et al. 2019; Cristaldi et al. 2020). The detected concentrations of microplastics in WWTP effluent ranged from 0 to 447 particles per litre, which is much lower. The varied extraction of samples, pretreatment procedures, and analytical procedures used in these investigations may be somewhat to blame for the significant diversity in microplastics contents in these WWTPs (Kay et al. 2018; Murphy et al. 2016). For example, using a finer mesh size might reveal higher microplastics content. Studies that lack chemical characterization also significantly increase the risk of measurement bias, particularly when separating natural from synthetic fibres (Blair et al. 2019; Lenaker et al. 2020). In certain studies, the amount of natural fibres was accordingly taken into account. To properly contrast the levels of microplastics in various searches, the sampling and analysis processes for microplastics must be standardized or harmonized (Adegoke et al. 2023). Catchment areas, population served, next-door land use, integrated sources of wastewater (residential, commercial, or industrial), etc., may all have an impact on the variations in concentrations of microplastic between plants (Di Nunno et al. 2021; Bellasi et al. 2020). Given that residential produces account for a substantial amount of the microplastics in the wastewater, human behaviours in the functioned catchment, such as residents' support for wearing synthetic clothing or using plastic goods, can significantly impact the quantity of microplastics in the wastewater (Sadia et al. 2022; Issac and Kandasubramanian 2021).

The served community and the overall quantity of microplastics found in the wastewater were positively connected, according to Mason et al.'s statistical analysis of 17 WWTPs in the United States in 2016. Nevertheless, Mintenig et al. (2017) discovered no connection between microplastics (500 mm) counts and population counterparts at the 12 German WWTPs under investigation. In order to fully understand the widespread use of microplastics, it may be beneficial to examine their presence in wastewater from regions with various economic status and habits in the future. This is based on a range of information from investigations done in various nations (Campanale et al. 2020). Researchers have discovered a link between the integrated sewer network and an upsurge in sewage fragments, which may be caused by neighbourhood land use and impurities associated with shipping, such as microplastics produced by tyre and brake wear. The Michielssen et al. (2016) study's findings also suggested that wastewater treatment plants (WWTPs) associated with centralized sewage systems may be to blame for large amounts of microplastics and other tiny manmade litter (SAL) in the influent. The amount of microplastics discharged into surface water is greatly decreased by integrated sewer, even if it may cause an increase in microplastics in the wastewater treatment plant.

Contrary to systems with discrete sewage networks, it has been discovered that in regions where the excess microplastics in mixed sewage lines are caused by runoff from roads, the emission to surface waterways will be reduced (Zillien et al. 2022; Schernewski et al. 2020). Microplastic pollution of wastewater can be substantially eliminated in the WWTP. The amount of microplastics in the effluent from WWTPs will vary according to the wastewater treatment techniques used (Reddy and Nair 2022). Microplastic concentrations in the effluent of WWTPs with tertiary treatment processes tend to be lower (0–51 particle/L) than those of WWTPs with only either primary or secondary treatment operations (9×104–447 particle/L) (Cristaldi et al. 2020). Studies also shown that some WWTPs' tertiary treatment did not further reduce the amount of microplastics in the runoff. In the effluent from WWTPs with only primary treatment, the concentration of microplastics has only been determined by one study (Talvitie et al. 2017). According to the same study, the concentration was almost an order of magnitude higher than that of WWTPs with secondary and tertiary remedies (Quach-Cu et al. 2018). The overall emissions of microplastics from WWTPs were nevertheless quite large (Xu et al. 2019), despite the comparatively low concentrations of microplastics in the discharge of these facilities due to the fact that the majority of these facilities handle millions of litres of wastewater each day. The total daily outflow of microplastics from the examined WWTPs was estimated to be 2×10^6 particles/day, or an average annual efflux of 5×10^7 m^3/year, based on calculations based on yearly outflow and wastewater content (Oveisy et al. 2022).

The daily discharge totals from some WWTPs in the USA and the Netherlands may even exceed 1×10^{10} particles. According to Conley et al. (2019) and Cristaldi et al. (2020), the average yearly effluxes of the high tiny particles discharge WWTPs were larger than 1×10^7 m^3, and their population equivalents were greater than 1×10^6. Focused microplastics treatment methods must be implemented immediately in order to stop the excessive levels of microplastics being released into the ecosystem from WWTPs. Customized microplastic treatment techniques must be developed immediately to stop the huge discharge of microplastics into aquatic systems as a result of the high volumes of microplastics produced from WWTPs (Krishnan et al. 2022).

12.3 Microplastics Properties in WWTPs

12.3.1 Shapes

Morphology is a significant criterion for classifying microplastics. According to Cristaldi et al. (2020), the crystalline structure of microplastics can influence how they communicate with other contaminants and microorganisms in wastewater as well as how they are eliminated in the WWTP. Fibre, which is much longer than broad, and particles, which are similar in diameter and in length, are two straightforward classifications for microplastics. The vast majority of the microplastics that were discovered in the wastewater were composed of the fibres, with a normal percentage of 52.7% (Herzke et al. 2021). Given how many fibres were discharged by household washing machine outcomes, the large amount of fibres in the wastewater may be somewhat attributed (Schöpel and Stamminger 2019).

12.3.2 Size and Mass

Currently, two methods are frequently employed to determine cluster sizes. One possibility is that microplastics are retained on sieves of different diameters. The varied forms of microplastics, as previously stated, limit this procedure's reliability (Reineccius et al. 2021). Another possibility is to use microscopic imaging approaches. It is challenging to quantify the precise size of microplastics utilizing a single number because they come in a variety of shapes. It was recommended to use traditional interfacial science criteria to get reliable and corresponding data on the true size of microplastic particles in a systematic way (Mariano et al. 2021; Nguyen and Tufenkji 2022).

The size categorization measures that were typically employed were 25, 100, and 500 mm. The proportion of microplastics larger than 500 mm in the input of the WWTP may rarely exceed 70% (Kiran et al. 2022). In comparison to over 60% of the microplastics in certain samples that were smaller than 100 mm, nearly 90% of the microplastics in the discharge were, on a typical basis, less than 500 mm (Iyare et al. 2020). However, the mesh size employed for collecting the samples may have an impact on the distribution of size of the microplastics. A high net size is likely to fail to capture the majority of minute particles (Lares et al. 2018).

12.3.3 Materials

The effluent and the influent of WWTPs contain over 30 different types of microplastic polymers, according to research done so far (Van Do et al. 2022). Polyester (PES, between 28 and 89 per cent), polyethylene (PE), polyethylene terephthalate (PET), and polyamide (PA, between 3 and 30%) were among the most common polymers found in the influence and wastewater of WWTPs. While PE is utilized in water containers, body and facial washes, packaging for food, and skin care products, PES, PET, and PA are all frequently utilized in polyester garments (Šaravanja et al. 2022). Amini Moghaddam et al. 2021, and Yan et al. 2015, report that the wastewater additionally includes polymers with the highest populations spanning among roughly 5% and 27%, including acrylate, alkyd, polypropylene (PP), polystyrene (PS), polyurethane (PU), PS acrylic, polyvinyl alcohol (PVAL), and polylactide (PLA). Other polymers were typically present in wastewater in amounts less than 5%, and occasionally, they may have been as low as 1%. Therefore, rather than studying every one of the accessible particulates, it is possible to start with the more common polymers (Chamas et al. 2020). According to the chemical and physical characteristics of the microplastics found in WWTPs, a large portion of them have been produced during normal everyday activities (Edo et al. 2020). Unknown are the general origins and locations of entry for microplastics into WWTPs. These finding merits further investigation since it could help in regulating the origins of microplastics to reduce related pollution (Liu et al. 2022).

12.4 Microplastic Detection

12.4.1 Attributes of Microplastics

In the environment, microplastics (MPs) are a typical solid form that is available in a variety of sizes, densities, colours, and polymer types. Furthermore, they have the capacity to mix toxins and plastic particles (Szymańska and Obolewski 2020). Only these characteristics can account for MP mobility in the environment. The majority of PPs are discovered in various aquatic bodies. As an example, small PPs have been discovered in estuaries all around China, particularly the Lake Taihu of the Yangtze delta and other estuaries, as well as the municipal water systems of Wuhan. On the other hand, there is a greater availability of larger PPs in the Siling Co Basin and the Bay of Jiangxi (Vaid et al. 2021). The massive polymers, however, shrink to the size of tiny PPs throughout the process of weathering. Since there are currently no data on the geographic distribution of PP sizes in the environment, the field of science is concerned and pushing for suitable solutions (Krauklis et al. 2022). Based on morphological studies, MPs can be divided into six classes: fibre, spherical, foam, sheet, fragment, and film. The natural world is where you can find most fibre-shaped MPs. Microfiber quantities in water usually vary from 0.02 to 25.8 fibres per litre, whereas microfiber sizes vary from 0.09 to 27.06 mm (Dey et al. 2021). Since cleanings take place throughout several production steps, the clothing sector is primarily responsible for the discharge of microfibers into the environment. Some countries with significant clothing sectors, such as Bangladesh, China, Vietnam, and India, may be contributing a lot of microfibers to the ecosystem (De Falco et al. 2019). However, artificial plastic bags and packaging supplies produce MPs with a film structure, while certain plastic-based products can degrade into foam-like particles of plastic. Different particles, aggressive blasting medium, and resin pellet system of transportation permeability are the main sources of sphere-type particles (Oliveira et al. 2020).

12.4.2 Microplastic Sampling in Wastewater Treatment Plants

12.4.2.1 Sampling Process of MPs

MPs from WWTPs may be examined using the lengthy procedure (Fig. 12.1). For the collection of samples from WWTP, mechanical filtering alongside multiple metallic mesh or sieve is typically used (Rusten and Ødegaard 2006). The quantity of MPs obtained determines the mesh/sieve size. Applying various sieve series, MPs can be separated into a number of size groups. For raw and treated wastewater, different sample sizes must be used (Turan et al. 2021). Due to the high concentration of organic load in the influent, which could quicker block screens and screens, a significant volume of influent is not recommended. However, an abundance of wastewater is required as there are not enough organic substances around. Carefully manage numerous samples, particularly if the investigation is centre on big particles (Cescon and Jiang 2020).

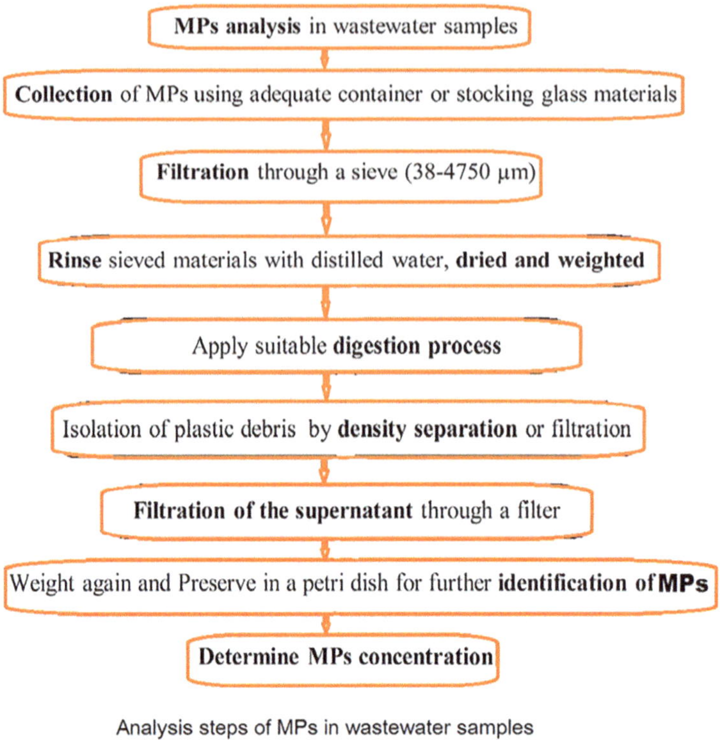

Fig. 12.1 Analysis steps of microplastics in wastewater samples (Alvim et al. 2020; Silva et al. 2018)

For instance, 10.5–13.5 L are needed to check for massive particles (>300 m), which constitute about 5% of the overall number of particles (De Gisi et al. 2016). To quantify small particles (20–100 m), which make up for approximately 70% of all observable particles, only 2 L are required. To perform a thorough investigation, it is advised to determine all MPs between 1 and 5000 m (Vu et al. 2015). The study's sample quantity variability has been greatly influenced by the analytical method's limit of quantification. It is more difficult to determine constant sampling amounts in all circumstances when homogeneity is absent (Armbruster and Pry 2008). The best strategies and techniques for collecting, handling, and analysing data imply that the collection of samples should be easy to do, repeatable, and aimed at avoiding contaminating.

12.4.3 Methods of Detection

MPs are difficult to differentiate with a single detection technique because of changes in size, structure, colour, and polymer type. Due to this, numerous detecting techniques are employed. Physical characterization (i.e., colour, size) and chemical characterization (i.e., composition, structure) are frequently used in conjunction to detect microplastics (Adhikari et al. 2022). Larger and some small coloured MPs are identified using visual analysis or unaided eyes and forceps. The bulk of tiny particles, meanwhile, is probably going to pass undetected all through processing. Because of this, the first identifying process is quick and simple. As a result, more investigation is needed to provide a complete comprehension of the MPs, which motivates the creation of MPs detecting Dey et al. (2021).

12.4.3.1 Visual Identification

12.4.3.1.1 Optical Techniques

Optical assessments of MP candour, colour, and form are commonly made with the unaided eye using optical microscopes. Six various kinds of microplastics, including fibres, bits, pellets, foam, granules, and sheets, were found in WWTP by Gies et al. (2018). Sheet and fragmentation were discovered to be the two MPs kinds that were most frequently present in samples of water (Ayyamperumal et al. 2022). This technique is praised for being straightforward, inexpensive, and quick when used for finding MPs between 1 and 5 mm. Notwithstanding the benefits that have already indicated, there are a few limitations on the optical method for recognizing MPs, which could result in results that are arbitrary and depending on the person conducting the experiment (Lv et al. 2021). The fewer in number the MPs are and the greater meddling there is, the less successful this strategy is. Because of this, eye examination is not advised as an independent technique for MPs identification and the existence of MPs must be confirmed using spectrometry procedures (Ivleva 2021).

12.4.3.2 Electron Microscopy

Even for tiny particles like nanoplastic, scanning electron microscopy (SEM) may produce remarkably clear, high-magnification images. High-resolution photographs show the fact that it is simple to discriminate among microplastics and biological debris due to the degree of roughness of the particle surface (Mariano et al. 2021). Microplastics can also be found using TEM, or transmission electron microscopy. The cross-validation the data from numerous users using an established method will assist in minimizing errors despite the fact there may be variability in the user's ability to recognize microplastics using TEM (Woo et al. 2021). Energy-dispersive X-ray spectroscopy (EDS) is used to conduct additional research in order to ascertain the elemental make-up of the object. Antioxidants and other components such as Al, Ca, Mg, Na, and Si are frequently captured as markers in order to determine microplastics using EDS (Azizah et al. 2016). The difference between inorganic and carbon-rich polymers can be determined by a particle's surface elemental composition. Due to the expensive machinery and the time and effort required for the collection and analysis, SEM and TEM can only process a limited number of samples (Ivleva 2021)

12.4.3.3 Polarizing Microscopy

In toxicity testing, polyethylene (PE) particles have been identified using polarized light microscopy. The crystalline make-up of the plastic can have an impact on the ability to transmit polarized light that can be tested (Sierra et al. 2020). Depending on the manufacturing process and the kind of polymer, the degree of crystallization differs even among the same polymer. However, tiny microplastic samples are necessary to allow enough polarized light to get through. This method is useless for specimens of microplastic that are opaque. (Woo et al. 2021).

12.4.3.4 Spectrometry Methods—Fourier Transform Infrared (FTIR) and Raman Spectroscopy

Chemically detecting MPs frequently makes use of vibrational spectroscopy methods such as Fourier transform infrared (FTIR). The sample's molecular stimulation generates spectrum fingerprints that can be contrasted to the repository spectra of other recognized materials (Barth 2007). The attenuate total reflection (ATR) of FTIR is typically used to locate MPs larger than 300 mm with a 1-min analysis frame. The infrared light's ability to penetrate the sample limits the transmission mode of FTIR, although the reflection mode is only effective with opaque and dense materials (Ling and Sommer 2015). Two distinct studies on reflection from a focal point array (FPA) and ATR-FTIR FTIR micro-spectroscopy (FPA-FTIR) came to the same conclusion that they were both suitable for usage in wastewaters and water/sediments, correspondingly (Veerasingam et al. 2021). Fast, precise, and immune to fluorescence interferences are FTIR analysis. The structure, quantity, and degree of resilience of microplastics are also provided for a better understanding of MPs routes (Chen et al. 2020). However, FTIR is unable to distinguish between black, opaque, and conditioned plastic particles smaller than 20 mm. FTIR is also regarded as a laborious and challenging technique because each sample must be examined separately under the probe (Birch et al. 2021).

12.4.3.5 Raman Spectroscopy

When combined with not harmful chemical analysis and microscopy, Raman spectroscopy is analogous to the FTIR technique. Raman spectroscopy, which has a narrower laser beam diameter than FTIR, can find microplastics as small as a few micrometres in size (Sobhani et al. 2019). Raman spectroscopy noncontact analysis has the benefit of preserving the microplastic material throughout and following testing for potential additional studies (Chakraborty et al. 2022). Understanding the outcomes of each technique when recognizing complicated microplastics can be more challenging due to changes in how they react and spectra of microplastics between FTIR and Raman spectroscopy (Araujo et al. 2018).

Raman spectroscopy makes it challenging to determine the intended polymer type since preservatives and colouring agents included in microplastics might obstruct the Raman signal. The microplastics' curved surface might potentially cause issues with precision. Similar drawbacks to FTIR apply here as well, such as its high cost and need for specialized research (Nava et al. 2021). Micro-Raman spectroscopy can only distinguish between materials if a suitable Raman reference spectrum is available for them in the database and if a probability of the material ambiguity has been previously entered (Shipp et al. 2017). Identification of the intended microplastic may be challenging because PE can be mistaken for sodium stearate or SDS; as a result, the chance of accidental positives should be taken into account (Kamemoto et al. 2010).

12.4.3.6 Thermal-Based Methods

Microplastics have recently been subjected to heat analysis techniques, which can monitor changes in the physical and chemical properties of polymers with heat resilience. The determination of polymers in the decomposition remnants of the material being studied serves as the basis for these procedures.

12.4.3.7 Differential Scanning Calorimetry

The technique of differential scanning calorimetry (DSC) is effective for examining thermal properties. To support the physical properties of polymers, it displays differences in dissolution, crystallization, transition temperatures, and corresponding enthalpy and entropy (Sturtevant 1987). DSC can be used to distinguish between various polymer kinds because each plastic product has unique properties. When DSC identifies microplastics with similar melting points, there are restrictions because of overlapping peaks. Using it, only particular microplastics that are primary, such PE and PP, can be recognized (Woo et al. 2021; Zhang et al. 2022).

12.4.3.8 Thermogravimetric Analysis

Thermogravimetry analysis is a different method for thermal analysis (TGA). By recording the weight loss of the sample while it heats up at a specified pace under predefined ambient and temperature circumstances, this method verifies the sample's qualitative and quantitative information by studying its reliance on time and temperature (Saldarriaga et al. 2015). In past studies, TGA and DSC were used to investigate microplastics in wastewater. Only PE and polypropylene (PP) could, however, be separated with certainty in these trials (Majewsky et al. 2016). Polyvinyl chloride (PVC), polyamide (PA), and polyester (PES) were not mentioned. Overlapping phase transition signals, such as those of polyethylene terephthalate (PET) and polyurethane (PU), are difficult to distinguish (Andler et al. 2022).

12.4.3.9 Pyrolysis-Gas Chromatography-Mass Spectrometry

An alternate technique for examining burned gases in polymers is pyrolysis-gas chromatography-mass spectrometry (Py-GC-MS). If a sample's programme does not match a recognized polymer benchmark, the data can be compared to determine this (Ma et al. 2014). Only a few samples of plastic (particle) are pyrolyzed at temperatures greater than those employed in TGA (e.g., 700 °C), after which they are divided and tested by GC-MS analysis (Sauer and Kliem 2010). These samples range in size from 0.3 to 7 mg. PVC, PS, polyvinyl acetate, poly(acrylonitrile butadiene styrene), and poly(acrylonitrile butadiene styrene) (PVA) are all detected in a bulk examination of silt and suspended solid particles. In addition, there is styrene-butadiene rubber (Liu et al. 2019). This technique is able to recognize microplastics generated from polystyrene derivatives. More study is needed to determine the molecular markers for various plastic types (Schirinzi et al. 2019).

12.4.3.10 Short-Wave Infrared (SWIR) Multispectral Imaging

Similar to Raman and FTIR research, the distinct short-wave infrared (SWIR) reflectance properties of plastics can be utilized to locate and distinguish MPs in a sample. To determine and classify particles abundances, a hyperspectral image reveals the spectrum at each pixel of the image (SWIR, near-infrared, Raman, or FTIR according to the apparatus) (Xu et al. 2019). For MPs to go through hyperspectrum imaging for categorization, spectroscopy techniques such as Raman, FTIR, and SWIR have been used. Full hyperspectral imaging is typically not required because of the compressed phase spectroscopy's wide range of bands of absorption (Karlsson et al. 2016). It has been demonstrated that hyperspectral imaging is less efficient than multispectral imaging, but only when wavelength bands are chosen to target important characteristics and to opposition polymers with the surrounding environment without first focusing on frequencies that are not helpful in making that choice (Lu et al. 2020; Adão et al. 2017). Without a spectrometer, multispectral imaging methods can be used with a cheap camera and a few filters. To scan dried seawater specimens for MP detection, we created a multispectral SWIR approach (Lapray et al. 2014). A SWIR camera and an infrared filter wheel were used to evaluate several wavelength ranges (Sensors Unlimited Inc., Princeton, NJ, USA, Model GA1280JSX) (Thorlabs, Newton, NJ, USA) (Banerji et al. 2019; Liobe and Wieners 2022).

12.4.4 Emerging Technologies

Strategies should be created to get around the current challenges in microplastic identification by creating new analytical tools and fusing new detection technologies with current instruments (Lv et al. 2021). One of the problems with microplastic analysis is the largest size that can be detected. For the currently employed analytical techniques, the minimal visible size limit is a few micrometres. Because smaller plastic particles might cause more dangerous side effects, it is more important than ever to understand the existence, distribution, destiny, and toxicity of nanoscale plastics (Park and Park 2021). As a result, new identification methods and techniques for nanoplastic sampling, extraction, purification, and concentration are need to be developed.

12.4.4.1 Atomic Force Microscopy (AFM)-IR and AFM-Raman

AFM is a potential contender for the study of nanoplastic when used in conjunction with IR or Raman spectroscopy. The probes used in AFM can interact with things both directly and indirectly and can produce images with nanometer resolution (Woo et al. 2021). When AFM is used in conjunction with IR or Raman spectroscopy, the chemical make-up of the target material can also be revealed. The Fourier transform is utilized to determine the frequency and amplitude of the vibration from the ring-down arrangement as a result of the sample's expansion in temperature carried on by IR absorption (Luo et al. 2020).

The target's IR spectrum of absorption and an image with a resolution in space of 50–100 nm may both be acquired concurrently using AFM-IR. It was feasible to obtain the AFM and IR spectra of 100 nm PS beads in earlier studies. Unfortunately, it takes a lot of time and effort to isolate a single nanoscale plastic particle target for AFM-IR analysis in an unknown material (Heidari 2018). The human search for plastic particles is one of the demanding labour analytical steps in conventional microplastic analysis methodologies. How rapidly organic and inorganic particles are removed throughout the separation and purification processes determines how long the pretreatment procedure will last. Small, transparent plastic particles are particularly difficult to detect manually (Liu et al. 2022; Thomas et al. 2020). These problems can be solved by automated FTIR/Raman mapping and Raman spectroscopy-based particle tracking; however, these methods necessitate expensive tools that not all microplastics research facilities are able to afford (Woo et al. 2021).

12.4.4.2 Nile Red Staining of Plastics

To solve the problem of minute, translucent particles, a straightforward staining technique could be used. Only a few instances have used Oil red EGN, Eosin B, Hostasol Yellow 3G, and Rose Bengal. The fluorescent dye 9-diethylamino-5H-benzo phenoxazine-5-one, on the other hand, is especially helpful in colouring very hydrophobic microplastics (Shruti et al. 2022). It is frequently employed to stain lipids that are physiologically neutral. Only strongly fluorescent Nile red is present, and it only binds to neutral lipids in a hydrophobic environment. The Nile red staining process has a quick staining time (10–30 min) and a high recovery efficiency (up to 96%); where required, a quick bleaching wash is also carried out (Greenspan and Fowler 1985). This constitutes a good technique for finding concealed microplastics and serves as an excellent first step before more in-depth spectroscopic investigation. By incorporating a fluorescence filter into the FTIR microscope after fluorescence microscopy, the same particle's spectrum identification may be quickly verified (Zhou et al. 2022). After Nile red staining, combining fluorescence microscopy with FTIR confirmation may make it less likely that microplastics will go unnoticed during the in situ identification of the sample. Additionally, it takes less time to determine every particle that resembled plastic than spectroscopy individually (Maes et al. 2017). The potential for co-staining with natural chemical compounds is one of the main drawbacks of using Nile red staining on in situ materials. So, before applying the Nile red stain, the sample must be thoroughly cleaned. It can be challenging and time-consuming to completely eliminate organic waste during the pretreatment stage (Kang et al. 2020). Attempts to remove organic waste from plastics using H_2O^2 or a density separation approach, such as Fenton reagent or NaCl, to get around this have failed since plastics have different densities0 (Thomas et al. 2020).

12.4.4.3 Near-IR Spectra Analysis Method

The use of near-IR (NIR) spectroscopy to evaluate microplastics has also been investigated. While FTIR investigates the spectrum between 600 and 4000 cm^{-1}, NIR evaluates the spectra across the range of 4000–15,000 cm^{-1} (Rani et al. 2023). NIR spectra analysis often makes use of X-H chemical vibration combinations such C-H, O-H, and N-H. NIR analysis is difficult to employ for quantitative analysis and insensitive. However, it can quickly go through and assess a large amount of plastic sample data. In order to determine the kind of a sample rather than its number, it is preferable to use this analysis technique (Zhang et al. 2022).

12.4.4.4 Vis-NIR Measurement

By counting the total quantity of light reflected from the surface of a sample over a wavelength spectrum of 350–2500 nm, vis-NIR spectroscopy can calculate the amount of reflectance for each frequency. This analysis technique can quantify microplastics since it considers the chemical make-up of the sample (Corradini et al. 2019). Microplastics, such as those made of low-density polyethylene (LDPE), PET, and PVC, which are frequently present in the environment, can be found using open vis-NIR spectrum data sources. It is critical to rely on human judgement because this is a visual recognition approach and biological fragments can still be confused for plastic (Stock et al. 2021).

12.4.4.5 Nano-Thermal Analysis

AFM pictures are created using nano-thermal evaluation (nano-TA), a local thermal measurement method with exceptional spatial resolution. With less than 100 nm of spatial resolution, this enables the understanding of the thermal behaviour of materials (Nguyen-Tri et al. 2020). When a site that is intriguing is chosen during a nano-TA examination, the probe travels to a fixed location on the sample surface. When it comes to the rigidity (hardness) of microplastics, the probe is highly sensitive. As soon as the sample surface achieves the glass transition temperature, the tip progressively ceases heating (Duvigneau et al. 2010). Since the sample relaxes during phase inversion, it is primarily utilized to explore the nanoscale surface characteristics of microplastics and to determine the glass transition temperature. It also describes a technique for putting the probe within the sample (Woo et al. 2021). Recent research has employed nano-TA to examine the features of microplastics impacted by advanced oxidation as well as the effects of TiO2-dyed microplastics on ageing. Understanding the thermal characteristics of separate locations, such as glass temperature, from the nano-TA study of microplastics is important in order to determine the chemical make-up and physical state of each microplastics region

(Luo et al. 2021). Phase transition temperatures have been investigated for both crystalline and semi-transparent polymers, including polycaprolactone (PCL), LDPE, polyoxymethylene (POM), PET, PS, polymethyl methacrylate (PMMA), and similar materials (Fakirov 2013).

12.5 Strategies to Minimize the Entry of Microplastic in WWTPs

It is important to understand that, in conjunction with the fact that these chemicals are already present in microplastics, a range of additives and plasticizers added during manufacturing can contribute to the dangerous effects of microplastics (Fauser et al. 2022). The surface of these microplastics also absorbs various pollutants, chemicals, and germs as soon as they are introduced into the surrounding environment. As a result, it is important to be worried about more than only microplastics because associated contaminants can pose serious health risks. Although there is no regulation of primary microplastics, especially in nations that are developing, they are primarily found in medicines and beauty products (Yuan et al. 2022). Microplastic removal becomes challenging due to their pervasiveness and small size, which often permits them to elude sieving/filtration procedures. The resulting effluent contains a significant amount of microplastics because traditional wastewater treatment plants are ineffective at eliminating them (Singh et al. 2021). Due to the release of this effluent into bodies of water on the surface and/or runoff from the surface, the supply of drinking water chain may become polluted with microplastics (Khant and Kim 2022).

Furthermore, the bulk of microplastics entrapped in sludge is just deposited on the ground without any extra processing (Wang et al. 2022). This results in the contamination of soil and groundwater through the process of percolation. The risk that human beings may ingest any microplastics present in freshwater is also increased by the fact that it is often viewed as safe for consumption and does not go through any form of treatment process. In countries with low or middle incomes, where comprehensive solid waste and wastewater administration is still difficult, the inclusion of new technologies for elimination of microplastics imposes extra demands on the supply chain. Consequently, the new possibilities for therapy must be both economical and simple to use (Lwanga et al. 2022).

12.5.1 Traditional Wastewater Treatment Technologies Can Remove Microplastics

The majority of the labour is done at the basic, secondary, and tertiary stages of the wastewater treatment process. In order to get rid of the large floating particles and grits, the preparatory phase is commonly included in the primary treatment. Screening, grit removal, oil and grease removal, and precipitation (with coagulation) are a few of the procedures used in basic treatment, which is generally the physical elimination of undesirable elements from the wastewater stream. The pollutants that may easily floating or settle with the aid of gravity are removed at this stage (Sathya et al. 2022). Secondary (biological) treatment tries to eliminate the wastewater's biological content using either the connected growth system or the detached proliferation system. The final stage of the management manipulates, known as tertiary or sophisticated treatment, uses methods such reverse osmosis, the ozonation ultrafiltration, which is microfiltration, and nanofiltration process. The effectiveness of removal in wastewater treatment plants is calculated using the percentage of microplastics contained in the effluent and influent water (i.e., the percentage of microplastic fragments per litre) (Kabir et al. 2023).

In wastewater treatment plants, 50–98% of the microplastics are eliminated throughout main treatment and 35–59% are eliminated during preliminary treatment, according to the stage-wise elimination of microplastics (Iyare et al. 2020). At this point, scraping and settlement of the microplastics that have become entrapped throughout gravity separation are the main approach for eliminating microplastics. Due to the fact that fibres rather than pieces are easier to remove in these procedures, the microplastic particles' structure is very crucial (Dey et al. 2021).

Modern techniques such as membrane bioreactors may help to some extent reduce the microplastic quantity, although they have a greater probability to be used in facilities that only discharge the effluent after secondary treatment. Additionally, it has been demonstrated that the overall elimination of microplastics can reach up to 97% with tertiary therapy and up to 88% without it. Over time, lower reactor output causes reduced effectiveness of microplastic removal (Krishnan et al. 2022).

12.6 Methods to Manage the Microplastic Detected in WWTPS

The focus of MPs pollution management strategies should be on (1) source control, which is often achieved by legislation and public education campaigns, and (2) remediation and clean up (to eliminate the MPs already present in water).

There are already a number of legislative regulations in place with the aim of minimizing MP release. Since 2017, MP beads have been prohibited by US law from being used in cosmetics. In a similar spirit, many other countries are debating implementing strong safeguards, including Australia, Canada, and the European Union (EU) (Yuan et al. 2022). Cosmetics are the principal source of primary MPs discharged into the environment. A lot of progresses have been made to reduce the manufacturing and utilization of plastics that are only used once, with a focus on plastic straws and plastic bags, in relation to plastic (a potential source of secondary MPs) (Lamichhane et al. 2023). In many locations, the use of single-use plastic bags is forbidden. As an example, the EU recently proposed in a news release that a Europe-wide strategy on plastics be developed as part of the transformation to a more circular economy. By 2030, all plastic packaging supplied on the EU market will be recycled, single-use plastic consumption will be reduced, and MP use will be reduced on purpose (Nielsen et al. 2023).

12.6.1 By Microorganism

Cole et al. (2014) found that plant and animal taxa levels of microplastic are rapidly increasing. These plastic particles are everywhere, even in particularly remote regions of the world, like the Antarctic Islands or the deep ocean. Therefore, it is possible to get rid of this plastic trash by letting MPs degrade. This process is carried out by microorganisms, which are more inclined to convert the waste plastic into the plant matter, methane, carbon dioxide, water, and a number of inorganic compounds. The procedure also heavily relies on catalysts from living beings (Nabi et al. 2023). The only kinds of plastic polymers that can induce microplastics to deteriorate are those with specific chemical and physical attributes. Additionally, environmental factors that affect biotic deterioration include temperature, air humidity, sunlight, ultraviolet radiation, and other light-related factors (Chamas et al. 2020).

12.6.2 Fungal Degradation

Numerous fungus has been shown to exploit plastics as a source of nutrition. Therefore, these organisms could be used to degrade MPs. Researchers did this by destroying PE-based MPs with marine fungi (Zalerion maritimum), which are common in Portuguese maritime waters. In the experiments, beneficial associations were observed as an upsurge in the proportion of microbiological weight with a decrease in the percentage of plastic mass after 14 days of contact with microplastic to the fungus in regulated condition (25 °C, dark circumstances, 120 rpm constant stirring) (Kim et al. 2022). According to Paço et al. (2017), in this instance, the plastic mass variation was 56.7% 2.9, the rate of clearance was over 43%, and the biomass variability was 82.0% 2.1. The study was also conducted in 2011 to determine whether endophytic fungus (*Pestalotiopsis microspora*) could deteriorate PUR MPs. PUR is viewed as a carbon source, yet in anaerobic conditions (25 °C), serine hydrolase is what breaks apart the polymer.

In addition, in a comparable study, Junqing Zhang and associates (Zhang et al. 2020) identified a fungus (Aspergillus flavus) termed PEDX3 from intestinal material from wax moths or honeycomb moths (Galleria mellonella) for PE repair within 28 days. The computed value of 3.9025 1.18% for the mass loss of high-density polyethylene (HDPE) has been confirmed by the FTIR investigation. Using reverse transcription-polymerase chain reaction analysis, the second oxidising substrate of the laccase-like multi-copper oxidase genes (AFLA 006190 and AFLA 053930) was also identified (Dey et al. 2021). The fungus may degrade MPs in a controlled environment with a particular temperature, stirring, and an abundance of an enzyme (serine hydrolases). The length of time for plastic degradation is greater, nevertheless, as a result of the decreased reaction rate. Pretreatment methods such as solvolysis, ozonolysis, and photo-oxidation have to be applied in order to begin polymer breakdown preceding fungal destruction (Temporiti et al. 2022).

12.6.3 Marine Microalga

MHET (2-hydroxyethyl terephthalate) and TPA (terephthalic acid), which are formed when this microalga is transformed into a possible deterioration chassis, may break down PET into these products. Aqueous microalga (Chlorophyta,

Rhodophyta, Cyanobacteria, and Bacillariophyta) should be investigated for their ability to degrade MPs with a potential for business in a comparable manner that marine microalga can discharge an enzyme that shreds up MPs (Mrigwani et al. 2022).

12.6.4 Periphytic Biofilm

In 2020, Sadaf Shabbir and his research associates created periphytic biofilm (EPS), an extracellular polymer material, in order to break down various polymeric MPs including PET, PE, and PP. Weight loss was used to measure plastic degradation during the tests to determine MPs when extra nutrients like glucose and peptone were present (Shabbir et al. 2020). The batch experiment was conducted in a dark room for 60 days at a pH of 7 and a temperature of 30 °C. When additional nutrients are present, a significant limit (roughly a twofold increase) is reached. In comparison to periphytic biofilm alone, the conjunction of glucose with biofilm raised the decomposition rate by 13.24–19.72%, 9.52–18.02%, and 5.95–14.02% for PET, PP, and PE, correspondingly. Therefore, periphytic biofilm's plastic disintegration rate is still minor contrasted to that of other breaking-down procedures. It is important to do study to find more effective catalysts, such as sucrose, lactose, and fructose, which are polysaccharide starches that can speed up the evaluation of destruction (Tarafdar et al. 2021).

12.6.5 By Adsorption

When MPs are removed from water utilizing adsorbent materials, an intriguing and cutting-edge technique may be biosorption. According to Bilal et al. (2018), biosorption, a simple physicochemical procedure requiring the van der Waals force or ion exchange, is how adsorbate attaches to the surface of adsorbents. Recent studies have shown that the coastal brown alga Fucus vesiculosus may be able to adsorb MPs due to the presence of alginic acid in its cell walls. The presence of the carboxylic functional group on the surface of brown algae on alginate polymer is closely linked to the algae's capacity to soak up substances. Thus, it can be concluded that the presence of the carboxylic group is what confers a capability of adsorbents to bind plastic.

12.6.6 By Magnetic Extraction

The analysis found that magnetic separation had a better chance of removing MPs from sewage. With the aid of magnetic seeds (Fe nanoparticles), oxalic acid (as a Fe di-sorbent), and external magnetic attraction, the MPs were isolated from the seeds. Iron-based nanoparticles have been used because of their ferromagnetic properties, inexpensive price, and higher exploitable specific surface area (Rhein et al. 2022).

The nanoparticles' surface were coated with hexadecyltrimethoxysilane to improve their hydrophobicity, which made them more receptive to water and made it possible for plastic particles to stick to them. More than 92% of cases allow for the elimination of a range of 10–20 m PS and PE beads (Ramage et al. 2022).

12.6.7 By Electrocoagulation

In addition to chemical and microbial coagulation, the electrocoagulation procedure is an approach to treatment in which cations are produced by metal electrodes in an electrical field. It has an abundance of benefits notably sludge reduction, automatic treatment, and efficient and economical operation, and it produces coagulants with metal electrodes that are more likely to come into touch with MPs. Using this technique, PE-based MPs (300–355 m) were also eliminated. A greater clearing percentage (99.24%) at pH 7.5 and current density 11 A/m^2 was seen throughout the experiments because there was more coagulant analysis at this neutral pH level. Additionally, it was claimed that the optimum effectiveness was attained within 40 min of usage and that a continual increase in current density did not result in improved removal effectiveness (Vinduja and Balasubramanian 2013). Water permeability has also had no effect on the removal's efficiency. The operational costs for MP removal were additionally determined and compiled, demonstrating that electrocoagulation can provide greater effectiveness at a lower cost than additional pollutant removal (Perren and Kozinets 2018). This process may be used economically to tackle MPs prior to release into the waters around them.

12.6.8 Bioengineering-Based Solutions

The other bioengineering-based solution is to explore innovative decomposition pathways for common plastics, such as different bacterial and fungi species, or to retrieve the necessary enzymes to ensure the breakdown by enzymes of plastics. Polyesters that are biodegradable can be hydrolyzed by carboxylesterases that are found external to cells (Picó and Barceló 2019). Specialized bacteria, including the marine fungus Zalerion maritimum for PE and Ideonella sakaiensis for PET, may degrade plastics. Although options seem encouraging, the removal of macro and MPs is still in its early stages at the laboratory scale. The development of techniques for the in situ decay of MPs through the use of microbes or using enhanced attenuation that occurs naturally employing native microflora is essential (Gao and Sun 2021). Since bioengineering-based remedies are crucial for the now and the future, much more study and innovation are still needed prior to they can be widely utilized. Biotechnology-based approaches are the exciting directions to follow in order to ameliorate what may be the most important environmental issue of our time. An increasingly circular and ecological plastic society is absolutely necessary (Degli Esposti et al. 2021).

12.7 Conclusion

To locate microplastics in intricate environmental matrices, several combinations of microplastic analytical methodologies will be used. It takes longer to find microplastics as their size gets smaller. Sub-micron analysis has grown increasingly important to assess the risks and consequences of microplastics on the environment and human health. As the demand for microplastic contamination surveillance increases, it will be crucial to improve present approaches and develop new procedures in order to reduce recognition time and effort. Future studies should concentrate on creating a fully or completely automated analysis methodology that allows for image analysis-based methods to identify the physical characteristics (such as size and form) of microplastics with chemical analysis to identify the plastic's component elements. Researchers should continue to identify and remove microplastics from environmental samples by introducing and perfecting novel dyeing processes, nanotechnology, and approaches to analysis.

Acknowledgement Dr. Enespa would like to thank the Department of Plant Pathology, Sri Mahesh Prasad Degree College, Lucknow, and Dr. Prem Chandra would like to acknowledge the School of Agriculture, IGNOU, New Delhi, for assistance in the preparation of this chapter manuscript.

References

Adão T, Hruška J, Pádua L, Bessa J, Peres E, Morais R, Sousa JJ (2017) Hyperspectral imaging: a review on UAV-based sensors, data processing and applications for agriculture and forestry. Remote Sens 9(11):1110

Adegoke KA, Adu FA, Oyebamiji AK, Bamisaye A, Adigun RA, Olasoji SO, Ogunjinmi OE (2023) Microplastics toxicity, detection, and removal from water/wastewater. Mar Pollut Bull 187:114546

Adhikari S, Kelkar V, Kumar R, Halden RU (2022) Methods and challenges in the detection of microplastics and nanoplastic: a mini-review. Polym Int 71(5):543–551

Alvim CB, Mendoza-Roca J, Bes-Piá A (2020) Wastewater treatment plant as microplastics release source—quantification and identification techniques. J Environ Manag 255:109739

Amini Moghaddam M, Di Martino A, Šopík T, Fei H, Císař J, Pummerová M, Sedlařík V (2021) Polylactide/polyvinylalcohol-based porous bioscaffold loaded with gentamicin for wound dressing applications. Polymers 13(6):921

Andler R, Tiso T, Blank L, Andreeßen C, Zampolli J, D'Afonseca V, Díaz-Barrera A (2022) Current progress on the biodegradation of synthetic plastics: from fundamentals to biotechnological applications. Rev Environ Sci Biotechnol 21(4):829–850

Andrady AL, Neal MA (2009) Applications and societal benefits of plastics. Philos Trans R Soc B Biol Sci 364(1526):1977–1984

Araujo CF, Nolasco MM, Ribeiro AM, Ribeiro-Claro PJ (2018) Identification of microplastics using Raman spectroscopy: latest developments and future prospects. Water Res 142:426–440

Armbruster DA, Pry T (2008) Limit of blank, limit of detection and limit of quantitation. Clin Biochem Rev 29(1):S49

Ateia M, Ersan G, Alalm MG, Boffito DC, Karanfil T (2022) Emerging investigator series: microplastic sources, fate, toxicity, detection, and interactions with micro pollutants in aquatic ecosystems—a review of reviews. Environ Sci Processes Impacts 24(2):172–195

Atugoda T, Vithanage M, Wijesekara H, Bolan N, Sarmah AK, Bank MS, Ok YS (2021) Interactions between microplastics, pharmaceuticals and personal care products: implications for vector transport. Environ Int 149:106367

Ayyamperumal R, Huang X, Li F, Chengjun Z, Chellaiah G, Gopalakrishnan G, Antony JK (2022) Investigation of microplastic contamination in the sediments of Noyyal River-southern India. J Hazard Mater Adv 8:100198

Azizah N, Hashim U, Arshad MM, Gopinath SC, Nadzirah S, Farehanim MA, Ayub RM (2016) Analysis study of single gold nanoparticle system of interdigitated device electrodes (ides) by using energy-dispersive X-ray (EDX). ARPN J Eng Appl Sci 11(14):8889–8892

Banerji S, Meem M, Majumder A, Dvonch C, Sensale-Rodriguez B, Menon R (2019) Single flat lens enabling imaging in the short-wave infra-red (SWIR) band. OSA Continuum 2(10): 2968–2974

Barth A (2007) Infrared spectroscopy of proteins. Biochim Biophys Acta 1767(9):1073–1101

Bellasi A, Binda G, Pozzi A, Galafassi S, Volta P, Bettinetti R (2020) Microplastic contamination in freshwater environments: a review, focusing on interactions with sediments and benthic organisms. Environments 7(4):30

Bilal M, Rasheed T, Sosa-Hernández JE, Raza A, Nabeel F, Iqbal HM (2018) Biosorption: an interplay between marine algae and potentially toxic elements—a review. Mar Drugs 16(2):65

Birch QT, Potter PM, Pinto PX, Dionysiou DD, Al-Abed SR (2021) Isotope ratio mass spectrometry and spectroscopic techniques for microplastics characterization. Talanta 224:121743

Bitter H, Lackner S (2021) Fast and easy quantification of semi-crystalline microplastics in exemplary environmental matrices by differential scanning calorimetry (DSC). Chem Eng J 423:129941

Blair RM, Waldron S, Phoenix VR, Gauchotte-Lindsay C (2019) Microscopy and elemental analysis characterization of microplastics in sediment of a freshwater urban river in Scotland, UK. Environ Sci Pollut Res 26:12491–12504

Campanale C, Massarelli C, Savino I, Locaputo V, Uricchio VF (2020) A detailed review study on potential effects of microplastics and additives of concern on human health. Int J Environ Res Public Health 17(4):1212

Cescon A, Jiang JQ (2020) Filtration process and alternative filter media material in water treatment. Water 12(12):3377

Chakraborty I, Banik S, Biswas R, Yamamoto T, Noothalapati H, Mazumder N (2022) Raman spectroscopy for microplastic detection in water sources: a systematic review. Int J Environ Sci Technol 15:1–14

Chamas A, Moon H, Zheng J, Qiu Y, Tabassum T, Jang JH, Suh S (2020) Degradation rates of plastics in the environment. ACS Sustain Chem Eng 8(9):3494–3511

Chandra P, Enespa, Singh DP (2020) Microplastic degradation by bacteria in aquatic ecosystem. In Microorganisms for sustainable environment and health. Elsevier, pp 431–467

Chen Y, Wen D, Pei J, Fei Y, Ouyang D, Zhang H, Luo Y (2020) Identification and quantification of microplastics using Fourier-transform infrared spectroscopy: current status and future prospects. Curr Opin Environ Sci Health 18:14–19

Cheng YL, Kim JG, Kim HB, Choi JH, Tsang YF, Baek K (2021) Occurrence and removal of microplastics in wastewater treatment plants and drinking water purification facilities: a review. Chem Eng J 410:128381

Cole M, Webb H, Lindeque PK, Fileman ES, Halsband C, Galloway TS (2014) Isolation of microplastics in biota-rich seawater samples and marine organisms. Sci Rep 4(1):1–8

Conley K, Clum A, Deepe J, Lane H, Beckingham B (2019) Wastewater treatment plants as a source of microplastics to an urban estuary: removal efficiencies and loading per capita over one year. Water Res X 3:100030

Corradini F, Bartholomeus H, Lwanga EH, Gertsen H, Geissen V (2019) Predicting soil microplastic concentration using vis-NIR spectroscopy. Sci Total Environ 650:922–932

Cristaldi A, Fiore M, Zuccarello P, Oliveri Conti G, Grasso A, Nicolosi I, Ferrante M (2020) Efficiency of wastewater treatment plants (Wwtps) for microplastic removal: a systematic review. Int J Environ Res Public Health 17(21):8014

Cunsolo S, Williams J, Hale M, Read DS, Couceiro F (2021) Optimising sample preparation for FTIR-based microplastic analysis in wastewater and sludge samples: multiple digestions. Anal Bioanal Chem 413:3789–3799

De Falco F, Di Pace E, Cocca M, Avella M (2019) The contribution of washing processes of synthetic clothes to microplastic pollution. Sci Rep 9(1):6633

De Gisi S, Lofrano G, Grassi M, Notarnicola M (2016) Characteristics and adsorption capacities of low-cost sorbents for wastewater treatment: a review. Sustain Mater Technol 9:10–40

Degli Esposti M, Morselli D, Fava F, Bertin L, Cavani F, Viaggi D, Fabbri P (2021) The role of biotechnology in the transition from plastics to bioplastics: an opportunity to reconnect global growth with sustainability. FEBS Open Biol 11(4):967–983

Dekiff JH, Remy D, Klasmeier J, Fries E (2014) Occurrence and spatial distribution of microplastics in sediments from Norderney. Environ Pollut 186:248–256

Dey TK, Uddin ME, Jamal M (2021) Detection and removal of microplastics in wastewater: evolution and impact. Environ Sci Pollut Res 28:16925–16947

Di Nunno F, Granata F, Parrino F, Gargano R, de Marinis G (2021) Microplastics in combined sewer overflows: an experimental study. J Mar Sci Eng 9(12):1415

Duis K, Coors A (2016) Microplastics in the aquatic and terrestrial environment: sources (with a specific focus on personal care products), fate and effects. Environ Sci Eur 28(1):1–25

Duvigneau J, Schonherr H, Vancso GJ (2010) Nanoscale thermal AFM of polymers: transient heat flow effects. ACS Nano 4(11):6932–6940

Dyachenko PA, Dolotov LE, Lazareva EN, Kozlova AA, Inozemtseva OA, Verkhovskii RA, Zharov VP (2020) Detection of melanoma cells in whole blood samples using spectral imaging and optical clearing. IEEE J Select Top Quantum Electron 27(4):1–11

Edo C, González-Pleiter M, Leganés F, Fernández-Piñas F, Rosal R (2020) Fate of microplastics in wastewater treatment plants and their environmental dispersion with effluent and sludge. Environ Pollut 259:113837

Erni-Cassola G, Gibson MI, Thompson RC, Christie-Oleza JA (2017) Lost, but found with Nile red: a novel method for detecting and quantifying small microplastics (1 mm to 20 μm) in environmental samples. Environ Sci Technol 51(23):13641–13648

Fakirov S (2013) Nano-and microfibrillar single-polymer composites: a review. Macromol Mater Eng 298(1):9–32

Fauser P, Vorkamp K, Strand J (2022) Residual additives in marine microplastics and their risk assessment—a critical review. Mar Pollut Bull 177:113467

Ferronato N, Torretta V (2019) Waste mismanagement in developing countries: a review of global issues. Int J Environ Res Public Health 16(6):1060

Fries E, Dekiff JH, Willmeyer J, Nuelle MT, Ebert M, Remy D (2013) Identification of polymer types and additives in marine microplastic particles using pyrolysis-GC/MS and scanning electron microscopy. Environ Sci Processes Impacts 15(10):1949–1956

Gao R, Sun C (2021) A marine bacterial community capable of degrading poly (ethylene terephthalate) and polyethylene. J Hazard Mater 416:125928

Gies EA, LeNoble JL, Noël M, Etemadifar A, Bishay F, Hall ER, Ross PS (2018) Retention of microplastics in a major secondary wastewater treatment plant in Vancouver, Canada. Mar Pollut Bull 133:553–561

Greenspan P, Fowler SD (1985) Spectrofluorometric studies of the lipid probe, Nile red. J Lipid Res 26(7):781–789

Harrison JP, Ojeda JJ, Romero-González ME (2012) The applicability of reflectance micro-Fourier-transform infrared spectroscopy for the detection of synthetic microplastics in marine sediments. Sci Total Environ 416:455–463

Heidari A (2018) Atomic force microscopy based infrared (AFM–IR) spectroscopy and nuclear resonance vibrational spectroscopy comparative study on malignant and benign human cancer cells and tissues under synchrotron radiation with the passage of time. J Appl Biotechnol Bioeng 5(3):142–148

Hernández Fernández J, Cano H, Guerra Y, Puello Polo E, Ríos-Rojas JF, Vivas-Reyes R, Oviedo J (2022) Identification and quantification of microplastics in effluents of wastewater treatment plant by differential scanning calorimetry (DSC). Sustainability 14(9):4920

Herzke D, Ghaffari P, Sundet JH, Tranang CA, Halsband C (2021) Microplastic fiber emissions from wastewater effluents: abundance, transport behavior and exposure risk for biota in an arctic fjord. Front Environ Sci 9:662168

Issac MN, Kandasubramanian B (2021) Effect of microplastics in water and aquatic systems. Environ Sci Pollut Res 28:19544–19562

Ivleva NP (2021) Chemical analysis of microplastics and nanoplastic: challenges, advanced methods, and perspectives. Chem Rev 121(19):11886–11936

Iyare PU, Ouki SK, Bond T (2020) Microplastics removal in wastewater treatment plants: a critical review. Environ Sci Water Res Technol 6(10):2664–2675

Kabir MS, Wang H, Luster-Teasley S, Zhang L, Zhao R (2023) Microplastics in landfill leachate: sources, detection, occurrence, and removal. Environ Sci Ecotechnol 16:100256

Kamemoto LE, Misra AK, Sharma SK, Goodman MT, Luk H, Dykes AC, Acosta T (2010) Near-infrared micro-Raman spectroscopy for in vitro detection of cervical cancer. Appl Spectrosc 64(3):255–261

Kang H, Park S, Lee B, Ahn J, Kim S (2020) Modification of a Nile red staining method for microplastics analysis: a Nile red plate method. Water 12(11):3251

Karlsson TM, Grahn H, van Bavel B, Geladi P (2016) Hyperspectral imaging and data analysis for detecting and determining plastic contamination in seawater filtrates. J Near Infrared Spectrosc 24(2):141–149

Kay P, Hiscoe R, Moberley I, Bajic L, McKenna N (2018) Wastewater treatment plants as a source of microplastics in river catchments. Environ Sci Pollut Res 25:20264–20267

Kedzierski M, Geslain E, Pedrotti ML, Ghiglione JF, Bruzaud S (2021) Pre-detection of microplastics using active thermography. Chemosphere 262:127648

Kehinde O, Ramonu OJ, Babaremu KO, Justin LD (2020) Plastic wastes: environmental hazard and instrument for wealth creation in Nigeria. Heliyon 6(10):e05131

Khant NA, Kim H (2022) Review of current issues and management strategies of microplastics in groundwater environments. Water 14(7):1020

Kim SH, Lee JW, Kim JS, Lee W, Park MS, Lim YW (2022) Plastic-inhabiting fungi in marine environments and PCL degradation activity. Antonie Van Leeuwenhoek 115(12):1379–1392

Kiran BR, Kopperi H, Venkata Mohan S (2022) Micro/nano-plastics occurrence, identification, risk analysis and mitigation: challenges and perspectives. Rev Environ Sci Biotechnol 21(1): 169–203

Krauklis AE, Karl CW, Rocha IB, Burlakovs J, Ozola-Davidane R, Gagani AI, Starkova O (2022) Modelling of environmental ageing of polymers and polymer composites—modular and multiscale methods. Polymers 14(1):216

Krishnan RY, Manikandan S, Subbaiya R, Karmegam N, Kim W, Govarthanan M (2022) Recent approaches and advanced wastewater treatment technologies for mitigating emerging microplastics contamination–a critical review. Sci Total Environ 858:159681

Kumar R, Verma A, Shome A, Sinha R, Sinha S, Jha PK, Vara Prasad PV (2021) Impacts of plastic pollution on ecosystem services, sustainable development goals, and need to focus on circular economy and policy interventions. Sustainability 13(17):9963

Kwon HJ, Hidayaturrahman H, Peera SG, Lee TG (2022) Elimination of microplastics at different stages in wastewater treatment plants. Water 14(15):2404

Lagarde F, Olivier O, Zanella M, Daniel P, Hiard S, Caruso A (2016) Microplastic interactions with freshwater microalgae: hetero-aggregation and changes in plastic density appear strongly dependent on polymer type. Environ Pollut 215:331–339

Lamichhane G, Acharya A, Marahatha R, Modi B, Paudel R, Adhikari A, Parajuli N (2023) Microplastics in environment: global concern, challenges, and controlling measures. Int J Environ Sci Technol 20(4):4673–4694
Lapray PJ, Wang X, Thomas JB, Gouton P (2014) Multispectral filter arrays: recent advances and practical implementation. Sensors 14(11):21626–21659
Lares M, Ncibi MC, Sillanpää M, Sillanpää M (2018) Occurrence, identification and removal of microplastic particles and fibers in conventional activated sludge process and advanced MBR technology. Water Res 133:236–246
Lenaker PL, Corsi SR, Mason SA (2020) Spatial distribution of microplastics in surficial benthic sediment of Lake Michigan and Lake Erie. Environ Sci Technol 55(1):373–384
Lenz R, Enders K, Stedmon CA, Mackenzie DM, Nielsen TG (2015) A critical assessment of visual identification of marine microplastic using Raman spectroscopy for analysis improvement. Mar Pollut Bull 100(1):82–91
Li J, Shan E, Zhao J, Teng J, Wang Q (2023) The factors influencing the vertical transport of microplastics in marine environment: a review. Sci Total Environ 870:161893
Ling C, Sommer AJ (2015) The advantages of an attenuated total internal reflection infrared microspectroscopic imaging technique for the analysis of polymer laminates. Microsc Microanal 21(3):626–636
Liobe JC, Wieners J (2022) Low SWaP SWIR cameras for multi-mode applications. Infrared Technol Appl XLVIII 12107:80–84
Liu J, Zhu X, Cao Z (2019) Poly (acrylonitrile–butadiene–styrene) as a special β-nucleating agent on the toughness of isotactic polypropylene. Polymers 11(11):1894
Liu W, Zhang J, Liu H, Guo X, Zhang X, Yao X, Zhang T (2021) A review of the removal of microplastics in global wastewater treatment plants: characteristics and mechanisms. Environ Int 146:106277
Liu Y, Wang B, Pileggi V, Chang S (2022) Methods to recover and characterize microplastics in wastewater treatment plants. Case Stud Chem Environ Eng 5:100183
Löder MGJ, Kuczera M, Mintenig S, Lorenz C, Gerdts G (2015) Focal plane array detector-based micro-Fourier-transform infrared imaging for the analysis of microplastics in environmental samples. Environ Chem 12(5):563–581
Lu B, Dao PD, Liu J, He Y, Shang J (2020) Recent advances of hyper spectral imaging technology and applications in agriculture. Remote Sens 12(16):2659
Luo H, Xiang Y, Zhao Y, Li Y, Pan X (2020) Nanoscale infrared, thermal and mechanical properties of aged microplastics revealed by an atomic force microscopy coupled with infrared spectroscopy (AFM-IR) technique. Sci Total Environ 744:140944
Luo H, Zeng Y, Zhao Y, Xiang Y, Li Y, Pan X (2021) Effects of advanced oxidation processes on leachates and properties of microplastics. J Hazard Mater 413:125342
Lv L, Yan X, Feng L, Jiang S, Lu Z, Xie H, Li C (2021) Challenge for the detection of microplastics in the environment. Water Environ Res 93(1):5–15
Lwanga EH, Beriot N, Corradini F, Silva V, Yang X, Baartman J, Geissen V (2022) Review of microplastic sources, transport pathways and correlations with other soil stressors: a journey from agricultural sites into the environment. Chem Biol Technol Agric 9(1):1–20
Ma XM, Lu R, Miyakoshi T (2014) Application of pyrolysis gas chromatography/mass spectrometry in lacquer research: a review. Polymers 6(1):132–144
Maes T, Jessop R, Wellner N, Haupt K, Mayes AG (2017) A rapid-screening approach to detect and quantify microplastics based on fluorescent tagging with Nile red. Sci Rep 7(1):44501
Majewsky M, Bitter H, Eiche E, Horn H (2016) Determination of microplastic polyethylene (PE) and polypropylene (PP) in environmental samples using thermal analysis (TGA-DSC). Sci Total Environ 568:507–511
Mariano S, Tacconi S, Fidaleo M, Rossi M, Dini L (2021) Micro and nanoplastics identification: classic methods and innovative detection techniques. Front Toxicol 3:636640

Michielssen MR, Michielssen ER, Ni J, Duhaime MB (2016) Fate of microplastics and other small anthropogenic litter (SAL) in wastewater treatment plants depends on unit processes employed. Environ Sci Water Res Technol 2(6):1064–1073
Miller ME, Hamann M, Kroon FJ (2020) Bioaccumulation and biomagnification of microplastics in marine organisms: a review and meta-analysis of current data. PLoS One 15(10):e0240792
Mintenig SM, Int-Veen I, Löder MG, Primpke S, Gerdts G (2017) Identification of microplastic in effluents of waste water treatment plants using focal plane array-based micro-Fourier-transform infrared imaging. Water Res 108:365–372
Mohanan N, Montazer Z, Sharma PK, Levin DB (2020) Microbial and enzymatic degradation of synthetic plastics. Front Microbiol 11:580709
Mrigwani A, Thakur B, Guptasarma P (2022) Conversion of polyethylene terephthalate into pure terephthalic acid through synergy between a solid-degrading cutinase and a reaction intermediate-hydrolysing carboxylesterase. Green Chem 24(17):6707–6719
Murphy F, Ewins C, Carbonnier F, Quinn B (2016) Wastewater treatment works (WWTW) as a source of microplastics in the aquatic environment. Environ Sci Technol 50(11):5800–5808
Nabi I, Zaheer M, Jin W, Yang L (2023) Biodegradation of macro-and micro-plastics in environment: a review on mechanism, toxicity, and future perspectives. Sci Total Environ 858:160108
Nava V, Frezzotti ML, Leoni B (2021) Raman spectroscopy for the analysis of microplastics in aquatic systems. Appl Spectrosc 75(11):1341–1357
Ngo PL, Pramanik BK, Shah K, Roychand R (2019) Pathway, classification and removal efficiency of microplastics in wastewater treatment plants. Environ Pollut 255:113326
Nguyen B, Tufenkji N (2022) Single-particle resolution fluorescence microscopy of nanoplastics. Environ Sci Technol 56(10):6426–6435
Nguyen-Tri P, Ghassemi P, Carriere P, Nanda S, Assadi AA, Nguyen DD (2020) Recent applications of advanced atomic force microscopy in polymer science: a review. Polymers 12(5):1142
Nielsen MB, Clausen LPW, Cronin R, Hansen SF, Oturai NG, Syberg K (2023) Unfolding the science behind policy initiatives targeting plastic pollution. Microplast Nanoplast 3(1):3
Oliveira J, Belchior A, da Silva VD, Rotter A, Petrovski Ž, Almeida PL, Gaudêncio SP (2020) Marine environmental plastic pollution: mitigation by microorganism degradation and recycling valorization. Front Mar Sci 7:567126
Oveisy N, Rafiee M, Rahmatpour A, Nejad AS, Hashemi M, Eslami A (2022) Occurrence, identification, and discharge of microplastics from effluent and sludge of the largest WWTP in Iran—south of Tehran. Water Environ Res 94(8):e10765
Paço A, Duarte K, da Costa JP, Santos PSM, Pereira R, Pereira ME (2017) Biodegradation of polyethylene microplastics by the marine fungus Zalerion maritimum. Sci Total Environ 586: 10–15
Park H, Park B (2021) Review of microplastic distribution, toxicity, analysis methods, and removal technologies. Water 13(19):2736
Perren R, Kozinets RV (2018) Lateral exchange markets: how social platforms operate in a networked economy. J Mark 82(1):20–36
Picó Y, Barceló D (2019) Analysis and prevention of microplastics pollution in water: current perspectives and future directions. ACS Omega 4(4):6709–6719
Quach-Cu J, Herrera-Lynch B, Marciniak C, Adams S, Simmerman A, Reinke RA (2018) The effect of primary, secondary, and tertiary wastewater treatment processes on antibiotic resistance gene (ARG) concentrations in solid and dissolved wastewater fractions. Water 10(1):37
Ramage SJ, Pagaling E, Haghi RK, Dawson LA, Yates K, Prabhu R, Devalla S (2022) Rapid extraction of high-and low-density microplastics from soil using high-gradient magnetic separation. Sci Total Environ 831:154912
Rani M, Ducoli S, Federici S, Depero LE (2023) Influx of near-infrared technology in microplastic community: a bibliometric analysis. Microplastics 2(1):107–121
Reddy AS, Nair AT (2022) The fate of microplastics in wastewater treatment plants: an overview of source and remediation technologies. Environ Technol Innov 28:102815

Reineccius J, Bresien J, Waniek JJ (2021) Separation of microplastics from mass-limited samples by an effective adsorption technique. Sci Total Environ 788:147881

Renner G, Schmidt TC, Schram J (2018) Analytical methodologies for monitoring micro (nano) plastics: which are fit for purpose? Curr Opin Environ Sci Health 1:55–61

Rhein F, Nirschl H, Kaegi R (2022) Separation of microplastic particles from sewage sludge extracts using magnetic seeded filtration. Water Res X 17:100155

Rusten B, Ødegaard H (2006) Evaluation and testing of fine mesh sieve technologies for primary treatment of municipal wastewater. Water Sci Technol 54(10):31–38

Sadia M, Mahmood A, Ibrahim M, Irshad MK, Quddusi AHA, Bokhari A, Show PL (2022) Microplastics pollution from wastewater treatment plants: a critical review on challenges, detection, sustainable removal techniques and circular economy. Environ Technol Innov 28: 102946

Saldarriaga JF, Aguado R, Pablos A, Amutio M, Olazar M, Bilbao J (2015) Fast characterization of biomass fuels by thermogravimetric analysis (TGA). Fuel 140:744–751

Šaravanja A, Pušić T, Dekanić T (2022) Microplastics in wastewater by washing polyester fabrics. Materials 15(7):2683

Sathya K, Nagarajan K, Carlin Geor Malar G, Rajalakshmi S, Raja Lakshmi P (2022) A comprehensive review on comparison among effluent treatment methods and modern methods of treatment of industrial wastewater effluent from different sources. Appl Water Sci 12(4):70

Sauer S, Kliem M (2010) Mass spectrometry tools for the classification and identification of bacteria. Nat Rev Microbiol 8(1):74–82

Schernewski G, Radtke H, Hauk R, Baresel C, Olshammar M, Osinski R, Oberbeckmann S (2020) Transport and behavior of microplastics emissions from urban sources in the Baltic Sea. Front Environ Sci 8:579361

Schirinzi GF, Llorca M, Seró R, Moyano E, Barceló D, Abad E, Farré M (2019) Trace analysis of polystyrene microplastics in natural waters. Chemosphere 236:124321

Schöpel B, Stamminger R (2019) A comprehensive literature study on microfibres from washing machines. Tenside Surfactants Detergents 56(2):94–104

Shabbir S, Faheem M, Ali N, Kerr PG, Wang LF, Kuppusamy S, Li Y (2020) Periphytic biofilm: an innovative approach for biodegradation of microplastics. Sci Total Environ 717:137064

Shipp DW, Sinjab F, Notingher I (2017) Raman spectroscopy: techniques and applications in the life sciences. Adv Opt Photon 9(2):315–428

Shruti VC, Pérez-Guevara F, Roy PD, Kutralam-Muniasamy G (2022) Analyzing microplastics with Nile red: emerging trends, challenges, and prospects. J Hazard Mater 423:127171

Sierra I, Chialanza MR, Faccio R, Carrizo D, Fornaro L, Pérez-Parada A (2020) Identification of microplastics in wastewater samples by means of polarized light optical microscopy. Environ Sci Pollut Res 27:7409–7419

Silva AB, Bastos AS, Justino CI, da Costa JP, Duarte AC, Rocha-Santos TA (2018) Microplastics in the environment: challenges in analytical chemistry—a review. Anal Chim Acta 1017:1–19

Singh S, Kalyanasundaram M, Diwan V (2021) Removal of microplastics from wastewater: available techniques and way forward. Water Sci Technol 84(12):3689–3704

Smith M, Love DC, Rochman CM, Neff RA (2018) Microplastics in seafood and the implications for human health. Curr Environ Health Rep 5:375–386

Sobhani Z, Al Amin M, Naidu R, Megharaj M, Fang C (2019) Identification and visualization of microplastics by Raman mapping. Anal Chim Acta 1077:191–199

Stock V, Laurisch C, Franke J, Dönmez MH, Voss L, Böhmert L, Sieg H (2021) Uptake and cellular effects of PE, PP, PET and PVC microplastic particles. Toxicol In Vitro 70:105021

Sturtevant JM (1987) Biochemical applications of differential scanning calorimetry. Annu Rev Phys Chem 38(1):463–488

Sun J, Dai X, Wang Q, Van Loosdrecht MC, Ni BJ (2019) Microplastics in wastewater treatment plants: detection, occurrence and removal. Water Res 152:21–37

Szymańska M, Obolewski K (2020) Microplastics as contaminants in freshwater environments: a multidisciplinary review. Ecohydrol Hydrobiol 20(3):333–345

Tagg AS, Sapp M, Harrison JP, Ojeda JJ (2015) Identification and quantification of microplastics in wastewater using focal plane array-based reflectance micro-FT-IR imaging. Anal Chem 87(12): 6032–6040

Talvitie J, Mikola A, Setälä O, Heinonen M, Koistinen A (2017) How well is microlitter purified from wastewater?—a detailed study on the stepwise removal of microlitter in a tertiary level wastewater treatment plant. Water Res 109:164–172

Tang KHD, Hadibarata T (2021) Microplastics removal through water treatment plants: its feasibility, efficiency, future prospects and enhancement by proper waste management. Environ Challenges 5:100264

Tarafdar A, Lee JU, Jeong JE, Lee H, Jung Y, Oh HB, Kwon JH (2021) Biofilm development of Bacillus siamensis ATKU1 on pristine short chain low-density polyethylene: a case study on microbe-microplastics interaction. J Hazard Mater 409:124516

Temporiti MEE, Nicola L, Nielsen E, Tosi S (2022) Fungal enzymes involved in plastics biodegradation. Microorganisms 10(6):1180

Thomas D, Schütze B, Heinze WM, Steinmetz Z (2020) Sample preparation techniques for the analysis of microplastics in soil—a review. Sustainability 12(21):9074

Thompson RC, Moore CJ, Vom Saal FS, Swan SH (2009) Plastics, the environment and human health: current consensus and future trends. Philos Trans R Soc B Biol Sci 364(1526): 2153–2166

Turan NB, Erkan HS, Engin GO (2021) Microplastics in wastewater treatment plants: occurrence, fate and identification. Process Saf Environ Prot 146:77–84

Tursi A, Baratta M, Easton T, Chatzisymeon E, Chidichimo F, De Biase M, De Filpo G (2022) Microplastics in aquatic systems, a comprehensive review: origination, accumulation, impact, and removal technologies. RSC Adv 12(44):28318–28340

Urbanek AK, Rymowicz W, Mirończuk AM (2018) Degradation of plastics and plastic-degrading bacteria in cold marine habitats. Appl Microbiol Biotechnol 102:7669–7678

Vaid M, Sarma K, Gupta A (2021) Microplastic pollution in aquatic environments with special emphasis on riverine systems: current understanding and way forward. J Environ Manag 293: 112860

Van Do M, Le TXT, Vu ND, Dang TT (2022) Distribution and occurrence of microplastics in wastewater treatment plants. Environ Technol Innov 26:102286

Veerasingam S, Ranjani M, Venkatachalapathy R, Bagaev A, Mukhanov V, Litvinyuk D, Vethamony P (2021) Contributions of Fourier transform infrared spectroscopy in microplastic pollution research: a review. Crit Rev Environ Sci Technol 51(22):2681–2743

Vinduja V, Balasubramanian N (2013) Electrocoagulation-integrated hybrid membrane processes for the treatment of tannery wastewater. Environ Sci Pollut Res 20:7441–7449

Vu TV, Delgado-Saborit JM, Harrison RM (2015) Particle number size distributions from seven major sources and implications for source apportionment studies. Atmos Environ 122:114–132

Wang C, O'Connor D, Wang L, Wu WM, Luo J, Hou D (2022) Microplastics in urban runoff: global occurrence and fate. Water Res 225:119129

Wojnowska-Baryła I, Bernat K, Zaborowska M (2022) Plastic waste degradation in landfill conditions: the problem with microplastics, and their direct and indirect environmental effects. Int J Environ Res Public Health 19(20):13223

Woo H, Seo K, Choi Y, Kim J, Tanaka M, Lee K, Choi J (2021) Methods of analyzing microsized plastics in the environment. Appl Sci 11(22):10640

Xu X, Jian Y, Xue Y, Hou Q, Wang L (2019) Microplastics in the wastewater treatment plants (WWTPs): occurrence and removal. Chemosphere 235:1089–1096

Yan J, Huang Y, Miao YE, Tjiu WW, Liu T (2015) Polydopamine-coated electrospun poly (vinyl alcohol)/poly (acrylic acid) membranes as efficient dye adsorbent with good recyclability. J Hazard Mater 283:730–739

Yousif E, Haddad R (2013) Photodegradation and photostabilization of polymers, especially polystyrene. SpringerPlus 2(1):1–32

Yuan Z, Nag R, Cummins E (2022) Human health concerns regarding microplastics in the aquatic environment-from marine to food systems. Sci Total Environ 823:153730

Zhang J, Gao D, Li Q, Zhao Y, Li L, Lin H, Zhao Y (2020) Biodegradation of polyethylene microplastic particles by the fungus Aspergillus flavus from the guts of wax moth Galleria mellonella. Sci Total Environ 704:135931

Zhang G, Tuo X, Zhai S, Zhu X, Luo L, Zeng X (2022) Near-infrared spectral characteristic extraction and qualitative analysis method for complex multi-component mixtures based on TRPCA-SVM. Sensors 22(4):1654

Zhou F, Wang X, Wang G, Zuo Y (2022) A rapid method for detecting microplastics based on fluorescence lifetime imaging technology (FLIM). Toxics 10(3):118

Zillien C, Posthuma L, Roex E, Ragas A (2022) The role of the sewer system in estimating urban emissions of chemicals of emerging concern. Rev Environ Sci Biotechnol 21:957–991

Report of Microplastic Presence in Drinking Water

13

Pankajkumar R. Waghmare and Mudasir A. Dar

Abstract

Recent studies have shown the presence of microplastics in freshwater reservoirs, commercially available bottled drinking water and in most common drinking water sources, such as tap water from various locations. Microplastics have received more attention because of their potential in carrying pathogenic organisms, heavy metals, and new contaminants, such as phthalate esters, into organisms, possibly impacting the environment and finally human health. The unavailability of effective and targeted water treatment technology to remove microplastics leads to their unintentional consumption by humans, resulting in a variety of health issues. Limited information on microplastic pollution in drinking water is available, as different studies have used various analytical techniques for detecting and extracting microplastics from water samples collected without the standard operating procedure. Therefore, reported studies showed huge variations while determining the microplastic concentration, resulting in difficulty in concluding the exact microplastic content in both tap and bottled water. Microplastics are one of the classes of emerging new contaminants; and very few global-level regulatory attempts have been made for the reduction of microplastic production and usage in various commercial products. However, research on microplastics is in naive phase, and more research is needed for adverse effect on human health. The present chapter focuses on some aspects of microplastics in drinking water,

P. R. Waghmare (✉)
State Key Laboratory of Microbial Technology, Shandong University, Qingdao, Shandong, P. R. China
e-mail: wpankaj97@sdu.edu.cn

M. A. Dar
Biofuels Institute, School of The Environment and Safety Engineering, Jiangsu University, Zhenjiang, Jiangsu, P. R. China

M. Shahnawaz et al. (eds.), *Microplastic Pollution*,
https://doi.org/10.1007/978-981-99-8357-5_13

including its genesis, fate of distribution in environment, human exposure, and effect on human health. The development of efficient microplastic removal technologies is a global concern for achieving the Sustainable Development Program goal 6 of clean drinking water to all human beings.

Keywords

Microplastics · Tap water · Bottled water · Phthalate esters · Human health · Standard operating procedure

13.1 Introduction

Since 1950, global plastic production has suddenly risen to 400 million tons and it is forecasted to reach up to 1100 million tons by 2050 (UN Environment Program 2017; Lim 2021). Plastic is most commonly used in many industrial fields and our daily lives because it is available at cheaper prices, lightweight, durable, and resistant to temperature and chemicals (Lim 2021; Zhao et al. 2023a). Approximately 60% of produced plastics are disposed among which a smaller part is recycled and a major part of disposed plastics directly reach landfills or natural environment due to improper management (Geyer et al. 2017). This mismanaged waste plastic is responsible for plastic pollution which severely affects quality of freshwater and marine water systems (Geyer et al. 2017; Bayarkhuu and Byun 2022). The abiotic and biotic processes are responsible for progressive weathering and fragmentation of plastics generates considerable small-size plastic as microplastics (MPs) (<5 mm) and nanoplastics (NPs) (1–100 nm (or up to 500 nm in some cases) (Gigault et al. 2018; Ng et al. 2018; Cheng et al. 2021; Cao et al. 2022). The presence of these small-size plastic fragments affects aquatic environment biota thus easily enters to food chain and drinking water indicating direct human exposure. Recent studies identified microplastics in human feces (Schwabl et al. 2019) and in blood (Leslie et al. 2022), which provides evidence of microplastic exposure to humans through consumption plastic contaminated processed or packaged water and drinking water. However, understanding the behavior of microplastics in an aqueous system is not easily predictable because of the lack of uniformity in analytical techniques used for environmental samples (Bayarkhuu and Byun 2022). Yong et al. (2020) reported distribution and accumulation of microplastics in living organism's tissue samples or vital organs such as digestion system, blood, liver, pancreas, heart, and brain.

The revised guidelines of European Drinking Water Directive were adopted on December 2020 and enforced from January 12, 2021. Therefore, the European Union (EU) Member States are morally obligated to follow and implement the provisions of the revised directive guidelines into national laws and regulations. Article 13 of the revised directive guidelines refers to a "watch list" of substances including new compounds like microplastics (MPs) which emphasizes its effect on human health via drinking water intended and potential risks to water supply system. This indicates that drinking water must be monitored to access microplastics risk to human health. In short, accurate analysis of microplastic concentrations in drinking water is the most crucial step to meet obligations objected in the revised directive guidelines (Bäuerlein et al. 2022). Clean drinking water is essential for healthy life, and reports on waterborne diseases positively increase the public concerns for clean and safe drinking water (Anadu and Harding 2000). Therefore, along with the economic development in the society, most of population is convinced for packed bottle drinking water instead of tap water (Luo et al. 2018). The various multinational companies have hugely invested in drinking water businesses and fulfilled consumer's drinking water demand by cost effective packaging material like plastic. Improper disposal of plastic in environment caused severe pollution problem resulting to microplastic generation.

The present chapter focuses on various aspects of microplastics in drinking water, including their origin and distribution, contaminant transport ability, effect on the human health, and finally environmental risks. The public awareness about microplastic pollution and development of microplastic removal technologies are the global concern for achieving the clean drinking water goal of a Sustainable Development of humankind.

13.2 Types of Microplastics

Microplastics are insoluble in water which decreases its bioavailability and thus hampers biological degradation. The prolong persistence of microplastics increases their chances of bio-magnification and bio-accumulation. Thus, microplastics can inflict extreme adverse physical and chemical effects on organisms (Law and Thompson 2014). The physical properties of microplastics (concentration, shape, and particle size) are associated with physical adverse effect while the chemical adverse effects are mostly related to the hazardous chemicals (plasticizers, dyes, triclosan) used during plastic synthesis.

Figure 13.1 and Table 13.1 illustrate different types of plastics having variations in chemical compositions, which impart special properties for various applications. However, this may create the problem of its disposal once these plastics get older and finally reach to waste dumping site or directly to environment if not properly disposed. The most commonly used signal-use plastics (water bottles, plastic bags) are made from polyethylene (Fig. 13.1a) a linear ethylene polymer. Polyethylene is further modified to produce polyethylene terephthalate (PET) (Fig. 13.1b), which is having better properties such as flexibility, hydrophobicity, durability, strength, higher crystallinity, and thermoplasticity (Trinh et al. 2020). Polystyrene (Fig. 13.1c) is an aromatic homopolymer discovered in 1839 by Eduard Simon in Germany. Thus, polystyrene is one of the oldest and the most widespread polymers in the world (Terashima 2015). Styrofoam (derived from polystyrene) is the most commonly used shock absorber packaging material while transportation of delicate items (Domininghaus 1998). Polypropylene (Fig. 13.1d) is a saturated linear polymer of propylene (Verbeek and Gavin 2020; Verbeek et al. 2020). Polypropylene has three different forms as an amorphous, crystalline and semicrystalline (Crawford

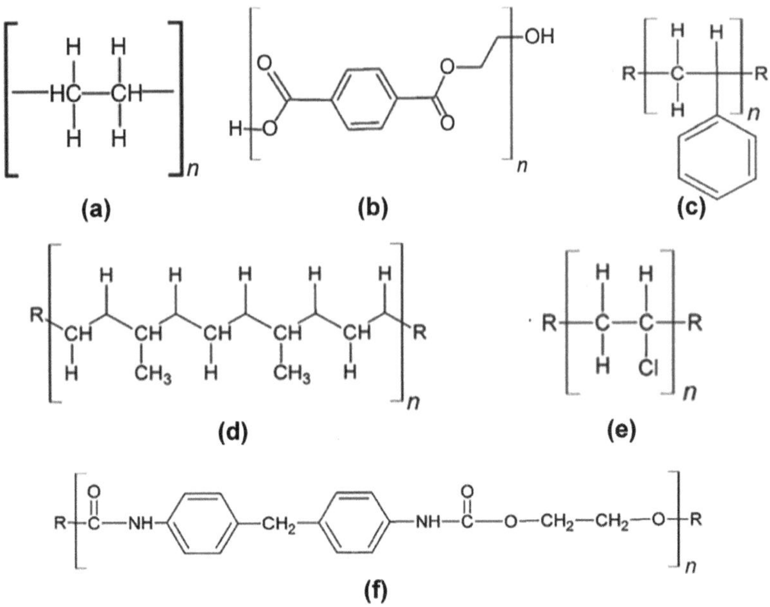

Fig. 13.1 Structure of different types of microplastics. (**a**) Polyethylene, (**b**) polyethylene terephthalate, (**c**) polystyrene, (**d**) polypropylene, (**e**) polyvinyl chloride, and (**f**) polyurethane

and Quinn 2017), due to which it has wide variety of applications. Since the emergence of COVID-19 pandemic, the demands of propylene-based personal protective equipment (PPE) kits drastically increased (Harussani et al. 2022). Globally, polyvinyl chloride (Fig. 13.1e) is the third largest plastic material produced and it is growing at CAGR at 3.2% up to 2030 (Ceresana 2014; Peng et al. 2020). Polyurethane (Fig. 13.1f) is heteropolymer of diisocyanate and diols (polyols), first synthesized by Dr. Otto Bayer in 1937 (Reghunadhan and Thomas 2017; De Souza et al. 2021). Polyurethane is composed of two segments as soft and hard depending on its crystalline nature (Akindoyo et al. 2016).

13.3 Origin of Microplastics

Microplastics are categorized as primary and secondary microplastics depending on size and fragmentation condition. Primary microplastics are produced in micro size, while secondary microplastics are generated in the environment due to fragmentation of large size by plastic physical and chemical processes such as mechanical abrasion and photo-degradation, respectively (Conkle et al. 2018). The environment is already been contaminated with microplastics originated from different sources; Burns and Boxall (2018) demonstrated the presence of microplastic in aquatic environment and presented detailed risk assessment of aquatic animals. The recent studies showed the exponential increase in microplastics in freshwater especially

Table 13.1 Chemical composition of microplastics, its monomer, synthesis process, properties, and applications

Chemical composition	Monomers	Synthesis process	Properties	Applications	References
Polyethylene	Ethylene (C_2H_4)	Cleavage of double bond in C_2H_4 then chain coupling	Hydrophobic	Bottles, Containers, Plastic bags	Arutchelvi et al. (2008)
Polyethylene terephthalate	Ethylene glycol (EG) and terephthalic acid (TPA)	Polymerization of EG and TPA in presence of catalyst	Flexible, hydrophobic, durable, strength, higher crystallinity, thermoplastic	Textile, packaging material (soft drink bottles, food containers for microwave)	Trinh et al. (2020); Djapovic et al. (2021)
Polystyrene	Styrene	Condensation of styrene and dehydrogenation in presence of benzene catalyst	Thermoplastic, Shock absorber	Shock absorbing packaging material, electronics (wire coating, CD), toys, personal consumable (toothbrushes)	Domininghaus (1998); Terashima (2015)
Polypropylene	Propylene	Propylene catalytic dehydrogenation	Lighter, stronger, transparent, resistant to heat, water vapors, oils, and microbial degradation	Injection molding, fibers, film, foam, personal protective equipment (PPE)	Crawford and Quinn (2017); Verbeek and Gavin (2020); Harussani et al. (2022)
Polyvinyl chloride	Vinyl chloride	Chlorination of acetylene by hydrochloric acid	Low cost, durable, lightweight, resistant to chemicals and fire	Building and construction industry (pipes, doors, imitation leather), medical devices	Crawford and Quinn (2017); Plastics Europe (2022)
Polyurethane	Diisocyanate and Diols	Prepolymerization of monomers and chain extension	Flexible, durable, high mechanical strength, resistant (moisture, chemicals)	Construction materials (adhesives, insulation), automobiles (UV coatings, car seats), medical devices (pacemaker, catheters, dialysis tubes, etc.)	Akindoyo et al. (2016); Reghunadhan and Thomas (2017); Heath et al. (2020); De Souza et al. (2021)

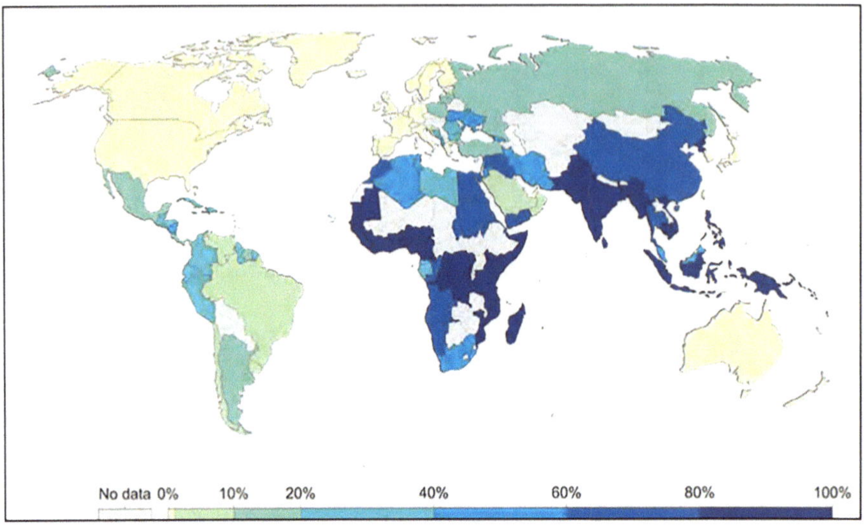

Fig. 13.2 A global perspective of plastic waste disposal mismanagement with relation to socio-economic factor—inequality. (Adapted from Corradini Santander (2021))

surface water (rivers, dams, canals) commonly used for drinking water (Novotna et al. 2019, 2023; Kirstein et al. 2021). Microplastics are frequently detected from different sources of drinking water but unfortunately there is very limited information is available showing its presence and its fate in drinking water (Eerkes-Medrano et al. 2019). Zhang et al. (2020) recently observed the presence of microplastics in 81% of 159 tap water samples from different countries and 93% of 259 packaged water bottle samples from 11 different brands. Shen et al. (2020) overviewed the microplastic sources and transport to raw water and treated drinking water including tap water and bottled water. Refilling of drinking water kiosks made from plastic material are also one of the major contributors for microplastic in drinking water supplied through decentralized water supply system (Pérez-Guevara et al. 2022). Uncontrolled production of plastic from petrochemical byproducts generated during crude-oil refinery resulted to very large-scale production of plastic at cheaper cost. Thus, continuous supply of fresh plastic from oil refinery industry caused sudden rise in single-use plastic used for various applications. Dumping of large volume of plastic waste and litter gets deposited in atmosphere also agriculture and municipal activities produce microplastic finally reach to drinking water resources (Eerkes-Medrano et al. 2019; Kirstein et al. 2021). Inadequate waste disposal and effluent treatment lead to leakage of 80% plastic waste to water bodies in environment (Okeke et al. 2022). A global perspective of plastic waste disposal mismanagement is shown in Fig. 13.2.

The large volumes of plastic wastes were diverted from landfills of developing countries to China and other countries having advanced and large-scale recycling capacities. Recently Government of China banned plastic import since 2018 due to adverse environmental impacts of plastic recycling (Angnunavuri et al. 2023). Therefore, the plastic exporting countries needed the exploration of domestic solution for plastic waste treatment (Wen et al. 2021). Additionally, other Asian and South-East Asian countries (Malaysia, Taiwan, Vietnam, and India) imposed bans single-use plastic and also restricted importation of waste plastics as a step to protect their environment. In developing countries, weak governance system, lack of awareness, higher poverty and population density, unplanned urbanization, and unavailability of information on local hydrology are cumulatively resulted to plastic pollution with exponential increase in plastic waste (Angnunavuri et al. 2023).

Barra et al. (2018) firmly postulated that large-scale plastic production is the principal driver for environmental degradation through waste plastic accumulation, natural resources pollution, climate change, and finally adverse effect on human health. This heavily impacted large group of population which are depended on natural resources like land and water to fulfil daily requirements and economic development (Verster et al. 2017). Microplastics have been detected in remote locations such as high mountain ranges, glacial ice, the Antarctic sea ice, and the Artic Central basin (Allen et al. 2019; Kelly et al. 2020; Kanhai et al. 2020).

Microplastic ingestion by aquatic animals leads to blockage in digestive system, starvation, growth retardation, also compromised fertility. The morphologies of microplastics such as microfiber, filaments, and granules act as good carrier agent for different microbial organisms. Along with this, various additive chemicals to microplastics are responsible for severe ecological toxicity. These additive chemicals are slowly released from plastic materials; Blackburn and Green (2022) reported bisphenol A phthalate, persistent organic pollutants, halogenated flame retardants, polybrominated diphenyl ethers, and polycyclic aromatic hydrocarbons. Recent report by Yin et al. (2023a) showed that microplastics carries pathogenic bacteria and phthalates associated with microplastics also enhance growth of these bacteria. Considering plastic pollution as the global-level serious issue, the classification of plastics and its associated chemicals to hazardous material class will be helpful for careful plastic waste management (Rochman et al. 2013). Table 13.2 enlists classification of microplastics with different shapes, size, and origins. Science Advice for Policy by European Academies (SAPEA) summarized adverse effects of microplastics on biota at different levels of biological organizations (Science Advice for Policy by European Academies 2019) as shown in Fig. 13.3.

13.4 Distribution and Detection of Microplastics

The drinking water is broadly referring to the freshwater systems present on the earth in the form of groundwater and free flowing surface water in rivers. Approximately 70% of the earth is consists of water, but only 2.5% freshwater is available among which only 0.014% of free liquid freshwater is found in lakes, rivers, and swamps (WorldAtlas 2018; King and Brown 2021). This shows that freshwater is limited natural resource and the most crucial factor responsible for maintaining existence of many living things on earth including human beings. An increase in human

Table 13.2 A glance of microplastics with different shapes, sizes, and origins (adopted from Corradini Santander 2021)

Shapes	Polymer	Origin	Size ranges	References
Tire wear particles	Poly butadiene Styrene butadiene Polysulfide Neoprene isoprene	Automobile tires	On-road driving (0.5–20 μm) Road runoff (1–100 μm)	Wagner et al. (2018)
Synthetic fibers	Polyamide Polyester Polyacrylonitrile Polyethylene Polypropylene	Textiles Fishing nets Ropes	Laundry (washing) (25–>3000 μm) Laundry (drying) (19–4000 μm) Industrial uses (<15–20 μm)	Carney Almroth et al. (2018) O'Brien et al. (2020) Du et al. (2020)
Fragments	Polypropylene Polystyrene Polyethylene Polyvinyl chloride	Extrusion leftovers Disposable plastics Building materials Plastic tools	Extrusion (0.001–0.21 μm) Others (<5 mm)	Fadare et al. (2020)
Pellets	Polyethylene Polymethyl methacrylate Polyester Polyvinyl chloride	Microbeads Glitter Face cleansers (face scrubs and face washes)	Cosmetics (5–0.1 mm) (~200 μm)	Miraj et al. (2021) O'Connor et al. (2019) Yurtsever (2019) Bhavsar and Gore (2023)
Films	Polyethylene Polypropylene Polyamide Polyethylene terephthalate	Agricultural mulches Plastic bags Varnish Paint	(5–1 mm)	Qi et al. (2020)
Bioplastics	Polylactic acid Polyglycolic acid Polybutylene succinate Poly(vinyl alcohol)	Extrusion leftovers Disposable plastics Agricultural mulches	ND	Fojt et al. (2020) Shruti and Kutralam-Muniasamy (2019)

development activities in the freshwater already gets polluted and over-exploited and causes biodiversity losses. Thus, drinking water management became an important problem along with sanitation, hygiene, and climate change.

The world's freshwater is already being contaminated by plastic waste, microplastics, and plastic-associated chemicals due to wind's continuous flow and

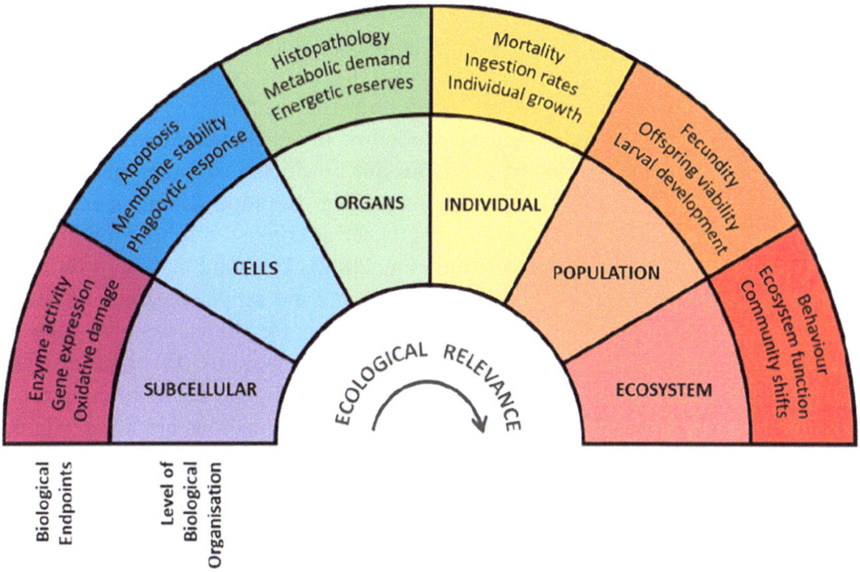

Fig. 13.3 Adverse effect of microplastics on biota at different levels of biological organizations (European Commission 2019; Science Advice for Policy by European Academies 2019)

surface runoff from plastic dumping site (Bäuerlein et al. 2022). Single-use plastic composed of polyethylene compound ranks the most for contaminating fresh and drinking water with microplastics (Koelmans et al. 2019) due to considerable scale of anthropogenic activities in surface water systems. Mason et al. (2018) and Ma et al. (2019) identified significant rise in microplastic in freshwater contaminated with wastewater effluents, incited for more scientific investigation studies on microplastic contamination to freshwater and drinking water resources. Mason et al. (2018) and Kosuth et al. (2018) conducted global studies in 18 countries across Africa, Asia, North America, South America, Europe, and Middle East detected diverse distribution of microplastics in package drinking water, tap water, and commercial drinks (soft drinks and beers).

Plastic packaging materials used for drinking water are the main source of microplastics in drinking water (Winkler et al. 2022). Studies have found microplastics in raw and treated water, which can evade filtration systems and enter domestic water supply systems (Pivokonsky et al. 2018). As per the World Wide Fund for Nature (2019), average person unintentionally consumes 1972 microplastic every week from food and beverages, among which 88% (1769 microplastics) are from drinking water alone. Quantitative exposure analysis in the European Union region estimated an annual consumption of approximately 35.7 lakhs microplastics per person per year from bottle drinking water (Danopoulos et al. 2020). PET materials containing drinking water are influenced by bottle aging, mechanical stress, and photo-degradation through direct sun exposure (Taheri et al. 2023). Along with microplastics produced from packaging material also can introduce novel contaminants such as bisphenol A and phthalates into drinking water and food for leachates (Cho et al. 2018; Angnunavuri et al. 2022; Zhong et al. 2023). These chemicals slowly release and become the most potential toxic secreted contaminants which enter drinking water through drinking water supplies as microplastics remained undetected. The facility of advanced water purification system is lacking in many developing nations due to which removal of microplastics from drinking water is not possible. Micro/granular filtration techniques can efficiently remove microplastics from drinking water as compared to traditional filtration systems (Wang et al. 2022a; Barbier et al. 2022). The combination of different processes (coagulation, flocculation, sedimentation, sand/activated carbon filtration, and floatation) in one system can be more efficient for the removal of 88% of microplastics from water as compared to stand-alone systems (Pivokonsky et al. 2018; Li et al. 2020). Detection of microplastics in water by routine monitoring is not possible because the standardized microplastic analysis methods are not available. Therefore, steps to be taken for microplastic pollution mitigation at the organizational level are presently inadequate.

Microplastics are ubiquitous in nature resulted to plastic pollution means that can potentially impacts living organisms. Microplastics will persist in the environment for a long time. We need to find out origin and toxicity against living animals and plants. We also need to study fate and transport of microplastics to drinking water, especially in developing countries. This can be helpful for the improvement of existing water treatment methods.

13.5 Exposure of Microplastics to Humans

Microplastics derived from the different sources of plastic materials are now becoming unavoidable from day-to-day life activities; however, the potential consequences of microplastics on human health are remained to be explored. Microplastics have the grave impact on human health because it persists in environment for longer duration and toxicity of additional chemicals in leachates; it acts as a carrier for microbes, organic pollutants, and heavy metals (Lee et al. 2014; Enders et al. 2015; Yin et al. 2023a). There are different modes through which human beings gets exposed to microplastic. Revel et al. (2018) suggested the major routes of microplastic exposure to human such as ingestion of food and drinking water, inhalation with polluted air, and direct skin contact. It has been estimated that approximately 50,000 microplastics are annually consumed per person through drinking water (Cox et al. 2019). Many previous studies investigated that microplastics are mainly derived from the different chemical composition plastic such as polyethylene, polypropylene, polystyrene, polyurethane, polyvinyl chloride (PVC), and polyethylene terephthalate (PET) (Bhatt et al. 2023). The exposure of polyethylene and polystyrene microplastics to humans through modes of food and drinking water is in the form of plastic films used in single-use polythene plastic bags and commercial plastic water bottles used as packaging materials; the major source of microplastics is synthetic fabric, textile, and reusable plastic products made from polypropylene and polyethylene terephthalate (Allen et al. 2019; Plastics Europe 2022). Along with this, humans may directly expose to suspended microplastics produced through plastic combustion at household waste and at landfill waste disposal site, plastic recycling industrial effluents, and discharges into environment (Dris et al. 2017).

13.6 Impacts of Microplastics on Human Health

Ingestion or inhalation of microplastics by humans, the microplastics are easily translocated and accumulated into different tissues and organs (Kumar et al. 2023). The exposure of microplastics through inhalation and ingestion initiates natural immune response mechanisms such as sneezing, mucociliary escalator, lymphatic transport, phagocytosis, and excretion (Prata 2018) although some microplastics skips from body's natural clearance response and finally reaches to the respiratory and digestive system. Immune system identifies microplastics as foreign material and initiates counter-immune response by releasing cellular messenger molecules, various proteases, and reactive oxygen species (ROS) which triggers inflammation to respiratory and digestive system organs (Wright and Kelly 2017; Gasperi et al. 2018). It has been reported that microplastics retain its surface morphology for about 6 months in the lung fluid system. Additionally, different microplastic particles form clusters due to which it persists in human lungs for longer period and may cause inflammation and damage internal cell layers due to continuous ROS generation (Law et al. 1990). The chronic inflammation and irritation due to microplastic exposure may eventually cause cancer resulting from lung fibrosis (Valavanidis et al. 2013; Revel et al. 2018; Fan et al. 2022). Polystyrene microplastics have been investigated for many toxic effects, like the induction of apoptosis in granulosa cells, apoptosis, and fibrosis in the ovary (An et al. 2021), blood-testis barrier disruption and spermatogenesis dysfunction (Wei et al. 2021), hepatotoxicity (Cheng et al. 2022; Yin et al. 2023b), cirrhotic liver (Horvatits et al. 2022), and impairment of learning and memory (Wang et al. 2022b). Table 13.3 shows detected of different polymer microplastics in human specimen. Microplastics released in the environment can enter the food chain and reach to human intestine. Intestinal epithelial cells can internalize microplastics by translocation depending on its size, concentration, and persistent time. Importantly, exposure of microplastics at higher concentrations can significantly induce immune system through secretion of a panel of inflammatory cytokines which are directly linked with human inflammatory bowel disease (IBD) (Fig. 13.4) (Chen et al. 2023). However, several animal studies and tissue models suggested microplastic and nanoplastics as the possible cancer causative agents, due to lack of relevant human intestinal models till date effect of these particles on human intestine not yet validated (Chen et al. 2023). Tamargo et al. (2022) identified role of microplastics in the modifications of human gut and colonic microbiota communities (Fig. 13.4). Microplastics decreases the beneficial probiotic bacteria species *Blautia*, *Faecalibacterium*, *Bifidobacterium*, and *Lactobacilli* while

Table 13.3 Evidence of microplastic concentration and polymer in human specimen (Adapted from Angnunavuri et al. 2023)

Human specimen	Maximum concentration or particle size	MP polymer	References
Blood	7.1[a]	Poly methyl methacrylate	Leslie et al. (2022)
Lungs	3.96[b] 1–10 μm	PET Polystyrene	Jenner et al. (2022) Goodman et al. (2021)
Breast milk	2.75[c]	Polyethylene, PVC, high-density Polyethylene	Ragusa et al. (2022)
Cirrhotic liver	3.4[b]	Polystyrene, PET, polypropylene, PVC	Horvatits et al. (2022)
Broncho-alveolar lavage fluid	9.18[e]	Microplastic fiber	Baeza-Martínez et al. (2022)
Sputum	567[d]	Mix MPs	Huang et al. (2022)
Placenta	7–10 μm	Polyethylene	Ragusa et al. (2021)
Feces	10.19[g]	Polypropylene	Wibowo et al. (2021)
Feces	41.8[b]	PET, polyamide	Yan et al. (2022)
Infant feces	82,000[f]	PET, polycarbonate	Zhang et al. (2021)

[a] μg/mL
[b] particles/g
[c] g/mL,
[d] particles/10 mL
[e] particles/100 mL
[f] ng/g feces
[g] μg/g feces

enhancing pathogenic microbial species (*Staphylococcus*, *Proteobacteria*, *Prevotella*) (Zhao et al. 2023b).

In addition to microplastics, numerous types of plasticizers such as phthalate esters, including diethyl phthalate, dioctyl phthalate, dimethyl phthalate, and di-(2-ethylhexyl) phthalate, are released in the form of leachates (Zhong et al. 2023; Yuan et al. 2023). Cho et al. (2018) reported that bisphenol A (BPA) is the additional chemical used as a plasticizer can induce the oxidative stress in humans and damage to erythrocytes, hepatocytes, and placental and endometrial stromal cells of testis. Prolong exposure of BPA eventually induces immune system and upregulates gene expression of inflammatory cytokines and estrogen receptor ER-α which may result into endocrine system-related diseases (Pérez-Albaladejo et al. 2020; Flaws et al. 2020). The plasticizer BPA and phthalates released from microplastics have been identified as a potential endocrine disrupting chemicals (EDCs). Both plasticizers mimic or interfere with reproduction processes regulated

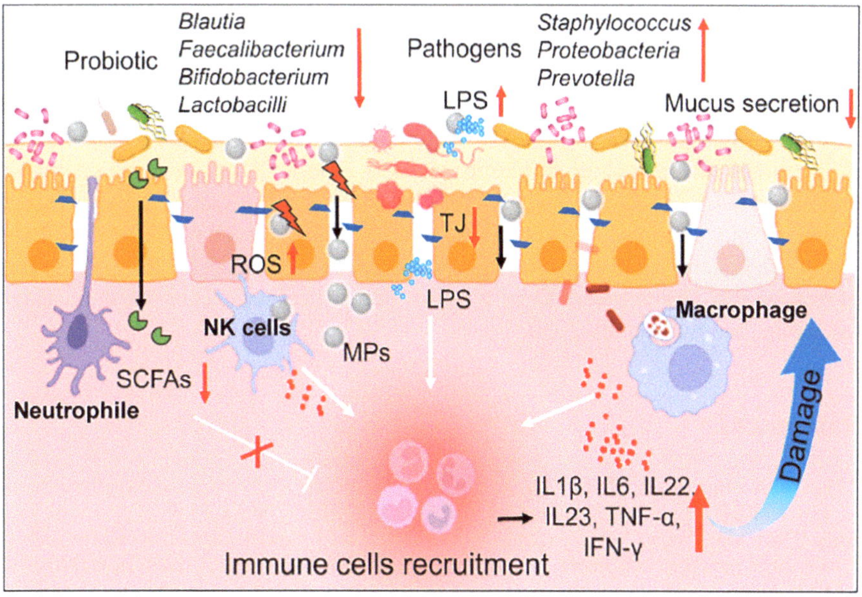

Fig. 13.4 Illustration about development of inflammatory bowel disease (IBD) due to microplastic exposure. Microplastics (MPs), short-chain fatty acids (SCFAs), lipopolysaccharide (LPS), tight junction (TJ), reactive oxygen species (ROS), natural killer cells (NK cells), interleukins (IL), and interferon (IFN) (adapted from Zhao et al. 2023b)

by estrogens and androgens. Overexposure to these EDCs may cause the reproductive dysfunctions which includes reduced fertility, pregnancy loss, and infertility (Flaws et al. 2020; Zhao et al. 2023a). The microplastic impact mechanism on living cell's signal transduction not yet illustrated, Wei et al. (2021) showed ROS-mediated impairment of mTORC1 and mTORC2. The effect of microplastic on other signal transduction pathways in living cells can be studied by the new powerful approaches like proteomic and phosphoproteomic analysis at high-throughput manner.

13.7 Conclusion and Future Prospects

The presence of microplastics in drinking water has created potential threat and serious concern arisen due to human activities. Understanding the origin, fate, and transportation of microplastics to drinking water can be helpful for the advancement of drinking water treatment plants. There is a significant knowledge gap in identifying material characterization, toxicity assessment, and responsibility for microplastic contamination. Along with the first step of microplastic pollution recognition, the social awareness also plays an essential role through "science-society-policy interface." These awareness programs can effectively convey alert message about the adverse effects of all forms of microplastics and nanoplastic contaminants. Microplastic degradation or decomposition by action of microbes (fungi and bacteria) is still not yet possible. So at present it is very difficult to eliminate or reduce number of microplastics entered in the food chain. The biodegradable and recyclable lignocellulosic biomass-derived polymers can be the best alternative fossil fuel-based plastic for achieving circular economy with zero plastic waste. The multiple international level campaigns and efforts are going on to reduce plastic usage and recycling of plastic products. The regulation control on plastic production is the main issue need to be included in these campaigns which can address proper plastic waste disposal management and adoption of recycling targets. The special attention on microplastics in drinking water is necessary to meet targets of Sustainable Development Goal 6 which seeks to ensure clean drinking water, sanitation, and hygiene by 2030.

References

Akindoyo JO, Beg MDH, Ghazali S et al (2016) Polyurethane types, synthesis and applications—a review. RSC Adv 6:114453–114482. https://doi.org/10.1039/C6RA14525F

Allen S, Allen D, Phoenix VR et al (2019) Atmospheric transport and deposition of microplastics in a remote mountain catchment. Nat Geosci 12:339–344. https://doi.org/10.1038/s41561-019-0335-5

An R, Wang X, Yang L et al (2021) Polystyrene microplastics cause granulosa cells apoptosis and fibrosis in ovary through oxidative stress in rats. Toxicology 449:152665. https://doi.org/10.1016/j.tox.2020.152665

Anadu EC, Harding AK (2000) Risk perception and bottled water use. J Am Water Works Assoc 92:82–92. https://doi.org/10.1002/j.1551-8833.2000.tb09051.x

Angnunavuri PN, Attiogbe F, Dansie A, Mensah B (2022) Evaluation of plastic packaged water quality using health risk indices: a case study of sachet and bottled water in Accra, Ghana. Sci Total Environ 832:155073. https://doi.org/10.1016/j.scitotenv.2022.155073

Angnunavuri PN, Attiogbe F, Mensah B (2023) Particulate plastics in drinking water and potential human health effects: current knowledge for management of freshwater plastic materials in Africa. Environ Pollut 316:120714. https://doi.org/10.1016/j.envpol.2022.120714

Arutchelvi J, Muniyasamy S, Arkatkar A et al (2008) Biodegradation of polyethylene and polypropylene. Indian J Biotechnol 7:9–22

Baeza-Martínez C, Olmos S, González-Pleiter M et al (2022) First evidence of microplastics isolated in European citizens' lower airway. J Hazard Mater 438:129439. https://doi.org/10.1016/j.jhazmat.2022.129439

Barbier J-S, Dris R, Lecarpentier C et al (2022) Microplastic occurrence after conventional and nanofiltration processes at drinking water treatment plants: preliminary results. Front Water 4: 886703. https://doi.org/10.3389/frwa.2022.886703

Barra R, Leonard SA, Whaley C, Bierbaum R (2018) Plastics and the circular economy. Scientific and Technical Advisory Panel to the Global Environment Facility, Washington, DC

Bäuerlein PS, Hofman-Caris RCHM, Pieke EN, Ter Laak TL (2022) Fate of microplastics in the drinking water production. Water Res 221:118790. https://doi.org/10.1016/j.watres.2022.118790

Bayarkhuu B, Byun J (2022) Optimization of coagulation and sedimentation conditions by turbidity measurement for nano- and microplastic removal. Chemosphere 306:135572. https://doi.org/10.1016/j.chemosphere.2022.135572

Bhatt A, Prajapati D, Gupte A, Gupte S (2023) Microplastic pollution: sources, environmental hazards, and mycoremediation as a sustainable solution. In: Satyanarayana T, Deshmukh SK (eds) Fungi and fungal products in human welfare and biotechnology. Springer Nature Singapore, Singapore, pp 127–156

Bhavsar PS, Gore AH (2023) Extraction and analysis of microplastic beads from personal care products. CAC 19:184–189. https://doi.org/10.2174/1573411019666221020094611
Blackburn K, Green D (2022) The potential effects of microplastics on human health: What is known and what is unknown. Ambio 51(3):518-530. https://doi.org/10.1007/s13280-021-01589-9
Burns EE, Boxall ABA (2018) Microplastics in the aquatic environment: evidence for or against adverse impacts and major knowledge gaps: microplastics in the environment. Environ Toxicol Chem 37:2776–2796. https://doi.org/10.1002/etc.4268
Cao Y, Lin H, Zhang K et al (2022) Microplastics: a major source of phthalate esters in aquatic environments. J Hazard Mater 432:128731. https://doi.org/10.1016/j.jhazmat.2022.128731
Carney Almroth BM, Åström L, Roslund S et al (2018) Quantifying shedding of synthetic fibers from textiles; a source of microplastics released into the environment. Environ Sci Pollut Res Int 25:1191–1199. https://doi.org/10.1007/s11356-017-0528-7
Ceresana (2014) Global demand for PVC to rise by about 3.2%/year to 2021. In: Additives for polymers, 2014:10–11. https://doi.org/10.1016/S0306-3747(14)70175-7
Chen Y, Williams AM, Gordon EB et al (2023) Biological effects of polystyrene micro- and nano-plastics on human intestinal organoid-derived epithelial tissue models without and with M cells. Nanomed Nanotechnol Biol Med 50:102680. https://doi.org/10.1016/j.nano.2023.102680
Cheng YL, Kim J-G, Kim H-B et al (2021) Occurrence and removal of microplastics in wastewater treatment plants and drinking water purification facilities: a review. Chem Eng J 410:128381. https://doi.org/10.1016/j.cej.2020.128381
Cheng W, Li X, Zhou Y et al (2022) Polystyrene microplastics induce hepatotoxicity and disrupt lipid metabolism in the liver organoids. Sci Total Environ 806:150328. https://doi.org/10.1016/j.scitotenv.2021.150328
Cho YJ, Park SB, Park JW et al (2018) Bisphenol A modulates inflammation and proliferation pathway in human endometrial stromal cells by inducing oxidative stress. Reprod Toxicol 81: 41–49. https://doi.org/10.1016/j.reprotox.2018.06.016
Conkle JL, Báez Del Valle CD, Turner JW (2018) Are we underestimating microplastic contamination in aquatic environments? Environ Manag 61:1–8. https://doi.org/10.1007/s00267-017-0947-8
Corradini Santander FA (2021) Development and application of tests for microplastic detection in soil. Wageningen University
Cox KD, Covernton GA, Davies HL et al (2019) Human consumption of microplastics. Environ Sci Technol 53:7068–7074. https://doi.org/10.1021/acs.est.9b01517
Crawford CB, Quinn B (2017) Physiochemical properties and degradation. In: Microplastic pollutants. Elsevier, pp 57–100
Danopoulos E, Twiddy M, Rotchell JM (2020) Microplastic contamination of drinking water: a systematic review. PLoS One 15:e0236838. https://doi.org/10.1371/journal.pone.0236838
De Souza FM, Kahol PK, Gupta RK (2021) Introduction to polyurethane chemistry. In: Gupta RK, Kahol PK (eds) ACS symposium series. American Chemical Society, Washington, DC, pp 1–24
Djapovic M, Milivojevic D, Ilic-Tomic T et al (2021) Synthesis and characterization of polyethylene terephthalate (PET) precursors and potential degradation products: toxicity study and application in discovery of novel PETases. Chemosphere 275:130005. https://doi.org/10.1016/j.chemosphere.2021.130005
Domininghaus H (1998) Synthetische Kunststoffe. In: Die Kunststoffe und ihre Eigenschaften. Springer Berlin Heidelberg, Berlin, pp 95–1174
Dris R, Gasperi J, Mirande C et al (2017) A first overview of textile fibers, including microplastics, in indoor and outdoor environments. Environ Pollut 221:453–458. https://doi.org/10.1016/j.envpol.2016.12.013
Du F, Cai H, Zhang Q et al (2020) Microplastics in take-out food containers. J Hazard Mater 399: 122969. https://doi.org/10.1016/j.jhazmat.2020.122969

Eerkes-Medrano D, Leslie HA, Quinn B (2019) Microplastics in drinking water: a review and assessment. Curr Opin Environ Sci Health 7:69–75. https://doi.org/10.1016/j.coesh.2018.12.001

Enders K, Lenz R, Stedmon CA, Nielsen TG (2015) Abundance, size and polymer composition of marine microplastics ≥10 μm in the Atlantic Ocean and their modelled vertical distribution. Mar Pollut Bull 100:70–81. https://doi.org/10.1016/j.marpolbul.2015.09.027

European Commission (2019) Environmental and health risks of microplastic pollution. Publications Office, LU, Belgium

Fadare OO, Wan B, Guo L-H, Zhao L (2020) Microplastics from consumer plastic food containers: are we consuming it? Chemosphere 253:126787. https://doi.org/10.1016/j.chemosphere.2020.126787

Fan Z, Xiao T, Luo H et al (2022) A study on the roles of long non-coding RNA and circular RNA in the pulmonary injuries induced by polystyrene microplastics. Environ Int 163:107223. https://doi.org/10.1016/j.envint.2022.107223

Flaws J, Damdimopoulou P, Patisaul HB, et al (2020) Plastics, EDCs and health. A Guide for public interest organizations and policy-makers on endocrine disrupting chemicals & plastics. Endocrine Society

Fojt J, David J, Přikryl R et al (2020) A critical review of the overlooked challenge of determining micro-bioplastics in soil. Sci Total Environ 745:140975. https://doi.org/10.1016/j.scitotenv.2020.140975

Gasperi J, Wright SL, Dris R et al (2018) Microplastics in air: are we breathing it in? Curr Opin Environ Sci Health 1:1–5. https://doi.org/10.1016/j.coesh.2017.10.002

Geyer R, Jambeck JR, Law KL (2017) Production, use, and fate of all plastics ever made. Sci Adv 3: e1700782. https://doi.org/10.1126/sciadv.1700782

Gigault J, Halle AT, Baudrimont M et al (2018) Current opinion: what is a nanoplastic? Environ Pollut 235:1030–1034. https://doi.org/10.1016/j.envpol.2018.01.024

Goodman KE, Hare JT, Khamis ZI et al (2021) Exposure of human lung cells to polystyrene microplastics significantly retards cell proliferation and triggers morphological changes. Chem Res Toxicol 34:1069–1081. https://doi.org/10.1021/acs.chemrestox.0c00486

Harussani MM, Sapuan SM, Rashid U et al (2022) Pyrolysis of polypropylene plastic waste into carbonaceous char: priority of plastic waste management amidst COVID-19 pandemic. Sci Total Environ 803:149911. https://doi.org/10.1016/j.scitotenv.2021.149911

Heath DE, Guelcher SA, Cooper SL (2020) Polyurethanes. In: Biomaterials science. Elsevier, pp 103–107

Horvatits T, Tamminga M, Liu B et al (2022) Microplastics detected in cirrhotic liver tissue. eBioMedicine 82:104147. https://doi.org/10.1016/j.ebiom.2022.104147

Huang S, Huang X, Bi R et al (2022) Detection and analysis of microplastics in human sputum. Environ Sci Technol 56:2476–2486. https://doi.org/10.1021/acs.est.1c03859

Jenner LC, Rotchell JM, Bennett RT et al (2022) Detection of microplastics in human lung tissue using μFTIR spectroscopy. Sci Total Environ 831:154907. https://doi.org/10.1016/j.scitotenv.2022.154907

Kanhai LDK, Gardfeldt K, Krumpen T et al (2020) Microplastics in sea ice and seawater beneath ice floes from the Arctic Ocean. Sci Rep 10:5004. https://doi.org/10.1038/s41598-020-61948-6

Kelly A, Lannuzel D, Rodemann T et al (2020) Microplastic contamination in east Antarctic sea ice. Mar Pollut Bull 154:111130. https://doi.org/10.1016/j.marpolbul.2020.111130

King J, Brown C (2021) Africa's living rivers: managing for sustainability. Daedalus 150:240–259. https://doi.org/10.1162/daed_a_01882

Kirstein IV, Gomiero A, Vollertsen J (2021) Microplastic pollution in drinking water. Curr Opin Toxicol 28:70–75. https://doi.org/10.1016/j.cotox.2021.09.003

Koelmans AA, Mohamed Nor NH, Hermsen E et al (2019) Microplastics in freshwaters and drinking water: critical review and assessment of data quality. Water Res 155:410–422. https://doi.org/10.1016/j.watres.2019.02.054

Kosuth M, Mason SA, Wattenberg EV (2018) Anthropogenic contamination of tap water, beer, and sea salt. PLoS One 13:e0194970. https://doi.org/10.1371/journal.pone.0194970

Kumar LR, Yadav B, Drogui P, Tyagi RD (2023) Microplastics and nanoplastics in drinking water and food chain. In: Current developments in biotechnology and bioengineering. Elsevier, pp 183–200

Law KL, Thompson RC (2014) Microplastics in the seas. Science 345:144–145. https://doi.org/10.1126/science.1254065

Law BD, Bunn WB, Hesterberg TW (1990) Solubility of polymeric organic fibers and manmade vitreous fibers in gambles solution. Inhal Toxicol 2:321–339. https://doi.org/10.3109/08958379009145261

Lee H, Shim WJ, Kwon J-H (2014) Sorption capacity of plastic debris for hydrophobic organic chemicals. Sci Total Environ 470–471:1545–1552. https://doi.org/10.1016/j.scitotenv.2013.08.023

Leslie HA, Van Velzen MJM, Brandsma SH et al (2022) Discovery and quantification of plastic particle pollution in human blood. Environ Int 163:107199. https://doi.org/10.1016/j.envint.2022.107199

Li Y, Li W, Jarvis P et al (2020) Occurrence, removal and potential threats associated with microplastics in drinking water sources. J Environ Chem Eng 8:104527. https://doi.org/10.1016/j.jece.2020.104527

Lim X (2021) Microplastics are everywhere—but are they harmful? Nature 593:22–25. https://doi.org/10.1038/d41586-021-01143-3

Luo Q, Liu Z, Yin H et al (2018) Migration and potential risk of trace phthalates in bottled water: a global situation. Water Res 147:362–372. https://doi.org/10.1016/j.watres.2018.10.002

Ma P, Wei Wang M, Liu H et al (2019) Research on ecotoxicology of microplastics on freshwater aquatic organisms. Environ Pollut Bioavail 31:131–137. https://doi.org/10.1080/26395940.2019.1580151

Mason SA, Welch VG, Neratko J (2018) Synthetic polymer contamination in bottled water. Front Chem 6:407. https://doi.org/10.3389/fchem.2018.00407

Miraj SS, Parveen N, Zedan HS (2021) Plastic microbeads: small yet mighty concerning. Int J Environ Health Res 31:788–804. https://doi.org/10.1080/09603123.2019.1689233

Ng E-L, Huerta Lwanga E, Eldridge SM et al (2018) An overview of microplastic and nanoplastic pollution in agroecosystems. Sci Total Environ 627:1377–1388. https://doi.org/10.1016/j.scitotenv.2018.01.341

Novotna K, Cermakova L, Pivokonska L et al (2019) Microplastics in drinking water treatment—Current knowledge and research needs. Sci Total Environ 667:730–740. https://doi.org/10.1016/j.scitotenv.2019.02.431

Novotna K, Pivokonska L, Cermakova L et al (2023) Continuous long-term monitoring of leaching from microplastics into ambient water—A multi-endpoint approach. J Hazard Mater 444:130424. https://doi.org/10.1016/j.jhazmat.2022.130424

O'Brien S, Okoffo ED, O'Brien JW et al (2020) Airborne emissions of microplastic fibres from domestic laundry dryers. Sci Total Environ 747:141175. https://doi.org/10.1016/j.scitotenv.2020.141175

O'Connor D, Pan S, Shen Z et al (2019) Microplastics undergo accelerated vertical migration in sand soil due to small size and wet-dry cycles. Environ Pollut 249:527–534. https://doi.org/10.1016/j.envpol.2019.03.092

Okeke ES, Olagbaju OA, Okoye CO et al (2022) Microplastic burden in Africa: a review of occurrence, impacts, and sustainability potential of bioplastics. Chem Eng J Adv 12:100402. https://doi.org/10.1016/j.ceja.2022.100402

Peng B-Y, Chen Z, Chen J et al (2020) Biodegradation of polyvinyl chloride (PVC) in *Tenebrio molitor* (Coleoptera: Tenebrionidae) larvae. Environ Int 145:106106. https://doi.org/10.1016/j.envint.2020.106106

Pérez-Albaladejo E, Solé M, Porte C (2020) Plastics and plastic additives as inducers of oxidative stress. Curr Opin Toxicol 20–21:69–76. https://doi.org/10.1016/j.cotox.2020.07.002

Pérez-Guevara F, Roy PD, Elizalde-Martínez I et al (2022) Human exposure to microplastics from urban decentralized pay-to-fetch drinking-water refill kiosks. Sci Total Environ 848:157722. https://doi.org/10.1016/j.scitotenv.2022.157722

Pivokonsky M, Cermakova L, Novotna K et al (2018) Occurrence of microplastics in raw and treated drinking water. Sci Total Environ 643:1644–1651. https://doi.org/10.1016/j.scitotenv.2018.08.102

Plastics Europe (2022) Plastics—the facts 2022. An analysis of European plastics production, demand, conversion and end-of-life management. In: Plastics Europe. https://plasticseurope.org/knowledge-hub/plastics-the-facts-2022/. Accessed 10 June 2023

Prata JC (2018) Airborne microplastics: consequences to human health? Environ Pollut 234:115–126. https://doi.org/10.1016/j.envpol.2017.11.043

Qi Y, Ossowicki A, Yang X et al (2020) Effects of plastic mulch film residues on wheat rhizosphere and soil properties. J Hazard Mater 387:121711. https://doi.org/10.1016/j.jhazmat.2019.121711

Ragusa A, Svelato A, Santacroce C et al (2021) Plasticenta: first evidence of microplastics in human placenta. Environ Int 146:106274. https://doi.org/10.1016/j.envint.2020.106274

Ragusa A, Notarstefano V, Svelato A et al (2022) Raman microspectroscopy detection and characterisation of microplastics in human breastmilk. Polymers 14:2700. https://doi.org/10.3390/polym14132700

Reghunadhan A, Thomas S (2017) Polyurethanes. In: Polyurethane polymers. Elsevier, pp 1–16

Revel M, Châtel A, Mouneyrac C (2018) Micro(nano)plastics: a threat to human health? Curr Opin Environ Sci Health 1:17–23. https://doi.org/10.1016/j.coesh.2017.10.003

Rochman CM, Browne MA, Halpern BS et al (2013) Classify plastic waste as hazardous. Nature 494:169–171. https://doi.org/10.1038/494169a

Schwabl P, Köppel S, Königshofer P et al (2019) Detection of various microplastics in human stool: a prospective case series. Ann Intern Med 171:453–457. https://doi.org/10.7326/M19-0618

Science Advice for Policy by European Academies (2019) A scientific perspective on microplastics in nature and society. Science Advice for Policy by European Academies, Berlin

Shen M, Song B, Zhu Y et al (2020) Removal of microplastics via drinking water treatment: current knowledge and future directions. Chemosphere 251:126612. https://doi.org/10.1016/j.chemosphere.2020.126612

Shruti VC, Kutralam-Muniasamy G (2019) Bioplastics: missing link in the era of microplastics. Sci Total Environ 697:134139. https://doi.org/10.1016/j.scitotenv.2019.134139

Taheri S, Shoshtari-Yeganeh B, Pourzamani H, Ebrahimpour K (2023) Investigating the pollution of bottled water by the microplastics (MPs): the effects of mechanical stress, sunlight exposure, and freezing on MPs release. Environ Monit Assess 195:62. https://doi.org/10.1007/s10661-022-10697-2

Tamargo A, Molinero N, Reinosa JJ et al (2022) PET microplastics affect human gut microbiota communities during simulated gastrointestinal digestion, first evidence of plausible polymer biodegradation during human digestion. Sci Rep 12:528. https://doi.org/10.1038/s41598-021-04489-w

Terashima T (2015) Polystyrene (PSt). In: Kobayashi S, Müllen K (eds) Encyclopedia of polymeric nanomaterials. Springer Berlin Heidelberg, Berlin, pp 2077–2091

Trinh NH, Jaafar M, Viet CX, Zubir SA (2020) Palm kernel oil polyol based shape memory polyurethane: effect of polycaprolactone and polyethylene glycol as soft segment. Mater Res Express 7:025704. https://doi.org/10.1088/2053-1591/ab6f3c

UN Environment Program (2017) Visual feature: beat plastic pollution. http://unep.org/interactive/beat-plastic-pollution/. Accessed 27 Apr 2023

Valavanidis A, Vlachogianni T, Fiotakis K, Loridas S (2013) Pulmonary oxidative stress, inflammation and cancer: respirable particulate matter, fibrous dusts and ozone as major causes of lung carcinogenesis through reactive oxygen species mechanisms. IJERPH 10:3886–3907. https://doi.org/10.3390/ijerph10093886

Verbeek CJR, Gavin C (2020) Grafting functional groups onto biodegradable thermoplastic polyesters. In: Gutiérrez TJ (ed) Reactive and functional polymers, vol 2. Springer International Publishing, Cham, pp 245–281

Verbeek JH, Rajamaki B, Ijaz S et al (2020) Personal protective equipment for preventing highly infectious diseases due to exposure to contaminated body fluids in healthcare staff. Cochrane Database Syst Rev 4(4):CD011621. https://doi.org/10.1002/14651858.CD011621.pub4

Verster C, Minnaar K, Bouwman H (2017) Marine and freshwater microplastic research in South Africa: marine and freshwater microplastics research in South Africa. Integr Environ Assess Manag 13:533–535. https://doi.org/10.1002/ieam.1900

Wagner S, Hüffer T, Klöckner P et al (2018) Tire wear particles in the aquatic environment—A review on generation, analysis, occurrence, fate and effects. Water Res 139:83–100. https://doi.org/10.1016/j.watres.2018.03.051

Wang Q, Li Y, Liu Y et al (2022a) Effects of microplastics accumulation on performance of membrane bioreactor for wastewater treatment. Chemosphere 287:131968. https://doi.org/10.1016/j.chemosphere.2021.131968

Wang S, Han Q, Wei Z et al (2022b) Polystyrene microplastics affect learning and memory in mice by inducing oxidative stress and decreasing the level of acetylcholine. Food Chem Toxicol 162:112904. https://doi.org/10.1016/j.fct.2022.112904

Wei Y, Zhou Y, Long C et al (2021) Polystyrene microplastics disrupt the blood-testis barrier integrity through ROS-mediated imbalance of mTORC1 and mTORC2. Environ Pollut 289:117904. https://doi.org/10.1016/j.envpol.2021.117904

Wen Z, Xie Y, Chen M, Dinga CD (2021) China's plastic import ban increases prospects of environmental impact mitigation of plastic waste trade flow worldwide. Nat Commun 12:425. https://doi.org/10.1038/s41467-020-20741-9

Wibowo AT, Nugrahapraja H, Wahyuono RA et al (2021) Microplastic contamination in the human gastrointestinal tract and daily consumables associated with an Indonesian farming community. Sustainability 13:12840. https://doi.org/10.3390/su132212840

Winkler A, Fumagalli F, Cella C et al (2022) Detection and formation mechanisms of secondary nanoplastic released from drinking water bottles. Water Res 222:118848. https://doi.org/10.1016/j.watres.2022.118848

World Wide Fund for Nature (2019) No plastic in nature: assessing plastic ingestion from nature to human. World Wide Fund for Nature, Gland

WorldAtlas (2018) How much of the earth is covered by freshwater? In: WorldAtlas. https://www.worldatlas.com/articles/how-much-of-the-earth-is-covered-by-freshwater.html. Accessed 1 May 2023

Wright SL, Kelly FJ (2017) Plastic and human health: a micro issue? Environ Sci Technol 51:6634–6647. https://doi.org/10.1021/acs.est.7b00423

Yan Z, Liu Y, Zhang T et al (2022) Analysis of microplastics in human feces reveals a correlation between fecal microplastics and inflammatory bowel disease status. Environ Sci Technol 56:414–421. https://doi.org/10.1021/acs.est.1c03924

Yin H, Chen R, Wang H et al (2023a) Co-occurrence of phthalate esters and perfluoroalkyl substances affected bacterial community and pathogenic bacteria growth in rural drinking water distribution systems. Sci Total Environ 856:158943. https://doi.org/10.1016/j.scitotenv.2022.158943

Yin K, Wang D, Zhang Y et al (2023b) Polystyrene microplastics promote liver inflammation by inducing the formation of macrophages extracellular traps. J Hazard Mater 452:131236. https://doi.org/10.1016/j.jhazmat.2023.131236

Yong CQY, Valiyaveettil S, Tang BL (2020) Toxicity of Microplastics and Nanoplastics in Mammalian Systems. Int J Environ Res Public Health. 7(5):1509. https://doi.org/10.3390/ijerph17051509. PMID: 32111046; PMCID: PMC7084551.

Yuan F, Sun Y, Jiang X et al (2023) Dioctyl phthalate enhances volatile fatty acids production from sludge anaerobic fermentation: insights of electron transport and metabolic functions. Sci Total Environ 859:160102. https://doi.org/10.1016/j.scitotenv.2022.160102

Yurtsever M (2019) Tiny, shiny, and colorful microplastics: are regular glitters a significant source of microplastics? Mar Pollut Bull 146:678–682. https://doi.org/10.1016/j.marpolbul.2019.07.009

Zhang Y, Diehl A, Lewandowski A et al (2020) Removal efficiency of micro- and nanoplastics (180 nm–125 μm) during drinking water treatment. Sci Total Environ 720:137383. https://doi.org/10.1016/j.scitotenv.2020.137383

Zhang J, Wang L, Trasande L, Kannan K (2021) Occurrence of polyethylene terephthalate and polycarbonate microplastics in infant and adult feces. Environ Sci Technol Lett 8:989–994. https://doi.org/10.1021/acs.estlett.1c00559

Zhao T, Shen L, Ye X et al (2023a) Prenatal and postnatal exposure to polystyrene microplastics induces testis developmental disorder and affects male fertility in mice. J Hazard Mater 445:130544. https://doi.org/10.1016/j.jhazmat.2022.130544

Zhao Y, Liu S, Xu H (2023b) Effects of microplastic and engineered nanomaterials on inflammatory bowel disease: a review. Chemosphere 326:138486. https://doi.org/10.1016/j.chemosphere.2023.138486

Zhong X, Yi X, Cheng F et al (2023) Leaching of di-2-ethylhexyl phthalate from biodegradable and conventional microplastics and the potential risks. Chemosphere 311:137208. https://doi.org/10.1016/j.chemosphere.2022.137208

Detection of Microplastic in Seafood

14

Mubashir J. Mintoo and Ruqiya Kousar

Abstract

The study of microplastics and their impact on aquatic organisms is a rapidly growing field, as researchers work to understand the presence of these synthetic polymers in underwater habitats. Investigating microplastics in aquatic environments remains challenging due to their diverse interactions with biota and varying synthetic organic polymers. Human activity has caused microplastics to contaminate the marine environment. These tiny particles of plastic are so widespread that they are ingested by many wildlife species, such as fish and shellfish. Microplastics' physical and chemical toxicity is a worry since they are linked to manufacturing-related chemicals and can absorb toxins from the environment. Our knowledge is significantly limited in the areas of microplastic toxicity's effects and distribution. Given the nutritional importance of seafood consumption and the incomplete knowledge of the effects of microplastics on human health, it is imperative to fill these research gaps. So, in the present chapter an attempt was made to overview the microplastic detection in seafood under the following lines: (1) to pool all the reports of microplastic detection in the seafood, (2) to highlight the consequences of the consumption of seafood with microplastic, (3) to discuss the possible way to remove the microplastic from the seafood before consumption, and (4) to suggest future strategies to minimize the entry of microplastic in seafood.

Mubashir J. Mintoo and Ruqiya Kousar contributed equally with all other contributors.

M. J. Mintoo (✉)
Department of Cell and Molecular Pharmacology and Experimental Therapeutics, Medical University of South Carolina, Charleston, SC, USA

R. Kousar
Department of Botany and Environment Studies, DAV University, Jalandhar, Punjab, India

M. Shahnawaz et al. (eds.), *Microplastic Pollution*,
https://doi.org/10.1007/978-981-99-8357-5_14

Keywords

Microplastic · Mollusk · Crustaceans · Toxicity · Inflammation

14.1 Introduction

According to the Food and Agricultural Organization (FAO), there will be a significant shift in the fish supply chain by 2030, with aquaculture taking over as the leading source of fish worldwide (Kobayashi et al. 2015). The FAO predicts that the total production of fish from both aquaculture and capture fisheries will reach 201 million tons, a substantial increase from the 171 million metric tons of seafood produced globally in 2016 (Food and Agriculture Organization of the United Nations 2018, 2020). In other words, aquaculture is expected to surpass capture fisheries as the primary fish supplies, resulting in a higher overall fish production level. However, it is important to manage the production of marine-based foods with care to ensure that they are safe for human consumption. A prerequisite for establishing food security is having access to enough healthful food for everyone (Pinstrup-Andersen 2009). The attainment of the predicted rise in global fish production hinges on addressing the issue of food safety related to plastic particles found in seafood. The annual global production of plastic surpasses 300 million tons, and this amount continues to increase (Mendes et al. 2021). Since many plastic polymers are not readily biodegradable, they pose a potential hazard to both human and environmental well-being (Singh et al. 2021; Rahman and Bhoi 2021).

Plastics have infiltrated our daily routines and can be found in a variety of items such as clothing, coatings, transport vehicles, and cleaning products (Barboza and Gimenez 2015; Geyer et al. 2017). These materials are attractive due to their economical price, durability, lightweight, and versatility, leading to a seemingly endless list of applications. The negative impacts of plastic are becoming increasingly noticeable as substantial amounts are making their way into rivers and oceans, damaging marine ecosystems, and affecting related industries.

Microplastics (MP) refer to tiny plastic fragments that measure less than 5 mm in length and can cause damage to our oceans and the creatures living in them (Sharma and Chatterjee 2017). Plastic waste includes all types of debris, from large, easily visible, and removable pieces to virtually invisible small particles. Primary MPs are distinct from secondary MPs, which are generated from the degradation of larger plastic waste in the marine environment. Primary MPs originate from two sources: either they are deliberately incorporated into products as abrasive agents in personal care products such as shower gels and creams, or they are formed from the fragmentation of larger plastic objects during their production, use, or maintenance, such as tire abrasion from driving or textile erosion from washing (An et al. 2020).

This chapter will examine the potential human health hazards of plastic particles present in various types of seafood that are commonly consumed by people such as cephalopods, finfish, crustaceans, echinoderms, bivalves, and gastropods. The goal of this chapter is to give a complete picture where MPs come from, what happens to them in the marine environment, and how they move. It also seeks to carefully examine the prevalence of MPs in frequently consumed seafood and assess the potential health concerns they represent to people. Finally, the study seeks to identify the need for further research and make recommendations to assess the impact of MPs on both ocean health and human food security.

In the environment, MPs are now pervasive and can be discovered in fish, shellfish, crabs, mollusks, and even mammals. A recent study published in Environmental Health Perspectives examined the levels of MP contamination in seafood and estimated the yearly intake of MP by humans (Danopoulos et al. 2020). It is unclear whether consuming MPs is harmful to human health because the particles may contain microbes, chemicals that have been adsorbed, and potentially dangerous plastic components. According to studies, mollusks brought in from Asia tended to be the most contaminated, with an annual maximum human MP intake from seafood estimated at 53,864 particles (Nicole 2021). Regulatory developments on MPs are still in the nascent, information-gathering stage at the federal level, and even in states that are beginning to act. For example, California has begun development of regulations addressing MPs as a water quality standard (California State Water Resources Control Board 2019). A restriction on MPs that are deliberately added has been proposed by the European Chemicals Agency (ECHA) (European Chemicals Agency 2019). There is still much study to be done on whether exposure to MPs justifies regulatory action.

MPs have various shapes and sizes, are composed of different polymers and additives, and originate from diverse sources that release them into the environment. Polystyrene, polyethylene terephthalate, polyvinylchloride, polypropylene, polyethylene, and polyamide are the main MP polymers found in the marine environment. There are two types of MPs: primary and secondary (Rochman et al. 2019). Primary MPs are intentionally manufactured for several commercial purposes, such as microbeads, plastic pellets, road markings, and city dust, (Boucher and Friot 2017). In contrast, secondary MPs result from parent materials such as discarded plastic products in various forms such as fragmented pieces, fibers, or thin films and textile items, which break down into smaller particles after being released into different environmental compartments (Horton and Dixon 2018). Knowing the difference between the two types of MPs is important, as they can have very different sources and characteristics, and this information is crucial for creating effective ways to lessen their effect on the environment and human health.

The predominant form of MPs that are commonly found in fish is classified as fibers, which make up between 58 and 87% of all plastic shapes found (Hossain et al. 2023; Güven et al. 2017). Fragments, films, and threads are common types of MPs in fish and shellfish, unlike spherical MPs, which are scarce (Gamarro et al. 2020). While the exact chemical effects of MP ingestion on marine organisms are not yet fully understood, the physical impacts can include oxidative stress, inflammation, and even potential hunger (Pirsaheb et al. 2020). Ingested MPs can cause abrasions and lesions in the digestive tract, leading to internal damage and reduced feeding activity (Jabeen et al. 2018). Marine organisms can also be exposed to toxic chemicals through MPs, which are commonly found in plastics (Koelmans 2015). As a result, the ingestion of MPs is a growing concern for the health and well-being of marine ecosystems and the animals that depend on them.

Human health can be impacted by plastics in aquatic environments, which can lead to behavioral changes and reduced growth rates, but there is no information on how MPs in seafood might affect human health. According to Lusher and colleagues, over 220 different types of marine creatures, ranging from tiny zooplankton to large marine mammals, have been found to have ingested plastic (Lusher et al. 2017). Some researchers have shown that this number has increased to at least 690 species. This includes a wide range of animals such as bivalves, crustaceans, fish, sea turtles, and seabirds (Gall and Thompson 2015). Ingesting plastics can harm the health and survival of these organisms, which is a global concern for marine ecosystems.

14.1.1 Distribution and Movement of MPs Within Marine Ecosystems

MPs in marine environments can originate from various sources, including land-based activities and the breakdown of larger plastic items (Auta et al. 2017). These tiny particles can then be transported by rivers and ocean currents, and their fate can vary depending on environmental factors. MPs can accumulate in certain areas of the seafloor due to the influence of ocean currents. Moreover, MPs can sink from the surface to the bottom of the ocean along with marine snow, which is a natural phenomenon of organic matter and debris falling to the sediment (Porter et al. 2018).

While MPs are known to settle on the seafloor, recent studies have shown that extreme weather events, such as flooding and typhoons, can also play a significant role in redistributing and increasing the abundance of MPs in coastal marine ecosystems (Chen et al. 2021; Daniel et al. 2022). Furthermore, fishing gear and other macroplastic debris can also break down into smaller particles through biological, chemical, and physical processes (Browne et al. 2010). The intricate interaction among these factors underscores the pressing requirement for sustained research on the destiny and conveyance of MPs in marine ecosystems. The development of efficacious approaches to attenuate the repercussions of microplastics on human and environmental health is imperative.

14.2 Detection in Microplastics in Seafood

14.2.1 Presence of MPs in Bivalve a Mollusk

Bivalves, a taxonomic subclass of mollusks belonging to the class Bivalvia, exhibit extensive global consumption by the human population. In 2018, global bivalve production reached 15.2 million metric tons, with China, Peru, and Japan being the top-producing countries (Wijsman et al. 2019). Due to their rich content of omega-3 fatty acids, protein, minerals, and essential vitamins, bivalves are highly coveted as a nutritious dietary option. The most commonly consumed bivalves include clams, oysters, and mussels, which can be eaten raw, cooked, or canned (Wright et al. 2018). In terms of nutritional value, 100 g of cooked clams contain approximately 12.8 g of protein, while the same amount of cooked oysters contains around 7 g of protein (Joseph 2022a, b). However, bivalves are also known to accumulate contaminants such as heavy metals and harmful algae toxins, which can be harmful to human health if consumed in large amounts (Pavón et al. 2022; Lassudrie et al. 2020). As a result, there are regulations in place in many countries to monitor and control bivalve harvesting and processing practices to ensure that they are safe for human consumption.

Mollusks are a highly nutritious food source with significant economic value and are relatively simple to cultivate. Mollusks, particularly bivalves, are highly prized in aquaculture for their unique ability to filter and remove microbes and organic matter from the water (Gifford et al. 2004). This valuable trait enhances their desirability as a sustainable and healthy food source. Due to their wide distribution, certain bivalves, including M. edulis, are recognized as biological indicators for monitoring MP pollution in coastal regions (Azizi et al. 2018). This is due to their sensitivity to environmental changes, making them useful for assessing the health of coastal ecosystems. Bivalves are the most extensively studied type of seafood when it comes to MP uptake from the environment. This is because their filtering behavior puts them at a greater risk of ingesting MPs from the water column. The early investigations were centered primarily on blue mussels, which provided insights into the presence of MPs in both wild and market-bought samples (Beyer et al. 2017). However, other bivalve species, such as clams, oysters, and scallops, have also been the subject of investigations into MP uptake. These investigations have revealed that MPs can be found in various parts of bivalves, including their digestive systems, gills, and tissues, which has raised concerns about the potential health risks of consuming contaminated bivalves (Baechler et al. 2020).

MP fibers are frequently found in bivalves, such as mussels, clams, oysters, and scallops. In fact, fibers have been reported to account for the majority of MPs found in these species in various regions of the world, such as China (Ding et al. 2021), British Columbia, and Oregon (Baechler et al. 2020). One possible explanation for the prevalence of MP fibers in bivalves is that they are more difficult to expel from the digestive tract compared to other MP morphologies.

The abundance of microplastics (MPs) found in bivalves is subject to multifaceted influences, encompassing environmental contamination levels, species-specific filtration capabilities, and the specific tissues selected for examination during investigation. Studies indicate that bivalves sampled from areas with highly contaminated areas or those close to urban sources of MPs typically contain higher levels of MPs (Bom and Sá 2022). Conversely, certain investigations have suggested that there is a lack of discernible disparity in microplastic (MP) exposure levels across various origins. The ingestion of MPs by bivalves may have implications for both the health of the species and the safety of human consumption, as MPs may accumulate in the tissues of the bivalve and be transferred to humans upon consumption.

There are differences in the levels of MPs found in bivalves purchased from markets versus those caught in the wild, with the former containing lower concentrations. Similarly, farmed mussels have been found to have lower amounts of MPs than wild mussels (Cho et al. 2019; Lam et al. 2023). The use of depuration procedures has been shown to reduce the number of MPs in bivalves, which is significant for their preparation as food. MP concentrations in marine biota may vary depending on the season, as observed in oysters during the summer months.

The size and morphology of microplastics (MPs) exert a significant influence on the selective ingestion and retention of particles by bivalves. Smaller particles tend to accumulate more, and gut retention times vary across bivalve species and age. However, most studies sample whole organisms, making it difficult to determine the location of MPs in their tissues.

MPs can impact bivalve physiology directly and indirectly affect their habitats, leading to a decrease in food resources and transfer of organic pollutants. These effects can negatively impact filtration activity, feeding behavior, and reproduction (Zhang et al. 2020). It is worth noting that most studies on the topic have used uniform MP particles, which may not accurately represent the exposure regime in the environment. Several studies have identified the potential risk of human consumption due to the accumulation of MP fibers on the foot and mantle of blue mussels.

MP fibers have been shown to gather on the mantle and foot of blue mussels, according to some research.

14.2.2 Presence of MPs in Echinoderms

Echinoderms are a phylum of marine invertebrates that encompass sea cucumbers, urchins, and sea stars (Flammang 2006). These creatures are harvested as food sources in many countries, particularly in Asia where they are considered delicacies. Sea urchin gonads, or roe, are highly valued for their rich, buttery taste and are a popular ingredient in sushi (Chowdhury and Yahya 2012). Sea cucumbers, on the other hand, are used in traditional Chinese medicine and are also consumed as a food source. In some cultures, sea stars are also consumed as a delicacy (Slater 2015).

The global market for echinoderms is significant, with estimates suggesting that the value of the trade in sea urchin roe alone is in the billions of dollars annually. In Japan, for example, the demand for sea urchin roe is so high that it has led to overfishing and concerns about the sustainability of this practice. Similarly, in some parts of Asia, the overexploitation of sea cucumbers has led to population declines, with some species now listed as endangered (Eddy et al. 2015).

In terms of nutritional value, echinoderms are rich in protein and other nutrients. Sea urchin roe is a valuable dietary source of omega-3 fatty acids that are recognized for their ability to provide a range of health benefits (Alrajab and Shulgina 2022). Sea cucumbers are also highly valued for their potential health benefits, which include properties that may help reduce inflammation and combat cancer (Janakiram et al. 2015). However, there are also concerns about the presence of heavy metals and other contaminants in echinoderms, particularly those harvested from polluted waters (Bonsignore et al. 2018).

The presence and concentrations of microplastics (MPs) detected in echinoderms exhibit variations that are contingent upon factors such as organism size, geographical location, and species identity. Research has found that sea urchins tend to contain less MPs than starfish, possibly due to their feeding habits (Feng et al. 2020). MPs in echinoderms can lead to physical damage, such as blockages in the digestive system or entanglement, as well as changes in behavior and metabolism.

If echinoderms like sea urchins and sea cucumbers are consumed as food, the presence of MPs in these organisms could pose a potential threat to human health, particularly in Asia where they are frequently eaten. Despite growing concerns about the presence of MPs in these marine creatures, the number of studies that have examined the abundance of MPs in echinoderms is relatively low. One investigation, which studied heart urchins (Brissopsis lyrifera), found that 40% of them had MPs in their soft tissue, mainly as flakes, and only one particle per individual on average (Bour et al. 2018). However, since heart urchins are not commonly consumed, this finding might be less relevant. Another study conducted in the northern coastal region of China revealed that MPs were present in 90% of individuals across four species of sea urchins. On average, these creatures had five particles per individual, primarily in the form of fibers (Feng et al. 2020).

Research conducted in the Yellow Sea found that the gut of sea urchins contained significantly more MPs than their gonads and coelomic fluid, although the difference was not consistent across all species when adjusted for weight. Studies also show that several sea cucumber species, including those commonly eaten in Japan and China, have ingested MPs. MPs have been found in various parts of sea cucumbers, with cellophane microfibers being the most common type in farmed sea cucumbers from China's Bohai Sea and Yellow Sea (Feng et al. 2020).

There is still limited knowledge about the long-term effects of MPs on echinoderms and their ecosystems. However, more research is needed to determine the potential impact of MP contamination on human health for both sea urchins and sea cucumbers.

14.2.3 Presence of MPs in Cephalopods

Cephalopods are a group of marine animals that include octopuses, squids, and cuttlefish. These creatures are consumed in various parts of the world, particularly in coastal regions. They are a valuable source of protein, vitamins, and minerals and are low in fat, making them a healthy food choice (Mouritsen and Styrbæk 2018). Here are some facts and figures about cephalopods in the human diet:

- Cephalopods have constituted a staple in human consumption for millennia, becoming a significant component of diverse global cuisines (Mouritsen and Styrbæk 2018).
- In some countries, such as Japan and Spain, cephalopods are a staple in the diet, and a wide variety of dishes are prepared using these animals (Mouritsen et al. 2009; Mouritsen and Styrbæk 2018).
- Squid and octopus are the most commonly consumed cephalopods in the world, followed by cuttlefish.
- Cephalopods are a good source of protein, with a 4 ounce serving of cooked squid containing approximately 18 g of protein (WebMD Editorial Contributors 2022).
- They are also low in fat, with most varieties containing less than 2 g of fat per 4 ounce serving (WebMD Editorial Contributors 2022).
- Cephalopods are rich in vitamins and minerals, particularly vitamin B12, which is important for the functioning of the nervous system, and minerals such as zinc and copper.
- Cephalopods account for around 2.5% of seafood production. The global cephalopod market exhibited a valuation surpassing three million tons in 2019, with projections indicating sustained growth in the forthcoming years (Ospina-Alvarez et al. 2022).

However, there are concerns about overfishing and the impact of fishing on cephalopod populations, particularly in some parts of the world. As with all seafood, it is important to choose sustainable sources of cephalopods to ensure the long-term viability of these animals as a food source.

MP contamination in cephalopods has gained attention due to the potential risks it poses to human health as cephalopods are widely consumed globally. Several studies have investigated the presence of MPs in cephalopods, such as cuttlefish and squid, but there is still much to be understood regarding the extent and impact of this contamination. In particular, there is a need for further research on the transfer of MPs from cephalopods to humans and the potential health effects.

In one study by Oliveira et al. (2020), MP levels were measured in different parts of cuttlefish (Sepia officinalis), but the study did not include tissue that is relevant for human consumption (Oliveira et al. 2020). Another study found that 18% of Indian squid (Uroteuthis duvaucelii) were contaminated with MPs, with an average of 0.008 particles per gram of edible tissue and 0.2 particles per individual (Peng et al. 2020).

Further investigation is required to improve our comprehension of the origins, distribution, and potential consequences of MPs in cephalopods. Furthermore, there is a need to examine the potential impacts of MPs on human consumption of these organisms.

14.2.4 Presence of MPs in Crustaceans

Crustaceans, a class of arthropods predominantly found in aquatic environments, are widely utilized as a significant food resource in human alimentation. Here are some facts, figures, and more information about crustaceans in human diet:

- Types of Crustaceans: The most commonly consumed crustaceans include shrimp, crab, lobster, crayfish, and prawns.
- Nutritional Value: Crustaceans are a rich source of high-quality protein, omega-3 fatty acids, vitamins, and minerals. They are also low in fat and calories, making them a healthy choice for a balanced diet.
- Production and Consumption: In 2009, the global production of crustaceans was estimated to be around 11.2 million tons, with China being the largest producer and consumer of crustaceans (Bondad-Reantaso et al. 2012).
- Culinary Use: Crustaceans are commonly used in a wide range of culinary dishes, including seafood salads, soups, stews, and curries. They are also popular in sushi and sashimi, where raw or cooked shrimp, crab, and lobster are used as fillings.
- Allergies: Some people are allergic to crustaceans, and even small amounts can trigger an allergic reaction. Symptoms of a crustacean allergy can include itching, hives, difficulty breathing, and, in severe cases, anaphylaxis (Davis et al. 2020).
- Sustainability: The increasing demand for crustaceans has raised concerns about the sustainability of their production. Overfishing, habitat destruction, and climate change have all contributed to the decline in crustacean populations. Sustainable practices, such as responsible fishing, aquaculture, and recycling of shells and other waste products, can help to reduce the environmental impact of crustacean production.

Crustaceans are a nutritious and widely consumed group of aquatic animals. They are used in a variety of culinary dishes and provide high-quality protein, vitamins, and minerals (Prester 2011). However, concerns about the sustainability of their production and the risk of allergic reactions highlight the importance of responsible consumption and production practices.

Research has predominantly focused on the presence of microplastics (MPs) in the gastrointestinal tract of organisms, which is not typically a part of the human diet (Lim 2021). Nevertheless, shellfish, including crabs and mollusks, represent an exception as they are commonly consumed whole or with the gut removed (Lundebye et al. 2022). The potential ingestion of MPs from other tissues, such as muscle, hinges on their capacity to traverse the intestinal barrier and accumulate in edible tissues (Zeytin et al. 2020).

Previous investigations concerning frequently consumed crustaceans by humans have primarily focused on wild-caught specimens rather than their cultivated counterparts. These studies have typically investigated environmental pollutants, rather than specifically looking at the safety of seafood (Lundebye et al. 2022). For example, there have been numerous studies conducted on the langoustine species Nephrops norvegicus, a commercially valuable crustacean. As per a study, Nephrops norvegicus specimens from the Clyde Sea region exhibited a higher prevalence of single-strand MP fibers in their stomachs compared to those collected from the North Sea and Minch, where a considerably lower proportion of these fibers was observed (Welden and Cowie 2016). Economically significant species, including shrimp, prawns, and spiny spider crabs (Maja squinado), have also been discovered to harbor MPs (Devriese et al. 2015; Welden et al. 2018; Cau et al. 2019).

Coastal areas where crustaceans are caught may be contaminated with MPs, which come from sources like plastic on land. For example, Mediterranean shrimp (Aristeus antennatus) exhibited an average MP content of 39.2%, whereas specimens captured in proximity to urban centers demonstrated a complete prevalence of 100% MPs (Carreras-Colom et al. 2018). In the Clyde Sea, both isolated populations of N. norvegicus (40%) and those collected near Glasgow (84%) had the same percentage of MPs (Welden and Cowie 2016).

When analyzing crustaceans in their natural habitats, researchers often find fibers and fragments as the most common types of MPs, with fiber bundles being present in various species (McGoran et al. 2020). Although the majority of research has examined the presence of MPs in the stomachs of crustaceans, there is growing interest in investigating other tissues such as gills and muscles that could potentially impact human exposure. Removing the stomach before cooking and consumption does not necessarily eliminate the risk of ingesting MPs. In fact, studies have found MP contamination in various tissues of wild Portunus gracilimanus and P. trituberculatus fish, indicating that it is not restricted to the digestive system (Zhang et al. 2019).

14.2.5 Presence of MPs in Finfish

Finfish term is used to refer to a specific group of fish that are biologically distinct from other aquatic life such as crustaceans (e.g., crayfish), mollusks (e.g., cuttlefish), echinoderms (e.g., starfish), and other creatures (e.g., jellyfish) that are sometimes mistakenly referred to as "fish" even though they are not true fish. The term "true fish" is used to distinguish this biologically distinct group of fish from these other aquatic creatures. Finfish are the most common type of fish and include a wide range of species such as salmon, tuna, cod, herring, and many others (Cooke 2016). Finfish are found in a variety of aquatic environments, including oceans, lakes, and rivers, and are an important food source for many people around the world. They also play a significant role in recreational and commercial fishing industries. In fish, MPs have been found in a range of organs and tissues, including the liver, muscle, and gut (Abbasi et al. 2018).

Several studies have detected MPs in both wild and farmed fish species (Lusher et al. 2013). For example, a study published in the journal of marine pollution bulletin found that MPs were present in 63% of the fish sampled from the English Channel, with an average of 1.9 particles per fish (Lusher et al. 2013). Similarly, in another study published in a peer-reviewed journal it was found that salmon from both Europe and North America contained MPs (Baechler et al. 2020).

The concentration of MPs in fish can vary depending on a range of factors, such as the type of fish, its diet, and the level of plastic pollution in the surrounding environment. In general, fish that live in areas with high levels of plastic pollution, such as near coastal urban areas or in the vicinity of plastic manufacturing facilities, are more likely to have higher concentrations of MPs in their tissues.

The effects of MPs on fish health and behavior are still not fully understood. Some studies have suggested that MPs can cause physical damage to fish organs and tissues, such as inflammation and oxidative stress, while others have found evidence of behavioral changes in fish exposed to MPs, such as altered feeding behavior and reduced activity levels (Usman et al. 2021; Galloway et al. 2017).

Additional studies are necessary to fully comprehend the potential consequences to human health that may result from the consumption of fish contaminated with MPs. Some studies have suggested that MPs may contain toxic chemicals that could accumulate in the tissues of fish and be passed on to humans who consume them. However, the levels of MPs and associated chemicals found in fish are generally considered to be low, and the risk to human health is currently thought to be low as well.

Overall, reducing the release of MPs into the environment is an important step in protecting the health of both aquatic organisms and humans. This objective can be accomplished through implementing measures such as plastic use reduction, enhancing waste management systems, and advocating for sustainable fishing practices.

14.3 Consequences of the Consumption of Seafood with MP

14.3.1 Impact of MPs on Human Health: Ingestion and Toxicity

MPs are absorbed depending on their size, shape, solubility, surface charge, and chemistry. According to the European Food Safety Authority (EFSA) findings, microplastic particles (MPs) with a diameter below 130 μm possess the potential to permeate human tissues, while particles measuring 1.5 μm or smaller have the ability to traverse capillaries (Yoo et al. 2011; EFSA Panel on Contaminants in the Food China 2016). Endocytotic and paracellular transport through epithelial tissues are two proposed pathways for the uptake of MPs (Wright and Kelly 2017). According to EFSA Panel on Contaminants in the Food China (2016), 90% of ingested MPs are thought to be eliminated from the body (EFSA Panel on Contaminants in the Food China 2016).

MPs can be harmful to humans in a number of ways, with the main mechanisms of toxicity believed to be oxidative damage and resulting inflammation. Exposure to MPs can also result in genotoxicity, apoptosis, and necrosis, which can cause tissue damage, fibrosis, and the development of cancer. The severity of these negative effects is dependent on the size of the MP particles, with studies suggesting that nanoparticles can cause more reactive oxygen species and are more likely to be transported throughout the body. The properties of the MPs, such as their size, shape, and surface properties, can also influence their toxicity (Xuan et al. 2023).

The consequences of the consumption of seafood with MPs are still being studied, but there are a number of potential risks that have been identified. These include:

- Oxidative stress: MPs have the capacity to elicit oxidative stress, a state characterized by an imbalance between antioxidants and free radicals within the body. Oxidative stress, known to cause cellular and tissue damage, has been correlated with various health issues, including cancer, cardiovascular diseases, and neurodegenerative disorders (Villines 2020).
- Neurotoxicity: MPs can also damage the nervous system. Studies have shown that MPs can enter the brain and cause inflammation and cell death. This can lead to a range of neurological problems, including learning and memory deficits, and behavioral changes (Villines 2020).
- Endocrine disruption: MPs can also disrupt the endocrine system, which is responsible for regulating hormones. This can lead to a range of health problems, including reproductive problems, obesity, and cancer (Villines 2020).
- Immune system suppression: MPs can also suppress the immune system. This can make people more susceptible to infections and other diseases (Villines 2020).
- Cancer: Some studies have shown that MPs can increase the risk of cancer. This is because MPs can carry toxins that can damage DNA and lead to the development of cancer cells (Villines 2020).

It is imperative to acknowledge that investigations into the health implications of MPs are still in their nascent stages, warranting further research to comprehensively comprehend the associated risks. However, the evidence that is available suggests that there are a number of potential health risks associated with the consumption of seafood with MPs.

14.4 Future Strategies to Minimize the Entry of MP in Seafood

Following are some of the strategies for minimizing the MP invasion in seafood:

- **Using less single-use plastics and disposing them properly:** This can help avoid plastic waste from reaching the ocean and becoming MPs.
- **Helping the seafood industry to remove plastic pollution from the coastal waters:** This can help protect marine life from eating or getting trapped by plastics.
- **Working with consumers to fund environmental-friendly actions:** This can help increase awareness and promote behaviors that lower plastic use and waste.
- **Focusing on the life cycle of MPs:** This can help find and solve the problems of MP sources, pathways, and impacts on the environment and human health.
- Doing more studies on how MPs affect seafood and human health: This can help improve the methods of finding, measuring, and preventing MP contamination.

14.5 Conclusions

It is essential to emphasize that the deleterious impacts of microplastics on human health remain incompletely elucidated, necessitating further research to advance our understanding in this domain. While studies have shown that nanoplastics may be more harmful than larger MPs, it is also possible that the combined exposure to a variety of MPs of different sizes and shapes may have complex effects on the body. Additionally, the ways in which MPs interact with other pollutants and environmental stressors, as well as individual differences in susceptibility, may also play a role in their overall impact on human health.

Acknowledgments I appreciate Dr. Shahnawaz for allowing me to write this chapter on "Detection of Microplastic in Sea Food." I gained a lot of insight from this project, thanks to his expertise and guidance. I hope this chapter will enhance the current research and awareness on this important issue.

References

Abbasi S, Soltani N, Keshavarzi B, Moore F, Turner A, Hassanaghaei M (2018) Microplastics in different tissues of fish and prawn from the Musa estuary, Persian Gulf. Chemosphere 205:80–87

Alrajab M, Shulgina LV (2022) Dietary product based on sea urchin caviar and Sardinops melanostictus fat. J Appl Biol Biotechnol 10:102–106

An L, Liu Q, Deng Y, Wennan W, Gao Y, Ling W (2020) Sources of microplastic in the environment. In: He D, Luo Y (eds) Microplastics in terrestrial environments: emerging contaminants and major challenges. Cham, Springer International Publishing

Auta HS, Emenike CU, Fauziah SH (2017) Distribution and importance of microplastics in the marine environment: a review of the sources, fate, effects, and potential solutions. Environ Int 102:165–176

Azizi G, Akodad M, Baghour M, Layachi M, Moumen A (2018) The use of Mytilus spp. mussels as bioindicators of heavy metal pollution in the coastal environment. A review. J Mater Environ Sci 9:1170–1181

Baechler BR, Stienbarger CD, Horn DA, Joseph J, Taylor AR, Granek EF, Brander SM (2020) Microplastic occurrence and effects in commercially harvested north American finfish and shellfish: current knowledge and future directions. Limnol Oceanography Lett 5:113–136

Barboza LGA, Gimenez BCG (2015) Microplastics in the marine environment: current trends and future perspectives. Mar Pollut Bull 97:5–12

Beyer J, Green NW, Brooks S, Allan IJ, Ruus A, Gomes T, Bråte ILN, Schøyen M (2017) Blue mussels (Mytilus edulis spp.) as sentinel organisms in coastal pollution monitoring: a review. Mar Environ Res 130:338–365

Bom FC, Sá F (2022) Are bivalves a source of microplastics for humans? A case study in the Brazilian markets. Mar Pollut Bull 181:113823

Bondad-Reantaso MG, Subasinghe RP, Josupeit H, Cai J, Zhou X (2012) The role of crustacean fisheries and aquaculture in global food security: past, present and future. J Invertebr Pathol 110: 158–165

Bonsignore M, Manta DS, Mirto S, Quinci EM, Ape F, Montalto V, Gristina M, Traina A, Sprovieri M (2018) Bioaccumulation of heavy metals in fish, crustaceans, molluscs and echinoderms from the Tuscany coast. Ecotoxicol Environ Saf 162:554–562

Boucher J, Friot D (2017) Primary microplastics in the oceans: a global evaluation of sources. IUCN, Gland

Bour A, Avio CG, Gorbi S, Regoli F, Hylland K (2018) Presence of microplastics in benthic and epibenthic organisms: influence of habitat, feeding mode and trophic level. Environ Pollut 243: 1217–1225

Browne MA, Galloway TS, Thompson RC (2010) Spatial patterns of plastic debris along estuarine shorelines. Environ Sci Technol 44:3404–3409

California State Water Resources Control Board (2019) Microplastics in drinking water. https://www.waterboards.ca.gov/drinking_water/certlic/drinkingwater/microplastics.html

Carreras-Colom E, Constenla M, Soler-Membrives A, Cartes JE, Baeza M, Padrós F, Carrassón M (2018) Spatial occurrence and effects of microplastic ingestion on the deep-water shrimp Aristeus antennatus. Mar Pollut Bull 133:44–52

Cau A, Avio CG, Dessì C, Follesa MC, Moccia D, Regoli F, Pusceddu A (2019) Microplastics in the crustaceans Nephrops norvegicus and Aristeus antennatus: flagship species for deep-sea environments? Environ Pollut 255:113107

Chen L, Li J, Tang Y, Wang S, Xiao L, Cheng Z, Zhang X, Pengfei W, Chang X, Xia Y (2021) Typhoon-induced turbulence redistributed microplastics in coastal areas and reformed plastisphere community. Water Res 204:117580

Cho Y, Shim WJ, Jang M, Han GM, Hong SH (2019) Abundance and characteristics of microplastics in market bivalves from South Korea. Environ Pollut 245:1107–1116

Chowdhury A, Yahya K (2012) Sustainable seafood production: Malaysian status and comparison with the world. In: Conference: international conference aquaculture Indonesia (ICAI)-2012, Semarang, Indonesia

Cooke M (2016) Animal welfare in farmed fish. Bus Benchmark Farm Anim Welf Invest Brief 23: 1–16

Daniel DB, Muhamed Ashraf P, Thomas SN (2022) Impact of 2018 Kerala flood on the abundance and distribution of microplastics in marine environment off Cochin, southeastern Arabian Sea, India. Reg Stud Mar Sci 53:102367

Danopoulos E, Jenner LC, Twiddy M, Rotchell JM (2020) Microplastic contamination of seafood intended for human consumption: a systematic review and meta-analysis. Environ Health Perspect 128:126002

Davis CM, Gupta RS, Aktas ON, Diaz V, Kamath SD, Lopata AL (2020) Clinical management of seafood allergy. J Allergy Clin Immunol Pract 8:37–44

Devriese LI, Van der Meulen MD, Maes T, Bekaert K, Paul-Pont I, Frère L, Robbens J, Dick A, Vethaak. (2015) Microplastic contamination in brown shrimp (Crangon crangon, Linnaeus 1758) from coastal waters of the southern North Sea and channel area. Mar Pollut Bull 98: 179–187

Ding J, Sun C, He C, Li J, Peng J, Li F (2021) Microplastics in four bivalve species and basis for using bivalves as bioindicators of microplastic pollution. Sci Total Environ 782:146830

Eddy SD, Brown NP, Harris LG (2015) Aquaculture of the green sea urchin Strongylocentrotus droebachiensis in North America. In: Echinoderm aquaculture, pp 175–209

EFSA Panel on Contaminants in the Food China (2016) Presence of microplastics and nanoplastics in food, with particular focus on seafood. EFSA J 14:e04501

European Chemicals Agency (2019) Registry of restriction intentions: 4-(1,1,3,3-tetramethylbutyl) phenol, ethoxylated [covering any of its individual isomers and/or combinations thereof (with or without mono-constituent or multi-constituent substances)]. https://echa.europa.eu/registry-of-restriction-intentions/-/dislist/details/0b0236e18244cd73

Feng Z, Wang R, Zhang T, Wang J, Huang W, Li J, Juntian X, Gao G (2020) Microplastics in specific tissues of wild sea urchins along the coastal areas of northern China. Sci Total Environ 728:138660

Flammang P (2006) Adhesive secretions in echinoderms: an overview. In: Smith AM, Callow JA (eds) Biological adhesives. Springer Berlin Heidelberg, Berlin

Food and Agriculture Organization of the United Nations (2018) The state of world fisheries and aquaculture 2018—meeting the sustainable development goals. Food and Agriculture Organization of the United Nations

Food and Agriculture Organization of the United Nations (2020) The state of world fisheries and aquaculture 2020: sustainability in action. Food and Agriculture Organization of the United Nations

Gall SC, Thompson RC (2015) The impact of debris on marine life. Mar Pollut Bull 92:170–179

Galloway TS, Cole M, Lewis C (2017) Interactions of microplastic debris throughout the marine ecosystem. Nat Ecol Evol 1:0116

Gamarro G, Esther JR, Elvevoll EO, Olsen RL (2020) Microplastics in fish and shellfish—a threat to seafood safety? J Aquat Food Prod Technol 29:417–425

Geyer R, Jambeck JR, Law KL (2017) Production, use, and fate of all plastics ever made. Sci Adv 3: e1700782

Gifford S, Dunstan RH, O'Connor W, Roberts T, Toia R (2004) Pearl aquaculture—profitable environmental remediation? Sci Total Environ 319:27–37

Güven O, Gökdağ K, Jovanović B, Kıdeyş AE (2017) Microplastic litter composition of the Turkish territorial waters of the Mediterranean Sea, and its occurrence in the gastrointestinal tract of fish. Environ Pollut 223:286–294

Horton AA, Dixon SJ (2018) Microplastics: an introduction to environmental transport processes. Wiley Interdiscip Rev Water 5:e1268

Hossain MB, Jimmy Y, Nur A-AU, Banik P, Jolly YN, Al Mamun M, Arai T, Albeshr MF (2023) Microplastics in surface water from a mighty subtropical estuary: first observations on occurrence, characterization, and contamination assessment. Environ Res 226:115594

Jabeen K, Li B, Chen Q, Lei S, Chenxi W, Hollert H, Shi H (2018) Effects of virgin microplastics on goldfish (Carassius auratus). Chemosphere 213:323–332

Janakiram NB, Mohammed A, Rao CV (2015) Sea cucumbers metabolites as potent anti-cancer agents. Mar Drugs 13:2909–2923

Joseph M (2022a) 7 benefits of clams (and full nutrition facts). https://www.nutritionadvance.com/clams-nutrition-benefits/

Joseph M (2022b) 9 health benefits of oysters (and full nutrition facts)

Kobayashi M, Msangi S, Batka M, Vannuccini S, Dey MM, Anderson JL (2015) Fish to 2030: the role and opportunity for aquaculture. Aquacult Econ Manag 19:282–300

Koelmans AA (2015) Modeling the role of microplastics in bioaccumulation of organic chemicals to marine aquatic organisms. A critical review. In: Bergmann M, Gutow L, Klages M (eds) Marine anthropogenic litter. Springer International Publishing, Cham

Lam TW, Ling YC, Tsui J, Cheng YL, Ma ATH, Fok L (2023) Microplastic contamination in edible clams from popular recreational clam-digging sites in Hong Kong and implications for human health. Sci Total Environ 875:162576

Lassudrie M, Hégaret H, Wikfors GH, da Silva PM (2020) Effects of marine harmful algal blooms on bivalve cellular immunity and infectious diseases: a review. Dev Comp Immunol 108: 103660

Lim XZ (2021) Microplastics are everywhere—but are they harmful. Nature 593:22–25

Lundebye, Anne-Katrine, Amy L. Lusher, and Michael S. Bank. 2022. 'Marine microplastics and seafood: implications for food security.' in Michael S. Bank, Microplastic in the environment: pattern and process (Springer International Publishing: Cham)

Lusher AL, McHugh M, Thompson RC (2013) Occurrence of microplastics in the gastrointestinal tract of pelagic and demersal fish from the English Channel. Mar Pollut Bull 67:94–99

Lusher A, Hollman P, Mendoza-Hill J (2017) Microplastics in fisheries and aquaculture: status of knowledge on their occurrence and implications for aquatic organisms and food safety (FAO)

McGoran AR, Clark PF, Smith BD, Morritt D (2020) High prevalence of plastic ingestion by Eriocheir sinensis and Carcinus maenas (Crustacea: Decapoda: Brachyura) in the Thames estuary. Environ Pollut 265:114972

Mendes J, Curralo A, Curado A, Lopes SI (2021) The sustainable smartbottle: a proposed design methodology to minimize plastic pollution. In: Advances in design and digital communication: proceedings of the 4th international conference on design and digital communication, Digicom 2020, November 5–7, 2020, Barcelos, Portugal, pp 280–92. Springer
Mouritsen OG, Styrbæk K (2018) Cephalopod gastronomy—a promise for the future. Front Commun 3:38
Mouritsen OG, Mouritsen JD, Nyberg T (2009) Sushi: food for the eye, the body & the soul. Springer
Nicole W (2021) Microplastics in seafood: how much are people eating? Environ Health Perspect 129:034001
Oliveira AR, Sardinha-Silva A, Andrews PLR, Green D, Cooke GM, Hall S, Blackburn K, Sykes AV (2020) Microplastics presence in cultured and wild-caught cuttlefish, Sepia officinalis. Mar Pollut Bull 160:111553
Ospina-Alvarez A, de Juan S, Pita P, Ainsworth GB, Matos FL, Pita C, Villasante S (2022) A network analysis of global cephalopod trade. Sci Rep 12:322
Pavón A, Riquelme D, Jaña V, Iribarren C, Manzano C, Lopez-Joven C, Reyes-Cerpa S, Navarrete P, Pavez L, García K (2022) The high risk of bivalve farming in coastal areas with heavy metal pollution and antibiotic-resistant bacteria: a Chilean Perspective. Front Cell Infect Microbiol 12:867446
Peng L, Dongdong F, Qi H, Lan CQ, Huamei Y, Ge C (2020) Micro- and nano-plastics in marine environment: source, distribution and threats—a review. Sci Total Environ 698:134254
Pinstrup-Andersen P (2009) Food security: definition and measurement. Food Security 1:5–7
Pirsaheb M, Hossini H, Makhdoumi P (2020) Review of microplastic occurrence and toxicological effects in marine environment: experimental evidence of inflammation. Process Saf Environ Prot 142:1–14
Porter A, Lyons BP, Galloway TS, Lewis C (2018) Role of marine snows in microplastic fate and bioavailability. Environ Sci Technol 52:7111–7119
Prester L (2011) Biogenic amines in fish, fish products and shellfish: a review. Food Addit Contam Part A Chem Anal Control Expo Risk Assess 28:1547–1560
Rahman MH, Bhoi PR (2021) An overview of non-biodegradable bioplastics. J Clean Prod 294: 126218
Rochman CM, Brookson C, Bikker J, Djuric N, Earn A, Bucci K, Athey S, Huntington A, McIlwraith H, Munno K, De Frond H, Kolomijeca A, Erdle L, Grbic J, Bayoumi M, Borrelle SB, Tina W, Santoro S, Werbowski LM, Zhu X, Giles RK, Hamilton BM, Thaysen C, Kaura A, Klasios N, Ead L, Kim J, Sherlock C, Ho A, Hung C (2019) Rethinking microplastics as a diverse contaminant suite. Environ Toxicol Chem 38:703–711
Sharma S, Chatterjee S (2017) Microplastic pollution, a threat to marine ecosystem and human health: a short review. Environ Sci Pollut Res 24:21530–21547
Singh R, Kumar N, Mehrotra T, Bisaria K, Sinha S (2021) Chapter 9—Environmental hazards and biodegradation of plastic waste: challenges and future prospects. In: Saxena G, Kumar V, Shah MP (eds.) Bioremediation for environmental sustainability. Elsevier
Slater M (2015) Use and exploitation of sea cucumbers. In: Echinoderm aquaculture
Usman S, Razis AFA, Shaari K, Amal MNA, Saad MZ, Isa NM, Nazarudin MF (2021) Polystyrene microplastics exposure: an insight into multiple organ histological alterations, oxidative stress and neurotoxicity in Javanese Medaka fish (Oryzias javanicus Bleeker, 1854). Int J Environ Res Public Health 18:9449
Villines Z (2020) Plastics in seafood: microplastics, pollution, safety, and more. Medical News Today. https://www.medicalnewstoday.com/articles/plastics-in-seafood#what-are-microplastics
WebMD Editorial Contributors (2022) Squid: is it good for you?
Welden NAC, Cowie PR (2016) Environment and gut morphology influence microplastic retention in langoustine, Nephrops norvegicus. Environ Pollut 214:859–865

Welden NA, Abylkhani B, Howarth LM (2018) The effects of trophic transfer and environmental factors on microplastic uptake by plaice, Pleuronectes plastessa, and spider crab, Maja squinado. Environ Pollut 239:351–358

Wijsman JWM, Troost K, Fang J, Roncarati A (2019) Global production of marine bivalves. Trends and challenges. In: Smaal AC, Ferreira JG, Grant J, Petersen JK, Strand Ø (eds) Goods and services of marine bivalves. Springer International Publishing, Cham

Wright SL, Kelly FJ (2017) Plastic and human health: a micro issue? Environ Sci Technol 51:6634–6647

Wright AC, Fan Y, Baker GL (2018) Nutritional value and food safety of bivalve molluscan shellfish. J Shellfish Res 37:695–708

Xuan L, Xiao L, Huang R (2023) The geno-toxicological impacts of microplastic (MP) exposure on health: mechanistic pathways and research trends from a Chinese perspective. Environ Sci Processes Impacts 25:26

Yoo J-W, Doshi N, Mitragotri S (2011) Adaptive micro and nanoparticles: temporal control over carrier properties to facilitate drug delivery. Adv Drug Deliv Rev 63:1247–1256

Zeytin S, Wagner G, Mackay-Roberts N, Gerdts G, Schuirmann E, Klockmann S, Slater M (2020) Quantifying microplastic translocation from feed to the fillet in European sea bass Dicentrarchus labrax. Mar Pollut Bull 156:111210

Zhang F, Wang X, Jiayi X, Zhu L, Peng G, Pei X, Li D (2019) Food-web transfer of microplastics between wild caught fish and crustaceans in East China Sea. Mar Pollut Bull 146:173–182

Zhang F, Man YB, Mo WY, Man KY, Wong MH (2020) Direct and indirect effects of microplastics on bivalves, with a focus on edible species: a mini-review. Crit Rev Environ Sci Technol 50: 2109–2143

Reports of Microplastic in Human Blood 15

Mubashir J. Mintoo and Ruqiya Kousar

Abstract

Plastic pollution is a growing concern with potential impacts on human health. Research has identified plastic particles in various environmental compartments, including sediment, seawater, and human feces. The blood compartment has been identified as a suitable matrix for human biomonitoring of plastic particles. Various analytical approaches have been developed to identify and quantify plastic particles in biological matrices, including mass concentrations of polymers using pyrolysis-GC/MS. In the present chapter, an attempt was made to review all literature available on the detection of microplastic in human blood under the following lines: (1) to enlist sources, types of microplastics, and possible ways of their entry in the human body, (2) to review the reports on microplastic detection in human blood, (3) to highlight the techniques used to detect the microplastic in human blood, (4) to discuss the potential health implications of microplastics in the body, and (5) to suggest the strategies to minimize the entry of microplastic in human body.

Keywords

Blood · Ecosystem · Saphenous vein · Polytetrafluoroethylene · Polyethylene (PE) beads

---

Mubashir J. Mintoo and Ruqiya Kousar contributed equally with all other contributors.

M. J. Mintoo (✉)
Department of Cell and Molecular Pharmacology and Experimental Therapeutics, Medical University of South Carolina, Charleston, SC, USA

R. Kousar
Department of Botany and Environment Studies, DAV University, Jalandhar, Punjab, India

M. Shahnawaz et al. (eds.), *Microplastic Pollution*,
https://doi.org/10.1007/978-981-99-8357-5_15

15.1 Introduction

Almost nine million premature deaths were ascribed to pollution in 2019 in a study published in lancet journal titled "Pollution and health: a progress update" (Fuller et al. 2022). Moreover, there is mounting evidence that environmental pollution, namely, high PM2.5, is associated with subpar cardiovascular outcomes. However, the precise makeup of particles in the PM2.5, or PM10 group, is yet unknown (Al-Kindi et al. 2020). Microplastics (MPs), a newly emerging pollutant found in these fractions, have recently attracted attention as a source of concern (Yu et al. 2022). MPs are tiny synthetic polymer particles that range in size from micro to nano. Although there is no universal agreement on the precise size range, most sizes fall between 1 and 5 mm (da Silva et al. 2020). Microplastics from plastic bags, water bottles, and cosmetics are present in numerous items that are bought every day. These contaminants may accumulate in the environment and be consumed by people through their food, water, or even air (Nielsen et al. 2020).

15.1.1 Sources of Microplastics

Microplastics can come from a variety of sources, including:

- Cosmetics and personal care products. Microplastics are often used as exfoliants in face washes, scrubs, and toothpaste.
- Clothing. Microfibers from synthetic clothing can be released into the environment when it is washed.
- Food packaging. Microplastics can leach from food packaging, such as plastic bottles and bags.
- Drinking water. Microplastics can enter drinking water through the breakdown of plastic pipes and other infrastructure.
- Air pollution. Microplastics can be released into the air from the burning of plastics.

15.1.2 Entry of Microplastics in the Human Body

Microplastics have been discovered in the stomach, lungs, liver, kidneys, and blood, among other areas of the human body. These particles can enter the body in several ways, including through the skin, inhalation of contaminated air, consumption of tainted food and water, and ingestion (Vethaak and Legler 2021). Microplastics have been identified in human feces, indicating that these particles can build up in the gut and may have an impact on the gut flora (Luqman et al. 2021). Human organs, such as the liver and kidney, have been found to contain microplastics, suggesting that they may be carried throughout the body and accumulate there (Vethaak and Legler 2021). Furthermore, recent findings have suggested the presence of microplastics in lymphatic fluid and human blood (Wright and Kelly 2017).

15.2 The Presence of Microplastics in the Saphenous Vein Tissue of Humans

Microplastics are small plastic particles commonly found in the environment, and recent studies have found that they can accumulate in human tissues. Blood samples were collected from two healthy people in a study that examined the effects of polyethylene (PE) beads with sizes ranging from 10 to 45 mm. (Çobanoğlu et al. 2021). According to this study, human peripheral blood cells exposed to PE beads for 48 h at 250–500 mg/mL developed nuclear buds, nucleoplasmic bridges, and micronuclei, all indicative of increased genomic instability (Çobanoğlu et al. 2021).

Although microplastics (MPs) have been studied in cell and tissue cultures, little is known about their effects on humans. However, recent data suggest a link between high MP levels and inflammatory illnesses, particularly bowel diseases (Yan et al. 2022). Patients with coronary heart disease and obstructed coronary arteries often require coronary artery bypass grafts (CABGs) to restore blood flow to their hearts. Saphenous vein autografts are frequently used as conduits in CABG surgeries. (Caliskan et al. 2020). CABG procedures, however, fail in 40–50% of cases after 10 years for a variety of reasons. Notably, no published research has investigated a possible connection between exposure to environmental MP and the results of CABG (Caliskan et al. 2020). Hence, the evaluation of MP infiltration and composition in vascular conduits is a crucial initial step. Furthermore, no published research has examined whether MPs can enter or pass through any type of biological barrier, including blood vessels. Jeanette and colleagues from Department of Biological and Marine Sciences, University of Hull, United Kingdom, reported the presence of microplastic in saphenous vein tissues. This study sought to determine whether microplastic (MP) contamination was present in human venous tissue samples, and if so, to what extent and under what conditions (Rotchell et al. 2023).

15.2.1 Use of FTIR Spectroscopy for Identification and Characterization of Microplastics

FTIR spectroscopy is a powerful tool for detection of microplastic. It can be used to identify the chemical composition of microplastics, even when they are present in small quantities. FTIR spectroscopy does not damage the sample being analyzed, as it is a non-destructive technique. This is beneficial for microplastic detection, as it enables the analysis of microplastics that are found in environmental samples including water, soil, and sediment. It has also been used to detect microplastics in food and beverages.

FTIR spectroscopy works by passing infrared radiation through a sample. The infrared radiation interacts with the molecules in the sample, causing them to vibrate. The frequency of the vibration is characteristic of the type of molecule. By measuring the frequencies of the vibrations, FTIR spectroscopy can identify the chemical composition of the sample. FTIR spectroscopy is a valuable technique for microplastic identification and will probably be instrumental for the advancement of microplastic research.

To identify and characterize any MPs present, the samples were digested and subjected to FTIR spectroscopy with a size restriction of 5 μm. According to a study by Rotchell et al., they identified 20 MP particles from five different MP polymer types, such as alkyd resin, PVAc, and nylon-EVA, in four out of five vein tissue samples (Rotchell et al. 2023). The majority (90%) of the particles identified were MPs with irregular morphology. Moreover, the study revealed that the MP concentrations in the tissue samples were not significantly different from those in the procedural blanks, suggesting airborne contamination during collection. However, the plastic polymer types in the tissue samples differed from those in the control samples. (Rotchell et al. 2023). The procedural blanks had four different MP polymer types, represented by $n = 13$ MP particles, with polytetrafluoroethylene (PTFE) being the most prevalent. This study presents uncorrected findings along with various methods of contamination adjustment and reports the greatest level of contamination control. These findings underscore the need for further research to ascertain the effects of MPs on vascular health by showing that MPs can be carried within human tissues, specifically blood vessels (Rotchell et al. 2023).

15.3 Microplastics in Human Blood

Heather et al. recently discovered and quantified the microplastics in human blood (Leslie et al. 2022). This study sheds light on the growing concern about plastic pollution and its potential impact on human health. It emphasizes the need for reliable and accurate analytical methods to measure plastic particles in biological matrices (Leslie et al. 2022). The study discusses the presence of plastic particles in various environmental compartments, such as sediment, seawater, and human feces. The study identifies the blood compartment as a suitable matrix for human biomonitoring of plastic particles. The study describes various analytical approaches for identifying and quantifying plastic particles in biological matrices, including mass concentrations of polymers using pyrolysis-GC/MS (Leslie et al. 2022).

15.3.1 Techniques for Detection of Microplastic in Human Blood

Pyrolysis-GC/MS: Identifying and quantifying microplastics in environmental samples can be done by pyrolysis-GC/MS, which is a powerful analytical technique. Pyrolysis is a process that breaks down organic materials into their constituent molecules, and GC/MS is a technique that can be used to separate and identify these molecules. In pyrolysis-GC/MS, a sample of microplastics is heated to a high temperature in a pyrolysis furnace. The microplastics undergo thermal decomposition into their component molecules, which are then transported by a gas flow into a gas chromatograph. The molecules are separated by the gas chromatograph based on their molecular weight and polarity, and the mass spectrometer then analyzes the separated molecules.

Pyrolysis-GC/MS is a sensitive and versatile technique that can be used to detect microplastics in a variety of environmental samples. It has been used to successfully detect microplastics in water, soil, sediment, and air. Pyrolysis-GC/MS is a valuable tool for microplastic research and is this approach may become essential for future microplastics surveillance and control.

Mass Spectrometer: The mass spectrometer identifies the molecules by measuring their mass-to-charge ratio. This information is then used to identify the type of plastic that the microplastics are made from. In a typical mass spectrometry (MS) procedure, a sample of varying physical states (solid, liquid, or gaseous) undergoes ionization, typically accomplished by subjecting it to electron bombardment. This process can result in the dissociation of certain molecules within the sample, leading to the formation of positively charged fragments, or alternatively, the direct generation of positively charged intact molecules without fragmentation. Subsequently, these ions or fragments are segregated based on their mass-to-charge ratio, typically achieved by accelerating and exposing them to an electric or magnetic field. In this setup, ions possessing identical mass-to-charge ratios will experience uniform deflection. Finally, the ions are identified by a detection system capable of registering charged particles, such as an electron multiplier. The outcomes are presented as spectra showcasing the signal intensity of detected ions in relation to their mass-to-charge ratios. By comparing the identified masses with established values for known masses (such as entire molecules), the atoms or molecules within the sample can be recognized. Alternatively, identification can be accomplished by examining the distinctive fragmentation pattern exhibited by the ions.

The objective of the research was to determine the concentrations and masses of five polymers with high production volumes (PMMA, PP, PS, PE, and PET) in the blood of healthy human subjects. The results of the study indicated that all five polymers were present in the blood samples, with PP and PE being the most abundant (Leslie et al. 2022). The study highlights the criticality of method development and validation in order to attain the requisite sensitivity and mitigate the occurrence of false positives, thereby ensuring reliable and accurate results. This highlights the need for reliable and accurate analytical methods to measure plastic particles in biological matrices (Leslie et al. 2022). The findings of the study revealed that 77% of the donors exhibited a detectable mass of plastic particles in their blood samples, with PET, PS, and PE being the most prevalent polymers encountered. The average sum concentration across all donors was determined to be 1.6 μg of total plastic particles per milliliter of blood sample (Leslie et al. 2022).

The study presented compelling evidence regarding the bioavailability of plastic particles for absorption into the human bloodstream, thereby indicating the necessity for additional research to comprehensively understand the extent of exposure and potential hazards associated with these substances in the human population. This study can serve as a valuable resource for researching and studying the impact of plastic pollution on human health. The study can be a starting point for further exploration of the topic and can help students to develop their research skills.

15.4 Potential Health Implications of Microplastics in Human Body

There is a mounting concern regarding the potential health ramifications associated with microplastics. Several studies have demonstrated that microplastics can induce inflammation, oxidative stress, and genotoxic effects in animal models. Nonetheless, further investigations are required to comprehensively grasp the complete spectrum of risks posed to human health.

Here are some of the potential health implications of microplastics in the human body:

- Inflammation: Microplastics have the potential to trigger inflammation within the body. Inflammation is a natural immune response that occurs in the presence of injury or infection. However, when inflammation becomes chronic, it can cause detrimental effects on cellular and tissue integrity. This chronic inflammation has been associated with the development of various chronic diseases, including cardiovascular disease, stroke, cancer, and diabetes.
- Oxidative stress: Microplastics have the potential to induce oxidative stress within the human body. Oxidative stress denotes a state of dysregulation in the equilibrium between the generation of unstable molecules called free radicals and the body's capacity to counteract their effects. These free radicals possess an inherent instability that enables them to inflict harm upon cells and tissues. The presence of oxidative stress has been associated with various persistent ailments, encompassing cardiovascular disease, cerebrovascular accidents, malignancies, and neurodegenerative disorders.
- Genotoxicity: Microplastics possess the capability to induce genotoxicity within the human body, leading to detrimental effects on the DNA structure. Genotoxicity encompasses the harmful alterations inflicted upon DNA. Such damage to DNA can result in the occurrence of mutations, thereby augmenting the susceptibility to the development of cancer.

It is crucial to emphasize that the investigation into the health consequences of microplastics is still in its nascent phase. Further research is imperative to comprehensively ascertain the complete scope of risks associated with microplastic exposure on human health.

15.5 Strategies to Minimize the Entry of Microplastic in Human Body

Plastic pollution in the ocean is a major environmental issue, as it harms marine animals and ecosystems. Students can research how plastic waste affects different marine species, such as sea turtles, whales, and fish. They could also look into how plastic pollution affects human health, by analyzing the impact of microplastics in drinking water and seafood.

1. **Materials that could replace plastic:** There are many alternatives to plastic that are more eco-friendly, such as glass, metal, and paper. Students could explore the advantages and disadvantages of these materials and investigate how they could be used to replace plastic in different industries, such as packaging and transportation.
2. **Policies and regulations around plastic waste reduction:** Numerous nations and urban centers have enacted measures to diminish the accumulation of plastic waste, including prohibitions on disposable plastics and levies imposed on plastic bags. Students could study these policies and evaluate their effectiveness in reducing plastic pollution. They could also compare different approaches taken by governments around the world and identify best practices for reducing plastic waste.
3. **New ways to measure plastic particles in biological samples:** Scientists are developing new techniques to measure plastic particles in biological samples, such as blood and urine. Students could research these methods and evaluate their accuracy and reliability. They could also investigate the potential health risks associated with plastic exposure, such as endocrine disruption and cancer.
4. **The role of corporations in reducing plastic waste:** Many companies are taking steps to reduce their plastic footprint, by using more sustainable materials and improving their recycling practices. Students could analyze the strategies used by these companies and evaluate their impact on the environment. They could also investigate the role of consumers in driving corporate sustainability, and the potential for collaboration between companies and governments to reduce plastic waste.
5. **Biodegradable plastics:** Biodegradable plastics are designed to decompose in the environment, reducing the amount of plastic waste. However, some biodegradable plastics may not break down completely and can still harm the environment. Students could research the different types of biodegradable plastics and evaluate their environmental impact. They could also investigate new technologies that could make biodegradable plastics more effective and sustainable.
6. **Ways to educate the public and promote sustainable practices:** Finally, students could explore different ways to educate the public about plastic pollution and promote sustainable practices. This could entail the formulation of educational initiatives, coordination of community gatherings, and collaboration with local businesses and organizations to curtail the generation of plastic waste. Students could also investigate the role of social media and other digital platforms in raising awareness about plastic pollution and promoting sustainability.

Acknowledgment I would like to thank Dr. Shahnawaz for inviting me to write this chapter on "Reports of Microplastic in human blood." It was a pleasure to work on this project and to learn from his expertise and guidance. I hope this chapter will contribute to the ongoing research and awareness on this important issue.

References

Al-Kindi SG, Brook RD, Biswal S, Rajagopalan S (2020) Environmental determinants of cardiovascular disease: lessons learned from air pollution. Nat Rev Cardiol 17:656–672

Caliskan E, Ramos D, de Souza A, Böning OJ, Liakopoulos Y-HC, John Pepper C, Gibson M, Perrault LP, Wolf RK, Kim K-B, Emmert MY (2020) Saphenous vein grafts in contemporary coronary artery bypass graft surgery. Nat Rev Cardiol 17:155–169

Çobanoğlu H, Belivermiş M, Sıkdokur E, Kılıç Ö, Çayır A (2021) Genotoxic and cytotoxic effects of polyethylene microplastics on human peripheral blood lymphocytes. Chemosphere 272: 129805

da Silva VH, Murphy F, Amigo JM, Stedmon C, Strand J (2020) Classification and quantification of microplastics (<100 μm) using a focal plane Array-Fourier transform infrared imaging system and machine learning. Anal Chem 92:13724–13733

Fuller R, Landrigan PJ, Balakrishnan K, Bathan G, Bose-O'Reilly S, Brauer M, Caravanos J, Chiles T, Cohen A, Corra L, Cropper M, Ferraro G, Hanna J, Hanrahan D, Howard H, Hunter D, Janata G, Kupka R, Lanphear B, Lichtveld M, Martin K, Mustapha A, Sanchez-Triana E, Sandilya K, Schaefli L, Shaw J, Seddon J, Suk W, Téllez-Rojo MM, Yan C (2022) Pollution and health: a progress update. Lancet Planetary Health 6:e535–e547

Leslie HA, van Velzen MJM, Brandsma SH, Dick Vethaak A, Garcia-Vallejo JJ, Lamoree MH (2022) Discovery and quantification of plastic particle pollution in human blood. Environ Int 163:107199

Luqman A, Nugrahapraja H, Wahyuono RA, Islami I, Haekal MH, Fardiansyah Y, Putri BQ, Amalludin FI, Rofiqa EA, Götz F, Wibowo AT (2021) Microplastic contamination in human stools, foods, and drinking water associated with Indonesian coastal population. Environments 8:138

Nielsen TD, Hasselbalch J, Holmberg K, Stripple J (2020) Politics and the plastic crisis: a review throughout the plastic life cycle. Wiley Interdiscip Rev Energy Environ 9:e360

Rotchell JM, Jenner LC, Chapman E, Bennett RT, Bolanle IO, Loubani M, Sadofsky L, Palmer TM (2023) Detection of microplastics in human saphenous vein tissue using μFTIR: a pilot study. PLoS One 18:e0280594

Vethaak AD, Legler J (2021) Microplastics and human health. Science 371:672–674

Wright SL, Kelly FJ (2017) Plastic and human health: a micro issue? Environ Sci Technol 51:6634–6647

Yan Z, Liu Y, Zhang T, Zhang F, Ren H, Zhang Y (2022) Analysis of microplastics in human feces reveals a correlation between fecal microplastics and inflammatory bowel disease status. Environ Sci Technol 56:414–421

Yu H, Zhang Y, Tan W, Zhang Z (2022) Microplastics as an emerging environmental pollutant in agricultural soils: effects on ecosystems and human health. Front Environ Sci 10:855292

Removal of Microplastic from Wastewater Treatment Plants

16

Babatunde Oluwafemi Adetuyi, Charles Oluwaseun Adetunji, Peace Abiodun Olajide, Olubanke Olujoke Ogunlana, John Tsado Mathew, Abel Inobeme, Oluwabukola Atinuke Popoola, Falana Yetunde Olaitan, Olalekan Akinbo, Oluwafemi Adebayo Oyewole, Eniola K.I.T., and Mohammed Bello Yerima

Abstract

Since water treatment facilities provide a workable alternative for expelling microplastics from the surroundings, this review examines the overall and stage-by-stage effectiveness of different water treatment techniques in expelling microplastics and the innovations for handling water treatment waste products.

B. O. Adetuyi · P. A. Olajide · J. T. Mathew
Biochemistry Unit, Faculty of Pure and Applied Sciences, Department of Natural Sciences, Precious Cornerstone University, Ibadan, Nigeria

C. O. Adetunji (✉)
Applied Microbiology, Biotechnology and Nanotechnology Laboratory, Department of Microbiology, Edo State University, Auchi, Edo State, Nigeria
e-mail: adetunji.charles@edouniversity.edu.ng

O. O. Ogunlana · A. Inobeme
Department of Biochemistry, Covenant University, Ota, Ogun State, Nigeria

O. A. Popoola · F. Y. Olaitan
Plant Improvement Unit, Agricultural Biotechnology Department, National Biotechnology Development Agency, Abuja, Nigeria

O. Akinbo
Centre of Excellence in Science, Technology, and Innovation, AUDA-NEPAD, Johannesburg, South Africa

O. A. Oyewole
Department of Microbiology, Federal University of Technology, Minna, Nigeria

E. K.I.T.
Department of Biological Sciences, Joseph Ayo Babalola University, Ikeji Arakeji, Osun State, Nigeria

M. B. Yerima
Department of Microbiology, Sokoto State University, Sokoto, Sokoto State, Nigeria

M. Shahnawaz et al. (eds.), *Microplastic Pollution*,
https://doi.org/10.1007/978-981-99-8357-5_16

Over 80 academic papers over the last decade were analyzed. Primary treatment at wastewater treatment plants has been claimed to remove between 16.5 and 98.4% of microplastics; however, this review demonstrates that the effectiveness of these technologies varies greatly. A preliminary treatment sewage treatment facility's aggregate microplastic sorption capacity ranges from 78.1 to 100%, with stage-by-stage performance varying from 7.0 to 99.9%. As much as 99.9% of microplastics are eliminated during the tertiary treatment process, drinking water treatment plants typically remove 1.8–54.5% of microplastics during the initial coagulation and sedimentation process, while additional treatment increases this to an overall 88.6%. Experiments show that using a higher than typical dose of flocculent can greatly boost microplastic removal by up to 62%. The majority of microplastics that cannot be extracted using traditional methods end up in sludge. Sludge and membranes are identified as significant contributors to microplastics in water treatment, notwithstanding the review's increasing focus on the ability of treating wastewater to remove microplastics from waterbodies. It advises improving water treatment by altering coagulation factors, introducing cutting-edge procedures to more efficiently remove microplastics, and possibly utilizing innovative sol-gel technologies. In order to lessen the chances of microplastics being released back into the environment, it proposes thermal pre-treatment of sludge.

Keywords

Efficiency · Microplastics · Water treatment · Membrane · Sludge

16.1 Introduction

It is now well-established that microplastics are widespread across the world's ecosystems, including those found in land, freshwater, and saltwater bodies (Wong et al. 2020). Microbeads, microfibers, and plastic pellets are all examples of primary microplastics, which come from the manufacture of products containing plastics of that size, whereas secondary microplastics result from the breakdown of bigger plastic items (Wong et al. 2020). Some of the most common sources of microplastics are synthetic textiles and cosmetics, specifically facial washes, exfoliants, and toothpaste. Microplastics in the environment are often called "secondary," meaning that they are the result of plastic waste or improper waste management. There are many other industries that contribute to this problem, but the packaging and construction industries produce the most plastic waste (Geyer et al. 2017). China, Indonesia, and the Philippines, all emerging nations, are thought to generate the largest proportion of improperly handled plastics that end up in our environment (Jambeck et al. 2015). According to estimates, between 60 and 99 million tons of plastic waste were unlawfully discarded globally in 2015. As human activity spreads further, this amount is anticipated to increase (Lebreton and Andrady 2019). Direct plastic discharge into waterways, sewage and sullage, and industrial wastewater are all pathways for the introduction of microplastics into aquatic ecosystems (Wong et al. 2020). Rainfall runoff is another potential route for microplastics to enter waterways and ultimately end up in rivers and seas (Choong et al. 2020). Water treatment has been identified in various investigations as a crucial step in the effective elimination of microplastics to get rid of the microplastics that are currently present in the ecosystem because it stops factory wastes or groundwater streams (Yang et al. 2019). The most practical method for removing microplastics is thought to be water treatment before other strategies, including bacterial degradation, reach maturity. Although biodegradable polymers degrade faster than conventional plastics, it seems that the microplastics they generate are still present in the environment for a sufficient amount of time to enter and bio-accumulate in nutrient cycling (Tang 2020).

Microplastics are compositionally complicated, with several different additives such as plasticizers, fillers, and stabilizers. The notion that they offer surface area for the adsorption of several environmental chemicals, such as drugs and hydrocarbons, further complicates their environmental and human health consequences (Tang 2021). The elimination of microplastics and nanoplastics from the environment has become an issue of paramount importance because of the latter's propensity to infiltrate cells and impair biological functioning (Tang 2020). Water and sludge treatment offers a quick and manageable way to get rid of microplastics, especially as long-term strategies for eliminating them from the environment have not yet been developed. Even though not all WTPs treat wastewater, most WTPs include WWTPs as part of their operations. Surface water, groundwater, and seawater are all common water supplies for drinking water treatment facilities, and all three are susceptible to microplastic contamination (Collivignarelli et al. 2018). WWTPs are facilities dedicated to processing municipal, commercial, and residential waste (Murphy et al. 2016). The toxic waste of treated water, such as effluents and sediment, contains microplastics that are later released back into the ecosystem (Mahon et al. 2017). Polyethylene (PE) microplastics have been identified at concentrations of up to 9000 per m^3 in the effluents of WWTPs (Mintenig et al. 2017). It has also been claimed that WWTP discharges contain synthetic fibers, primarily polyester, at a volume of 1000 m^3 (Mason et al. 2016). Furthermore, microplastics, especially PE, were found in WWTP sewage sludge (Mintenig et al. 2017). The discovery of microplastics at varying concentrations from 129 to 238 microplastics/g of dry-weight sludge in a sewage treatment facility in Iran provides grounds for concern that wastewater sediment could be a significant source of microplastics (Petroody et al. 2021).

With a microplastic elimination effectiveness of up to 99% (Carr et al. 2016), conventional water remediation can include primary and/or secondary treatment phases while also focusing microplastics for release. Pre-treatment is frequently utilized to get rid of substantial particulates such as silt, rocks, and timber before the main treatment starts because primary treatment is not always required. The purpose of secondary treatment, sometimes referred to as the activated sludge process, is to capture any residual organic elements that were neglected through primary treatment using an air circulation column or membrane bioreactor (MBR)

(Pal 2017). Secondary treatment is more common in WWTPs due to the higher proportion of organic compounds in sewage. A WWTP can install an intermediate tertiary treatment unit to further purify supplementary waste, resulting in an improved output (Pal 2017). Tertiary treatment techniques include chlorination, ozonation, UV radiation, disk filtration, quick sand filtration, and activated carbon granulation (Pal 2017). Similar to the tertiary treatment stage, the last stage of drinking water treatment entails filtration and disinfection. Technological advancements have been made to conventional water treatment to produce cleaner effluents; however, these advancements have not been directed toward microplastic removal. Standard membrane water treatment methods used in desalination processes concentrate microplastic on the membrane and in the retentate before discharging them directly into the ecosystem (Mejía et al. 2017). Membrane filtration systems, which are extensively employed in the intermediate and last phases of water processing and in residential filtration units, may become more contaminated with microplastics during cleaning or disposal (Landaburu-Aguirre et al. 2016).

Owing to the dual role that treatment of wastewater serves as a deterrent and a source of microplastics, this article aims to assess the approaches for removing microplastics from the effluent of water treatment facilities in addition to the microplastic elimination of different methods for treating water.

16.2 Microplastic Cleanup Technology for Water

16.2.1 Conventional Water Treatment

The typical aspects of conventional wastewater treatment include primary and secondary treatments (sometimes with tertiary treatment incorporated to treat intermediate output) (Pal 2017). First, a screen is used to catch any large floatables before they jam the principal treatment pipelines. The inlet is then directed into a granulator, which is basically a long, narrow reservoir, so that particles such as sandstones can drain out of the water more gradually (Ranade and Bhandari 2014). When influent that has been cleared of grit flows to the primary clarifiers, gravitation deposits the particulate, forming the main sludge. When settling suspended materials, it is usual practice to add coagulants and flocculants (Lv et al. 2019). Through mostly biological mechanisms, secondary treatment seeks to eliminate the dissolved organic detritus from the initial outflow (Ranade and Bhandari 2014). The biological process known as activated sludge consists of a bioreactor and a secondary clarifier. At the base of the bioreactor, perforated humidifiers are used to introduce pressurized gas. Oxygen from the air helps microorganisms thrive, decreasing the water's biochemical oxygen demand (BOD) by eating away at the dissolved organics that pollute it (Rao et al. 2012).

The primary effluent is routed through a secondary clarifier, where activated sludge is sedimented. The microbes can be recreated in the bioreactor where around 30% of the deposited sludge is regenerated, while the remaining intermediate sludge is transferred for processing and proper disposal (Rao et al. 2012). An alternative to the traditional activated sludge method, membrane bioreactors use microfiltration or ultrafiltration membranes to treat wastewater. Conventional activated sludge processes use a secondary clarifier to separate the activated sludge and provide high-quality effluent, but the membrane process eliminates the need for a clarifier altogether (Iorhemen et al. 2016). The nitrogen and phosphorus removal component of the process of activated sludge is the anaerobic-anoxic-aerobic (A2O) method. Phosphorus-accumulating organisms do their work in the anaerobic tank. Denitrification occurs in the anaerobic tank, transforming nitrate into nitrogen gas. Nitrification, the oxidation of ammonia to nitrate, and BOD elimination both occur in the aerobic tank (Zhou et al. 2011). Reusing the effluent from the aerobic tank in the anaerobic tank helps remove nitrate, and after being sedimented, activated sludge is returned to the anaerobic digester (Zhang et al. 2018).

In order to further eliminate the particulates, final sewage treatment often entails dissolved air floating or sifting using a variety of media, including sand filters, disk filters, bioactive filters, and gravity filters (Ranade and Bhandari 2014). To get rid of organic contaminants in the water, activated carbon filtration could be used as an extra step. Ultrafiltration and reverse osmosis membrane technology is another option, albeit it is not typically used in a typical WWTP (Hidayaturrahman and Lee 2019). In this overview, traditional WWTPs are described independently from those that use membrane technology for tertiary treatment (Rao et al. 2012).

In primary and secondary treatments, skimming and settling have the possibility of eliminating more than 99.9% of microplastics due to their ability to separate lighter microplastics that glide from denser microplastics that are stuck in solid flocculation (Carr et al. 2016). However, when microplastics in unstable flocs fail to deposit properly, they can avoid being removed through skimming and settling and instead resuspend in the water. Additionally, the development of biofilms on microplastics alters their surface characteristics and prevents their elimination during water treatment (Carr et al. 2016). During secondary treatment, the length of time wastewater is held and the concentration of nutrients in the wastewater both play a role in the production of biofilm (Sun et al. 2019). Carr et al. and Ziajahromi et al. examined the presence of microplastics in the secondary and tertiary effluents of wastewater treatment plants (WWTPs) in the Sydney Water Catchment Area, Australia, and the Sanitation Districts of Los Angeles County, California. The latter briefly noted that the Sydney population was responsible for the wastewater problem, while the former gave no indication of where the water came from. While the latter assessed the pre-existing plumbing and flows of facility conformity specimens and plumbed ultimate discharge streams, mesh filters were placed centrally at the effluents with a predefined flow rate limit (Ziajahromi et al. 2017). The quality of the influents and the technology used to treat the wastewater may also play a role. It has been found by Mason et al. (2016) that the population that a WWTP serves is directly proportional to the variation in the concentration of microplastics in the WWTP's discharge. Leslie et al. (2017) found that airflow during preliminary treatment eliminated 99% of microplastics in a WWTP. Although the recorded microplastic levels were substantially higher than those discovered by Carr et al. (2016), this information supported their finding that main and preliminary treatment successfully removed the majority of the microplastics. Microplastics were collected using 200-m, 100-m, and 20-m meshes in a previous work by Talvitie et al. (2015), who separated them into fiber and particle categories during many WWTP water treatment stages. According to the research, most of the fibers were removed during primary sedimentation, while the amount removed during secondary sedimentation was negligible. The findings were in agreement with those of Carr et al. (2016) and gave insight into the discrepancy retrieval effectiveness of distinct microplastics at various phases of water treatment, despite post-treatment microplastic concentrations (4.9 fibers/L and 8.6 particles/L).

Michielssen et al. (2016) observed that secondary treatment removed most of the SAL (95.6%) and that tertiary treatment employing granular sand filtration had a greater removal efficiency of 97.2%. The SAL in this study ranged in size from 20 m to 4.75 mm. In addition to microplastics, the secondary effluent contained 5.9 SAL per liter. The plant attains extraction efficiency similar to that of China's major water reclamation facility by using a carriage of aerated grit compartments, main depositional environment, anaerobic/anoxic/aerobic (A2O), intermediate deposition, and advanced treatment along with denitrification, ultrafiltration, ozonation, and ultraviolet. The facility has a 95.2% efficiency in removing microplastics (5000 m), with the aerated grit chamber removing 58.8%, the A2/O-secondary sedimentation removing 54.5%, and the advanced treatment removing 71.7% (Yang et al. 2019). Based on the results of the research, it was determined that an aerated grit chamber could effectively remove over 60% of microfibers.

Talvitie et al. (2017a) evaluated the elimination of microplastics from the secondary effluent of four WWTPs that use disk filters, dissolved air flotation, and instant sand filters for secondary effluent treatment and membrane bioreactors (MBRs) for main effluent treatment. Pea gravel and quartz are used in quick sand filters to get rid of harmful microbes, nutrients, and other suspended materials (Talvitie et al. 2017a). Membrane bioreactors (MBRs) immersed in tanks of water are used to remove nutrients, soluble organic materials, and suspended particles (Talvitie et al. 2017a). The study found that MBR (secondary treatment) was 99.9% effective at removing microplastics (>300–20 m), while disk filters ranged from 40.95 to 98.55% effective, dissolved air flotation was 95% effective, and quick sand filters were 97% effective (Talvitie et al. 2017a).

16.2.2 Drinking Water Treatment Plants (DWTPs)

In spite of the fact that additional treatment to decompose dissolved elements is not included, the purification of both drinking water and sewage is very comparable. This is because the raw water used to feed DWTPs frequently has a better accuracy and has a lower organic load than wastewater from urban and industrial sources. Alum is added to the raw water for coagulation and flocculation, and then, screening and grit removal take place. Similar to wastewater treatment, the mixture is then introduced to sedimentation tanks, where the heavier floc particles fall to the bottom. Instead of subsequent treatment, as is conducted in wastewater treatment, the water is filtered, and the purification unit is typically a sand filter with an activated carbon filter added. Such a process would be classified as tertiary treatment in a WWTP. Before being pumped into homes' plumbing systems, the filtrate is typically chlorinated to kill bacteria and fluoridated for taste.

The majority of research that has been done does not distinguish between the effectiveness of microplastic clearance at varying levels of treatment since the phases of the process at DWTPs are less specifically articulated as at WWTPs. These investigations, on the other hand, zero in on specific water treatment methods. Three different Czech Republic DWTPs used a combination of coagulation/flocculation, sedimentation/flotation, and filtration with sand and activated carbon to remove an average of 70–83% of microplastics (>100–1 m) from the water (Pivokonsky et al. 2018). The study discovered that up to 628 28 microplastics were detected per liter of recycled water, and up to 95% of them were less than 10 m, though not monitoring the variability in microplastic levels in each phase of treating wastewater. Although additional research is needed, it appears that microplastics that are light and buoyant can be removed via flotation (Pivokonsky et al. 2018). Based on an investigation on groundwater purification in Germany for consumption reasons, microplastic particle levels (150–50 m) varied from 1 to 7 particle/m^3 in untreated and processed water, with polyester representing the predominant prevalent microplastic observed. Fiber concentrations in raw and finished water samples were found to be anywhere from 3 to 46 fibers/m^3 (Mintenig et al. 2019). The raw groundwater was aerated and filtered in tanks that were partially made of polypropylene and partially made of epoxy resin for storage and reaction. The reason for our investigation, which examined water at several stages of treatment and observed no appreciable variation in microplastic levels, is due to the very low content of microplastics in the untreated water (Mintenig et al. 2019). However, it pointed out the potential for microplastics to be introduced during the treatment procedure.

Considering that settling is believed to eliminate the majority of the microplastics in WWTPs, Ma et al. (2019a) examined the effectiveness of various kinds and concentrations of coagulants in eliminating PE all through treatment of drinking water. As PE size of the particles reduced from 5000 to 500 m in diameter while retaining the same concentration of coagulant, they found that $AlCl_3 \cdot 6H_2O$ was much more efficient in eliminating PE than $FeCl_3 \cdot 6H_2O$. Despite improvements in removal efficiency with decreasing PE particle sizes, even at large dosages, coagulation employing $FeCl_3 \cdot H_2O$ yielded low PE removal efficiency, which was also validated in another investigation by Ma et al. (2019b). $AlCl_3 \cdot 6H_2O$ dose increased only marginally boosted removal effectiveness, notably of bigger PE particles. While introducing anionic polyacrylamide significantly increased the evacuation of smaller PE (diameter 500 m) at 5 mM $AlCl_3 \cdot 6H_2O$, the impact of using polyacrylamide (PAM) as a coagulation facilitator was minimal at small concentrations of $AlCl_3 \cdot 6H_2O$ (0.5 mM). In contrast to the high microplastic elimination attained with skimming and settling, Ma et al. (2019a) recorded a high microplastic removal rate of 61.19% utilizing coagulation (Carr et al. 2016). Ziajahromi et al. found that subsequent treatment reduced microplastics by an average of 68%; therefore, this number is more comparable (Ziajahromi et al. 2017). At 15 mg/L polyacrylamide concentration, it was shown that $FeCl_3 \cdot H_2O$ had a greater effect on removing PE with a diameter of 500 m than $AlCl_3{\cdot}6H_2O$ at similar concentrations (Ma et al. 2019a, b). Also, the ideal experimental conditions may not be applicable to real-world drinking water treatment because of the concentration and pH dependence of coagulants' efficacy in removing microplastics. By way of illustration, Ma et al. (2019a) employed anion PAM concentrations that were significantly higher than the suggested 1 mg/L (WHO 2011).

In a drinking wastewater treatment plant with a 150 million m^3/day capacity that uses coagulation/flocculation, deposition, sand filtration, ozonation, and particulate activated carbon filtration, the microplastics were observed. In the influent, they observed that about 84.4–86.7% of the microplastic particles were between 1 and 5 m in size, which is similar to the prevalence of microplastics 10 m in a Czech study (Pivokonsky et al. 2018). Coagulation and deposition eliminated 40.5–54.5% of microplastics from the untreated water, which is comparable to the outcomes of jar studies carried out by Ma et al. (2019a) utilizing $AlCl_3{\cdot}6H_2O$ and polyacrylamide. Sizes of microplastics reported by Ma et al. (diameter 500 m) and Wang et al. (2020) agree, suggesting that bigger microplastics >10 μm are effectively removed (2019a, b). Due to the sampling of microplastics larger than 100 μm, Wang et al. (2020)'s prediction may be wrong by over 500 μm. However, Wang et al. (2020) did not detail the coagulant type they employed. However, simulations revealed that in addition to the typical oxide colloid, a prototype polymer sphere with a diameter of 1–5 μm and a density of 1.3 g/cm^3 could also be eliminated from water using the conventional alum ($KAl(SO_4)_2{\cdot}12H_2O$) aggregation and water treatment methods (Skaf et al. 2020). Surfactant, which is known to greatly alter the dispersion of PE microfibers in water, did not appear to affect microsphere removal at 20 mg/L (Skaf et al. 2020).

Wang et al. (2020) research found that microplastics >5 m in size were more effectively removed by sand filtration, with removal rates ranging from 29.0 to 44.4%. Without any kind of sophisticated treatment, only about 59–70.5% of the microplastics were removed. This is consistent with the relatively low efficacy of removing microplastics (Pivokonsky et al. 2018). However, the use of ozone in wastewater treatment resulted in an 89.9% stage-wise elimination efficiency. After the ozonation effluent was filtered using granular activated carbon, between 56.8 and 60.9% of the microplastics were eliminated. Furthermore, when ozonation and particulate activated carbon filtering were coupled, 17.2–22.2% microplastic sorption capacity was attained (Wang et al. 2020). This result is consistent with the work of Zhang et al. (2020), who found that granular filtration can remove microplastics with an efficiency of at least 86.9%, with removal rates as high as 99.9% reported for particles measuring 105–125 m, although Wang et al. (2020)'s decreased removal effectiveness was likely attributable to insufficient microfiber removal (50%).

16.2.3 Membrane Water Treatment

Water with high levels of total dissolved solids can be treated using membrane technology alone, or in tandem with secondary treatment procedures such as the membrane bioreactor (MBR) (Pronk et al. 2019). Tertiary treatment of secondary effluents from wastewater treatment plants (WWTPs) is another common application. Reverse osmosis (RO) is used in an Australian tertiary wastewater treatment plant (WWTP) to minimize microplastics from 0.28 to 0.21 per liter (Ziajahromi et al. 2017). Despite the reduced pore size of the RO membrane compared to the ultrafiltration membrane of the bioreactor, this is still significantly higher than the 0.005 microplastic L1 reported by Talvitie et al. (2017a) for MBR. It is possible for microplastics to escape through the RO membrane due to pore irregularities, the presence of bigger pores on the membrane, or other flaws (Yan et al. 2015).

The use of tertiary treatment based on filtering for microplastic removal has been met with skepticism. Overall particle discharge was positively correlated with tertiary filtration, and fiber discharge in particular, while fragment discharge was not significantly correlated with tertiary filtration, suggesting that microbial residuals are released during tertiary filtration as a result of microbial growth in filtration media that mimics fibers (Mason et al. 2016). However, the data analysis did not reveal the type of filtering system used (Mason et al. 2016).

There has been some research into how to make membrane technology more effective and less expensive for treating water. To determine what filter type, filter angle, and filter pressure are best for removing microplastics, Beljanski et al. (2016b) built and tested a gravity-powered filtration system on a small scale (a bench). For the purpose of testing, a scenario mimicking a WWTP was created. A two-dimensional filter with an 80-m screen and a three-dimensional filter with a thicker multi-layer structure were compared. The research showed that the 2D filter clogged up more quickly than the 3D filter, but that backwashing was more effective in removing microplastics. The 3D filter could be backwashed to remove 95–100% of microfibers at pressures of 1.68 and 3.68 kPa, albeit the lower pressure required longer backwashing times (Beljanski et al. 2016a), but the test merely mentioned that the microplastics came from a facial scrub and that a maximum of 30 mg/L was loaded. However, the document omitted details about the microplastics' sizes both before and after use (Beljanski et al. 2016a).

Additionally, Li et al. (2018) have investigated the utilization of dynamic membrane technology for microplastic removal utilizing a bench-scale gravity-driven filtration machine fitted with a 90-m mesh. For secondary filtration of wastewater pollutants, a dynamic membrane accumulates a cake layer on top of an existing membrane. The research showed that within 20 min, the turbidity had dropped to below 1 NTU (nephelometric turbidity unit) thanks to the removal of microparticles (90% sized 90.5 m). Transmembrane pressures for dynamic membrane filtration are significantly lower than those for microfiltration and ultrafiltration (Li et al. 2018).

16.3 Management of Water Treatment Wastes to Reduce Microplastics

16.3.1 Membranes

Water filtration polymeric membranes contribute to microplastic pollution for two reasons: They are themselves made of plastic (polyamide, polycarbonate, and polytetrafluoroethylene), and they also trap microplastics in the process of purifying water. Market value for membrane filtration is expected to reach USD 7242 million by 2025 as demand rises (Global Info Research 2020). In order to keep up with demand, scientists have begun exploring the possibility of recycling used RO membranes for use in nanofiltration and ultrafiltration. Additionally, bio-polymers are gaining attention as a viable option for membrane manufacturing (Poerio et al. 2019). Polyethylene terephthalate (PET) from recycled plastic bottles has been shown to have promise as a raw material for solution electrospinning-based nanofiber filtration membranes, as was discussed by Zander et al. (2015). By testing the membranes with latex beads ranging in size from 30 to 2000 nm, they found that the successfully manufactured PET nanofiber membranes had a clearance rate of >99% (Zander et al. 2015).

The polyamide thin film composite RO membrane from a decommissioned desalination plant was recycled by Morón-López et al. (2019) into nanofiltration and ultrafiltration membranes for use in membrane bioreactors. Good gas permeability at low pressure was seen in testing of the modified membranes, indicating the viability of producing recycled membranes at a cheap cost for MBRs. Even more research has been done on the topic of recycling used RO membranes; for example, Mohamedou et al. (2010) looked into the possibility of using recycled RO membrane in nanofiltration processes, and da Silva et al. (2012) investigated the feasibility of using recycled RO membrane in RO procedures. Recycling RO membrane has been studied by Mohamedou et al. (2010), who found that it might be used to stabilize rocks and slow the growth of grass in backyard gardens. Research into polymeric membranes has largely concentrated on reusing and recycling membranes, with only a small amount of effort devoted to synthesizing membranes from waste plastic. Nonetheless, this work has helped reduce plastic input and, by extension, microplastics in the environment.

16.3.2 Filtration Concentrate

However, research into further processing of the retentate or concentrate of membrane filtering in order to remove microplastics is sparse. It was noted by Arola et al. (2019) that electrodialysis or shear-enhanced nanofiltration/RO concentration can be used to recover nutrients and struvite from the filtration waste streams of tertiary treatment in WWTPs, but no mention was made of the eradication of microplastics in these waste streams. Only membrane WTPs perform pre-treatment to reduce membrane fouling. Most microplastics can be removed with conventional pre-treatment methods such as chlorination, ozonation, dissolved air flotation, and coagulation-flocculation (Poerio et al. 2019). Despite this, microplastics remain even after pre-treatment and tend to accumulate in the retentate of subsequent filtration. Microfiltration, ultrafiltration, and nanofiltration used as an unconventional pre-treatment to RO also produce retentate that is likely to be contaminated with microplastics (Poerio et al. 2019).

Similarly, almost no research has been conducted on the microplastic composition of the reject streams of membrane-based WTPs. This makes it difficult to gauge the extent to which the reject streams contribute to the polluting of waterways with microplastics. While Arola et al. (2019) focused on microplastics, Mejía et al. (2017) demonstrated the potential of ultraviolet light and titanium dioxide in photo-oxidative removal of organic micropollutants such as diclofenac (>95%), ibuprofen (>95%), and naproxen (65%) from the concentrate of a nanofiltration.

16.3.3 Emerging Technologies of Microplastic Removal in Water Treatment

Microplastics are a persistent problem, and scientists are constantly looking for innovative ways to filter them out of wastewater and drinking water. In recent years, electrocoagulation has gained popularity as a less expensive alternative to traditional coagulation methods such as chemicals and microorganisms (Garcia-Segura et al. 2017).

The use of magnetic extraction as a water treatment technology for removing microplastics is another option to consider. Microplastics in water are magnetically extracted using Fe nanoparticles coated with hexadecyltrimethoxysilane (Grbic et al. 2019). Small (20 m) PE and polystyrene microplastics were extracted at a rate of 92% from saltwater, whereas the approach recovered 84% of medium-sized microplastics (200–1000 m) from freshwater and 78% from silt. Because of its potential utility for microplastic removal from sludge, it has been used to recover microplastics from sediment, albeit at a lesser efficiency due to the impeded movement of Fe nanoparticles by soil (Grbic et al. 2019).

This approach is simple, cheap, and successful no matter what the influent pH, temperature, or pollutant size and concentration are. Microplastics in wastewater can be bound, agglomerated, and separated through the production of alkoxy-silyl functionalized molecules with adhesive-like function (Herbort et al. 2018). In a similar vein, research is being done to see if particular organosilanes may be used in tandem with chemically induced agglomeration. As the organosilanes combine with the microplastics, an inclusion compound is formed, which entraps the trash in huge clumps that can then be skimmed out of the water. Results from laboratory-scale experiments conducted by Sturm and Schuhen (2019) demonstrated that, of the C1, C4, C8, and C10 silanes evaluated, C4 silane attained the greatest microplastic inclusion rate of 75% across the longest gelation time of 120 s. Higher inclusion rates were achieved because a longer gelation time allowed for the production of agglomerates. When the agglomeration process was fine-tuned at the pilot scale, it was possible to attain higher inclusion rates of >99%.

Although a 10-m mesh filter was used, the sizes of the microplastics retained were not specified. In contrast to the 71.7% success rate seen in the advanced therapy study by Yang et al. (2019), this rate is significantly higher. When bigger microplastics are excluded from the quantification, the biofilter's removal efficiency increases (Liu et al. 2020).

16.4 Conclusion

WTPs are essential barriers for separating microplastics from the environment while being a source and sink of microplastics. It has been observed that most microplastics in wastewater and surface waters can be removed by primary and secondary treatments at WWTPs and through coagulation and sedimentation at DWTPs. It has been discovered that the efficiency of microplastic removal can be further increased through the incorporation of tertiary treatment in WWTPs, the adoption of membrane filtering in DWTPs, the suitable selection of coagulants, and the optimization of water treatment. Unfortunately, water treatment facilities (WTPs) have not been able to completely eliminate microplastics in their final discharges, and the microplastics they remove are often released back into the environment via sludge, discarded membranes, and reject streams carrying filtration concentrate. The gradual elimination of microplastics from the environment can be achieved by concentrating on microplastic removal from these water treatment wastes. This review consequently advocates for the dissemination of the idea that WTPs can help alleviate microplastic pollution and that microplastics from WTPs can be properly handled so that WTPs are no longer a major source of microplastics. This article sheds light on how water treatment facilities have progressed recently and the various alternatives available for dealing with microplastics that become caught in the processes. Research into microplastics' ultimate destination in membrane-based WTPs and strategies for eliminating microplastics from sludge, membranes, and filter concentrate are all encouraged (Adetunji et al. 2019, 2020, 2021a, b, c, d, 2022a, b, c; Adetunji 2019).

References

Adetunji CO (2019) Environmental impact and Ecotoxicological influence of biofabricated and inorganic nanoparticle on soil activity. In: Panpatte D, Jhala Y (eds) Nanotechnology for agriculture. Springer, Singapore. https://doi.org/10.1007/978-981-32-9370-0_12

Adetunji CO, Panpatte DG, Bello OM, Adekoya MA (2019) Application of nanoengineered metabolites from beneficial and eco-friendly microorganisms as a biological control agents for plant pests and pathogens. In: Panpatte D, Jhala Y (eds) Nanotechnology for agriculture: crop production & protection. Springer, Singapore. https://doi.org/10.1007/978-981-32-9374-8_13

Adetunji CO, Oloke JK, Prasad G (2020) Effect of carbon-to-nitrogen ratio on eco-friendly mycoherbicide activity from *Lasiodiplodia pseudotheobromae* C1136 for sustainable weeds management in organic agriculture. Environ Dev Sustain 22:1977–1990. https://doi.org/10.1007/s10668-018-0273-1

Adetunji CO, Akram M, Olaniyan OT, Ajayi OO, Inobeme A, Olaniyan S, Hameed L, Adetunji JB (2021a) Targeting SARS-CoV-2 novel Corona (COVID-19) virus infection using medicinal plants. In: Dua K, Nammi S, Chang D, Chellappan DK, Gupta G, Collet T (eds) Medicinal plants for lung diseases. Springer, Singapore. https://doi.org/10.1007/978-981-33-6850-7_21

Adetunji CO, Michael OS, Varma A, Oloke JK, OseniKadiri MA, EbunoluwaBodunrinde R, Imtiaz A, Adetunji JB, Shahzad K, Jain A, Ubi BE, Majeed N, Ozolua P, Olisaka FN (2021b) Recent advances in the application of biotechnology for improving the production of secondary metabolites from quinoa. In: Varma A (ed) Biology and biotechnology of quinoa. Springer, Singapore. https://doi.org/10.1007/978-981-16-3832-9_17

Adetunji CO, Olaniyan OT, Akram M, Ajayi OO, Inobeme A, Olaniyan S, Khan FS, Adetunji JB (2021c) Medicinal plants used in the treatment of pulmonary hypertension. In: Dua K, Nammi S, Chang D, Chellappan DK, Gupta G, Collet T (eds) Medicinal plants for lung diseases. Springer, Singapore. https://doi.org/10.1007/978-981-33-6850-7_14

Adetunji CO, Ajayi OO, Akram M, Olaniyan OT, Chishti MA, Inobeme A, Olaniyan S, Adetunji JB, Olaniyan M, Awotunde SO (2021d) Medicinal plants used in the treatment of influenza A virus infections. In: Dua K, Nammi S, Chang D, Chellappan DK, Gupta G, Collet T (eds) Medicinal plants for lung diseases. Springer, Singapore. https://doi.org/10.1007/978-981-33-6850-7_19

Adetunji CO, Olaniyan OT, Adeyomoye O, Dare A, Adeniyi MJ, Alex E, Rebezov M, Isabekova O, Shariati MA (2022a) Smart sensing for COVID-19 pandemic. In: Pani SK, Dash S, dos Santos WP, Chan Bukhari SA, Flammini F (eds) Assessing COVID-19 and other pandemics and epidemics using computational modelling and data analysis. Springer, Cham. https://doi.org/10.1007/978-3-030-79753-9_9

Adetunji CO, Olaniyan OT, Adeyomoye O, Dare A, Adeniyi MJ, Alex E, Rebezov M, Petukhova E, Shariati MA (2022b) Internet of health things (IoHT) for COVID-19. In: Pani SK, Dash S, dos Santos WP, Chan Bukhari SA, Flammini F (eds) Assessing COVID-19 and other pandemics and epidemics using computational modelling and data analysis. Springer, Cham. https://doi.org/10.1007/978-3-030-79753-9_5

Adetunji CO, Olaniyan OT, Adeyomoye O, Dare A, Adeniyi MJ, Alex E, Rebezov M, Koriagina N, Shariati MA (2022c) Diverse techniques applied for effective diagnosis of COVID-19. In: Pani SK, Dash S, dos Santos WP, Chan Bukhari SA, Flammini F (eds) Assessing COVID-19 and other pandemics and epidemics using computational modelling and data analysis. Springer, Cham. https://doi.org/10.1007/978-3-030-79753-9_3

Arola K, Van der Bruggen B, Mänttäri M, Kallioinen M (2019) Treatment options for nanofiltration and reverse osmosis concentrates from municipal wastewater treatment: a review. Crit Rev Environ Sci Technol 49:2049–2116. https://doi.org/10.1080/10643389.2019.1594519

Beljanski A, Cole C, Fuxa F, Setiawan E, Singh H (2016a) Efficiency and effectiveness of a low-cost, self-cleaning microplastic filtering system for wastewater treatment plants. In: 30th national conference on undergraduate research (NCUR), pp 1388–1395

Beljanski A, Cole C, Fuxa F, Setiawan E, Singh H (2016b) Efficiency and effectiveness of a low-cost, self-cleaning microplastic filtering system for wastewater treatment plants. In: Proceedings of the 30th anniversary national conference on undergraduate research, pp 1388–1395

Carr SA, Liu J, Tesoro AG (2016) Transport and fate of microplastic particles in wastewater treatment plants. Water Res 91:174–182. https://doi.org/10.1016/j.watres.2016.01.002

Choong WS, Hadibarata T, Tang DKH (2020) Abundance and distribution of microplastics in the water and riverbank sediment in Malaysia—a review. Biointerface Res Appl Chem 11:11700–11712

Collivignarelli MC, Abbà A, Benigna I, Sorlini S, Torretta V (2018) Overview of the main disinfection processes for wastewater and drinking water treatment plants. Sustainability 10:1–21. https://doi.org/10.3390/su10010086

da Silva MK, Ambrosi A, dos Ramos GM, Tessaro IC (2012) Rejuvenating polyamide reverse osmosis membranes by tannic acid treatment. Sep Purif Technol 100:1–8. https://doi.org/10.1016/j.seppur.2012.07.027

Garcia-Segura S, Eiband MMSG, de Melo JV, Martínez-Huitle CA (2017) Electrocoagulation and advanced electrocoagulation processes: a general review about the fundamentals, emerging applications and its association with other technologies. J Electroanal Chem 801:267–299. https://doi.org/10.1016/j.jelechem.2017.07.047

Geyer R, Jambeck JR, Law KL (2017) Production, use, and fate of all plastics ever made. Sci Adv 3: e1700782. https://doi.org/10.1126/sciadv.1700782

Global Info Research (2020) Global membrane filtration market 2020 by manufacturers, regions, type and application, forecast to 2025. https://www.marketstudyreport.com/reports/global-membrane-filtration-market-2020-by-manufacturers-regions-type-and-application-forecast-to-2025?gclid=EAIaIQobChMIwdzMpNHJ6gIVgx0rCh2L8An-EAAYAiAAEgLw9PD_BwE. Assessed 12 Aug 2021

Grbic J, Nguyen B, Guo E, You JB, Sinton D, Rochman CM (2019) Magnetic extraction of microplastics from environmental samples. Environ Sci Technol Lett 6:68–72. https://doi.org/10.1021/acs.estlett.8b00671

Herbort AF, Sturm MT, Fiedler S, Abkai G, Schuhen K (2018) Alkoxy-silyl induced agglomeration: a new approach for the sustainable removal of microplastic from aquatic systems. J Polym Environ 26:4258–4270. https://doi.org/10.1007/s10924-018-1287-3

Hidayaturrahman H, Lee TG (2019) A study on characteristics of microplastic in wastewater of South Korea: identification, quantification, and fate of microplastics during treatment process. Mar Pollut Bull 146:696–702. https://doi.org/10.1016/j.marpolbul.2019.06.071

Iorhemen OT, Hamza RA, Tay JH (2016) Membrane bioreactor (MBR) technology for wastewater treatment and reclamation: membrane fouling. Membranes (Basel) 6:33. https://doi.org/10.3390/membranes6020033

Jambeck JR, Geyer R, Wilcox C, Siegler TR, Perryman M, Andrady A, Narayan R, Law KL (2015) Plastic waste inputs from land into the ocean. Science 347(80):768LP–771LP. https://doi.org/10.1126/science.1260352

Landaburu-Aguirre J, García-Pacheco R, Molina S, Rodríguez-Sáez L, Rabadán J, García-Calvo E (2016) Fouling prevention, preparing for re-use and membrane recycling. Towards circular economy in RO desalination. Desalination 393:16–30. https://doi.org/10.1016/j.desal.2016.04.002

Lebreton L, Andrady A (2019) Future scenarios of global plastic waste generation and disposal. Palgrave Commun 5:6. https://doi.org/10.1057/s41599-018-0212-7

Leslie HA, Brandsma SH, van Velzen MJM, Vethaak AD (2017) Microplastics en route: field measurements in the Dutch river delta and Amsterdam canals, wastewater treatment plants, north sea sediments and biota. Environ Int 101:133–142. https://doi.org/10.1016/j.envint.2017.01.018

Li L, Xu G, Yu H, Xing J (2018) Dynamic membrane for micro-particle removal in wastewater treatment: performance and influencing factors. Sci Total Environ 627:332–340. https://doi.org/10.1016/j.scitotenv.2018.01.239

Liu F, Nord NB, Bester K, Vollertsen J (2020) Microplastics removal from treated wastewater by a biofilter. Water 12:1–11. https://doi.org/10.3390/W12041085

Lv X, Dong Q, Zuo Z, Liu Y, Huang X, Wu WM (2019) Microplastics in a municipal wastewater treatment plant: fate, dynamic distribution, removal efficiencies, and control strategies. J Clean Prod 225:579–586. https://doi.org/10.1016/j.jclepro.2019.03.321

Ma B, Xue W, Ding Y, Hu C, Liu H, Qu J (2019a) Removal characteristics of microplastics by Fe-based coagulants during drinking water treatment. J Environ Sci 78:267–275. https://doi.org/10.1016/j.jes.2018.10.006

Ma B, Xue W, Hu C, Liu H, Qu J, Li L (2019b) Characteristics of microplastic removal via coagulation and ultrafiltration during drinking water treatment. Chem Eng J 359:159–167. https://doi.org/10.1016/j.cej.2018.11.155

Mahon AM, O'Connell B, Healy MG, O'Connor I, Officer R, Nash R, Morrison L (2017) Microplastics in sewage sludge: effects of treatment. Environ Sci Technol 51:810–818. https://doi.org/10.1021/acs.est.6b04048

Mason SA, Garneau D, Sutton R, Chu Y, Ehmann K, Barnes J, Fink P, Papazissimos D, Rogers DL (2016) Microplastic pollution is widely detected in US municipal wastewater treatment plant effluent. Environ Pollut 218:1045–1054. https://doi.org/10.1016/j.envpol.2016.08.056

Mejía AC, Velasco AC, Sánchez PZ, Cisneros BJ, Maciel-Cerda A (2017) Photooxidation treatment of the reject stream of a nanofiltration membrane system BT -membranes. In: Materials, simulations, and applications. Springer International Publishing, Cham, pp 105–111. https://doi.org/10.1007/978-3-319-45315-6_12

Michielssen MR, Michielssen ER, Ni J, Duhaime MB (2016) Fate of microplastics and other small anthropogenic litter (SAL) in wastewater treatment plants depends on unit processes employed. Environ Sci Water Res Technol 2:1064–1073. https://doi.org/10.1039/C6EW00207B

Mintenig SM, Int-Veen I, Löder MGJ, Primpke S, Gerdts G (2017) Identification of microplastic in effluents of waste water treatment plants using focal plane array-based micro-Fourier-transform infrared imaging. Water Res 108:365–372. https://doi.org/10.1016/j.watres.2016.11.015

Mintenig SM, Löder MGJ, Primpke S, Gerdts G (2019) Low numbers of microplastics detected in drinking water from ground water sources. Sci Total Environ 648:631–635. https://doi.org/10.1016/j.scitotenv.2018.08.178

Mohamedou EO, Penate Suarez DB, Vince F, Jaouen P, Pontie M (2010) New lives for old reverse osmosis (RO) membranes. Desalination 253:62–70. https://doi.org/10.1016/j.desal.2009.11.032

Morón-López J, Nieto-Reyes L, Aguado S, El-Shehawy R, Molina S (2019) Recycling of end-of-life reverse osmosis membranes for membrane biofilms reactors (MBfRs). Effect of chlorination on the membrane surface and gas permeability. Chemosphere 231:103–112. https://doi.org/10.1016/j.chemosphere.2019.05.108

Murphy F, Ewins C, Carbonnier F, Quinn B (2016) Wastewater treatment works(WwTW) as a source of microplastics in the aquatic environment. Environ Sci Technol 50:5800–5808. https://doi.org/10.1021/acs.est.5b05416

Pal P (2017) Industrial water treatment process technology. Butterworth-Heinemann

Petroody SSA, Hashemi SH, van Gestel CAM (2021) Transport and accumulation of microplastics through wastewater treatment sludge processes. Chemosphere 278:130471. https://doi.org/10.1016/j.chemosphere.2021.130471

Pivokonsky M, Cermakova L, Novotna K, Peer P, Cajthaml T, Janda V (2018) Occurrence of microplastics in raw and treated drinking water. Sci Total Environ 643:1644–1651. https://doi.org/10.1016/j.scitotenv.2018.08.102

Poerio T, Piacentini E, Mazzei R (2019) Membrane processes for microplastic removal. Molecules 24:4148. https://doi.org/10.3390/molecules24224148

Pronk W, Ding A, Morgenroth E, Derlon N, Desmond P, Burkhardt M, Wu B, Fane AG (2019) Gravity-driven membrane filtration for water and wastewater treatment: a review. Water Res 149:553–565. https://doi.org/10.1016/j.watres.2018.11.062

Ranade VV, Bhandari VM (2014) Industrial wastewater treatment, recycling and reuse. Butterworth-Heinemann, Oxford

Rao DG, Senthilkumar R, Byrne JA, Feroz S (2012) Wastewater treatment: advanced processes and technologies. CRC Press, London

Skaf DW, Punzi VL, Rolle JT, Kleinberg KA (2020) Removal of micron-sized microplastic particles from simulated drinking water via alum coagulation. Chem Eng J 386:123807. https://doi.org/10.1016/j.cej.2019.123807

Sturm M, Schuhen K (2019) Development of a novel technological approach for the reduction of microplastic pollution in seawater desalination plants and for sea salt extraction. In: Proceedings of the 17th international conference on chemistry and the environment, Thessaloniki, Greece. European Chemical Society, pp 672–673

Sun J, Dai X, Wang Q, van Loosdrecht MCM, Ni BJ (2019) Microplastics in wastewater treatment plants: detection, occurrence and removal. Water Res 152:21–37. https://doi.org/10.1016/j.watres.2018.12.050

Talvitie J, Heinonen M, Pääkkönen JP, Vahtera E, Mikola A, Setälä O, Vahala R (2015) Do wastewater treatment plants act as a potential point source of microplastics? Preliminary study in the coastal Gulf of Finland, Baltic Sea. Water Sci Technol 72:1495–1504. https://doi.org/10.2166/wst.2015.360

Talvitie J, Mikola A, Koistinen A, Setälä O (2017a) Solutions to microplastic pollution removal of microplastics from wastewater effluent with advanced waste water treatment technologies. Water Res 123:401–407. https://doi.org/10.1016/j.watres.2017.07.005

Tang KHD (2020) Ecotoxicological impacts of micro and nanoplastics on marine fauna. Exam Mar Biol Oceanogr 3:1–5. https://doi.org/10.31031/EIMBO.2020.03.000563

Tang KHD (2021) Interactions of microplastics with persistent organic pollutants and the ecotoxicological effects: a review. Trop Aquat Soil Pollut 1:24–34

Wang Z, Lin T, Chen W (2020) Occurrence and removal of microplastics in an advanced drinking water treatment plant (ADWTP). Sci Total Environ 700:134520. https://doi.org/10.1016/j.scitotenv.2019.134520

WHO (2011) Acrylamide in drinking water. https://www.who.int/water_sanitation_health/dwq/chemicals/acrylamide.pdf. Assessed 12 Aug 2021

Wong JKH, Lee KK, Tang KHD, Yap PS (2020) Microplastics in the freshwater and terrestrial environments: prevalence, fates, impacts and sustainable solutions. Sci Total Environ 719:137512. https://doi.org/10.1016/j.scitotenv.2020.137512

Yan H, Miao X, Xu J, Pan G, Zhang Y, Shi Y, Guo M, Liu Y (2015) The porous structure of the fully-aromatic polyamide film in reverse osmosis membranes. J Membr Sci 475:504–510. https://doi.org/10.1016/j.memsci.2014.10.052

Yang L, Li K, Cui S, Kang Y, An L, Lei K (2019) Removal of microplastics in municipal sewage from China's largest water reclamation plant. Water Res 155:175–181. https://doi.org/10.1016/j.watres.2019.02.046

Zander NE, Sweetser D, Cole DP, Gillan M (2015) Formation of nanofibers from pure and mixed waste streams using electrospinning. Ind Eng Chem Res 54:9057–9063. https://doi.org/10.1021/acs.iecr.5b02279

Zhang S, Li Y, Jiao W, Qiu L, Zhong J, Sun W (2018) Research on lateral flow treatment technology to enhance phosphorus removal of AAO process. In: Proceedings of the E3S web of conferences, p 4029 EDP sciences

Zhang Y, Diehl A, Lewandowski A, Gopalakrishnan K, Baker T (2020) Removal efficiency of micro- and nanoplastics (180 nm–125 μm) during drinking water treatment. Sci Total Environ 720:137383. https://doi.org/10.1016/j.scitotenv.2020.137383

Zhou Z, Wu Z, Wang Z, Tang S, Gu G, Wang L, Wang Y, Xin Z (2011) Simulation and performance evaluation of the anoxic/anaerobic/aerobic process for biological nutrient removal. Korean J Chem Eng 28:1233. https://doi.org/10.1007/s11814-010-0502-2

Ziajahromi S, Neale PA, Rintoul L, Leusch FDL (2017) Wastewater treatment plants as a pathway for microplastics: development of a new approach to sample wastewater-based microplastics. Water Res 112:93–99. https://doi.org/10.1016/j.watres.2017.01.042

Part IV

Entry of Microplastic in the Food Chain and Its Toxicity and Impact on Terrestrial Flora and Fauna

Entry of the Microplastics in Food Chain and Food Web

17

Somia Liaqat, Mubashar Hussain, and Jaweria Riaz

Abstract

The increased reliance on plastic products in day-to-day life has resulted in massive buildup of plastic waste in every ecosystem. Microplastics include plastic materials with size ranging from 1 micron to 5 mm and being the persistent and globally pervasive pollutants, microplastics pose considerable risk to ecosystems, biota, and food webs. Microplastics have been reported from freshwater, land, sea, and biota and are unequally distributed showing the highest concentrations in bio-productive seas, coastal waters, and subtropical gyres. The characteristics of microplastics, i.e., detection capability, motility, and relative concentrations, feeding strategy of organisms, and geographical overlap affect the encounter rate of microplastics and biota. Trophic transfer, inhalation, ingestion, and entanglement are the main routes by which microplastics enter the food webs. Several invertebrates, bivalves, crabs, and small fish have been reported to ingest microplastics, which are then transferred across the food chain and food web. Microplastics are also considered the carriers of pathogens and toxic chemicals. Numerous research studies have reported that microplastics negatively affect living organisms, i.e., reproductive reduction, feeding disruption, disturbed metabolism, and intestinal damage. Microplastics are also known to translocate, bioaccumulate, and release toxic chemicals into the body. This chapter focuses on the sources, entry, bioaccumulation, and biomagnification of microplastics in food chains and food webs in various ecosystems.

Keywords

Accumulation · biomagnification · ingestion · microplastics · trophic level

S. Liaqat (✉) · M. Hussain · J. Riaz
Department of Zoology, University of Gujrat, Gujrat, Punjab, Pakistan

M. Shahnawaz et al. (eds.), *Microplastic Pollution*,
https://doi.org/10.1007/978-981-99-8357-5_17

17.1 Introduction

The extensive synthesis and inefficient management of plastics result in increased levels of plastic waste, hence causing various environmental issues. Increasing awareness regarding the harmful effects has resulted in increased recycling rate of plastics, but still the rate of disposal is high as compared to recycling rate; i.e., about 250 million tons plastic waste was disposed of into the marine water in 2015 (Wright and Kelly 2017). Various physical, biological, and chemical drivers, i.e., UV rays, water, or wind erosion, act upon the plastic litter in the environment and degrade it into smaller fragments known as microplastics (Wright and Kelly 2017). Microplastics are also generated firsthand in multi-purpose cosmetics. About 1 g of microplastics is used by each American in personal care products every year (Bouwmeester et al. 2015). Recent research studies have reported the microplastics from freshwater, sea, land, and living organisms as well (M. Liu et al. 2018). Pollution caused by microplastics has raised global concern, and it is considered the second major problem in the field of ecology and environment (Horton et al. 2017). Researchers across the globe have been working to find solutions to this grave issue. The number of research regarding microplastics is continuously increasing, especially after 2014. Most of these research studies have reported microplastics from marine environment. According to a research study, an excess of 5 trillion plastic particles can be found drifting on the sea surface globally (Eriksen et al. 2014). Although, research studies on microplastic pollution in freshwater are fewer but the available data suggests that microplastic contamination is equally drastic in both freshwater and marine environment (Horton et al. 2017; Zhang et al. 2018a). These microplastics are engulfed by living organisms, i.e., shore crabs, small fish, filter-feeding bivalves, and invertebrates (Horton et al. 2017; Wright and Kelly 2017). These microplastics can potentially be transferred across the food chain from lower trophic level, i.e., zooplankton and phytoplankton, toward higher trophic levels and through the energy pyramid (Zhu et al. 2018). The deleterious consequences of microplastics on living organisms include intestinal damage, feeding disruption, disturbed metabolic processes, and 1reproductive reduction (Anbumani and Kakkar 2018; Lei et al. 2018a; Lei et al. 2018b; Zhu et al. 2018). The major source of plastic waste in oceans is due to human activities on land. Recent researches have declared microplastic pollution as emerging threat to terrestrial environment in which soil acts as a larger reservoir of plastics as compared to sea (Bläsing and Amelung 2018; Hurley and Nizzetto 2018). A research study from Switzerland reported the occurrence of microplastics (55.5 mg/kg) in 90% of floodplain soils (Scheurer and Bigalke 2018). Another study from Shanghai, China, reported mesoplastic and microplastic contamination in farmland soils of suburban areas (Liu et al. 2018). Microplastics that enter into the soil will be stored, translocated, eroded, degraded, and eventually leach down the soil to groundwater, hence threatening organisms and affecting human health (Hurley and Nizzetto 2018). The accumulation and fate of microplastics are influenced by the soil biota. For example, soil fauna can ingest microplastics resulting in smaller microplastics in their digestive tracts. Soil-digging animals, i.e., moles and gophers, result in the translocation and abrasion of microplastics into nanoplastics (Rillig et al. 2017). Microplastics can have negative effects on living organisms because of the additives and toxicants that get adsorbed on plastic surface. Numerous additives have been used in plastic production, i.e., monomers, stabilizers, plasticizers, and flame retardants (de Souza Machado et al. 2018). A number of toxicants including polycyclic aromatic hydrocarbons (PAHs), metals, polychlorinated biphenyls (PCBs), and organochlorine pesticides (HCH, DDT) get absorbed on the hydrophobic plastic surface (Horton et al. 2017). These additives and toxicants can have far-reaching effects on living organisms.

17.2 Microplastic Pollution in Soil

Numerous research studies have explored the occurrence of microplastics in aquatic environments, and only few studies have considered microplastic pollution in terrestrial environment. Soil acts as an interface among hydrosphere, lithosphere, biosphere, and atmosphere. Once entered into the soil, microplastics accumulate and finally reach at harmful levels, hence negatively affecting biodiversity and organisms (Chae and An 2018; de Souza Machado et al. 2018). In addition, microplastics potentially transfer pollutants, i.e., additives or the toxicants adsorbed on surface, thus posing hazards to biota. Terrestrial ecosystems are the major and critical contributors of plastic waste in aquatic environments. Soils are the main storage of microplastics and it is bigger storage point as compared to oceans (Nizzetto et al. 2016). According to another research study, microplastic contamination in soil is 423 times greater as compared to oceans (Horton et al. 2017). Recent research studies are focusing more on microplastic pollution in soils. Researchers from Hebei, China, separated microplastics from the soil to observe the surface features. The study reported mean abundance of microplastics as 317 n/kg (dry weight) and surface features indicated the occurrence of 75% granulate microplastics, whereas fragments accounted for 25% of the total abundance (Qian et al. 2016). The abundance ranges of microplastics from sewage sludge samples suggested higher abundance in floodplain soils of China (Li et al. 2018). The application of plastics in agriculture causes farmland microplastic pollution. The application of sludge, sewage, and mulching are the highest contributing factors toward microplastic pollution in agricultural settings (Nizzetto et al. 2016). A research from Shanghai reported that most of the microplastics in deep and shallow farmland soils were polypropylene (50.51%) and polyethylene (43.43%) (Liu et al. 2018). Additionally, an abundance range of 7100–42,960 particles/kg was reported from fifty samples of arable soils (Zhang and Liu 2018). In addition to other characteristics of soil, recently, many research studies are focusing on microplastic contamination in soils (Bläsing and Amelung 2018; Chae and An 2018).

17.3 The Fate and Buildup of Microplastics in Soil

After microplastics pile up in soils, most of the degradation occurs in top-most layer of soil due to the high temperature, ultraviolet radiation, and availability of oxygen (Chae and An 2018). Terrestrial organisms and soil microbiota increase the rate of microplastic degradation. Additionally, crop rotation and tillage are the agricultural processes that contribute to the degradation of plastic debris into microplastics. However, considering the biological and natural factors, the rate of microplastic degradation is very slow. A research study reported that one year of PP incubation in soil results in 0.4% weight loss, whereas there was no degradation in PVC fragments after 3 decades and 5-year incubation in soil under natural conditions (Ali et al. 2014).

Microplastics found in the top layer of topsoil get incorporated into the deeper topsoil or into plow layer with larger cracks by tillage or by the movement of soil organisms. Leaching is the most important physical process that drives the contaminants to groundwater. Nano- and microplastics although not observed in groundwater but transfer via biopores are a possible mechanism for groundwater contamination (Rillig et al. 2017). Small animals in soil can actively transport microplastic particles within the soil. A research study reported that collembolans can transport microplastic beads in laboratory soil environment (Maaß et al. 2017). The incidental and small-scale horizontal movement of microplastic particles, facilitates the entry of microplastics into deeper layers of soil. A research study reported the vertical translocation of PE microbeads by earthworms (*Lumbricus terrestris*) having a speed of 10 cm down in 21 days (Rillig et al. 2017).

Numerous terrestrial animals also ingest microplastics. Researchers have indicated that different types of microplastics including PVC, PP, PA, PS, and PE are ingested by and accumulated in *Caenorhabditis elegans,* soil nematode (Lei et al. 2018a; Lei et al. 2018b). The bio-transport of microplastics occurs because of adherence, ingestion/egestion, burrowing, and casts of earthworms. Micro- and macroplastics have been analyzed and accessed in soil, chicken feces, earthworm casts, gizzards, and crops (Huerta Lwanga et al. 2017). The available data indicate the capability of microplastics to enter terrestrial food webs. Intake of microplastics by soil biota has a major influence on the deposition of microplastics in soil.

17.4 Microplastics in Terrestrial Food Chains

The transfer of microplastics along food chain has been investigated indicating high concentration of microplastics in successive trophic levels, i.e., earthworm casts, soil, and chicken feces (Huerta Lwanga et al. 2017). *Hypoaspis aculeifer* and *Folsomia candida* have been used to investigate microplastic transport in the soil (Zhu et al. 2018). Microplastic contamination has been observed in beer, sugar, seafood, honey, and salt (Yang et al. 2015; Zhang et al. 2018a).

17.5 Aquaculture as a Source of Microplastics

Microplastics in aquaculture settings normally result from the aging and damage of commercial fishing gear, i.e., trammel nets, net cages, and fishing lines. The indirect transfer of microplastics can also occur due to rainfall, rivers, surface runoff, wind, and rivers. Net cages, fish nets, and buoyant materials are necessary plastic gears in marine aquaculture. These fishing gears break down into microplastics due to prolonged exposure to ultraviolet light (Long et al.) and mechanical abrasion by sand and waves (Song et al. 2017). A higher concentration of microplastics has been observed in cultured mussels as compared to wild mussels, and this may be because of breakage of polypropylene lines used in aquaculture settings (Collos and Harrison 2014). A study from the Xiangshan Bay of China reported that 36.8% and 55.7% of the microplastics in the sediment and seawater result from net cages, floating material, and derivatives of fish net employed in mariculture. The increasing demand of polyethylene foam due to its porous structure results in a particularly high concentration of PE foam in aquaculture environment. The utilization of antipredator nets in shellfish culture does not provide protection to shellfish, instead acts as a microplastic source (Vonk et al. 2015). The aging and wear and tear of fish nets in small-scale fishing operations and marine bottom trawling result in microplastic diffusion into oceans and their ingestion (Bischoff et al. 1985; Churchill 1989; Welden et al. 2018). In freshwater aquaculture, these microplastics get dispersed in rivers, fishponds (Bordós et al. 2019), and lakes (Free et al. 2014).

The microplastic dispersion in aquaculture systems results from atmospheric fallout and wind. A highest MP concentration in cultured oysters has been observed due to the semi-closed geography resulting in typhoons that increase MP concentration in the Gulf (Savoca et al. 2019). In addition, rainfall and storms elevate the concentration of microplastics in marine water (Moore et al. 2002; Zhang et al. 2013), whereas wind results in the fluctuation of MP concentration in freshwater lakes (Fischer et al. 2016; Kukulka et al. 2012).

The natural forces, i.e., runoff and rivers facilitate the transfer of microplastics to aquaculture systems. The microplastics in cosmetics are discharged into sewers that enter the rivers, hence resulting in the accumulation of microplastics in aquaculture settings. An elevated concentration of microplastics has been observed at the entrance of fishpond as compared to the outlet (Bordós et al. 2019). This might be due to microplastic deposition in the sediments of fishpond. Microplastics in soil enter directly into the rivers through runoff. Treated sewage from the sewage treatment units is the major microplastic source in rivers. Even though 98% of the MPS are removed from treated sewage, the rest of the 6.5×10^7 particles are discharged in outcoming water on daily basis (Murphy et al. 2016).

Thus, both direct and indirect sources contribute to microplastics in aquaculture systems. Direct sources of microplastics are the fragmentation of fishing gears due to UV exposure and mechanical breakdown by wind, waves, and sand. Indirect sources include the ways in which microplastics enter aquaculture settings. Most of the research has focused on the sources of microplastics. However, more research is needed to investigate the ways in which the entry of microplastics into aquaculture systems can be blocked.

17.6 Microplastics Transfer Across Food Chain in Aquaculture Settings

17.6.1 Distribution of Microplastics in Various Trophic Levels

According to the categories of trophic levels (Raymond 1942), the first trophic level comprises the producers that convert inorganic material into organic material, the second trophic level comprises the herbivores that feed upon producers, and the third trophic level consists of the carnivores that prey upon the herbivores. The last trophic level constitutes the decomposers that convert the organic matter into inorganic matter, hence providing resources for the producers. All these trophic levels collectively form food chains and food chains form complex food webs.

When microplastics enter the natural environment, they negatively affect the biota and may also enter the food chain resulting in bioaccumulation. A research study reported that attached charged polystyrene beads on algae (producers), Chlorella and Scenedesmus, inhibit the process of photosynthesis (Bhattacharya et al. 2010). Simultaneously, MP exposure inhibits the growth of the algae *Skeletonema costatum* (C. Zhang et al. 2017). Zooplanktons being the second trophic level ingest microplastics, and this significantly decreases algal feeding by zooplankton (Cole et al. 2013) resulting in oxidative stress (Liu et al. 2019). The spectroscopic and digestion methods have reported LDPE microplastics in wild zooplankton (Sun et al. 2017; Vickers 2017). Numerous researches have investigated the consumption of microplastics at higher trophic levels, i.e., whale (mammal), turtle (reptile), tadpole (amphibian), fish, bivalves and sea cucumbers (mollusks), crab and shrimp (crustaceans), and jellyfish (coelenterates) (Bobkov et al. 2021; Devriese et al. 2015; Duncan et al. 2019; Fossi et al. 2012; Graham and Thompson 2009; Hu et al. 2018; Lusher et al. 2013; Vedolin et al. 2018; Yu et al. 2018). At the top-most level in food chain, microplastics have been reported in human feces (Liebmann et al. 2018).

Microplastics have been reported to enter the food chains and get transferred across the trophic levels, and this ultimately threatens human health. However, a significantly higher concentration of microplastics has been used under laboratory conditions as compared to natural environments and the microplastic concentration in field organisms is lower than the level of damage. Therefore, the biological effects of microplastic exposure need further investigation.

17.6.2 Microplastic Biomagnification

A number of pollutants, i.e., microplastics, persistent organic pollutants (POPs), and heavy metals, can be transferred and magnified across food chains and food webs

(Benninger et al. 1998). Determination of direct or indirect ingestion of MPs is difficult under natural environmental conditions. Therefore, most of the findings are based on laboratory conditions that attempt to mimic natural conditions to study the transfer of MPs along various trophic levels. In a research study, mysid shrimp (large-scale predators) were incubated with copepods, zooplankton (prey), and Marenzelleria spp., which had taken fluorescent microplastics. The fluorescent signals were observed in mysid shrimp (Luo et al. 2014). In another study, *Mytilus edulis* were incubated with fluorescent microplastics and were fed to *Carcinus maenas*. The results showed elevated levels of microplastics in crab's hemolymph, and these MPs disappeared after 21 days following exposure (Farrell and Nelson 2013). Under laboratory conditions, the wild-captured *Scomber scombrus* were incubated with predatory *Halichoerus grypus*. The results indicated microplastics in gray seal and mackerel feces (Dris et al. 2016). The gray seals do not directly engulf microplastics, so mackerel might be the contributor of microplastics in feces.

The fate of microplastics taken up by aquatic animals depends on their shape and size. These microplastics might get trapped in digestive tract and can also be transferred from intestinal tract to other tissues. Microplastics up to 5 μm are transferred to the hepatopancreas of *Eriocheir sinensis* and liver of zebrafish, but 20 μm microplastics will not be transported to liver of zebrafish (Lu et al. 2016; Yu et al. 2018). Many bivalves can capture and trap 3–4 μm microplastic particles having 100% retention efficiency; however, the retention efficiency of 1 μm particles reduces by 50% (Gosling 2008).

The biomagnification of microplastics during transfer across the trophic levels is linked to their retention time in tissues, which is affected by the shape and size of microplastics. The retention time of microplastics in the intestine of copepod (*Calanus helgolandicus*) gets prolonged due to hunger (Cole 2013). Microplastics that enter the hemolymph of *M. edulis* have a retention time exceeding forty-eight days (Browne et al. 2008). Some experts have explained the relationship b/w microplastic size and the chances of their biomagnification. The retention time of microplastics is directly related to their size, and the higher the retention time, the greater risk of passing microplastics to higher trophic levels.

17.6.3 Microplastic in Aquatic Food Products

The indirect exposure of humans to microplastics occurs when commercial aquatic products exposed to microplastics are consumed by humans. Numerous edible aquatic animals ingest microplastics, i.e., crustaceans (crab, shrimp), bivalves (oysters, mussels), and fish (Red mullet, Atlantic cod, European sea bass, European pilchard, and Atlantic horse mackerel). It can be concluded that trophic transfer of microplastics can ultimately cause the entry of microplastics in human body (Dris et al. 2016).

17.7 Entry Points of Microplastics in Food Chains and Food Webs

The major entry points of microplastics into aquatic food chains are their inhalation, entanglement, and ingestion. The chances of microplastic ingestion by organisms increase due to their small size, resulting in accidental ingestion (Schuyler et al. 2012). The probability of microplastic ingestion is influenced by various factors, i.e., shape, concentration, size, chemical character, distribution, and animal's feeding habit. The animals have specialized organs for prey detection, and microplastic particles might be rejected if suitable prey is accessible.

17.7.1 Filter Feeding

A vast majority of marine organisms are filter feeders including benthic fauna and small planktonic invertebrates to larger fauna. These filter feeders feed upon the suspended organic matter, i.e., zooplankton, algae, detritus, and fish larvae. Ingestible microplastic size by grazers depends on various factors, i.e., gape size, feeding mode (raptorial or filter feeding), and the feeding mechanism. In case of filter feeders, the feeding apparatus sets the size of ingested prey and traps the suspended food particles (Riisgård and Larsen 2010). The filtering devices used by suspension feeders are not just strainers that filter out suspended organic matter. Filtering apparatus in vertebrates, invertebrates, and unicellular organisms varies among and between taxa, with different levels of sensory capability and adaptability. The rate of microplastic capture depends on particle concentration, particle type (density, shape, size), water viscosity, filtering efficiency, and quantity of water that is filtered. Other than direct contact, the capturing mechanisms include various factors, i.e., mechanoreception and chemoreception (Riisgård and Larsen 2010). In addition, variable plankton clearance rates have been observed depending on the type of offered prey, temperature, and salinity (Garrido et al. 2013; Kiørboe et al. 1982). The daily rates of clearance by marine invertebrates vary from microliters for unicellular organisms, e.g., ciliates, to milliliters for copepods, liters for bivalves, hundreds of liters for zooplankton, and even more for baleen whales.

The outcome and efficiency of filter feeding are usually estimated by two parameters, i.e., clearance and ingestion rate. The direct estimation of ingestion rate involves the observation of ingested microplastics in the body, whereas indirect estimation constitutes the prey disappearance from the experimental medium with the passage of time. The feeding parameters of planktons have been estimated by using plastic particles as surrogates (Børsheim 1984; Huntley et al. 1983; Nygaard et al. 1988). The results indicated preference of copepods, i.e., *Calanus* for algae as compared to the polystyrene beads (Donaghay and Small 1979; Fernandez 1979; Huntley et al. 1983). The probability of ingestion of suspended particle thus depends upon the concentration of prey and clearance rate.

Considering the filter-feeding organisms, the routine encounter with microplastics under natural conditions might be very low. However, the situation is different for waters that contain high concentration of microplastics even for the organisms that have very low efficiency and clearance rate. The experimentally determined 24-h clearance rates for copepods vary from ca. 10 to <200 mL (Engström et al. 2000; Frost 1972; Setälä et al. 2009). It was further estimated by the authors that these microplastics could be passed to higher trophic level when zooplanktons are ingested by juvenile and adult salmon.

Invertebrates that have high water filtering capacity, longer lifespan (bivalves), and large filter feeders (whales) more frequently encounter microplastics as compared to zooplankton. Bivalves are the key entrance points for microplastics in marine food webs. They are the suspension feeders that form link between benthic and pelagic ecosystems, acting as a source of prey for marine vertebrates. It has been theoretically estimated that blue mussel beds in Baltic Sea can filter volume of water that equals whole sea basin in a year (Kautsky and Kautsky 2000). The quantity of microplastics in bivalve mollusks varies from below 0.5 particles (Baltic Sea and Eastern Atlantic) to more than 100 particles per individual (Western Atlantic) (Mathalon and Hill 2014; Railo 2017; Vandermeersch et al. 2015). These large filter feeders ingest micro-litter along with the prey. A research study compared the two semi-enclosed basins; i.e., Sea of Cortez and Mediterranean Sea indicated higher health hazards to fin whales in areas with relatively high pollution (Fossi et al. 2016).

17.7.2 Respiratory Intake

Ventilation or respiratory intake acts as a means of concentrating microplastics from surrounding water into the body. The shoe crab (*Carcinus maenas*) has been reported to respire polystyrene beads that accumulate on its gills surface (Watts et al. 2014). Similarly, for Baltic clams (*Macoma balthica*) and Blue mussels (*Mytilus trossulus*) after 24-h incubation, however, beads concentration was much higher in the digestive tract (Setälä et al. 2016).

17.7.3 Entanglement

The entanglement of numerous organisms with smaller and larger plastic fibers has been reported (Cole et al. 2013; Laist 1997; Taylor et al. 2016). These microplastics are in the feeding or swimming organs of invertebrates, entangling larger animals, and in the gapes of bivalves. Microplastic entanglement has been observed for the field-collected animals, i.e., hermit crabs and sea pens from deep seas (Taylor et al. 2016). The plastics on external surface of these animals get transferred to higher trophic levels when ingested by predators along with plastic fragments.

17.7.4 Trophic Transfer

After ingestion, the microplastics can be egested or retained in the body of organism. The transfer of microplastics across trophic levels has been reported in *Carcinus maenas* that indirectly ingested 0.5–10 μm microplastics in *Mytilus edulis* (Farrell and Nelson 2013; Watts et al. 2014); *Gasterosteus aculeatus* (fish) that ingested microplastic-containing insect larvae; *Neomysis integer* ingested fluorescent polystyrene (10 μm) microplastics ingested by mesozooplankton (Setälä et al. 2014); and *Nephrops norvegicus* ingested polypropylene ropes in fish (Murray and Cowie 2011). The transport of microplastics and associated POPs has also been reported from *Artemia nauplii* to *Danio rerio* (Batel et al. 2016). The quantity of microplastics in feces of *Stercorarius skua* was greater as compared to surface-feeding *Fulmarus glacialis* (Hammer et al. 2016).

The trophic transfer of microplastics occurs as the microplastics are consumed along with the prey, i.e., plastics adhered to the body surface of prey (trapped in the body or setae of copepod appendages) or to algae (Bhattacharya et al. 2010; Gutow et al. 2016). Plastic fragments have been reported in the GIT of marine organisms, i.e., invertebrates (Murray and Cowie 2011), fish (Lusher et al. 2013), turtles (Nelms et al. 2016), and seabirds (Van Franeker and Law 2015). This is due to the gut blockage caused by the coalesced plastic fibers that cannot be moved by peristaltic action. Prolonged gut retention times have been observed in *Carcinus maenas* in which microplastics lodge between the microvillae lining the stomach. The gut retention times of copepods increase due to starvation; e.g., the polystyrene (10 μm) retention time of *Calanus helgolandicus* increases from 2 hours to 7 days (Cole et al. 2013).

Various persistent pollutants, i.e., PCBs, methylmercury, and PAHs, have been observed to biomagnify in the food chains of aquatic ecosystems (Blais et al. 2007). Accumulation of these chemicals in higher trophic levels is due to their hydrophobic nature, causing their deposition in the fatty tissues. Until now, the quantitative biomagnification of microplastics has not been reported in seawater food webs. Biomagnification depends on the instability and retention time of plastics in organisms; hence, biomagnification occurs when plastics are ingested and retained in the body. Food availability largely affects the retention time of plastics (Cole et al. 2013; Watts et al. 2014) and shape (Murray and Cowie 2011), but will be largely affected by plastic size (Galloway 2015).

17.8 Fate of Microplastics in Marine Ecosystems

This section deals with the effects of biologically mediated ecological processes and trophic dynamics on the fate of microplastics and how they influence marine ecosystem, food webs, and biota.

17.8.1 Biological Transport of Microplastics

The microplastics adhered to, respired by, or consumed by an organism will cause biologically mediated transport resulting in lateral and vertical movement of microplastics in various habitats, i.e., sediments, water, and columns. The migratory routes, movement, and gut transit times are the factors that influence transport distance of microplastics.

The transport of microplastics from sea surface to deep water occurs mainly due to the daily vertical migrations of numerous taxa (Clark et al. 2016; Cole et al. 2016). The organisms might take up plastics from the surface, and then, egestion takes place thousand meters below surface, e.g., a 2–3 mm copepod swims at the speed of 30–90 mh^{-1} (Enright 1977), having approximately 2 h of gut evacuation time (Cole et al. 2013), and can potentially transport microplastics at depths of about 60–180 m. A research study reported microplastics in the GIT of 11% of the fishes caught from the Northeast Atlantic (Lusher et al. 2016). Even though the depth at which these plastics were consumed is not known, many identified species were surface feeders and undergo daily vertical migrations. This suggests that intake of microplastics might have occurred at the surface, whereas egestion occurred at depth.

The geographical distribution of plastics is considered regarding their physical presence, and abiotic processes, i.e., oceanic currents, rivers, and winds, are the prominent factors contributing to their wide distribution (Sherman and Van Sebille 2016). Migratory species also play an important part in long-distance transfer of persistent chemicals, i.e., methylmercury, PCBs, and DDT (Blais et al. 2007). Persistent organochlorines accumulation has been reported in the tissues of migratory fishes, i.e., salmon and trout, and are released in freshwater during spawning (Krümmel et al. 2003; Mu et al. 2004). Microplastics have also been reported from the intestines of numerous migratory species, i.e., cetaceans (Lusher et al. 2015), turtles (Nelms et al. 2016), and seabirds (Van Franeker and Law 2015). The egestion of plastics after the death of an animal, in scat or feces, results in their deposition in marine, terrestrial, or freshwater habitats, far away from the sites of their ingestion (Sileo et al. 1990).

17.8.2 Microplastic Incorporation in Biological Matrices

In the marine ecosystem, biofilms of epibionts, microorganisms, and plants rapidly colonize microplastics. The characteristics of these biofilms depend on the type of polymer, and hence, these microbial complexes can be used to trace the journey of the microplastics in seawater (Galloway et al. 2017). Biofilm development can potentially alter the polymer characteristics, i.e., increase in mass (Lobelle and Cunliffe 2011; Rummel et al. 2017; Zettler et al. 2013) and changing the chemical signal. It has been proposed that biofilm formation can result in the sinking of buoyant plastics, and this largely depends on the density and size of plastic (Ye and Andrady 1991).

In bivalves, the rejection or feeding on suspended particles depends on the size of particles and the filtration organs. It has been reported that the identification and sorting of unpalatable material in bivalves occurs lectin-containing mucus covering the feeding organs (Pales Espinosa et al. 2010). Mussels (*Mytilus edulis*) have also been observed to reject nanopolystyrene particles (Ward and Kach 2009).

The ingested microplastics pass through intestinal tract via the peristaltic action. In the GIT, microplastics may become trapped in the gut, absorbed in the lining of gut, or get incorporated in feces and egested. The assumed post-ingestion fate of these microplastics is the passage through gut and repackaging in fecal pellets. This repackaging potentially alters the properties of plastic in water column (Cole and Galloway 2015) and provides an alternative route through which microplastics are transported in marine ecosystem (Clark et al. 2016).

Marine aggregates and sinking feces are important processes by which nutrients and carbon from the euphotic zone get transferred deeper into the oceans (Turner 2015). It has been observed that feces of anchovies in aquaculture systems at the coast of Peru act as important contributors of sediments, having an average > 1 km d^{-1} fecal sinking rate (Staresinic et al. 1983). Considering this situation, the microplastics in these fecal pellets settle in benthos in a very short period of time. Alteration in the buoyancy of biological matrix has been observed due to marine aggregates (Long et al. 2015) and microplastic incorporation in fecal pellets (Cole et al. 2016). It has been concluded by numerous carbon flux studies that slow-sinking feces might become repackaged via coprophagy by larger zooplankton or decomposed by microbes instead of sinking to the seabed (Turner 2002). The coprophagy of microplastic-containing feces represents the route via which plastics can potentially re-enter the marine food web. This has been documented by the ingestion of microplastic-containing feces of small copepod (*Centropages typicus*), by larger copepod (*Calanus helgolandicus*) (Cole et al. 2016). It can be concluded that microplastic-containing fecal pellets are more susceptible to fragmentation, hence more available to the detritivores.

17.8.3 Destiny of Microplastics in Benthos

Sediments in benthic environment act as important deposits for microplastics including both low-density and high-density plastics. Biological matrices facilitate the movement of low-density plastics in benthos, whereas the high-density plastics readily settle down in sediment from the water column. The highly polluted benthic sediments consist of 3% microplastics (Carson et al. 2011), with estimated 4 billion plastic fibers per square km in Indian seamount sediments (Woodall et al. 2014). Microplastics in these sediments are available to benthic organisms, i.e., shellfish (Rochman et al. 2015), *Nephrops norvegicus,* and Norwegian lobster (Murray and Cowie 2011). A number of researches have focused on the capacity of benthic organisms, i.e., polychaetes (Besseling et al. 2013; Green et al. 2016; Wright et al. 2013), echinoderms (Graham and Thompson 2009), and bivalves (Sussarellu et al. 2016) to ingest microplastics, hence negatively affecting their health (reduced energy reserves and reduced bioturbation activity). Benthic fauna potentially changes the microplastic's properties and moves them into deeper sediments. It is experimentally observed in clams and polychaetes that transferred microplastic fibers (<1 mm) 1.7–5.1 cm deep in the time period of 3 weeks (Näkki et al. 2017). However, the redistribution of microplastics by these organisms under natural conditions is largely unknown, as well as whether or not bioturbation causes permanent burial of these plastic fibers.

17.9 Conclusions

In conclusion, this chapter sheds light on the alarming entry of microplastics into the food chain and food web. From soil pollution to aquatic ecosystems, microplastics pose a significant threat to terrestrial and marine organisms. The transfer across trophic levels, biomagnification, and diverse entry points highlights the complexity of this issue. Understanding the fate of microplastics in marine ecosystems and their incorporation into biological matrices is crucial for effective mitigation strategies. Urgent action is needed to minimize microplastic pollution and protect both human health and the environment.

References

Ali MI, Ahmed S, Robson G, Javed I, Ali N, Atiq N, Hameed A (2014) Isolation and molecular characterization of polyvinyl chloride (PVC) plastic degrading fungal isolates. J Basic Microbiol 54(1):18–27

Anbumani S, Kakkar P (2018) Ecotoxicological effects of microplastics on biota: a review. Environ Sci Pollut Res 25(15):14373–14396

Batel A, Linti F, Scherer M, Erdinger L, Braunbeck T (2016) Transfer of benzo [a] pyrene from microplastics to Artemia nauplii and further to zebrafish via a trophic food web experiment: CYP1A induction and visual tracking of persistent organic pollutants. Environ Toxicol Chem 35(7):1656–1666

Benninger L, Suayah I, Stanley DJ (1998) Manzala lagoon, Nile delta, Egypt: modern sediment accumulation based on radioactive tracers. Environ Geol 34(2):183–193

Besseling E, Wegner A, Foekema EM, Van Den Heuvel-Greve MJ, Koelmans AA (2013) Effects of microplastic on fitness and PCB bioaccumulation by the lugworm Arenicola marina (L.). Environ Sci Technol 47(1):593–600

Bhattacharya P, Lin S, Turner JP, Ke PC (2010) Physical adsorption of charged plastic nanoparticles affects algal photosynthesis. J Phys Chem C 114(39):16556–16561

Bischoff RJ, Gould JL, Rubenstein DI (1985) Tail size and female choice in the guppy (Poecilia reticulata). Behav Ecol Sociobiol 17(3):253–255

Blais JM, Macdonald RW, Mackay D, Webster E, Harvey C, Smol JP (2007) Biologically mediated transport of contaminants to aquatic systems. Environ Sci Technol 41(4):1075–1084

Bläsing M, Amelung W (2018) Plastics in soil: analytical methods and possible sources. Sci Total Environ 612:422–435

Bobkov YV, Walker WB III, Cattaneo AM (2021) Altered functional properties of the codling moth Orco mutagenized in the intracellular loop-3. Sci Rep 11(1):1–16

Bordós G, Urbányi B, Micsinai A, Kriszt B, Palotai Z, Szabó I et al (2019) Identification of microplastics in fish ponds and natural freshwater environments of the Carpathian basin, Europe. Chemosphere 216:110–116

Børsheim KY (1984) Clearance rates of bacteria-sized particles by freshwater ciliates, measured with monodisperse fluorescent latex beads. Oecologia 63(2):286–288

Bouwmeester H, Hollman PC, Peters RJ (2015) Potential health impact of environmentally released micro-and nanoplastics in the human food production chain: experiences from nanotoxicology. Environ Sci Technol 49(15):8932–8947

Browne MA, Dissanayake A, Galloway TS, Lowe DM, Thompson RC (2008) Ingested microscopic plastic translocates to the circulatory system of the mussel, Mytilus edulis (L.). Environ Sci Technol 42(13):5026–5031

Carson HS, Colbert SL, Kaylor MJ, McDermid KJ (2011) Small plastic debris changes water movement and heat transfer through beach sediments. Mar Pollut Bull 62(8):1708–1713

Chae Y, An Y-J (2018) Current research trends on plastic pollution and ecological impacts on the soil ecosystem: a review. Environ Pollut 240:387–395

Churchill JH (1989) The effect of commercial trawling on sediment resuspension and transport over the Middle Atlantic Bight continental shelf. Cont Shelf Res 9(9):841–865

Clark JR, Cole M, Lindeque PK, Fileman E, Blackford J, Lewis C et al (2016) Marine microplastic debris: a targeted plan for understanding and quantifying interactions with marine life. Front Ecol Environ 14(6):317–324

Cole M (2013) Microplastic swallowing zooplankton. Environ Sci Technol 47:6646–6655

Cole M, Galloway TS (2015) Ingestion of nanoplastics and microplastics by Pacific oyster larvae. Environ Sci Technol 49(24):14625–14632

Cole M, Lindeque P, Fileman E, Halsband C, Goodhead R, Moger J, Galloway TS (2013) Microplastic ingestion by zooplankton. Environ Sci Technol 47(12):6646–6655

Cole M, Lindeque PK, Fileman E, Clark J, Lewis C, Halsband C, Galloway TS (2016) Microplastics alter the properties and sinking rates of zooplankton faecal pellets. Environ Sci Technol 50(6):3239–3246

Collos Y, Harrison PJ (2014) Acclimation and toxicity of high ammonium concentrations to unicellular algae. Mar Pollut Bull 80(1-2):8–23

Devriese LI, Van der Meulen MD, Maes T, Bekaert K, Paul-Pont I, Frère L et al (2015) Microplastic contamination in brown shrimp (Crangon crangon, Linnaeus 1758) from coastal waters of the Southern North Sea and Channel area. Mar Pollut Bull 98(1-2):179–187

Donaghay P, Small L (1979) Food selection capabilities of the estuarine copepod Acartia clausi. Mar Biol 52(2):137–146

Dris R, Gasperi J, Saad M, Mirande C, Tassin B (2016) Synthetic fibers in atmospheric fallout: a source of microplastics in the environment? Mar Pollut Bull 104(1-2):290–293

Duncan EM, Broderick AC, Fuller WJ, Galloway TS, Godfrey MH, Hamann M et al (2019) Microplastic ingestion ubiquitous in marine turtles. Glob Chang Biol 25(2):744–752

Engström J, Koski M, Viitasalo M, Reinikainen M, Repka S, Sivonen K (2000) Feeding interactions of the copepods Eurytemora affinis and Acartia bifilosa with the cyanobacteria Nodularia sp. J Plankton Res 22(7):1403–1409

Enright J (1977) Copepods in a hurry: sustained high-speed upward migration 1. Limnol Oceanogr 22(1):118–125

Eriksen M, Lebreton LC, Carson HS, Thiel M, Moore CJ, Borerro JC et al (2014) Plastic pollution in the world's oceans: more than 5 trillion plastic pieces weighing over 250,000 tons afloat at sea. PLoS One 9(12):e111913

Farrell P, Nelson K (2013) Trophic level transfer of microplastic: Mytilus edulis (L.) to Carcinus maenas (L.). Environ Pollut 177:1–3

Fernandez F (1979) Particle selection in the nauplius of Calanus pacificus. J Plankton Res 1(4): 313–328

Fischer EK, Paglialonga L, Czech E, Tamminga M (2016) Microplastic pollution in lakes and lake shoreline sediments–a case study on Lake Bolsena and Lake Chiusi (Central Italy). Environ Pollut 213:648–657

Fossi MC, Marsili L, Baini M, Giannetti M, Coppola D, Guerranti C et al (2016) Fin whales and microplastics: the Mediterranean Sea and the Sea of Cortez scenarios. Environ Pollut 209:68–78

Fossi MC, Panti C, Guerranti C, Coppola D, Giannetti M, Marsili L, Minutoli R (2012) Are baleen whales exposed to the threat of microplastics? A case study of the Mediterranean fin whale (Balaenoptera physalus). Mar Pollut Bull 64(11):2374–2379

Free CM, Jensen OP, Mason SA, Eriksen M, Williamson NJ, Boldgiv B (2014) High-levels of microplastic pollution in a large, remote, mountain lake. Mar Pollut Bull 85(1):156–163

Frost B (1972) Effects of size and concentration of food particles on the feeding behavior of the marine planktonic copepod Calanus pacificus 1. Limnol Oceanogr 17(6):805–815

Galloway TS (2015) Micro-and nano-plastics and human health. In: Marine anthropogenic litter. Springer, Cham, pp 343–366

Galloway TS, Cole M, Lewis C (2017) Interactions of microplastic debris throughout the marine ecosystem. Nat Ecol Evol 1(5):1–8

Garrido S, Cruz J, Santos AMP, Re P, Saiz E (2013) Effects of temperature, food type and food concentration on the grazing of the calanoid copepod Centropages chierchiae. J Plankton Res 35(4):843–854

Gosling E (2008) Bivalve molluscs: biology, ecology and culture. Wiley

Graham ER, Thompson JT (2009) Deposit-and suspension-feeding sea cucumbers (Echinodermata) ingest plastic fragments. J Exp Mar Biol Ecol 368(1):22–29

Green DS, Boots B, Sigwart J, Jiang S, Rocha C (2016) Effects of conventional and biodegradable microplastics on a marine ecosystem engineer (Arenicola marina) and sediment nutrient cycling. Environ Pollut 208:426–434

Gutow L, Eckerlebe A, Giménez L, Saborowski R (2016) Experimental evaluation of seaweeds as a vector for microplastics into marine food webs. Environ Sci Technol 50(2):915–923

Hammer S, Nager R, Johnson P, Furness R, Provencher J (2016) Plastic debris in great skua (Stercorarius skua) pellets corresponds to seabird prey species. Mar Pollut Bull 103(1-2): 206–210

Horton AA, Walton A, Spurgeon DJ, Lahive E, Svendsen C (2017) Microplastics in freshwater and terrestrial environments: evaluating the current understanding to identify the knowledge gaps and future research priorities. Sci Total Environ 586:127–141

Hu L, Chernick M, Hinton DE, Shi H (2018) Microplastics in small waterbodies and tadpoles from Yangtze River Delta, China. Environ Sci Technol 52(15):8885–8893

Huerta Lwanga E, Mendoza Vega J, Ku Quej V, Chi J, Sanchez del Cid L, Chi C et al (2017) Field evidence for transfer of plastic debris along a terrestrial food chain. Sci Rep 7(1):14071

Huntley M, Barthel K-G, Star J (1983) Particle rejection by Calanus pacificus: discrimination between similarly sized particles. Mar Biol 74(2):151–160

Hurley RR, Nizzetto L (2018) Fate and occurrence of micro (nano) plastics in soils: knowledge gaps and possible risks. Curr Opin Environ Sci Health 1:6–11

Kautsky L, Kautsky N (2000) The Baltic Sea, including Bothnian Sea and Bothnian Bay. Seas Millenn Environ Eval 1:121–133

Kiørboe T, Møhlenberg F, Nicolajsen H (1982) Ingestion rate and gut clearance in the planktonic copepod Centropages hamatus (Lilljeborg) in relation to food concentration and temperature. Ophelia 21(2):181–194

Krümmel E, Macdonald R, Kimpe L, Gregory-Eaves I, Demers M, Smol J et al (2003) Delivery of pollutants by spawning salmon. Nature 425(6955):255–256

Kukulka T, Proskurowski G, Morét-Ferguson S, Meyer DW, Law KL (2012) The effect of wind mixing on the vertical distribution of buoyant plastic debris. Geophys Res Lett 39(7):7601

Laist DW (1997) Impacts of marine debris: entanglement of marine life in marine debris including a comprehensive list of species with entanglement and ingestion records. In: Marine debris. Springer, pp 99–139

Lei L, Liu M, Song Y, Lu S, Hu J, Cao C et al (2018a) Polystyrene (nano) microplastics cause size-dependent neurotoxicity, oxidative damage and other adverse effects in Caenorhabditis elegans. Environ Sci Nano 5(8):2009–2020

Lei L, Wu S, Lu S, Liu M, Song Y, Fu Z et al (2018b) Microplastic particles cause intestinal damage and other adverse effects in zebrafish Danio rerio and nematode Caenorhabditis elegans. Sci Total Environ 619:1–8

Li X, Chen L, Mei Q, Dong B, Dai X, Ding G, Zeng EY (2018) Microplastics in sewage sludge from the wastewater treatment plants in China. Water Res 142:75–85

Liebmann B, Köppel S, Königshofer P, Bucsics T, Reiberger T, Schwabl P (2018) Assessment of microplastic concentrations in human stool: final results of a prospective study. Paper presented at the Conference on nano and microplastics in technical and freshwater systems

Liu M, Lu S, Song Y, Lei L, Hu J, Lv W et al (2018) Microplastic and mesoplastic pollution in farmland soils in suburbs of Shanghai, China. Environ Pollut 242:855–862

Liu Z, Yu P, Cai M, Wu D, Zhang M, Huang Y, Zhao Y (2019) Polystyrene nanoplastic exposure induces immobilization, reproduction, and stress defense in the freshwater cladoceran Daphnia pulex. Chemosphere 215:74–81

Lobelle D, Cunliffe M (2011) Early microbial biofilm formation on marine plastic debris. Mar Pollut Bull 62(1):197–200

Long M, Moriceau B, Gallinari M, Lambert C, Huvet A, Raffray J, Soudant P (2015) Interactions between microplastics and phytoplankton aggregates: impact on their respective fates. Mar Chem 175:39–46

Lu Y, Zhang Y, Deng Y, Jiang W, Zhao Y, Geng J et al (2016) Uptake and accumulation of polystyrene microplastics in zebrafish (Danio rerio) and toxic effects in liver. Environ Sci Technol 50(7):4054–4060

Luo W, Verweij RA, van Gestel CA (2014) Determining the bioavailability and toxicity of lead contamination to earthworms requires using a combination of physicochemical and biological methods. Environ Pollut 185:1–9

Lusher AL, Hernandez-Milian G, O'Brien J, Berrow S, O'Connor I, Officer R (2015) Microplastic and macroplastic ingestion by a deep diving, oceanic cetacean: the True's beaked whale Mesoplodon mirus. Environ Pollut 199:185–191

Lusher AL, Mchugh M, Thompson RC (2013) Occurrence of microplastics in the gastrointestinal tract of pelagic and demersal fish from the English Channel. Mar Pollut Bull 67(1-2):94–99

Lusher AL, O'Donnell C, Officer R, O'Connor I (2016) Microplastic interactions with North Atlantic mesopelagic fish. ICES J Mar Sci 73(4):1214–1225

Maaß S, Daphi D, Lehmann A, Rillig MC (2017) Transport of microplastics by two collembolan species. Environ Pollut 225:456–459

Mathalon A, Hill P (2014) Microplastic fibers in the intertidal ecosystem surrounding Halifax Harbor, Nova Scotia. Mar Pollut Bull 81(1):69–79

Moore CJ, Moore SL, Weisberg SB, Lattin GL, Zellers AF (2002) A comparison of neustonic plastic and zooplankton abundance in southern California's coastal waters. Mar Pollut Bull 44(10):1035–1038

Mu H, Ewald G, Nilsson E, Sundin P, Wesén C (2004) Fate of chlorinated fatty acids in migrating sockeye salmon and their transfer to arctic grayling. Environ Sci Technol 38(21):5548–5554

Murphy F, Ewins C, Carbonnier F, Quinn B (2016) Wastewater treatment works (WwTW) as a source of microplastics in the aquatic environment. Environ Sci Technol 50(11):5800–5808

Murray F, Cowie PR (2011) Plastic contamination in the decapod crustacean Nephrops norvegicus (Linnaeus, 1758). Mar Pollut Bull 62(6):1207–1217

Näkki P, Setälä O, Lehtiniemi M (2017) Bioturbation transports secondary microplastics to deeper layers in soft marine sediments of the northern Baltic Sea. Mar Pollut Bull 119(1):255–261

Nelms SE, Duncan EM, Broderick AC, Galloway TS, Godfrey MH, Hamann M et al (2016) Plastic and marine turtles: a review and call for research. ICES J Mar Sci 73(2):165–181

Nizzetto L, Futter M, Langaas S (2016) Are agricultural soils dumps for microplastics of urban origin? ACS Publications

Nygaard K, Børsheim KY, Thingstad TF (1988) Grazing rates on bacteria by marine heterotrophic microflagellates compared to uptake rates of bacterial-sized monodisperse fluorescent latex beads. Mar Ecol Prog Ser 44(2):159–165

Pales Espinosa E, Hassan D, Ward JE, Shumway SE, Allam B (2010) Role of epicellular molecules in the selection of particles by the blue mussel, Mytilus edulis. Biol Bull 219(1):50–60

Qian Z, Zhang H, Yang Z, Yuan L, Yong X, Chuancheng F et al (2016) Separation of microplastics from a coastal soil and their surface microscopic features. Chin Sci Bull 61(14):1604–1611

Railo S (2017) Microlitter in Mytilus trossulus and its environment in the Northern Baltic Sea: wastewater as point source pollution. MSc thesis. University of Helsinki

Raymond LL (1942) The trophic-dynamic aspect of ecology. Ecology 23(4):399–417

Riisgård HU, Larsen PS (2010) Particle capture mechanisms in suspension-feeding invertebrates. Mar Ecol Prog Ser 418:255–293

Rillig MC, Ziersch L, Hempel S (2017) Microplastic transport in soil by earthworms. Sci Rep 7(1): 1–6

Rochman CM, Tahir A, Williams SL, Baxa DV, Lam R, Miller JT et al (2015) Anthropogenic debris in seafood: plastic debris and fibers from textiles in fish and bivalves sold for human consumption. Sci Rep 5(1):1–10

Rummel CD, Jahnke A, Gorokhova E, Kühnel D, Schmitt-Jansen M (2017) Impacts of biofilm formation on the fate and potential effects of microplastic in the aquatic environment. Environ Sci Technol Lett 4(7):258–267

Savoca S, Capillo G, Mancuso M, Faggio C, Panarello G, Crupi R et al (2019) Detection of artificial cellulose microfibers in Boops boops from the northern coasts of Sicily (Central Mediterranean). Sci Total Environ 691:455–465

Scheurer M, Bigalke M (2018) Microplastics in Swiss Floodplain soils. Environ Sci Technol 52(6): 3591–3598

Schuyler Q, Hardesty BD, Wilcox C, Townsend K (2012) To eat or not to eat? Debris selectivity by marine turtles. PLoS One 7(7):e40884

Setälä O, Fleming-Lehtinen V, Lehtiniemi M (2014) Ingestion and transfer of microplastics in the planktonic food web. Environ Pollut 185:77–83

Setälä O, Norkko J, Lehtiniemi M (2016) Feeding type affects microplastic ingestion in a coastal invertebrate community. Mar Pollut Bull 102(1):95–101

Setälä O, Sopanen S, Autio R, Erler K (2009) Grazing and food selection of the calanoid copepods Eurytemora affinis and Acartia bifilosa feeding on plankton assemblages containing Dinophysis spp. Boreal Environ Res 14(5):837

Sherman P, Van Sebille E (2016) Modeling marine surface microplastic transport to assess optimal removal locations. Environ Res Lett 11(1):014006

Sileo L, Sievert P, Samuel MD, Fefer SI (1990) Prevalence and characteristics of plastic ingested by Hawaiian seabirds. Paper presented at the Second International Conference on Marine Debris

Song, Y., Hong, S., & Jang, M. (2017). Han, G. M., Jung, S. W., Shim, W. J. Combined effects of UV exposure duration and mechanical abrasion on microplastic fragmentation by polymer type, Environ Sci Technol, 8, 4368

de Souza Machado AA, Kloas W, Zarfl C, Hempel S, Rillig MC (2018) Microplastics as an emerging threat to terrestrial ecosystems. Glob Chang Biol 24(4):1405–1416

Staresinic N, Farrington J, Gagosian RB, Clifford CH, Hulburt EM (1983) Downward transport of particulate matter in the Peru coastal upwelling: role of the anchoveta, Engraulis ringens. In: Coastal upwelling its sediment record. Springer, pp 225–240

Sun X, Li Q, Zhu M, Liang J, Zheng S, Zhao Y (2017) Ingestion of microplastics by natural zooplankton groups in the northern South China Sea. Mar Pollut Bull 115(1-2):217–224

Sussarellu R, Suquet M, Thomas Y, Lambert C, Fabioux C, Pernet MEJ et al (2016) Oyster reproduction is affected by exposure to polystyrene microplastics. Proc Natl Acad Sci U S A 113(9):2430–2435

Taylor M, Gwinnett C, Robinson LF, Woodall LC (2016) Plastic microfibre ingestion by deep-sea organisms. Sci Rep 6(1):1–9

Turner JT (2002) Zooplankton fecal pellets, marine snow and sinking phytoplankton blooms. Aquat Microb Ecol 27(1):57–102

Turner JT (2015) Zooplankton fecal pellets, marine snow, phytodetritus and the ocean's biological pump. Prog Oceanogr 130:205–248

Van Franeker JA, Law KL (2015) Seabirds, gyres and global trends in plastic pollution. Environ Pollut 203:89–96

Vandermeersch G, Van Cauwenberghe L, Janssen CR, Marques A, Granby K, Fait G et al (2015) A critical view on microplastic quantification in aquatic organisms. Environ Res 143:46–55

Vedolin MC, Teophilo C, Turra A, Figueira RCL (2018) Spatial variability in the concentrations of metals in beached microplastics. Mar Pollut Bull 129(2):487–493

Vickers NJ (2017) Animal communication: when I'm calling you, will you answer too? Curr Biol 27(14):R713–R715

Vonk SM, Hollander DJ, Murk AJ (2015) Was the extreme and wide-spread marine oil-snow sedimentation and flocculent accumulation (MOSSFA) event during the Deepwater Horizon blow-out unique? Mar Pollut Bull 100(1):5–12

Ward JE, Kach DJ (2009) Marine aggregates facilitate ingestion of nanoparticles by suspension-feeding bivalves. Mar Environ Res 68(3):137–142

Watts AJ, Lewis C, Goodhead RM, Beckett SJ, Moger J, Tyler CR, Galloway TS (2014) Uptake and retention of microplastics by the shore crab Carcinus maenas. Environ Sci Technol 48(15): 8823–8830

Welden NA, Abylkhani B, Howarth LM (2018) The effects of trophic transfer and environmental factors on microplastic uptake by plaice, Pleuronectes plastessa, and spider crab, Maja squinado. Environ Pollut 239:351–358

Woodall LC, Sanchez-Vidal A, Canals M, Paterson GL, Coppock R, Sleight V et al (2014) The deep sea is a major sink for microplastic debris. R Soc Open Sci 1(4):140317

Wright SL, Kelly FJ (2017) Plastic and human health: a micro issue? Environ Sci Technol 51(12): 6634–6647

Wright SL, Rowe D, Thompson RC, Galloway TS (2013) Microplastic ingestion decreases energy reserves in marine worms. Curr Biol 23(23):R1031–R1033

Yang D, Shi H, Li L, Li J, Jabeen K, Kolandhasamy P (2015) Microplastic pollution in table salts from China. Environ Sci Technol 49(22):13622–13627

Ye S, Andrady AL (1991) Fouling of floating plastic debris under Biscayne Bay exposure conditions. Mar Pollut Bull 22(12):608–613

Yu P, Liu Z, Wu D, Chen M, Lv W, Zhao Y (2018) Accumulation of polystyrene microplastics in juvenile Eriocheir sinensis and oxidative stress effects in the liver. Aquat Toxicol 200:28–36

Zettler ER, Mincer TJ, Amaral-Zettler LA (2013) Life in the "plastisphere": microbial communities on plastic marine debris. Environ Sci Technol 47(13):7137–7146

Zhang C, Chen X, Wang J, Tan L (2017) Toxic effects of microplastic on marine microalgae Skeletonema costatum: interactions between microplastic and algae. Environ Pollut 220:1282–1288

Zhang G, Liu Y (2018) The distribution of microplastics in soil aggregate fractions in southwestern China. Sci Total Environ 642:12–20

Zhang K, Shi H, Peng J, Wang Y, Xiong X, Wu C, Lam PK (2018a) Microplastic pollution in China's inland water systems: a review of findings, methods, characteristics, effects, and management. Sci Total Environ 630:1641–1653

Zhang Y, Beesoon S, Zhu L, Martin JW (2013) Biomonitoring of perfluoroalkyl acids in human urine and estimates of biological half-life. Environ Sci Technol 47(18):10619–10627

Zhu D, Bi Q-F, Xiang Q, Chen Q-L, Christie P, Ke X et al (2018) Trophic predator-prey relationships promote transport of microplastics compared with the single Hypoaspis aculeifer and Folsomia candida. Environ Pollut 235:150–154

Insect Interactions with Microplastics Under Different Ecosystems

18

Somia Liaqat, Mubashar Hussain, Kiran Aftab, Suleman Hussain Shah, and Muhammad Faheem Malik

Abstract

Microplastic pollution (<5 mm in size) is undoubtedly considered a massive threat to living organisms across the globe in almost all ecosystems, i.e., aquatic and terrestrial ecosystems. These microplastic particles are transferred across food chains and food webs through inhalation or ingestion by organisms. Various persistent organic pollutants and metals are known to absorb on microplastic surfaces. Insects, being the most abundant life form on the earth, are major drivers of ecosystem services. Their immense contribution to ecosystem stability, sustainability, and resilience has been acclaimed in the literature. Plastic pollution and insect interactions have been reported consistently in the last two decades globally from both aquatic and terrestrial ecosystems. Many insect groups chew and ingest plastic films including PP, PE, and PVC. *Plodia interpunctella* and *Tenebrio molitor* larvae are known to ingest and degrade polyethylene and Styrofoam, respectively. Insect pollinators transport microplastic particles to flowers, hence interfering with the pollination process. Termites and ants are known to transport microplastics in the soil. These microplastics are being ingested by insects, and vertebrate predators of insects transfer the toxic chemicals to higher trophic levels. Microplastics affect biota adversely by disturbing metabolism, reducing survival by impairing growth and development, and declining reproduction potential. This may affect endocrinal activity by interfering with hormones. Many plastic fragments can mimic hormones, which can potentially disrupt the ecological roles of many insects. In terrestrial ecosystems, microplastics can potentially interfere with the natural biocontrol of insect pests by hormonal dysregulation in the predator species, i.e., ladybird beetles, prolonging the duration of larval instars. Polyvinylchloride microplastics

S. Liaqat (✉) · M. Hussain · K. Aftab · S. H. Shah · M. F. Malik
Department of Zoology, University of Gujrat, Gujrat, Punjab, Pakistan

M. Shahnawaz et al. (eds.), *Microplastic Pollution*,
https://doi.org/10.1007/978-981-99-8357-5_18

have been reported to negatively affect the emergence, weight, and development of *Chironomus riparius*. The exposure to microplastics also alters various behaviors, i.e., oviposition and foraging behavior in different species of insects.

Keywords

Ecosystem services · Ingestion · Microplastics · Pollination · Insect interaction

18.1 Introduction

Microplastics include plastic particles with size of less than 5 mm (Van Cauwenberghe et al. 2015). The synthesis of plastic in the early 1900s was considered a global success, and the additional properties of plastics, i.e., chemical resistance, lightness, and durability, have resulted in the increased utilization of plastic in the last century (Antelava et al. 2019; Hopewell et al. 2009; Piringer and Baner 2008). The major purpose of most plastic products is to replace wood, glass, metal, and ceramic products (Wong et al. 2015). This has resulted in the widespread and unchecked use of plastic, hence resulting in increased pollution, concerning researchers across the globe (Penghui et al. 2021). There is an exponential growth in the buildup of plastic debris (Geyer et al. 2017). China has been reported to be the largest contributor to plastic pollution contributing 60 million tons of plastic debris annually (Jambeck et al. 2015; Qu et al. 2019). Indonesia is the top producer of plastic debris in Southeast Asia, followed by Thailand and Vietnam (Finnegan and Gouramanis 2021).

Researchers have confirmed the negative effects of microplastics in marine ecosystem (Bai et al. 2021; Gola et al. 2021; Lin 2016; Pauna et al. 2019; Troost et al. 2018) and estuarine environment (Athey et al. 2020; Bessa et al. 2018; Forero-López et al. 2021, 2022; Pan et al. 2021; Sruthy and Ramasamy 2017). Nonetheless, studies in the past few decades have highlighted the havoc of microplastics in freshwater only (Eerkes-Medrano and Thompson 2018; Gupta et al. 2022; Pinheiro et al. 2017; Wu et al. 2022). No doubt, microplastics are the emerging freshwater contaminants (Dijkstra et al. 2014) affecting aquatic life adversely, particularly immature aquatic insects. Aquatic insects provide important ecosystem services as predators, primary consumers, and detritivores (Idris et al. 2020; Zain et al. 2022).

18.2 Sources and Transportation of Microplastics in the Aquatic Ecosystems

Freshwater bodies, i.e., rivers, are the major passageways that transport plastic waste from land to the seas. Polluted rivers contribute 90% of plastic debris in oceans (Lebreton et al. 2017). The contribution of plastic waste to oceans by rivers varies across regions due to entrapment of plastic waste on beach, in traps, or in trees along riverside (Kooi et al. 2018). The fate of this entrapped plastic is ultimately the riverbank, and after being submerged in water, it affects the environment and ecosystem functioning (Barboza et al. 2018).

18.3 Microplastic: Insect Interactions in Aquatic Insects

Microplastics have been reported in the digestive tracts of aquatic invertebrates (Lusher et al. 2017). These microplastics remain in the bodies of invertebrates for few days to few weeks before excretion (Straub et al. 2017). This retention of plastic inside the living organisms can potentially cause their transfer across the food chain and to other ecosystems as well. However, consistent results have not been obtained for the effects of microplastics on biota. Some organisms are capable of egesting the microplastics, while others are severely affected, i.e., survival rate (Ziajahromi et al. 2018) and reproductive and feeding abilities.

Insects, being the most diverse and abundant group on the earth, are adapted to inhabit a wide range of habitats and utilize varied food sources. Their high morphological and physiological diversity has contributed to high species diversity, hence contributing to maintaining the essential ecosystem services. The interaction between insects and microplastics is unavoidable due to their omnipresent nature. This chapter covers the interactions of microplastics with insects in aquatic and terrestrial ecosystems. Numerous microplastics have been reported in the larval cases of caddisfly (Ehlers et al. 2019). Primary microplastics resembled those used in cosmetics and accounted for 15.2% of the total microplastics (Tanaka and Takada 2016). The deterioration of primary microplastics in the larval cases results in the production of secondary microplastics (Cole et al. 2011). Microplastics having different characteristics, i.e., color, shape, size, and polymer type, have been observed (Cho et al. 2021). The caddisfly larvae contain microplastics that are available in the waterbody in which they live; thus, the larval case analysis of caddisfly can provide information about microplastic diversity in the waterbody (Ehlers et al. 2019). This suggests that caddisfly cases can be used as the bioindicators of microplastic pollution. The lower fraction of minerals and sand grains in larval cases ass compared to the sediments makes it easier to analyze the filter from caddisfly cases. For example, an orange PP film (73 μm long) and a yellow PP sphere (43 μm diameter) have been separated from caddisfly cases, which would have been overlooked in sediment analysis (Ehlers et al. 2019).

C. quinquefasciatus larvae consume microplastics and act as vectors to transfer them to terrestrial and aerial habitats (Al-Jaibachi et al. 2018). PE microplastics delay larval development, which causes longer permanence in water and hence higher vulnerability to predators (Malafaia et al. 2020). The toxic effects of these pollutants might cause larval death before their emergence as adults and can also affect the fitness of adults. Whatever the case, there will be a significant impact on the overall population.

The PE microplastics significantly affect the nutritional status and energy metabolism of the larvae. The effects include reduced levels of triglycerides, total proteins, and carbohydrate levels (Yin et al. 2018). Microplastic-induced nutritional deficiency has also been reported in other animals, i.e., marine worms (Wright et al. 2013), crabs (Watts et al. 2015), and marine jacopever fish (Yin et al. 2018).

The degradative and biosynthetic ability of intestinal microbiota of insects plays significant role in their nutritional ecology. Microplastic-induced damage to the intestinal microbiota of *C. quinquefasciatus* larvae impairs the assimilation process (Minard et al. 2013). This nutritional deficiency is a major cause of the delay in larval development.

18.4 Microplastic Degradation by Insects

The ants and termites produce biopores in the soils that help in the transport of microplastics deep into the soil, and this results in slower turnover due to decreased microbial populations at certain depths (Buks et al. 2020; de Souza Machado et al. 2020; Oliveira et al. 2019). The soil Collembolans, formerly classified in class Insecta, promote nutrient recycling in soil ecosystem by the decomposition of litter and soil detritus (Neher and Barbercheck 2019). Exposure to microplastics alters the gut microbiota of these collembolans, hence indirectly affecting the decomposition rate by inhibiting their growth and reproduction (Vaccari et al. 2022; Xiang et al. 2019; Zhu et al. 2018).

18.5 Pollination by Insects

Pollination is important for biodiversity maintenance and ecosystem sustainability. A major proportion of flowering plants and agricultural crops directly rely on pollinators, mainly insects, to produce fruits and seeds. Microplastics have been reported in the flowers of various plant species (Oliveira et al. 2019; Shah et al. 2023; Trestrail et al. 2020). The translocation of latex beads from the transmission tracts of styles to the ovary has been identified (Carvallo and Muñoz-Michea 2023), which suggests the interference of pollen-sized microplastics. These microplastic particles mimic the pollen grains, which disrupt the process of natural pollination ultimately contributing to the biodiversity loss.

18.6 Seed Dispersal

Insects disperse seeds of many crops and natural flora without affecting the viability of seeds. Insects facilitate seed dispersal during their foraging activity where seeds are transported from one plant to another with the help of exoskeleton, legs, hairs, spines, mouthparts, or hooks on their bodies (Basset et al. 2021; Kumar and Paul 2021). Microplastics when abundantly present in the terrestrial environment potentially compete with seeds for being carried by insects instead of pollen grains. Such interference of microplastics in seed can reduce seed production and biodiversity.

18.7 Bioaccumulation of Microplastic Habitats

Various insect groups ingest microplastics, which can pass from the larvae to adults through metamorphosis (Al-Jaibachi et al. 2019; Parenti et al. 2020; Simakova et al. 2022). Wild insects act as prey for vertebrate groups, i.e., fish, bats, spiders, and birds, eventually transferring microplastics across the food chain. Moreover, insect-rearing systems fed with contaminated agricultural or industrial products are used as food in animal culturing systems.

The hydrophobic nature of microplastics causes the adsorption of various metals, polybrominated biphenyls, aromatic hydrocarbons, DDT, and polychlorinated biphenyls onto the surface of microplastics, hence accumulation in food webs (Borga et al. 2022; Fremlin et al. 2020; Volschenk et al. 2019). Plasticizers, i.e., bisphenol A and phthalates, added to the plastics to obtain desired properties, adversely affect the reproduction and endocrine system of crustaceans causing acute toxicity (Burgos-Aceves et al. 2021; Dao et al. 2022; Ramakrishna et al. 2021). The physiology, behavior, and ecology of insects largely depend on hormones, and disruption of their hormonal systems by hormone analogs can adversely affect their growth and reproduction that has long-term effects via food chains and food webs.

18.8 Pest Control

The natural predators, pathogens, and parasitoids of various insect pests keep their populations below the economic threshold level. All members of Neuroptera and Odonata are insect predators, and many families of other insect orders, i.e., Diptera, Coleoptera, Hemiptera, and Hymenoptera, act as biological control agents of insect pests. Moreover, some insects can also be used to control weeds; e.g., snout beetles have been successfully used to control water hyacinths in Lake Victoria (Oliveira et al. 2019). These natural biological control agents provide important ecosystem service in the wild and agroecosystems. Interference in the normal hormonal regulation in these insects can have indirect effects on their role as biological control agents. Prolonged duration of larval instars due to hormonal imbalance in insect predators can potentially increase their predation. For example, the native ladybird instars are prone to predation by invasive ladybird (*Harmonia axyridis*), hence increasing the mortality rate of these instars (Katsanis et al. 2013; Kenis et al. 2017; Labrie et al. 2023).

18.9 Soil Formation and Composition

Ants and termites are among the most important soil invertebrates that help in the decomposition of organic matter (Anitha 2020; Menta and Remelli 2020; Walker et al. 2022). The symbiotic gut microbiota of termites enables them to contribute to the breakdown of plant litter including lignin and cellulose (Arora et al. 2022;

Bhujbal et al. 2021). Subterranean termites can decompose plastic litter as well (Amit et al. 2022).

18.10 Effect of PE Microplastics on Emergence Responses of *Chironomus* spp.

Microplastic particles of varying sizes affect the emergence of various *Chironomus* spp. Exposure to PE microplastics (10–27 μm) resulted in 17.5% emergence rate in *C. tepperi*, due to ingestion of microplastic particles (Townsend and Seen 2012; Ziajahromi et al. 2018). The microplastics of a specific polymer type have a significant effect on the emergence rate of *Chironomus* spp. Adverse effects of polyamide microplastics (10–180 μm) have been observed for *C. riparius*, where no individual was found at the higher density (10,100 microplastic particles per kg sediment) (Khosrovyan and Kahru 2020). Microplastic exposure to *C. sancticaroli* affected their emergence time irrespective of sex when the concentration of microplastics was higher and substrate was unavailable (Castro et al. 2022). Hence, the effects on Chironomids depend on the concentration, size, and type of the polymer. Toxicity can also be influenced by the availability of substrate.

18.11 Microplastic Sources in the Terrestrial Ecosystems

The plastics used in protection nets, pipes, greenhouses, sludge, manure, and compost and coated fertilizers used in crop and animal production systems are an eminent source of microplastic pollution in agricultural soils (Lwanga et al. 2022; Miles et al. 2017; Weithmann et al. 2018; Weng et al. 2013; Zhou et al. 2020; Ziajahromi et al. 2017). Soil invertebrates also contribute to plastic transportation via electrostatic attraction forces, cutaneous transport, ingestion, and accumulation in biogenic structures (Dong et al. 2020; Lwanga et al. 2022).

18.12 Insect: Microplastic Interactions in Terrestrial Ecosystem

The comparative effect of Cd and microplastic exposure on the survival of two Drosophila genotypes indicated no effects of microplastic exposure on the survival of CS10 genotype (XXY females) and 995 genotype (XX females) as compared to controls (Lu et al. 2018). However, exposure to both Cd and microplastics did not affect the survival rate, while only Cd exposure led to significant increase the mortality rate (Zhang et al. 2020).

The motor activity of Drosophila has been accessed through climbing assay (Barone and Bohmann 2013). The motor activity significantly decreased in the treatment group Cd + MP(S) and Cd + MP(L) as compared to the Cd-alone treatment group (Zhang et al. 2020). It can be concluded that smaller size (0.1 μm) in the

Cd + MP(S) treatment group is responsible for highest toxicity and reduced motor activity observed in this group.

Gut damage has been observed in the flies exposed to microplastics alone, only Cd, and combined exposure to Cd and microplastics. Increased damage was observed for the combined exposure of both microplastics and Cd (Zhang et al. 2020). Position effect variegation (PEV) has been used to translocate the white eye gene from the euchromatic region of X-chromosome to the heterochromatic region near the centromere of X-chromosome. It can be concluded that combined exposure of microplastics with heavy metals can have significant effects on the gene expression of these insects.

18.13 PE Microplastics and Sex-Specific Toxicity in Drosophila

It has been reported that fruit flies can distinguish food without microplastics and with microplastics. All the flies except 7% of males preferred the control food (no microplastics) and avoided polystyrene microplastic-treated food, irrespective of the concentration of microplastics (El Kholy and Al Naggar 2023).

The fruit flies given the sucrose solution containing higher concentration (0.05 or 0.5 μg/mL) of PS microplastics ate less food as compared to the group fed upon sucrose solution containing lower microplastic concentration (0.005 μg/mL) and control group (El Kholy and Al Naggar 2023). It can be concluded that the concentration of microplastics in food directly affects the quantity of food intake.

The survival rate of both male and female flies was significantly higher for the control group exposed to food media containing varying concentrations of PS microplastics. Males showed higher sensitivity (all dead) to PS microplastics when exposed for longer time duration (14 days), whereas 20% of females survived after 20 days of exposure (El Kholy and Al Naggar 2022).

Resistance to starvation is an indication of an organism's fitness. Female flies are more resistant to starvation and survived, whereas all males died in 24-h starvation period, regardless of exposure to any concentration of PS microplastics (El Kholy and Al Naggar 2023).

Microplastic exposure affects the defecation rate in flies as significantly lower defecation rate (fecal pellets) was observed in the female flies fed with treated food media containing PS (0.5 μg/mL) microplastics as compared to the control group and flies fed with treated media containing 0.005 μg/mL PS microplastics (El Kholy and Al Naggar 2023). Fecal pellets of both male and female fruit flies contained PS microplastics. Uncontrollable release of fecal pellets has been observed in the male and female flies exposed to higher concentrations (0.5 μg/mL) of PS microplastics (Junaid et al. 2023).

18.14 Microplastics and Insect Behavior

Microplastics affect the response of female fungus gnats (*Bradysia difformis*) searching a substrate for oviposition. It has been experimentally confirmed that *B. difformis* females avoided the lentil plants grown in the soil containing 5% HDPE and preferred HDPE-free systems for oviposition (Rondoni et al. 2021). The loss of attractiveness due to microplastics can be compensated by adequate watering and the addition of a fungal mycelium (Rondoni et al. 2021).

Honeybees have been reported to consume microplastic-free sucrose solutions faster than the microplastics containing sucrose solution possibly due to alteration of viscosity of two solutions. The foraging behavior of bees does not help them distinguish and avoid the microplastics containing food (Buteler et al. 2022).

18.15 Conclusion

The production and utilization of microplastics across the globe have resulted in their release in every ecosystem. Due to the relatively small size and widespread distribution of microplastics, their exposure to living organisms is inevitable. Insects, being the most diverse invertebrate group found in every ecosystem, have been widely studied. The exposure of insects to microplastics negatively affects the important ecosystem services provided by insects, i.e., bioremediation, pollination, seed dispersal, pest control, and soil formation, mainly due to nutrient deficiency and dysregulation of hormones. Microplastics can also incorporate into the larval cases prolonging the larval stages of insects and cause biochemical changes leading to negative effects on biochemical mechanisms. Numerous research studies have documented the impacts of microplastics of varied chemical nature on different species. Insects can be used as the bioindicators of microplastic pollution in an ecosystem. They also act as model organisms to study the impacts of microplastics on the major physiological and biochemical processes of organisms.

References

Al-Jaibachi R, Cuthbert RN, Callaghan A (2018) Up and away: ontogenic transference as a pathway for aerial dispersal of microplastics. Biol Lett 14(9):20180479

Al-Jaibachi R, Cuthbert RN, Callaghan A (2019) Examining effects of ontogenic microplastic transference on Culex mosquito mortality and adult weight. Sci Total Environ 651:871–876

Amit K, Kalleshwaraswamy C, Sharma R, Sharma P, Poonia A (2022) Biodegradation of plastic using termites and their gut microbiota: a mini review. IOP Conf Ser Earth Environ Sci 1057: 012016

Anitha K (2020) Soil macro fauna: a retrospection with reference to soil formation and soil health. J Pharmacogn Phytochem 9(4S):592–594

Antelava A, Damilos S, Hafeez S, Manos G, Al-Salem SM, Sharma BK, Kohli K, Constantinou A (2019) Plastic solid waste (PSW) in the context of life cycle assessment (LCA) and sustainable management. Environ Manag 64:230–244

Arora J, Kinjo Y, Šobotník J, Buček A, Clitheroe C, Stiblik P, Roisin Y, Žifčáková L, Park YC, Kim KY (2022) The functional evolution of termite gut microbiota. Microbiome 10(1):78

Athey SN, Albotra SD, Gordon CA, Monteleone B, Seaton P, Andrady AL, Taylor AR, Brander SM (2020) Trophic transfer of microplastics in an estuarine food chain and the effects of a sorbed legacy pollutant. Limnol Oceanogr Lett 5(1):154–162

Bai Z, Wang N, Wang M (2021) Effects of microplastics on marine copepods. Ecotoxicol Environ Saf 217:112243

Barboza LGA, Vethaak AD, Lavorante BR, Lundebye A-K, Guilhermino L (2018) Marine microplastic debris: an emerging issue for food security, food safety and human health. Mar Pollut Bull 133:336–348

Barone MC, Bohmann D (2013) Assessing neurodegenerative phenotypes in Drosophila dopaminergic neurons by climbing assays and whole brain immunostaining. J Vis Exp (74):e50339

Basset Y, Jorge LR, Butterill PT, Lamarre GP, Dahl C, Ctvrtecka R, Gripenberg S, Lewis OT, Barrios H, Brown JW (2021) Host specificity and interaction networks of insects feeding on seeds and fruits in tropical rainforests. Oikos 130(9):1462–1476

Bessa F, Barría P, Neto JM, Frias JP, Otero V, Sobral P, Marques J (2018) Occurrence of microplastics in commercial fish from a natural estuarine environment. Mar Pollut Bull 128: 575–584

Bhujbal SK, Kumar M, Vijay VK, Kumar V, Ghosh P (2021) Potential of termite gut microbiota for biomethanation of lignocellulosic wastes: current status and future perspectives. Rev Environ Sci Biotechnol 20(2):419–438

Borga K, McKinney MA, Routti H, Fernie KJ, Giebichenstein J, Hallanger I, Muir DC (2022) The influence of global climate change on accumulation and toxicity of persistent organic pollutants and chemicals of emerging concern in Arctic food webs. Environ Sci: Processes Impacts 24(10): 1544–1576

Buks F, Van Schaik NL, Kaupenjohann M (2020) What do we know about how the terrestrial multicellular soil fauna reacts to microplastic? Soil Discussions 2020:1–44

Burgos-Aceves MA, Abo-Al-Ela HG, Faggio C (2021) Impact of phthalates and bisphenols plasticizers on haemocyte immune function of aquatic invertebrates: a review on physiological, biochemical, and genomic aspects. J Hazard Mater 419:126426

Buteler M, Alma AM, Stadler T, Gingold AC, Manattini MC, Lozada M (2022) Acute toxicity of microplastic fibers to honeybees and effects on foraging behavior. Sci Total Environ 822: 153320

Carvallo GO, Muñoz-Michea V (2023) Microplastics clog reproduction in a monkeyflower species. bioRxiv 2023.2005.2021.541634

Castro GB, Bernegossi AC, Felipe MC, Ogura AP, de Lima e Silva MR, Corbi JJ (2022) Polyethylene microplastics and substrate availability can affect emergence responses of the freshwater insect Chironomus sancticaroli. Ecotoxicology 31(4):679–688

Cho Y, Shim WJ, Jang M, Han GM, Hong SH (2021) Nationwide monitoring of microplastics in bivalves from the coastal environment of Korea. Environ Pollut 270:116175

Cole M, Lindeque P, Halsband C, Galloway TS (2011) Microplastics as contaminants in the marine environment: a review. Mar Pollut Bull 62(12):2588–2597

Dao T-S, Nguyen V-T, Baduel C, Bui M-H, Tran VT, Pham T-L, Bui B-T, Dinh KV (2022) Toxicity of di-2-ethylhexyl phthalate and tris (2-butoxyethyl) phosphate to a tropical microcrustacean (Ceriodaphnia cornuta) is higher in Mekong River water than in standard laboratory medium. Environ Sci Pollut Res 29(26):39777–39789

de Souza Machado AA, Horton AA, Davis T, MaaB S (2020) Microplastics and their effects on soil function as a life-supporting system. In: Microplastics in terrestrial environments: emerging contaminants and major challenges, pp 199–222

Dijkstra K-DB, Monaghan MT, Pauls SU (2014) Freshwater biodiversity and aquatic insect diversification. Annu Rev Entomol 59:143–163

Dong Y, Gao M, Song Z, Qiu W (2020) As (III) adsorption onto different-sized polystyrene microplastic particles and its mechanism. Chemosphere 239:124792

Eerkes-Medrano D, Thompson R (2018) Occurrence, fate, and effect of microplastics in freshwater systems. In: Microplastic contamination in aquatic environments. Elsevier, pp 95–132
Ehlers SM, Manz W, Koop JH (2019) Microplastics of different characteristics are incorporated into the larval cases of the freshwater caddisfly Lepidostoma basale. Aquat Biol 28:67–77
El Kholy S, Al Naggar Y (2022) Polystyrene microplastic beads caused cellular alterations in midgut cells and sex-specific toxic effects on survival, starvation resistance, and excretion of the model insect Drosophila melanogaster
El Kholy S, Al Naggar Y (2023) Exposure to polystyrene microplastic beads causes sex-specific toxic effects in the model insect Drosophila melanogaster. Sci Rep 13(1):204
Finnegan AMD, Gouramanis C (2021) Projected plastic waste loss scenarios between 2000 and 2030 into the largest freshwater-lake system in Southeast Asia. Sci Rep 11(1):1–12
Forero-López AD, Rimondino GN, Truchet DM, Colombo C, Buzzi NS, Malanca FE, Spetter CV, Fernández-Severini MD (2021) Occurrence, distribution, and characterization of suspended microplastics in a highly impacted estuarine wetland in Argentina. Sci Total Environ 785: 147141
Forero-López A, Brugnoni L, Abasto B, Rimondino G, Lassalle V, Ardusso M, Nazzarro M, Martinez A, Spetter C, Biancalana F (2022) Plastisphere on microplastics: in situ assays in an estuarine environment. J Hazard Mater 440:129737
Fremlin KM, Elliott JE, Green DJ, Drouillard KG, Harner T, Eng A, Gobas FA (2020) Trophic magnification of legacy persistent organic pollutants in an urban terrestrial food web. Sci Total Environ 714:136746
Geyer R, Jambeck JR, Law KL (2017) Production, use, and fate of all plastics ever made. Sci Adv 3(7):e1700782
Gola D, Tyagi PK, Arya A, Chauhan N, Agarwal M, Singh S, Gola S (2021) The impact of microplastics on marine environment: a review. Environ Nanotechnol Monit Manag 16:100552
Gupta D, Choudhary D, Vishwakarma A, Mudgal M, Srivastava A, Singh A (2022) Microplastics in freshwater environment: occurrence, analysis, impact, control measures and challenges. Int J Environ Sci Technol 1–32
Hopewell J, Dvorak R, Kosior E (2009) Plastics recycling: challenges and opportunities. Philos Trans R Soc B Biol Sci 364(1526):2115–2126
Idris NH, Shafie NS, Snawi MMM, Rozali MFA, Bahri AS, Shukor AM (2020) Macroinvertebrates checklist in river intakes of Pergau Lake, Jeli, Kelantan. IOP Conf Ser Earth Environ Sci 549: 012048
Jambeck JR, Geyer R, Wilcox C, Siegler TR, Perryman M, Andrady A, Narayan R, Law KL (2015) Plastic waste inputs from land into the ocean. Science 347(6223):768–771
Junaid M, Liu S, Chen G, Liao H, Wang J (2023) Transgenerational impacts of micro (nano) plastics in the aquatic and terrestrial environment. J Hazard Mater 443:130274
Katsanis A, Babendreier D, Nentwig W, Kenis M (2013) Intraguild predation between the invasive ladybird Harmonia axyridis and non-target European coccinellid species. BioControl 58(1): 73–83
Kenis M, Adriaens T, Brown PM, Katsanis A, Martin GS, Branquart E, Maes D, Eschen R, Zindel R, Van Vlaenderen J (2017) Assessing the ecological risk posed by a recently established invasive alien predator: Harmonia axyridis as a case study. BioControl 62:341–354
Khosrovyan A, Kahru A (2020) Evaluation of the hazard of irregularly-shaped co-polyamide microplastics on the freshwater non-biting midge Chironomus riparius through its life cycle. Chemosphere 244:125487
Kooi M, Besseling E, Kroeze C, Van Wezel AP, Koelmans AA (2018) Modeling the fate and transport of plastic debris in freshwaters: review and guidance. In: Freshwater microplastics: emerging environmental contaminants? pp 125–152
Kumar G, Paul DK (2021) Study of seed dispersal by the Indian sloth bear (Melursus ursinus) in Nawada Forest division, Bihar (India). Indian J Ecol 48(1):98–101
Labrie G, Meseguer R, Lucas E (2023) Stage-specific vulnerability of Harmonia axyridis (Coleoptera: Coccinellidae) to intraguild predation. Eur J Entomol 120:70

Lebreton LCM, van der Zwet J, Damsteeg J-W, Slat B, Andrady A, Reisser J (2017) River plastic emissions to the world's oceans. Nat Commun 8(1):15611
Lin VS (2016) Research highlights: impacts of microplastics on plankton. Environ Sci Processes Impacts 18(2):160–163
Lu K, Qiao R, An H, Zhang Y (2018) Influence of microplastics on the accumulation and chronic toxic effects of cadmium in zebrafish (Danio rerio). Chemosphere 202:514–520
Lusher A, Welden N, Sobral P, Cole M (2017) Sampling, isolating and identifying microplastics ingested by fish and invertebrates. Anal Methods 9(9):1346–1360
Lwanga EH, Beriot N, Corradini F, Silva V, Yang X, Baartman J, Rezaei M, van Schaik L, Riksen M, Geissen V (2022) Review of microplastic sources, transport pathways and correlations with other soil stressors: a journey from agricultural sites into the environment. Chem Biol Technol Agric 9(1):1–20
Malafaia G, da Luz TM, Guimarães ATB, da Costa Araújo AP (2020) Polyethylene microplastics are ingested and induce biochemical changes in Culex quinquefasciatus (Diptera: Culicidae) freshwater insect larvae. Ecotoxicol Environ Contam 15(1):79–89
Menta C, Remelli S (2020) Soil health and arthropods: from complex system to worthwhile investigation. Insects 11(1):54
Miles C, DeVetter L, Ghimire S, Hayes DG (2017) Suitability of biodegradable plastic mulches for organic and sustainable agricultural production systems. Hortscience 52(1):10–15
Minard G, Mavingui P, Moro CV (2013) Diversity and function of bacterial microbiota in the mosquito holobiont. Parasit Vectors 6(1):1–12
Neher DA, Barbercheck ME (2019) Soil microarthropods and soil health: intersection of decomposition and pest suppression in agroecosystems. Insects 10(12):414
Oliveira M, Ameixa OM, Soares AM (2019) Are ecosystem services provided by insects "bugged" by micro (nano) plastics? TrAC Trends Anal Chem 113:317–320
Pan Z, Liu Q, Jiang R, Li W, Sun X, Lin H, Jiang S, Huang H (2021) Microplastic pollution and ecological risk assessment in an estuarine environment: the Dongshan Bay of China. Chemosphere 262:127876
Parenti C, Binelli A, Caccia S, Della Torre C, Magni S, Pirovano G, Casartelli M (2020) Ingestion and effects of polystyrene nanoparticles in the silkworm Bombyx mori. Chemosphere 257: 127203
Pauna V, Buonocore E, Renzi M, Russo G, Franzese P (2019) The issue of microplastics in marine ecosystems: a bibliometric network analysis. Mar Pollut Bull 149:110612
Penghui L, Wang X, Su M, Zou X, Duan L, Zhang H (2021) Characteristics of plastic pollution in the environment: a review. Bull Environ Contam Toxicol 107:577–584
Pinheiro C, Oliveira U, Vieira M (2017) Occurrence and impacts of microplastics in freshwater fish. J Aquac Mar Biol 5(6):00138
Piringer OG, Baner AL (2008) Plastic packaging: interactions with food and pharmaceuticals. Wiley, Hoboken
Qu S, Guo Y, Ma Z, Chen W-Q, Liu J, Liu G, Wang Y, Xu M (2019) Implications of China's foreign waste ban on the global circular economy. Resour Conserv Recycl 144:252–255
Ramakrishna M, Girigoswami A, Chakraborty S, Girigoswami K (2021) Bisphenol A—An overview on its effect on health and environment. Biointerface Res Appl Chem 12:105–119
Rondoni G, Chierici E, Agnelli A, Conti E (2021) Microplastics alter behavioural responses of an insect herbivore to a plant-soil system. Sci Total Environ 787:147716
Shah S, Ilyas M, Li R, Yang J, Yang F-L (2023) Microplastics and nanoplastics effects on plant–pollinator interaction and pollination biology. Environ Sci Technol 57:6415
Simakova A, Varenitsina A, Babkina I, Andreeva Y, Bagirov R, Yartsev V, Frank Y (2022) Ontogenetic transfer of microplastics in bloodsucking mosquitoes Aedes aegypti L. (Diptera: Culicidae) is a potential pathway for particle distribution in the environment. Water 14(12):1852
Sruthy S, Ramasamy E (2017) Microplastic pollution in Vembanad Lake, Kerala, India: the first report of microplastics in lake and estuarine sediments in India. Environ Pollut 222:315–322

Straub S, Hirsch PE, Burkhardt-Holm P (2017) Biodegradable and petroleum-based microplastics do not differ in their ingestion and excretion but in their biological effects in a freshwater invertebrate Gammarus fossarum. Int J Environ Res Public Health 14(7):774
Tanaka K, Takada H (2016) Microplastic fragments and microbeads in digestive tracts of planktivorous fish from urban coastal waters. Sci Rep 6(1):34351
Townsend AT, Seen AJ (2012) Historical lead isotope record of a sediment core from the Derwent River (Tasmania, Australia): a multiple source environment. Sci Total Environ 424:153–161
Trestrail C, Walpitagama M, Hedges C, Truskewycz A, Miranda A, Wlodkowic D, Shimeta J, Nugegoda D (2020) Foaming at the mouth: ingestion of floral foam microplastics by aquatic animals. Sci Total Environ 705:135826
Troost TA, Desclaux T, Leslie HA, van Der Meulen MD, Vethaak AD (2018) Do microplastics affect marine ecosystem productivity? Mar Pollut Bull 135:17–29
Vaccari F, Forestieri B, Papa G, Bandini F, Huerta-Lwanga E, Boughattas I, Missawi O, Banni M, Negri I, Cocconcelli PS (2022) Effects of micro and nanoplastics on soil fauna gut microbiome: an emerging ecological risk for soil health. Curr Opin Environ Sci Health 30:100402
Van Cauwenberghe L, Devriese L, Galgani F, Robbens J, Janssen CR (2015) Microplastics in sediments: a review of techniques, occurrence and effects. Mar Environ Res 111:5–17
Volschenk C, Gerber R, Mkhonto M, Ikenaka Y, Yohannes Y, Nakayama S, Ishizuka M, Van Vuren J, Wepener V, Smit N (2019) Bioaccumulation of persistent organic pollutants and their trophic transfer through the food web: human health risks to the rural communities reliant on fish from South Africa's largest floodplain. Sci Total Environ 685:1116–1126
Walker AE, Robertson MP, Eggleton P, Bunney K, Lamb C, Fisher AM, Parr CL (2022) Indirect control of decomposition by an invertebrate predator. Funct Ecol 36:2943
Watts AJ, Urbina MA, Corr S, Lewis C, Galloway TS (2015) Ingestion of plastic microfibers by the crab Carcinus maenas and its effect on food consumption and energy balance. Environ Sci Technol 49(24):14597–14604
Weithmann N, Möller JN, Löder MG, Piehl S, Laforsch C, Freitag R (2018) Organic fertilizer as a vehicle for the entry of microplastic into the environment. Sci Adv 4(4):eaap8060
Weng Y-X, Jin Y-J, Meng Q-Y, Wang L, Zhang M, Wang Y-Z (2013) Biodegradation behavior of poly (butylene adipate-co-terephthalate)(PBAT), poly (lactic acid)(PLA), and their blend under soil conditions. Polym Test 32(5):918–926
Wong S, Ngadi N, Abdullah TAT, Inuwa IM (2015) Current state and future prospects of plastic waste as source of fuel: a review. Renew Sust Energ Rev 50:1167–1180
Wright SL, Rowe D, Thompson RC, Galloway TS (2013) Microplastic ingestion decreases energy reserves in marine worms. Curr Biol 23(23):R1031–R1033
Wu C, Xiong X, Hamidian AH, Zhang Y, Xu X (2022) A review on source, occurrence, and impacts of microplastics in freshwater aquaculture systems in China. Water Biol Secur 1: 100040
Xiang Q, Zhu D, Chen Q-L, O'Connor P, Yang X-R, Qiao M, Zhu Y-G (2019) Adsorbed sulfamethoxazole exacerbates the effects of polystyrene (~ 2 μm) on gut microbiota and the antibiotic resistome of a soil collembolan. Environ Sci Technol 53(21):12823–12834
Yin L, Chen B, Xia B, Shi X, Qu K (2018) Polystyrene microplastics alter the behavior, energy reserve and nutritional composition of marine jacopever (Sebastes schlegelii). J Hazard Mater 360:97–105
Zain NM, Fauzi N, Subki NS, Ghazali ZZ (2022) Occurrence of microplastics in immature aquatic insects of Gua Musang tributaries in Kelantan. IOP Conf Ser Earth Environ Sci 1102:012047
Zhang Y, Wolosker MB, Zhao Y, Ren H, Lemos B (2020) Exposure to microplastics cause gut damage, locomotor dysfunction, epigenetic silencing, and aggravate cadmium (Cd) toxicity in Drosophila. Sci Total Environ 744:140979
Zhou B, Wang J, Zhang H, Shi H, Fei Y, Huang S, Tong Y, Wen D, Luo Y, Barceló D (2020) Microplastics in agricultural soils on the coastal plain of Hangzhou Bay, East China: multiple sources other than plastic mulching film. J Hazard Mater 388:121814

Zhu D, Chen Q-L, An X-L, Yang X-R, Christie P, Ke X, Wu L-H, Zhu Y-G (2018) Exposure of soil collembolans to microplastics perturbs their gut microbiota and alters their isotopic composition. Soil Biol Biochem 116:302–310

Ziajahromi S, Neale PA, Rintoul L, Leusch FD (2017) Wastewater treatment plants as a pathway for microplastics: development of a new approach to sample wastewater-based microplastics. Water Res 112:93–99

Ziajahromi S, Kumar A, Neale PA, Leusch FD (2018) Environmentally relevant concentrations of polyethylene microplastics negatively impact the survival, growth and emergence of sediment-dwelling invertebrates. Environ Pollut 236:425–431

Microplastic as a Pathogenic Vector to Cause Diseases in Marine Biota

19

Abel Inobeme, Mohd. Shahnawaz, Charles Oluwaseun Adetunji, John Tsado Mathew, Babatunde Oluwafemi Adetuyi, Oluwabukola Atinuke Popoola, Falana Yetunde Olaitan, Olalekan Akinbo, Olatunji Matthew Kolawole, Oluwafemi Adebayo Oyewole, Eniola K.I.T., and Mohammed Bello Yerima

Abstract

In addition to being an eyesore, microplastic pollution has serious ecological consequences for aquatic life, terrestrial life, and human beings. According to studies, microplastics can operate as vectors for pollutants, soaking up and storing them in marine habitats, species, and ultimately food webs. Given that people will

A. Inobeme
Department of Biochemistry, Covenant University, Ota, Ogun State, Nigeria

M. Shahnawaz
Department of Botany, University of Ladakh, Kargil Campus, Khumbathang, Kargil, Ladakh UT, India

Present Address: Department of Botany, Shri Pandit Prem Nath Dogra Government Degree College Samba, Samba, Jammu and Kashmir UT, India

C. O. Adetunji (✉)
Applied Microbiology, Biotechnology and Nanotechnology Laboratory, Department of Microbiology, Edo State University Uzairue, Auchi, Edo State, Nigeria
e-mail: adetunji.charles@edouniversity.edu.ng

J. T. Mathew · B. O. Adetuyi
Department of Natural Sciences, Biochemistry Unit, Faculty of Pure and Applied Sciences, Precious Cornerstone University, Ibadan, Nigeria

O. A. Popoola · F. Y. Olaitan
Plant Improvement Unit, Agricultural Biotechnology Department, National Biotechnology Development Agency, Abuja, Nigeria

O. Akinbo
Centre of Excellence in Science, Technology, and Innovation, AUDA-NEPAD, Pretoria, South Africa

M. Shahnawaz et al. (eds.), *Microplastic Pollution*,
https://doi.org/10.1007/978-981-99-8357-5_19

inevitably be exposed to microplastics, it is important to take stock of what we know about their possible consequences, exposure pathways, and toxicity to human health. Therefore, the purpose of this review was to shed light on the human health risks associated with microplastic's pollutant sorption and bioaccumulation, as well as to identify the most prevalent pollutant classes absorbed by microplastic and the classes of pollutants that are bioaccumulated by microplastic in marine ecosystem organisms. Potential factors affecting microplastic's sorption and bioaccumulation of contaminants in marine habitats were also discussed. The review also shed light on the most common forms of microplastic, the overall amount of microplastic, and its geographic distribution throughout the world's waterways. In the literature review, it was found that microplastic sorption of contaminants was influenced by microplastic properties, chemical interactions, and water parameters. There is still no proof that microplastics pose a direct medical harm to people, despite several studies documenting the risks of microplastic-associated subunits, additives, and contaminants.

Keywords

Adsorption · Bioaccumulation · Contaminant · Microplastic · Pollutant · Vector

Abbreviations

4MBC	4-Methylbenzylidene camphor
AMX	Amoxicillin
BaP	Benzo [a]pyrene
CBZ	Carbamazepine
CIP	Ciprofloxacin hydrochloride
EDC	Endocrine-disrupting chemical
EE2	17α-Ethinylestradiol
EVA	Ethylene-vinyl acetate
FTIR	Fourier transform infrared spectroscopy
HBCD	Hexabromocyclododecane
HCH	Hexachlorocyclohexane

O. M. Kolawole
Department of Microbiology, Faculty of Life Sciences, University of Ilorin, Ilorin, Nigeria

O. A. Oyewole
Department of Microbiology, Federal University of Technology, Minna, Nigeria

E. K.I.T.
Department of Biological Sciences, Joseph Ayo Babalola University, Ikeji Arakeji, Osun State, Nigeria

M. B. Yerima
Department of Microbiology, Sokoto State University, Sokoto, Sokoto State, Nigeria

HDPE	High-density polyethylene
HeCB	Hexachlorobenzene
LDPE	Low-density polyethylene
NP	Nonylphenol
PA	Polyamide
PAH	Polycyclic aromatic hydrocarbon
PB	Polybutylene
PBAT	Polybutylene adipate terephthalate
PBDE	Polybrominated diphenyl ether
PCB	Polychlorinated biphenyl
PE	Polyethylene
PeCB	Pentachlorobenzene
PES	Polyethersulfone
PET	Polyethylene terephthalate
PFAA	Perfluoroalkyl acids
PFOS	Perfluorooctanesulfonate
PFOSA	Perfluorooctanesulfonamide
PMMA	Poly (methyl methacrylate)
POP	Persistent organic pollutant
PP	Polypropylene
PPCP	Pharmaceuticals personal care product
PS	Polystyrene
PU	Polyurethane
PVA	Polyvinyl acetate
PVC	Polyvinyl chloride
SBR	Styrene butadiene rubber
SDZ	Sulfadiazine
TC	Tetracycline
TCS	Triclosan
TMP	Trimethoprim
TYL	Tylosin
USEPA	US Environmental Protection Agency

19.1 Introduction

The plastics industry sprouted in the 1920s, but it really took off in the 1940s. The amount of plastic produced worldwide in 2014 was 20 times what it was in 1964 (Neufeld et al. 2016). In 2016 and 2018, the world's yearly production was 330 and 360 million metric tons, respectively (PlasticsEurope 2017, 2019). Despite growing concern about plastic pollution and a number of initiatives intended to reduce that pollution, annual plastic output continues to rise. In 2010, anywhere from 4.8 to 12.7 million metric tons of land-based plastic garbage from 192 coastal countries entered the ocean (Jambeck et al. 2015). In addition, harmful secondary microplastic is produced as marine plastic debris degrades and breaks down. From 60 million to

99 million metric tons of microplastic trash was produced in 2015 (Lebreton and Andrady 2019; Ryberg et al. 2019). The literature on microplastic pollution focuses primarily on three aspects: the detection of microplastics in food, the accumulation of microplastics in ecosystems at concentrations that are statistically significant, and the potential hazards that microplastics pose to human health and the natural environment (Akarsu et al. 2020). Most man-made compounds will inevitably be present in the environment, but their overabundance poses risks and necessitates careful consideration when it appears as pollution. Awareness of microplastic pollution's origin, path, degradation, sink, and effects is crucial for tackling the issue. Microplastics can be found anywhere, including the ocean, lakes, and rivers (Yu et al. 2019a). Water purification facilities, garbage dumps, farms, factories, and homes all contribute to microplastic pollution (Conley et al. 2019; Belzagui et al. 2019). In turn, toxins such as polycyclic aromatic hydrocarbons (PAH), heavy metals, and antibiotics, as well as microplastic, build up in the ocean, the polar regions, the ice sheet, and the food chain (Yu et al. 2018). Living things are adversely affected by microplastic trash, which is further broken down into nanoplastic and consumed in large quantities. If we are going to make headway in the fight against microplastic pollution, we need to learn as much as we can about its origins, pathways, degradation, and potential countermeasures (Li et al. 2020). Garbage patches like the ones found in the western North Atlantic Ocean and the Caribbean Sea are only a few examples of the huge areas of surface convergence induced by ocean dynamics, which have led to the natural accumulation of up to 580,000 plastic bits per square kilometer (Maximenko et al. 2012). The food supply for people and other aquatic animals is reduced as a result of these effects (Wong et al. 2020). Also, microorganisms, including diseases, were found to be linked to microplastic (Brandon et al. 2018). *Bacillus* spp., *Paenibacillus* spp., Actinobacteria, and Firmicutes were all identified in microplastic (Park and Kim 2019). Some microplastic-associated microorganisms can break down microplastic, despite its ability to remain in the environment and resist destruction (Yuan et al. 2020). Depending on the bacteria and enzymes they contain, bacterial consortiums with varying microplastic breakdown abilities exist (Tsiota et al. 2018). Microplastics have been found to be colonized by bacteria from the Rhodobacteraceae, Hyphomonadaceae, and Sphingomonadaceae families (Oberbeckmann et al. 2018). The biodegradation of marine microplastic that has been exposed to weathering and external physicochemical conditions can be aided by the employment of microplastic-degrading bacteria (Oberbeckmann and Labrenz 2020). Though it is challenging to biodegrade, microplastic provides a favorable environment for microbial colonization and growth (Rujnic-Sokele and Pilipovic 2017). Many types of microplastics have been found to harbor harmful bacteria and fungi, such as Aspergillus flavus, due to their porous and irregular surfaces (Auta et al. 2018). It has also been reported that worms can break down plastic (Yang et al. 2015b). Despite the fact that microplastic can biodegrade over time, older main or secondary microplastic is continually having its structural integrity compromised or fragmented in the environment, resulting in nanoplastic, which is even more bioavailable to organisms. In addition, microplastic is a vector of hazardous pollutants such as POP and heavy metals, bringing toxins into the ecosystem via the food chain

(Zhang et al. 2020). As a result of sorption and bioaccumulation, microplastic can boost the bioavailability of contaminants to ecosystems and organisms (Guzzetti et al. 2018). Controlling the production of microplastics, cleaning it up, getting communities involved, working on a regional level, and doing research are all potential answers to this problem (Ma et al. 2020b). A prime example is the European Union's ban on main microplastics and single-use plastics, as well as its promotion of the controlled release of marine trash. Individual initiatives, such as recycling and consumption control, have been advocated by the United Nations to reduce microplastic pollution (Barceló and Picó 2020). Research into the use of water purification systems and the microbial destruction or biosynthesis of plastic-like material are two further avenues being explored in the fight against microplastic pollution (Zhang and Chen 2020). Microplastic contamination can be reduced, but only if people are made aware of its dangers. In recent years, researchers' focus has switched from determining the causes of microplastic to studying its direct and indirect effects because of the massive amount of published data on the topic of microplastic's global prevalence and distribution. Therefore, this study primarily focused on the health concerns of microplastic to people and other marine ecosystem organisms, the amount and distribution of microplastic, and the various factors that influenced pollutant sorption or bioaccumulation by microplastic.

19.2 Types, Abundance, and Distribution of Microplastic

It is possible to classify microplastics according to their size, shape, density, and composition when they were first produced. Microplastics are any plastic particles that are typically smaller than 5 millimeters in size. Chemical, mechanical, and biological deterioration all contribute to the fracturing of bigger plastics in the environment (Ivleva et al. 2017). Primary microplastic refers to microplastic made into micro-sized particles, such microbeads, and secondary microplastic refers to microplastic made through the breakdown of bigger plastic components. Microplastic used in cosmetics, medicine delivery vehicles, and other engineering and industrial settings is called "primary microplastic" (Auta et al. 2017). Secondary microplastic, in comparison, starts out as a relatively large piece of plastic during production but is broken down and divided into smaller bits due to complicated weathering interactions (Rocha-Santos and Duarte 2015).

19.2.1 Fibers

Numerous studies have found that human activities such wastewater outflow, industrial discharge, mariculture, and settlement have had a significant impact on microplastic abundance heterogeneity. Studies conducted in the estuary of the Yangtze and along the coast of the East China Sea found that fiber was the most commonly collected morphotype (Zhao et al. 2014). Microplastic is thought to be introduced to the estuary and coastal areas by the influx from the river to the sea. Human activities, both on land and in the ocean, are the primary cause of microplastic proliferation. Microplastic pollution in China's Xiangshan Bay is mostly attributable to the widespread use of mariculture. More than half of the fiber-type microplastic discovered by Chen et al. (2018) was traced back to mariculture. Plankton net tows were used to gather seawater samples from nine different locations. Polyethylene (PE), polypropylene (PP), and polystyrene (PS) microplastic were found throughout the study period. Similar results, with PET, PP, and PE polymers accounting for the bulk of the fiber at another mariculture location in China's Maowei Sea, were found (Zhu et al. 2019). There was already a high concentration of microplastic in the Maowei Sea, but the neighboring riverine influx greatly increased that concentration. Dai et al. (2018) conducted an in-depth analysis of the causes and effects of microplastic contamination in the Bohai Sea as a result of human activity. Overall, the researchers found that 75% of the microplastics they collected belonged to the fiber morphotype. Higher concentrations of fibers were found in the middle and lower parts of the water column. It was speculated that a stainless steel bucket was superior to other sample collection methods, such as mantra nets or neustonic trawls, since it more easily trapped fiber. Zhu et al. (2018) found that film and fiber morphotypes predominated in the surface water of the northern Yellow Sea. The average number of microplastics per cubic meter of water was 545,282. Microplastic analysis of surface water revealed that PE was the most common plastic component. It was hypothesized that floating microplastic entered the North Yellow Sea via the Chinese coastal current and the Korean coastal current. The majority of the plastic contamination in this region originated from human waste and fishing operations. The Jiaozhou Bay, China, confined environment has been studied extensively for its high microplastic quantity and distribution because of its proximity to human activity. Generally speaking, the microplastic in these samples has the morphotype of a black or blue fiber. Most of the polymers studied here belonged to PA, PVA, PP, PE, or PET families. Microplastic flux into the bay was found to be significantly increased by the riverine intake at the estuary of the Haibo. There were additional elements, such as residual current and sediment mobility, that had a major effect on the distribution and content of microplastics (Zheng et al. 2019). The western part of Asia was also found to be polluted with microplastics. The average concentration of microplastic in marine waters in Qatar was 0.71 particle/m^3, for instance. The research found that fiber and granule morphotypes predominated (Castillo et al. 2016). It was hypothesized that oil rig construction and shipping operations were to blame for the pollution. Aliabad et al. (2019) studied the microplastic abundance in surface water from the seven selected stations in Chabahar Bay, Oman, in the Arabic region. Microplastics were mostly collected in white, blue, and red. Fiber, a morphology of secondary microplastics, was found to dominate the region's low microplastic abundance. There was conclusive evidence that PE and PP constituted over 60% of all samples analyzed. Kazour et al. (2019) published the first study on microplastic abundance in the Mediterranean Sea region; their research was conducted along the coast of Lebanon in the so-called Levantine Basin. This region was identified as having a significant concentration of microplastic pollution. Fibers accounted for over half of the microplastics found in the water samples, with polyethylene terephthalate (PE) being the most common type. Microplastic pollution in Mersin Bay, Turkey, was explained by Akarsu et al. (2020). Sewage treatment plants were observed to have a high discharge rate in the study area. That is why we anticipated that fibers would be a major source of microplastic. Microplastic richness in the water column was also found to vary greatly between the American and European continents. Microplastics have been documented by researchers in San Francisco Bay, California, USA (Sutton et al. 2016). The wastewater discharge sample predominantly contained fibers. In addition, water and sediment samples from Tampa Bay, Florida, revealed an average microplastic abundance (4.5 particles/m^3). Of the microplastics found in the water samples, 88% were fibers, followed by fragments and flakes (McEachern et al. 2019). It was thought that wastewater treatment discharge was the entry point for microplastic into the aquatic ecosystem. Microplastic concentrations were highest close to the manufacturing hub. The Baltic Sea, Gulf of Bothnia, and Skagerrak/Kattegat were among the Swedish marine areas studied for their microplastic richness. Microplastic was extracted from the ocean using either a manta trawl or in situ pump filtration. Comparatively, the microplastic content was found to be 0.10 particles/m^3 when using the pump filtering method, but only 0.04 particles/m^3 when using the manta trawling approach. For PE and PP water samples, fiber was the most common morphotype (Schönlau et al. 2020). Only fibers longer than 1 mm were considered in this analysis; those with a transparency grade below 10 were left out. As black and blue are common colors for ropes and fishing gear in the examined area, it is probable that this is where the synthetic fibers came from in all the samples. Marine waters off the coast of South Africa were also found to contain microplastic. The majority of microplastics are of the fiber morphotype, and Nel and Froneman (2015) observed a concentration of 257 to 1215 particles/m^3 of microplastics in the water column. Sewage effluent from land was thought to be the primary cause. Generally speaking, the microplastic found to be polluting the environment was of the fiber or fragment morphotype, was made of PE or PP, and was a product of anthropogenic, urban, fisheries, or marine activity. The primary hues were various shades of blue and black. The fibers could be either tangled or untangled, straight, or bent. Suva, the capital of the Republic of Fiji, has some of the cleanest water, sediment, and fish samples in all of Oceania (Ferreira et al. 2020). Planktonic circular tows were used to gather microplastic from water samples. The amount of microplastic in water samples from Laucala Bay, Suva Harbour, and Navakavu, Vanua Levu, was less than 0.24 items/m^3. The majority of MPs were determined to be fibers (60.2% 6.9%) and fragments (26.9% 3.5%), with the smallest percentages being films (9.4% 2.0%) and microbeads (3.5% 1.9%).

19.2.2 Fragments

The manta net was pulled horizontally at the surface of the Bohai Sea for 15 min at speeds ranging from 1.5 to 2.0 knots for each transect, and the results showed that low-density polymers such as PP, PE, and PS were present in the waters of all

11 sites. In this study, fragments made up 46% of the total plastic collected and were the most common morphotype. Evidence suggests that fishing and marine farming in the Bohai Sea are major contributors to microplastic pollution in the region (Zhang et al. 2017). The initial Yellow Sea observations showed a relatively modest amount of microplastic, with fragment morphotype dominating by a factor of 42%. Based on results from micro-Fourier transform infrared spectroscopy (micro-FTIR), the two most abundant polymers in microplastics were found to be polypropylene (PP) and polyethylene (PE), making up 55.93% and 32.20%, respectively. Microplastics were found in high concentrations near coastal cities, suggesting that the Yellow Sea coastal current played a crucial role in the dispersal of these particles (Sun et al. 2018). Microplastic was identified in a variety of forms in the maritime waters of Malaysia. Khalik et al. (2018) provided an explanation for the discrepancy in microplastic abundance between two geographical regions with contrasting human activities. One area is a port in Kuantan, Malaysia, while the other is a rural community in Kuala Nerus, Terengganu. Kuantan Port is one of the largest multicargo ports in Malaysia, while Kuala Nerus Port is specialized in commercial fishing and tourism. The microplastic abundance in Kuala Nerus was higher than in Kuantan port (0.13–0.69 pieces/L vs. 0.14–0.15 pieces/L). It was hypothesized that a contained system, like a port, preventing the free flow of plastic pollution from the open sea was responsible for the widespread presence of microplastics. PET, PS, PA, PVC, PP, and PE are the six polymers that were found in both environments. There is evidence of microplastic in Faafu Atoll, Maldives, thanks to the work of Saliu et al. (2018). Neuston nets were used to collect surface seawater from 12 different places. This region benefited from a low population density and a flood of visitors brought on by its popular tourist attractions. However, there were just 0.32 0.15 particles of microplastic per cubic meter of water on Faafu Atoll. Up to 52% of all particles in the studies were determined to be microplastics, with a large proportion of them being of the fragment morphotype. Savoca et al. (2019) additionally addressed the presence of microplastic in the Mediterranean Sea's Tyrrhenian seas. A Bongo net 60 cm in diameter was used to collect water samples. Fragmentary microplastics made of polyvinyl chloride (PVC) and low-density polyethylene (LDPE) predominated in ocean water samples. More than half (56.25%) of the bits of plastic were blue. Microplastic levels vary considerably among South Asian countries. Preliminary research by Koongolla et al. (2018) in Southern Sri Lanka found that 70% of the water samples they collected contained microplastic. Microplastic composition was dominated by polyethylene (PE) and polypropylene (PP), while fragment morphotype was the most common. The Maldives, like many other countries whose economy rely heavily on tourism, has seen an increase in microplastic pollution in its water supply. Microplastic was found by Ruiz-Orejón et al. (2018) in the coastal waters of the Balearic Islands, which are part of the Spanish marine waters. A mantra trawl net was used to collect the water samples. The main size range of the plastic fragments recovered was between 0.5 and 1.3 mm, and this morphotype made up more than 90% of the plastic materials. The prevalence of microplastics was significantly higher than that of mesoplastics and macroplastics. Microplastic was found in the water column at a high density along

the coast, especially within the first 10 km. Microplastic pollution was also found in the maritime waters of Italy, as reported by Baini et al. (2018) in the coastal waters of Tuscany. A manta trawl was used to gather surface samples from four different transects off the coast of Tuscany. There were relatively few microplastics present overall (0.26 item/m^3), with polyethylene (PE) and polypropylene (PP) being the most common types. Most of the particles in the water samples were fragments (62%), followed by filaments (29%); the most numerous were smaller than 1 mm (52%). Microplastic abundance was also detected in several parts of Africa. The prevalence of microplastics, for instance, has been investigated in King Harbour, Jamaica. To collect these samples, a conventional trawl net was used. Microplastic concentrations ranged from zero to 5.73 particles per cubic meter. Many of the samples had a fragmented morphology, and their diameters ranged from 1 to 2.5 mm. Microplastics have been observed in a rainbow of hues, with transparency accounting for 35% of the total. For this morphotype of fragment, the FTIR analysis revealed a PE and PP composition (Rose and Webber 2019). The first study to investigate microplastic in Kuwaiti waterways was carried out by Saeed et al. (2020). According to their results, microplastic abundance in water samples was lower compared with other matrices such as sediment and fish. Microplastics in seawater samples were collected using Neuston nets. Only 15 of the 44 sampling points turned up any microplastics at all, despite the fact that each trawl covered around a kilometer. Raman spectroscopy was used to determine that the polymers being examined belonged to the PP and PE classes. Fragment morphotype was assigned to six of the microplastics, while filament morphotype was assigned to the rest. Variants of different sort researchers have found microplastics in high numbers in water samples taken from Nansha Island, China, in the South China Sea. Microplastic content in 15 water samples was 1733 items/m^3, which is quite high. The microbead morphotype was determined to be the most common, accounting for 76.5% of all microplastics found (Nie et al. 2019). Tsang et al. (2017) discovered a low concentration of microplastics such as PP and PE in Hong Kong's maritime waters. It was found that pellet morphotype microplastics made up nearly all of the microplastics studied (96.8%). The outflow from the Pearl River, a neighboring river, or from raw wastewater, storm runoff, as well as other ground activities, were assumed to be the source of the plastic pollution. Many scientists have noticed a parallel pattern in Southeast Asia, citing the prevalence of microplastics along the northern shore of Surabaya as an example. Measured microplastic concentrations in Lamong Bay, Kenjeran Beach, and the Wonorejo shoreline ranged from 0.38 to 0.61 items/L. Location in close proximity to both people and shipping contributed significantly. There was a significant amount of microplastic found, with 58.4% being of the foam kind. Among the seven polymers discovered from the site of study, polystyrene was the most commonly observed (Cordova et al. 2019).

Several variables are influencing microplastics' ability to absorb pollutants. China, the United States, the United Kingdom, Italy, and South Korea are among the major countries that have published on microplastic sorbing pollutants, and this is despite the fact that microplastic pollution is a worldwide problem (Yu et al. 2019a). Since microplastic is being recognized as a vector or carrier of hazardous substances including POP and heavy metals, there has been an uptick in research on microplastic sorption (Kwon et al. 2017).

19.3 Physical Properties of Microplastic as a Factor

Microplastic's ability to absorb pollutants has been shown to depend on its physical and chemical characteristics (Ma et al. 2020a). Microplastics of lighter colors were found to have lower amounts of PAH and PCB, as well as PAH of lower molecular weight. To combat microplastic sorption, scientists can look into developing environmentally friendly colorants by using the hypothesized association between color and chemical composition (Ma et al. 2020a). Additionally, it has been observed that higher-density microplastics have lower levels of PAH, PCB, and phenanthrene (Fries and Zarfl 2012). It is difficult to overlook the deterioration of microplastic since it not only increases in surface area but also generates oxygen groups, permeability, ruggedness, charge, and polarity (Fries and Zarfl 2012). The sorption capacity of aged microplastic can be greatly enhanced by the formation of hydrogen bonds between the oxygen-containing functional groups (Liu et al. 2018). Ni and Cu were detected on PVC after being exposed to them for 502 days in an experiment conducted by Kedzierski et al. (2018). After a year, polybutylene adipate terephthalate (PBAT) showed Si, Al, Fe, and Mg, while PET showed no change (Kedzierski et al. 2018). Furthermore, previous investigations have shown conflicting results when looking at how microplastic age affects pollutant sorption ability. Pollutant sorption capacity was shown to be lower in weathered or aged microplastic, as reported by Ma et al. (2020a) However, it has also been found that aged microplastic has a higher pollutant sorption capacity than virgin microplastic (Yu et al. 2019a). The sorption ability of a microplastic particle is affected by a number of parameters, including color, density, and surface shape or weathering, and the order in which these factors are presented does not reflect their relative importance.

19.4 Microplastic Chemical Composition as a Contributing Element

The sorption capacity of microplastic is affected not only by its physical qualities but also by its chemical properties (Li et al. 2018). When the chemical structure is porous or has a big surface area, it can absorb more. For instance, whereas polyethylene (PE) had the largest sorption capacity, polyethylene terephthalate (PET) had the highest sorption rate. Due to the former's greater surface area, wider space among polymer chains, and higher free volume, the pollutant had a greater sorption capacity and was more easily diffused into PE (Rochman et al. 2013). While PET has a faster sorption rate than PE, PE traps more impurities into structural voids within the polymer, possibly as a result of PET's reduced surface area and crystalline polymeric structure, which prevents diffusion into the substance (Teuten et al. 2007).

Pollutants' chemical interactions with microplastics, such as electrostatic attraction, van der Waals force, hydrophobic interaction, - interaction, and hydrogen bonding, all contribute to the wide range of their affinities with microplastics (Ma et al. 2020a). Hydrogen bonding between PA and antibiotics, for instance, resulted in high affinity due to the presence of amide groups in both molecules (Li et al. 2018). For another, when comparing polystyrene with carboxyl groups (PS-COOH) and amine groups (PS-NH2), the latter was found to be toxic to sea urchin (Paracentrotus lividus) embryos via inducing apoptosis, cell death, oxidative stress, and disruption of cell membranes (Vega and Epel 2004). The most extensively researched microplastic-associated pollutants at the moment are hydrophobic POP because of their close connection and hydrophobicity-driven sorption with microplastic (Liu et al. 2019b). Hydrophobic organic pollutants include polychlorinated biphenyls (PCBs), polycyclic aromatic hydrocarbons (PAHs), and endocrine-disrupting chemicals (EDCs), such as [Ma]. Common examples of such interactions are microplastics and hydrophobic chemical compounds like PAH in aquatic environments (Tan et al. 2019). Organic contaminants had greater attractions for rubber polymers than glass polymers in terms of crystallinity, and water-insoluble organic pollutants frequently preferred amorphous to crystalline forms (Rochman et al. 2013). Partitioning, rather than electrostatic force and hydrophobicity, has a preponderant role in the interaction of hydrophilic POP, such as perfluorooctanesulfonate (PFOS) and perfluorooctanesulfonamide (PFOSA), and microplastic. Since the sulfonamide functional group renders PFOSA non-ionic, but the anionic PFOS is less attractive to PE, and vice versa, it stands to reason that the presence of the functional group would make PFOSA more appealing to PE (Wang et al. 2015). Microplastics' chemical interactions with pollutants, such as hydrophobicity and hydrophilicity, vary depending on the type of microplastic. For instance, non-polar PFOSA was drawn to non-polar PE by hydrophobic interactions, but was less drawn to benzene-containing PS because of lower free volume, steric hindrance, and bonding rotation (Wang et al. 2015). Whereas PE can only generate van der Waals interactions, PS has been shown by Yu et al. (2019b) to be capable of generating planarity, hydrophobic, and - interactions. There was no difference in PAH sorption between polypropylene and polyethylene, another type of plastic that was reported to be a sorption component. Further research is needed to identify further patterns and connections between the physical or chemical features of a microplastic and a given pollutant that contribute to its sorption capacity. Microplastic also attracts benzene derivatives rather than pure benzene due to a greater hydrophobic contact (Huffer and Hofmann 2016). Hydrophobic, covalent hydrogen, and halogen interactions were observed to draw five bisphenol analogs to the PVC microplastic. It is true that electrostatic forces existed, but they were more susceptible to modification by their immediate surroundings (Wu et al. 2016). PCB was shown to sorb onto PE at a higher rate (higher uptake diffusion and partition coefficients) than it did onto PS or PVC. However, as the cohesive density and molar volume of PCB increased, so did the decrease in the sorption capacity with PVC at higher chlorinated congeners (Besseling et al. 2016). There was also no temporal or spatial correlation between microplastic and metals in terms of affinity or sorption ability (Yu et al. 2019b).

19.5 Water Properties as a Factor

Microplastic's ability to bind pollutants has been shown to be influenced by environmental factors and water characteristics such as salinity, pH, and dissolved organic matter. It is crucial to consider how both dissolved and suspended organic matter affect the in situ sorption capability and dispersal of ocean microplastics (Besseling et al. 2016). Wu et al. state that dissolved organic matter can affect sorption by doing one of three things: (1) competing with other sorbents, (2) interacting with pollutants after they have been sorption onto other sorbents, or (3) engaging with pollutants via complexation or hydrophobic interactions (Wu et al. 2016). Because of the greater affinity between medicines and dissolved organic matter, the sorption capacity of PE was decreased for three PPCP: triclosan (TCS), 17-ethinyl estradiol (EE2), and 4-methylbenzylidene camphor (4MBC). PE microplastic sorption of carbamazepine (CBZ) was not affected by the presence of dissolved organic materials (Wu et al. 2016). Considering the acidification of the marine ecosystem, studying the effect of pH on microplastic sorption is crucial. In contrast to PP and PE, whose sorption capacities fluctuate little with pH, PS and PVC have a higher sorption capacity for the antibiotic tylosin (TYL) because the proportion of TYL+ is larger as the pH falls (Guo et al. 2018). As pH falls, the microplastic surface is protonated, increasing the production of anionic PFOS through its sorption onto PS and PE. When compared to PS, the PE surface is sometimes simpler to protonate (Wang et al. 2015). Nonetheless, the sorption capacity of HDPE reduces for Pb, Ni, Co, and Cd when the pH lowers, whereas it increases for Cr and remains constant for Cu. Despite the fact that it has been demonstrated that the pH of the nearby water affects the sorption of contaminants by microplastic, for instance by protonation (Guo et al. 2018), a more thorough comparative or analytical technique is required to understand the associations between pH and microplastic and pollutants. In terms of sorption behavior, salinity has both beneficial and detrimental consequences. Sorption capacities, such as those for high PCB or lubricating oil in PE and PS, have been found to increase with increasing salinity (Wang et al. 2015). However, dichlorodiphenyltrichloroethane (DDT) and ciprofloxacin have also been shown to have lower sorption capabilities at higher salinities due to greater competition for the sorption site (Liu et al. 2018). In addition, phenanthrene's response to salt was obscure (Bakir et al. 2014). However, the salinity impact on metal sorption by microplastic did not show a consistent pattern. The sorption capacities of Cd, Co, and Ni decreased with increasing salinity, but those of Cr and, strangely enough, Cu and Pb remained unchanged. According to published research, free ions and competition for sorption sites are primarily responsible for the effects of salinity on sorption capacity. Chemical interactions may play a significant role in the pollutant sorption by microplastics since the covalent bonds on a microplastic are tied to both the physical parameters of the extracellular environment (i.e., pH and salinity) and the physical characteristics of the polymer (such as substance, color, and age). Therefore, the varied physical characteristics of the microplastic and its surroundings have an influence on its sorbability due to chemical interaction. In time or across different pollutant types, for example, variations in surface functional groups, polarization, or chemical interactions such as water insolubility eventually nevertheless had an impact on the dissolution rate of microplastics in relation to the relevant pollutant (Liu et al. 2018). Research investigating the complex formation between microplastic, contaminants, and the ecosystem may reveal new tangential features of microplastic's ability to bind contaminants.

19.6 Microplastic Risks to the Coastal Environment and Its Inhabitants

Most people are aware that huge plastic trash causes problems for marine ecosystems and the economy (via things such as damage to aquaculture, energy production, fishing, and shipping as well as the loss of esthetic appeal and the health of species) (Lusher et al. 2015). Microplastic has a substantially larger surface area-to-volume ratio than macroplastic, allowing it to sorb and desorb and, therefore, bioaccumulate quantitatively more dangerous chemicals, despite the fact that both can spread diseases and release plasticizers or additives (Avio et al. 2017). Benthic invertebrates, fish, deep ocean biota, and larger mammals from a variety of trophic levels have all been discovered to contain microplastics in their digestive tracts. Growing concerns about the dangers to aquatic biota are prompted by the fact that ingested microplastic is transmitted up the food chain (Wright et al. 2013a). In the ocean, microplastics can build up and cause problems for marine life and ecosystems (Guzzetti et al. 2018). Microplastic poses a significant hazard to marine life, but we cannot begin to comprehend the full scope of those threats unless we have a firm handle on the elements that influence microplastic's bioavailability (Kokalj et al. 2018). In addition, the representative model species chosen in microplastic toxicological studies should be chosen carefully according to the aims of the study, as various organisms consume microplastic in different ways depending on their individual physiology and morphology (Ma et al. 2020a). Despite this, research has accumulated to show that the kinds of organisms and the physicochemical characteristics of microplastic impact the bioavailability of microplastic to organisms by altering the distribution and ease of access of microplastic as nourishment for microbes, which, in turn, affects the microplastic-associated hazards or degree of harm faced by different species (Setala et al. 2016). Multiple pathways are compromised by microplastic, posing threats to marine ecosystems and creatures. Since microplastic works as a reservoir or vector in a typical marine ecosystem, it accumulates chemical contaminants, transports them over great distances, and makes pollutants highly bioavailable to species when consumed (Liao and Yang 2020). Without the addition of pollutants, microplastic poses a wide range of indirect and direct threats to a wide variety of organisms (Vajargah et al. 2019). Microplastic has external effects on the ecosystem, such as affecting the distribution of food and light, which can have consequences for species beyond their internal health (Cole et al. 2016). It has been documented that microplastics can be transported through marine species (Windsor et al. 2019). Depending on the species and the location, microplastic translocation through biological membranes might have different effects. When microplastics are ingested or breathed in, they can travel to the digestive system, gills, lungs, blood, hemolymph, liver, and kidneys (Abbasi et al. 2018). In addition, microplastic leads to digestive tract obstruction, decreased feeding, or satiation, structural deterioration, nutritional or growth issues, inflammation, and a changed metabolic profile. Ineffective egestion and intestinal injury were also linked to microplastic. The monogononta rotifer Brachionus koreanus, for instance, had longer retention durations when exposed to microplastic (Jeong et al. 2016). Similarly, Caenorhabditis elegans exposed to microplastics experienced increased mortality, accumulation, decreased intestinal Ca2+ level, and increased expression of the glutathione S-transferase, all of which contributed to chronic intestinal damage (Lei et al. 2018). It was also revealed that marine species and ecosystems felt the external effects of microplastic. Marine plankton's ability to breathe and produce food is stunted when microplastics contaminate the water's surface (Amin et al. 2020). In addition, microplastics can disrupt the regular pattern of food distribution among marine organisms because they are discharged alongside feces, which is a key food source (Cole et al. 2011). Still, there were reports of inconsistencies. Ingestion of microplastic by the Antarctic krill Euphausia superba resulted in only mild acute toxicity and no mortalities (Dawson et al. 2018). Potential processes for the transfer of POP and metals include sorption, aggregation, ingestion, retention, egestion, reingestion, and release of chemicals. Live animals such as lugworms, marine amphipods, mussels, rainbow fish, and microalgae have been found to transmit and bioaccumulate organic contaminants by sorption on microplastic. These pollutants include PAH, PCB, EDC, and PBDE (Wang et al. 2020). While some research has found no deleterious impacts of microplastic, reports on the dangers of microplastic to the environment should not be disregarded.

Microplastics can bioaccumulate a wide variety of contaminants. One of the most pressing contemporary environmental issues is microplastic pollution in the ocean. All the way from herbivores and secondary consumers to top predators, reports of microplastic ingestion have emerged (Wang et al. 2020). There may be negative and long-lasting effects of microplastic in the ocean, according to previous research. For instance, larger tiny particles may have detrimental physical impacts on living beings if ingested (Wright et al. 2013b). The introduction of toxic substances into the ecosystem via the food chain has been hypothesized as one more negative consequence of microplastic. Hydrophobic organic compounds and other chemical contaminants in saltwater may be absorbed by microplastics. Due to its high sorption capacity, plastics may have gathered toxins from nearby polluted water or from chemical additives used in their production. Similarly, when environmental conditions shift away from encouraging sorption, pollutants in the ingested microplastic may be desorbed and released into the biological systems of living animals. Because of this, microplastics can act as both a pollutant source and a pollution sink (Liu et al. 2019a). As for humans, reports of microplastic-pollutant complexes releasing their contents into the surrounding biological tissue reflect a condition similar to that observed in other organisms that have bioaccumulated hazardous pollutants. The topic of whether or not microplastic contributes considerably to the intake and accumulation of environmental contaminants through the marine food chain is thus raised. It has been shown that microplastic can operate as vectors for the transmission of chemical contaminants from saltwater to organisms and that various different forms of environmental pollutants have been identified in living species alongside plastic debris. The hydrophobic plastic surfaces make hydrophobic organic compounds the most prominent class of molecules to transfer in this fashion. Hydrophobic organic chemicals, such as PCB, PBDE, PAH, and hexabromocyclododecane (HBCD), may be absorbed by plastics such as PS, PE, and PP and transported to living things. New research also indicates that microplastics might act as a carrier for toxic metals in the ocean. Exposure of zebrafish (Danio rerio) to Ag-incubated PE microbeads dramatically raised the proportion of intestinal Ag levels, suggesting that the microplastic altered the uptake route of a metal contaminant in a model fish species (Khan et al. 2015). Pollutants bound by microplastics have caused similar ecotoxicological reactions in the past. Compared to lugworms (Arenicola marina) exposed to clean PVC particles, those exposed to PVC particles containing TCS or nonylphenol (NP) had reduced survival, feeding rate, and phagocytic activity (Browne et al. 2013). Researchers using Japanese medaka (Oryzias latipes) as model organisms found that consuming LDPE microplastic-containing environmental contaminants absorbed from San Diego Bay was associated with liver toxicity and pathological reactions. Beachhoppers (Platorchestia smithi) that consumed PAH-contaminated PE beads exhibited decreased jump height and survival rate, suggesting that the microplastic pollution they were exposed to altered their behavior (Tosetto et al. 2016). In addition, microplastic may serve as a vehicle for the spread of disease. Among the 28 bacterial species found on North Adriatic microplastic by Viršek et al. (2017) is the pathogenic bacterium Aeromonas salmonicida, which has a significant effect on fish aquaculture. Microplastic itself can act as a vector for transmission between different trophic levels. Microplastic is easily eaten by zooplanktons, the primary food supply for many tertiary consumers (Botterell et al. 2019). Microplastics consumed may be passed up the food chain via predation (Farrell and Nelson 2013), or they may be regested and subsequently re-ingested (Mazurais et al. 2015). Few studies have attempted to quantify microplastic's potential function in the trophic transmission of pollutants. Bioaccumulation in lugworms exposed to PCB-contaminated PE was higher than in the control group exposed to PCB alone (Besseling et al. 2013), which corroborated a previous assessment of the bioavailability model presented by Teuten et al. (2007). Great shearwaters (Puffinus gravis) and short-tailed shearwaters (Puffinus tenuirostris) were shown to have higher amounts of PCB and PBDE in their adipose tissue and eggs when they swallowed more plastic particles (Tanaka et al. 2013). It is important to keep in mind, too, that microplastics could not be the only conduit for transferring pollutants up the food chain. High PCB levels in northern fulmar (Fulmarus glacialis) are more likely to be provided by their prey than by the plastic trash they consume, as shown by a comparison with the concentrations of plastic debris in their stomachs (Herzke et al. 2016). Microplastic may also serve as a scavenger, removing toxins from people, in contrast to its vector role. Researchers used 13C-labeled PBDE to demonstrate that sandhopper (Talitrus saltator) microplastic intake could transfer and remove pollutants, suggesting a partial balance between the beneficial and negative impacts (Scopetani et al. 2018). Herring (Clupea harengus membras) caught in the Baltic Sea had higher quantities of hydrophobic chemical compounds in their muscle tissue, and more plastic particles than usual were found in their stomachs (Ogonowski et al. 2017). Most model evaluations, however, suggested that microplastic ingestion was not a major route by which chemicals absorbed from seawater entered the ecosystem (Koelmans et al. 2016), but it is still unclear how much pollution might be expected to build in the food web using microplastic as a vector. Plastics' potential as vectors may also be affected by the chemical bonds they share with other materials (Koelmans et al. 2016). Microplastic-bound hydrophobic organic chemicals are taken up differently by species with similar gut retention durations as fish, as measured by the desorption rates in artificial gut fluid. In addition, hexachlorocyclohexane (HCH) absorption by PE microplastic ingestion can be minimal compared with absorption via other exposure routes, such as water ventilation and food consumption. Microplastics may not be directly toxic, but they can raise the body's overall intake of harmful chemicals such as pentachlorobenzene (PeCB) and hexachlorobenzene (HeCB) (Lee et al. 2019). Therefore, it will be important for future research to zero in on the particular pieces of plastic trash that may have stronger chemical affinities. When compared to more conventional materials, PBAT's vector effect as an organic pollutant is stronger due to its enhanced sorption and desorption capabilities (Zuo et al. 2019).

19.7 Pollutant Bioaccumulation of Microplastic and Its Factors

Bioaccumulation is affected by desorption, the internal environment of the organism, and time of retention [60]. Other bioaccumulation parameters, such as microplastic size and hydrophobicity, were analogous to those that influence desorption (Ma et al. 2016). Many studies have looked into bioaccumulation and reported their findings (Barboza et al. 2018a). In a recent study, Barboza et al. found that the content of mercury in the gills and livers of young European seabass Dicentrarchus labrax was twice as when exposed to a mixture of microplastics and mercury, compared with exposure to mercury alone (Barboza et al. 2018b). Bioaccumulation of PBDEs, PCBs, and PAHs was also reported to increase in the presence of PE microplastic. PE microplastic was found to cause PBDE bioaccumulation in the marine amphipod Allorchestes compressa (Chua et al. 2014). In addition, the hydrophobicity of PCB congeners in the presence of PS resulted in varying degrees of bioaccumulation capacity in the marine lugworm Arenicola marina. The bioaccumulation of phenanthrene in the planktonic crustacean Daphnia magna was also found to be elevated in the presence of smaller PS microplastic (50 nm). Yet, another investigation found no difference in bioaccumulation or EC50 for Daphnia magna immobility between the microplastic PA/bisphenol mixture and the bisphenol-only setting (Rehse et al. 2018). The possibility exists that the PA material acts as a very small carrier of bisphenol. The EC50 for Daphnia magna immobilization did not differ between the microplastic PE/phenanthrene mixture and the phenanthrene-only configuration, as was previously reported (Frydkjaer et al. 2017). Researchers such as Batel et al. (2016) observed that in a simulated aquatic food chain consisting of zebrafish Danio rerio and brine shrimp nauplii Artemia sp., microplastic transferred benzo [a] pyrene (BaP) from shrimp nauplii to zebrafish. Tourinho et al. (2019) proposed four scenarios in which microplastic could cause pollutant bioaccumulation in organisms: (1) strong sorption and strong desorption, (2) strong sorption and low desorption, (3) low sorption and strong desorption, and (4) low sorption and low desorption. Microplastics, first, serve as a vector for bioaccumulation due to their high sorption and low desorption strengths. At high sorption levels, pollutants are released back into the environment by chemical desorption, leading to greater bioaccumulation in the organism. Under gut circumstances, the desorption rate was reported to be very high. For instance, increased venlafaxine concentrations (increased desorption and bioaccumulation) in the liver of the pond fish Misgurnus anguillicaudatus was linked to a 50-fold increase in microplastic (high sorption) (Qu et al. 2018). Similarly, Bakir et al. (2014) found that pollutant desorption in the stomach was 30 times higher than in saltwater (Nor and Koelmans 2019). Second, even though microplastic is pre-sorbing with the pollutant, the scenario with strong sorption and low desorption may reduce bioavailability. Mussels and zebrafish, which both have low bioaccumulation of pre-sorbed fluoranthene and Ag via PE, are examples that are similar to this scenario (Magara et al. 2018). Third, in the low sorption and strong desorption scenario, the sorbed percentage is low but a large number of contaminants are desorbed. This situation is similar to the poor sorption of PBDE and bifenthrin on PE, both of which are similarly bioaccumulated in amphipods and the midge Chironomus tepperi or are nearly totally bioaccumulated in them [125]. Lastly, poor bioaccumulation is often the result of low sorption and low desorption. Mussels of the genus Mytilus were shown to have little fluoranthene bioaccumulation due to the microplastic's poor sorption of fluoranthene (Ziajahromi et al. 2019). Pollutants in this setting are normally ingested; therefore, microplastics would not affect the distribution or bioavailability of pollutants. However, modest bioaccumulation might result from ingesting microplastics with low contamination levels. Nor and Koelmans (2019) demonstrated that PCB was transmitted from tainted meals to clean PE microplastic in a gut-like environment. The significant sorption of pollutants in the gut of the organism supports a fifth scenario in which unsorbed or low-contaminated microplastic is swallowed and later egested in a sorbed condition. In addition, depuration may have an influence later on, making it a crucial experimental step. For instance, discrepancies can arise, especially in comparative research, if the microplastics still present in the organism's stomach at the end of the interval or experiment are included or excluded (Tourinho et al. 2019). After depuration, the fluoranthene/microplastic mixture setup revealed greater toxicity to the mussel Mytilus spp. than the fluoranthene-only setup did before the experiment began (Paul-Pont et al. 2016). Pomatoschistus microps fish showed delayed mortality when exposed to a pyrene/PE microplastic mixture (Oliveira et al. 2013). However, most studies have not revealed significant bioaccumulation of contaminants by microplastic, according to the current literature. Rather than microplastics, the food pathway was found to have a more direct effect on bioaccumulation (Koelmans et al. 2016).

19.8 Risks of Pollutant Bioaccumulation to the Marine Ecosystem and Humans

Microplastics causes danger to the marine environment. In addition to the inherent concerns of microplastic, microplastic-pollutant complexes also pose problems because the pollutants they attract. Some studies suggested that pollutants desorb in the digestive tract before being transported to the circulatory system, intestinal epithelia, tissues, and organs such as the liver. Physiological stress, morphological abnormalities, immobilization, neurotoxicity, oxidative damage, enzymatic dysregulation, reduced growth rate, reduced chlorophyll concentration, altered immune response, genotoxicity, and mortality are all effects of microplastic and the bioaccumulation of pollutants caused by microplastic-pollutant complexes. The accumulation of heavy metals in the brain and muscles leads to neurotoxicity, oxidative damage, and enzymatic dysregulation; for instance, the Ag accumulates in the intestinal mucous, mucosal epithelium, muscle layer, and serosa (Khan et al. 2017). Another case in point is the translocation of bioaccumulated phenanthrene and its byproducts into the tissue, stomach, and carapace of daphnids, where they induced toxicity and immobilization of the joint system (Ma et al. 2016). On marine fish and microalgae, researchers found evidence of microplastic and other pollutants' combined effects (Kim et al. 2017). However, there are contrasting results on microplastic's biological harm. Rochman et al. (2014), who tested virgin microplastic's toxicity on a fish cell line, found that it had no effect. No death, toxicity, weight loss, or bioaccumulation was observed after microplastic consumption by the Antarctic krill Euphausia superba, according to another study. Short-term effects of bisphenol A (BPA) and polycyclic aromatic hydrocarbons (PAH) on Daphnia magna were also reported to be mitigated at lower exposure concentrations.

Dangers to people can be exposed to microplastics in two ways: through inhalation and through the diet, specifically through fish and shellfish. As a result of the measured abundances, it was calculated that 100,000 microplastics per year may be found in the shellfish consumed by people in China (Wright and Kelly 2017). In addition, 15 different brands of commercial sea salt were found to have an average of 700 microplastics/kg of microplastic, mostly PET and PE (Yang et al. 2015a). PET and PP microplastics were found in all eight human feces samples tested in a recent study (Schwabl et al. 2019). Food intended for human consumption is contaminated with microplastic, as shown in the current literature, although more recent investigations on the direct clinical implications of this problem are lacking. Over the course of more than three decades, scientists have studied microplastic's impact on animal health. Microplastic may be able to evade the clearance system, travel deep into the lungs of humans and other mammals, become entrenched or stuck there, and cause either chronic or acute inflammation, according to available research (Wright and Kelly 2017). There was a report back in the day that said microfibers were poisonous to lung cells and cancerous (Omenn et al. 1986). Researchers have found that microplastic can have indirect effects on the immune system by migrating from living cells to human organ systems such as the lymphatic and circulatory systems (Rieux et al. 2005). More recent reports on the health effects of microplastic are restricted to the use of model animals and tend to focus on the poisoning of organisms by the additives and pollutants leached from the microplastic rather than the microplastic itself (Costa et al. 2016). Using the human intestinal epithelial cell line and co-cultures that imitate intestinal M-cells and goblet cells, one study analyzed the in vitro gastrointestinal uptake of microplastic and found that it did not affect cell differentiation or activation in any way (Stock et al. 2019). Studying Cr-contaminated PLA microplastic in the human digestive system, for instance, was done using a whole-GI in vitro technique, a systematic model that comprised the oral, stomach, small intestine, and large intestine digestive phases (Liao and Yang 2020). Cr concentrations were high throughout the digestive process. Microplastic contains a variety of pollutants, monomers, and additives, some of which include persistent organic pollutants (POP) and endocrine-disrupting chemicals (EDC), which have been found in the human body at levels greater than would be predicted from gut exposure alone (Vethaak and Leslie 2016). Further, phthalate, a plasticizer, can be harmful to the liver and the reproductive system if consumed in excess of the US Environmental Protection Agency (USEPA) criteria (6 g/L) (Lambert et al. 2014). Styrene, benzoyl peroxide, zeolites, and azobisisobutyronitrile are just a few of the many plastic additives that pose a risk to human health (Todd et al. 2003). Microplastic in the water absorbs a variety of environmental pollutants, including the carcinogenic PAH. These substances, now known to pose a threat to human health, accumulate in the fat of animals and are ingested by people (Hwang et al. 2020). Microplastic-pollutant complexes are harmful to living species, including humans, as shown by the literature review. This toxicity is due in part to the leaching of sorbed pollutants and the usage of certain additives and monomers. There is a pressing need for a more recent investigation into the possibility of microplastic causing any medical condition directly against human patients, despite the fact that there are older studies describing the effect or existence of microplastic in humans and some relatively new studies on the effect of microplastic conducted using model organisms or human cell lines as the reference (Stock et al. 2019). As long as there is no definitive proof that plastic has a negative impact on humans, it is possible that plastic manufacture, use, and ocean trash leakage can go unchecked.

19.9 Gaps in Microplastic Research and Future Outlook

As a result of microplastic pollution, the safety of food produced for human consumption is being jeopardized. However, there is a dearth of data regarding microplastic's effects on the human body's internal systems. Further, the literature search highlighted the significance of approach variations in data comparability and reliability. Pre-incubation or pre-sorption, depuration, coupled co-exposure to a microplastic and pollutant mixture, pollutant-only effect study, and dose-response analysis are only a few examples of methodological considerations. Furthermore, research has utilized various microplastic sampling and analytical methodologies due to the absence of standardized approaches (Jiang et al. 2020). Data comparability and consistency may be compromised by variations in microplastic definitions, sampling strategies, and analyses used by various scientists. Further, contradictory results in the historical literature imply that the impact of microplastic aging on pollutant sorption capacity relies on the specific pollutant or microplastic and the way their chemical bonds, forces, or interactions change over time. Some research has found that microplastic has no deleterious impacts, but other research has found that it poses serious risks to the environment. Because each organism's internal environment is unique, there are also unanswered questions about the bioaccumulation factors of microplastic-bound contaminants. Insightful scientific data and facts, public awareness, stakeholder involvement, and the use of environmentally benign, alternative polymeric material are some of the potential future ways that can decrease microplastic contamination (Gong et al. 2019). To validate the biological, physical, or chemical features responsible for the sorption capacity of a microplastic with a particular pollutant, however, more research is needed. Recent studies have shown that microplastics can absorb pollutants in part due to the existence of biofilms on their surfaces. A material's maximal sorption capacity may also be a limiting factor in terms of its toxicity. Furthermore, the association patterns of microplastic and contaminants can be linked to environmental factors such as pH and salinity using a comprehensive comparative or statistical method. New indirect aspects of pollutant sorption by microplastic may be uncovered through research into the chemical interactions between microplastics, contaminants, and the environment. In addition, future research needs to zero in on the plastic trash that may have greater chemical affinities than the others. As an organic contaminant, for instance, PBAT has a more potent vector effect than more conventional materials due to its larger sorption and desorption capabilities. Moreover, future tests should look into the possibility of unsorbed or low-contaminated microplastic to be egested in a sorbed condition and carry pollutants out of the gut of organisms, as an alternative to that of presorbed microplastic. Future microplastic research is likely to focus on standardizing the techniques involved in microplastic research and gaining a better knowledge of the processes associated with pollutant sorption and bioaccumulation by microplastic. Concern over microplastic pollution in environmental engineering, ecology, and materials science has been ongoing for some time. Microplastic pollution is a growing problem, and it will take a combination of research into plastic compositions, plastic characterization, plastic source and sink tracking, microplastic-environment interaction comprehension, policy implementation, biotechnological approaches, infrastructure improvements, and technological breakthroughs to solve the problem (Hale et al. 2020). Recycling and compostable packaging are two examples of source reduction strategies that could be implemented, as are measures to limit the escape of microplastics from waste treatment facilities and the use of plastics made from renewable resources (Amelia et al. 2020). To reduce microplastic contamination, we need systematic, cutting-edge study and strategy.

19.10 Conclusions

Anthropogenic, urban, fishing, and marine activities often contribute to microplastic pollution in the marine environment, and these microplastics are typically of the fiber or fragment type, containing PE or PP material. Size, color, density, shape, and mobility of microplastic, as well as species, shape, and physiology of organisms, are currently known variables of microplastic bioavailability toward organisms. Environmental parameters such as dissolved organic matter, pH, and salinity, as well as the kind of pollutant and the presence of biofilms, are also known to affect the pollutant sorption of microplastic. The global microplastic pollution disaster can be alleviated in part by better understanding the aspects of bioavailability, sorption, and bioaccumulation already recognized to contribute to it. Hydrophobic organic compounds, metals, and environmental pollutants (POP, PAH, and EDC) can be transported by sorption, aggregation, ingestion, retention, egestion, reingestion, and release. Reading about how microplastics absorb pollutants led us to discover that chemical interaction is the most important and direct factor. New indirect aspects of pollutant sorption by microplastic may be uncovered through research into the chemical interactions between microplastics, contaminants, and the environment. Overall, it is still unclear how much pollution is expected to build in the food chain using microplastic as a vector. For instance, the plastics' chemical affinities may affect their vector role; hence, future research should concentrate on the plastic debris that may have higher chemical affinities than the others. The sorption and bioaccumulation (i.e., desorption) of pollutants in the presence of microplastic are regulated by a wide variety of parameters, including the species of organisms present and the type of pollutants (Adetunji et al. 2019, 2020, 2021a, b, c, d, 2022a, b, c; Adetunji 2019). Clearer and more complete understanding of the elements controlling the sorption and bioaccumulation behaviors remains a research problem. Production, consumption, and plastic waste leakage to the oceans in the future may be predictable if progressive and impartial research is combined with management.

References

Abbasi S, Soltani N, Keshavarzi B, Moore F, Turner A, Hassanaghaei M (2018) Microplastics in different tissues of fish and prawn from the Musa Estuary, Persian Gulf. Chemosphere 205:80–87

Adetunji CO (2019) Environmental impact and ecotoxicological influence of biofabricated and inorganic nanoparticle on soil activity. In: Panpatte D, Jhala Y (eds) Nanotechnology for agriculture. Springer, Singapore. https://doi.org/10.1007/978-981-32-9370-0_12

Adetunji CO, Panpatte DG, Bello OM, Adekoya MA (2019) Application of nanoengineered metabolites from beneficial and eco-friendly microorganisms as a biological control agents for plant pests and pathogens. In: Panpatte D, Jhala Y (eds) Nanotechnology for agriculture: crop production & protection. Springer, Singapore. https://doi.org/10.1007/978-981-32-9374-8_13

Adetunji CO, Oloke JK, Prasad G (2020) Effect of carbon-to-nitrogen ratio on eco-friendly mycoherbicide activity from *Lasiodiplodia pseudotheobromae* C1136 for sustainable weeds management in organic agriculture. Environ Dev Sustain 22:1977–1990. https://doi.org/10.1007/s10668-018-0273-1

Adetunji CO, Akram M, Olaniyan OT, Ajayi OO, Inobeme A, Olaniyan S, Hameed L, Adetunji JB (2021a) Targeting SARS-CoV-2 novel corona (COVID-19) virus infection using medicinal plants. In: Dua K, Nammi S, Chang D, Chellappan DK, Gupta G, Collet T (eds) Medicinal plants for lung diseases. Springer, Singapore. https://doi.org/10.1007/978-981-33-6850-7_21

Adetunji CO, Michael OS, Varma A, Oloke JK, Kadiri O, Akram M, Bodunrinde RE, Imtiaz A, Adetunji JB, Shahzad K, Jain A, Ubi BE, Majeed N, Ozolua P, Olisaka FN (2021b) Recent advances in the application of biotechnology for improving the production of secondary metabolites from quinoa. In: Varma A (ed) Biology and biotechnology of quinoa. Springer, Singapore. https://doi.org/10.1007/978-981-16-3832-9_17

Adetunji CO, Olaniyan OT, Akram M, Ajayi OO, Inobeme A, Olaniyan S, Khan FS, Adetunji JB (2021c) Medicinal plants used in the treatment of pulmonary hypertension. In: Dua K, Nammi S, Chang D, Chellappan DK, Gupta G, Collet T (eds) Medicinal plants for lung diseases. Springer, Singapore. https://doi.org/10.1007/978-981-33-6850-7_14

Adetunji CO, Ajayi OO, Akram M, Olaniyan OT, Chishti MA, Inobeme A, Olaniyan S, Adetunji JB, Olaniyan M, Awotunde SO (2021d) Medicinal plants used in the treatment of influenza A virus infections. In: Dua K, Nammi S, Chang D, Chellappan DK, Gupta G, Collet T (eds) Medicinal plants for lung diseases. Springer, Singapore. https://doi.org/10.1007/978-981-33-6850-7_19

Adetunji CO, Olaniyan OT, Adeyomoye O, Dare A, Adeniyi MJ, Alex E, Rebezov M, Isabekova O, Shariati MA (2022a) Smart sensing for COVID-19 pandemic. In: Pani SK, Dash S, dos Santos WP, Chan Bukhari SA, Flammini F (eds) Assessing COVID-19 and other pandemics and epidemics using computational modelling and data analysis. Springer, Cham. https://doi.org/10.1007/978-3-030-79753-9_9

Adetunji CO, Olaniyan OT, Adeyomoye O, Dare A, Adeniyi MJ, Alex E, Rebezov M, Petukhova E, Shariati MA (2022b) Internet of Health Things (IoHT) for COVID-19. In: Pani SK, Dash S, dos Santos WP, Chan Bukhari SA, Flammini F (eds) Assessing COVID-19 and other pandemics and epidemics using computational modelling and data analysis. Springer, Cham. https://doi.org/10.1007/978-3-030-79753-9_5

Adetunji CO, Olaniyan OT, Adeyomoye O, Dare A, Adeniyi MJ, Alex E, Rebezov M, Koriagina N, Shariati MA (2022c) Diverse techniques applied for effective diagnosis of COVID-19. In: Pani SK, Dash S, dos Santos WP, Chan Bukhari SA, Flammini F (eds) Assessing COVID-19 and other pandemics and epidemics using computational modelling and data analysis. Springer, Cham. https://doi.org/10.1007/978-3-030-79753-9_3

Akarsu C, Kumbur H, Gökdağ K, Kıdeyş AE, Sanchez-Vidal A (2020) Microplastics composition and load from three wastewater treatment plants discharging into Mersin Bay, north eastern Mediterranean Sea. Mar Pollut Bull 150:110776

Aliabad MK, Nassiri M, Kor K (2019) Microplastics in the surface seawaters of Chabahar Bay, Gulf of Oman (Makran coasts). Mar Pollut Bull 143:125–133

Amelia TSM, Sukri SNF, Jaapar AN, Md Amin R, Bhubalan K (2020) Uptake and egestion of polyhydroxyalkanoate microbeads in the marine copepod Nitokra lacustris pacifica. J Sustain Sci Manag 15:45–53

Amin RM, Sohaimi ES, Anuar ST, Bachok Z (2020) Microplastic ingestion by zooplankton in Terengganu coastal waters, southern South China Sea. Mar Pollut Bull 150:110616

Auta HS, Emenike CU, Fauziah SH (2017) Distribution and importance of microplastics in the marine environment: a review of the sources, fate, effects, and potential solutions. Environ Int 102:165–176

Auta HS, Emenike CU, Jayanthi B, Fauziah SH (2018) Growth kinetics and biodeterioration of polypropylene microplastics by *Bacillus* sp. and *Rhodococcus* sp. isolated from mangrove sediment. Mar Pollut Bull 127:15–21

Avio CG, Gorbi S, Regoli F (2017) Plastics and microplastics in the oceans: from emerging pollutants to emerged threat. Mar Environ Res 128:2–11
Baini M, Fossi MC, Galli M, Caliani I, Campani T, Finoia MG, Panti C (2018) Abundance and characterization of microplastics in the coastal waters of Tuscany (Italy): the application of the MSFD monitoring protocol in the Mediterranean Sea. Mar Pollut Bull 133:543–552
Bakir A, Rowland SJ, Thompson RC (2014) Transport of persistent organic pollutants by microplastics in estuarine conditions. Estuar Coast Shelf Sci 140:14–21
Barboza LGA, Vethaak AD, Lavorante BR, Lundebye AK, Guilhermino L (2018a) Marine microplastic debris: an emerging issue for food security, food safety and human health. Mar Pollut Bull 133:336348
Barboza LGA, Vieira LR, Branco V, Figueiredo N, Carvalho F, Carvalho C, Guilhermino L (2018b) Microplastics cause neurotoxicity, oxidative damage and energy-related changes and interact with the bioaccumulation of mercury in the European seabass, Dicentrarchus labrax (Linnaeus 1758). Aquat Toxicol 195:49–57
Barceló D, Picó Y (2020) Case studies of macro- and microplastics pollution in coastal waters and rivers: is there a solution with new removal technologies and policy actions? Case Stud Chem Environ Eng 2:100019
Batel A, Linti F, Scherer M, Erdinger L, Braunbeck T (2016) Transfer of benzo a pyrene from microplastics to Artemia nauplii and further to zebrafish via a trophic food web experiment: CYP1A induction and visual tracking of persistent organic pollutants. Environ Toxicol Chem 35:1656–1666
Belzagui F, Crespi M, Álvarez A, Gutiérrez-Bouzán C, Vilaseca M (2019) Microplastics' emissions: microfibers' detachment from textile garments. Environ Pollut 248:1028–1035
Besseling E, Wegner A, Foekema EM, van den Heuvel-Geve MJ, Koelmans AA (2013) Effects of microplastic on fitness and PCB bioaccumulation by the lugworm Arenicola marina (L.). Environ Sci Technol 47:593–600
Besseling E, Quik JTK, Sun M, Koelmans AA (2016) Fate of nano- and microplastic in freshwater systems: a modeling study. Environ Pollut 220:540
Botterell ZL, Beaumont N, Dorrington T, Steinke M, Thompson RC, Lindeque PK (2019) Bioavailability and effects of microplastics on marine zooplankton: a review. Environ Pollut 245:98–110
Brandon AM, Gao SH, Tian R, Ning D, Yang SS, Zhou J, Wu WM, Criddle CS (2018) Biodegradation of polyethylene and plastic mixtures in mealworms (larvae of Tenebrio molitor) and effects on the gut microbiome. Environ Sci Technol 52:6526–6533
Browne MA, Niven SJ, Galloway TS, Rowland SJ, Thompson RC (2013) Microplastic moves pollutants and additives to worms, reducing functions linked to health and biodiversity. Curr Biol 23:2388–2392
Castillo A, Al-Maslamani I, Obbard JP (2016) Prevalence of microplastics in the marine waters of Qatar's exclusive economic zone (EEZ). In: MICRO 2016. Fate and impact of microplastics in marine ecosystems from the coastline to the open sea. Elsevier, Amsterdam, p 112
Chen M, Jin M, Tao P, Wang Z, Xie W, Yu X, Wang K (2018) Assessment of microplastics derived from mariculture in Xiangshan Bay, China. Environ Pollut 242:1146–1156
Chua EM, Shimeta J, Nugegoda D, Morrison PD, Clarke BO (2014) Assimilation of polybrominated diphenyl ethers from microplastics by the marine amphipod, Allorchestes compressa. Environ Sci Technol 48:8127–8134
Cole M, Lindeque P, Halsband C, Galloway TS (2011) Microplastics as contaminants in the marine environment: a review. Mar Pollut Bull 62:2588–2597
Cole M, Lindeque PK, Fileman E, Clark J, Lewis C, Halsband C, Galloway TS (2016) Microplastics alter the properties and sinking rates of zooplankton faecal pellets. Environ Sci Technol 50:3239–3246
Conley K, Clum A, Deepe J, Lane H, Beckingham B (2019) Wastewater treatment plants as a source of microplastics to an urban estuary: removal efficiencies and loading per capita over one year. Water Res X 3:100030

Cordova MR, Purwiyanto AIS, Suteja Y (2019) Abundance and characteristics of microplastics in the northern coastal waters of Surabaya, Indonesia. Mar Pollut Bull 142:183–188

Costa JP, Santos PSM, Duarte AC, Rocha-Santos T (2016) (Nano) plastics in the environment – sources, fates and effects. Sci Total Environ 566-567:15–26

Dai Z, Zhang H, Zhou Q, Tian Y, Chen T, Tu C, Fu C, Luo Y (2018) Occurrence of microplastics in the water column and sediment in an inland sea affected by intensive anthropogenic activities. Environ Pollut 242:1557–1565

Dawson A, Huston W, Kawaguchi S, King C, Cropp R, Wild S, Eisenmann P, Townsend K, Nash SB (2018) Uptake and depuration kinetics influence microplastic bioaccumulation and toxicity in Antarctic krill Euphausia superba. Environ Sci Technol 52:3195–3201

Farrell P, Nelson K (2013) Trophic level transfer of microplastic: Mytilus edulis (L.) to Carcinus maenas (L.). Environ Pollut 177:1–3

Ferreira M, Thompson J, Paris A, Rohindra D, Rico C (2020) Presence of microplastics in water, sediments and fish species in an urban coastal environment of Fiji, a Pacific small island developing state. Mar Pollut Bull 153:110991

Fries E, Zarfl C (2012) Sorption of polycyclic aromatic hydrocarbons (PAHs) to low and high density polyethylene (PE). Environ Sci Pollut Res Int 19:1296–1304

Frydkjaer CK, Iversen N, Roslev P (2017) Ingestion and egestion of microplastics by the Cladoceran daphnia magna: effects of regular and irregular shaped plastic and sorbed phenanthrene. Bull Environ Contam Toxicol 99:655–661

Gong W, Jiang M, Han P, Liang G, Zhang T, Liu G (2019) Comparative analysis on the sorption kinetics and isotherms of fipronil on nondegradable and biodegradable microplastics. Environ Pollut 254:112927

Guo XT, Pang JW, Chen SY, Jia HZ (2018) Sorption properties of tylosin on four different microplastics. Chemosphere 209:240–245

Guzzetti E, Sureda A, Tejada S, Faggio C (2018) Microplastic in marine organism: environmental and toxicological effects. Environ Toxicol Pharmacol 64:164–171

Hale RC, Seeley ME, La Guardia MJ, Mai L, Zeng EY (2020) A global perspective on microplastics. J Geophys Res Oceans 125:e2018JC014719

Herzke D, Anker-Nilssen T, Nøst TH, Götsch A, Christensen-Dalsgaard S, Langset M et al (2016) Negligible impact of ingested microplastics on tissue concentrations of persistent organic pollutants in northern fulmars off coastal Norway. Environ Sci Technol 50:1924–1933

Huffer T, Hofmann T (2016) Sorption of non-polar organic compounds by microsized plastic particles in aqueous solution. Environ Pollut 214:194–201

Hwang J, Choi D, Han S, Jung SY, Choi J, Hong J (2020) Potential toxicity of polystyrene microplastic particles. Sci Rep 10:7391

Ivleva NP, Wiesheu AC, Niessner R (2017) Microplastic in aquatic ecosystems. Angew Chem Int Ed 56:1720–1739

Jambeck JR, Geyer R, Wilcox C, Siegler TR, Perryman M, Andrady A, Narayan R, Law KL (2015) Plastic waste inputs from land into the ocean. Science 347:768–771

Jeong CB, Won EJ, Kang HM, Lee MC, Hwang DS, Hwang UK, Zhou B, Souissi S, Lee SJ, Lee JS (2016) Microplastic size-dependent toxicity, oxidative stress induction, and p-JNK and p-p38 activation in the monogonont rotifer (Brachionus koreanus). Environ Sci Technol 50:8849–8857

Jiang Y, Yang F, Zhao Y, Wang J (2020) Greenland Sea gyre increases microplastic pollution in the surface waters of the Nordic seas. Sci Total Environ 712:136484

Kazour M, Jemaa S, Issa C, Khalaf G, Amara R (2019) Microplastics pollution along the Lebanese coast (eastern Mediterranean Basin): occurrence in surface water, sediments and biota samples. Sci Total Environ 696:133933

Kedzierski M, D'Almeida M, Magueresse A, Le Grand A, Duval H, Cesar G, Sire O, Bruzaud S, Tilly VL (2018) Threat of plastic ageing in marine environment. Adsorption/desorption of micropollutants. Mar Pollut Bull 127:684–694

Khalik WMAWM, Ibrahim YS, Anuar ST, Govindasamy S, Baharuddin NF (2018) Microplastics analysis in Malaysian marine waters: a field study of Kuala Nerus and Kuantan. Mar Pollut Bull 135:451–457
Khan FR, Syberg K, Shashoua Y, Bury NR (2015) Influence of PE microplastic beads on the uptake and localization of silver in zebrafish (Danio rerio). Environ Pollut 206:73–79
Khan FR, Boyle D, Chang E, Bury NR (2017) Do PE microplastic beads alter the intestinal uptake of Ag in rainbow trout Oncorhynchus mykiss? Analysis of the microplastics vector effect using in vitro gut sacs. Environ Pollut 231:200–206
Kim D, Chae Y, An YJ (2017) Mixture toxicity of nickel and microplastics with different functional groups on Daphnia magna. Environ Sci Technol 51:12852–12858
Koelmans AA, Bakir A, Burton GA, Janssen CR (2016) Microplastic as a vector for chemicals in the aquatic environment: critical review and model-supported reinterpretation of empirical studies. Environ Sci Technol 50:3315–3326
Kokalj AJ, Kunej U, Skalar T (2018) Screening study of four environmentally relevant microplastic pollutants: uptake and effects on Daphnia magna and Artemia franciscana. Chemosphere 208: 522–529
Koongolla JB, Andrady AL, Kumara PTP, Gangabadage CS (2018) Evidence of microplastics pollution in coastal beaches and waters in southern Sri Lanka. Mar Pollut Bull 137:277–284
Kwon JH, Chang S, Hong SH, Shim WJ (2017) Microplastics as a vector of hydrophobic contaminants: importance of hydrophobic additives. Integr Environ Assess Manag 13:494–499
Lambert S, Sinclair C, Boxall A (2014) Occurrence, degradation, and effect of polymer-based materials in the environment. In: Whitacre DM (ed) Reviews of environmental contamination and toxicology 227. Springer, Switzerland
Lebreton L, Andrady A (2019) Future scenarios of global plastic waste generation and disposal. Palgrave Commun 5:6
Lee H, Lee HJ, Kwon JH (2019) Estimating microplastic-bound intake of hydrophobic organic chemicals by fish using measured desorption rates to artificial gut fluid. Sci Total Environ 651: 162–170
Lei L, Wu S, Lu S, Liu M, Song Y, Fu Z, Shi H, Raley-Susman KM, He D (2018) Microplastic particles cause intestinal damage and other adverse effects in zebrafish Danio rerio and nematode Caenorhabditis elegans. Sci Total Environ 619:1–8
Li J, Zhang K, Zhang H (2018) Adsorption of antibiotics on microplastics. Environ Pollut 237:460–467
Li Y, Zhang H, Tang C (2020) A review of possible pathways of marine microplastics transport in the ocean. Anthr Coasts 3:6–13
Liao Y, Yang J (2020) Microplastic serves as a potential vector for Cr in an in-vitro human digestive model. Sci Total Environ 703:134805
Liu G, Zhu Z, Yang Y, Sun Y, Yu F, Ma J (2018) Sorption behavior and mechanism of hydrophilic organic chemicals to virgin and aged microplastics in freshwater and seawater. Environ Pollut 246:26–33
Liu X, Shi H, Xie B, Dionysiou DD, Zhao Y (2019a) Microplastics as both a sink and a source of Bisphenol A in the marine environment. Environ Sci Technol 5:10188–10196
Liu X, Xu J, Zhao Y, Shi H, Huang CH (2019b) Hydrophobic sorption behaviors of 17beta-estradiol on environmental microplastics. Chemosphere 226:726–735
Lusher A, Milian GH, O'Brien J, Berrow S, O'Connor I, Officer R (2015) Microplastic and macroplastic ingestion by a deep diving, oceanic cetacean: the True's beaked whale Mesoplodon Mirus. Environ Pollut 199:85–191
Ma Y, Huang A, Cao S, Sun F, Wang L, Guo H, Ji R (2016) Effects of nanoplastics and microplastics on toxicity, bioaccumulation, and environmental fate of phenanthrene in fresh water. Environ Pollut 219:166–173
Ma H, Pu S, Liu S, Bai Y, Mandal S, Xing B (2020a) Microplastics in aquatic environments: toxicity to trigger ecological consequences. Environ Pollut 261:114089

Ma ZF, Ibrahim YS, Lee YY (2020b) Microplastic pollution and health and relevance to the Malaysia's roadmap to zero single-use plastics 2018–2030. Malays J Med Sci 27:1–6
Magara G, Elia AC, Syberg K, Khan FR (2018) Single contaminant and combined exposures of PE microplastics and fluoranthene: accumulation and oxidative stress response in the blue mussel, Mytilus edulis. J Toxicol Environ Health A 81:1–13
Maximenko N, Hafner J, Niier P (2012) Pathways of marine debris derived from trajectories of Lagrangian drifters. Mar Pollut Bull 65:51–62
Mazurais D, Ernande B, Quazuguel P, Severe A, Huelvan C, Madec L, Mouchel O, Soudant P, Robbens J, Huvet A, Zambonino-Infante J (2015) Evaluation of the impact of PE microbeads ingestion in European sea bass (Dicentrarchus labrax) larvae. Mar Environ Res 112:78–85
McEachern K, Alegria H, Kalagher AL, Hansen C, Morrison S, Hastings D (2019) Microplastics in Tampa Bay, Florida: abundance and variability in estuarine waters and sediments. Mar Pollut Bull 148:97–106
Nel HA, Froneman PW (2015) A quantitative analysis of microplastic pollution along the south-eastern coastline of South Africa. Mar Pollut Bull 101:274–279
Neufeld L, Stassen F, Sheppard R, Gilman T (2016) The new plastics economy—rethinking the future of plastics. World economic forum. Ellen MacArthur Foundation, McKinsey & Company. http://www.ellenmacarthurfoundation.org/publications Ngo PL, Pramanik BK, Shah K, Roychand R (2019) Pathway, classification and removal efficiency of microplastics in wastewater treatment plants. Environ Pollut 255:113326
Nie H, Wang J, Xu K, Huang Y, Yan M (2019) Microplastic pollution in water and fish samples around Nanxun reef in Nansha Islands, South China Sea. Sci Total Environ 696:134022
Nor NHM, Koelmans AA (2019) Transfer of PCBs from microplastics under simulated gut fluid conditions is biphasic and reversible. Environ Sci Technol 53(4):1874–1883
Oberbeckmann S, Labrenz M (2020) Marine microbial assemblages on microplastics: diversity, adaptation, and role in degradation. Annu Rev Mar Sci 12:209–232
Oberbeckmann S, Kreikemeyer B, Labrenz M (2018) Environmental factors support the formation of specific bacterial assemblages on microplastics. Front Microbiol 8:2709
Ogonowski M, Wenman V, Danielsson S, Gorokhova E (2017) Ingested microplastic is not correlated to HOC concentrations in Baltic Sea herring. CEST [internet]. Rhodes, Greece
Oliveira M, Ribeiro A, Hylland K, Guilhermino L (2013) Single and combined effects of microplastics and pyrene on juveniles (0+ group) of the common goby Pomatoschistus microps (Teleostei Gobiidae). Ecol Indic 34:641–647
Omenn GS, Merchant J, Boatman E, Dement JM, Kuschner M, Nicholson W, Peto J, Rosenstock L (1986) Contribution of environmental fibers to respiratory cancer. Environ Health Perspect 70: 51–56
Park SY, Kim CG (2019) Biodegradation of micro-polyethylene particles by bacterial colonization of a mixed microbial consortium isolated from a landfill site. Chemosphere 222:527–533
Paul-Pont I, Lacroix C, Fernandez GC, Hegaret H, Lambert C, Le Goïc N, Frere L, Cassone AL, Sussarellu R, Fabioux C, Guyomarch J, Albentosa M, Huvet A, Soudant P (2016) Exposure of marine mussels Mytilus spp. to polystyrene microplastics: toxicity and influence on fluoranthene bioaccumulation. Environ Pollut 216:724–737
PlasticsEurope (2017) Plastics – the facts 2017. An analysis of European plastics production, demand and waste data. PlasticsEurope. https://www.plasticseurope.org/en/resources/publications/274-plastics-facts-2017. Accessed 17 Aug 2020
PlasticsEurope (2019) Plastics – the facts 2019. An analysis of European plastics production, demand and waste data. PlasticsEurope. https://www.plasticseurope.org/en/resources/publications/1804-plastics-facts-2019. Accessed 17 Aug 2020
Qu H, Ma R, Wang B, Zhang Y, Yin L, Yu G, Deng S, Huang J, Wang Y (2018) Effects of microplastics on the uptake, distribution and biotransformation of chiral antidepressant venlafaxine in aquatic ecosystem. J Hazard Mater 359:104–112

Rehse S, Kloas W, Zarfl C (2018) Microplastics reduce short-term effects of environmental contaminants. Part I: effects of bisphenol A on freshwater zooplankton are lower in presence of PA particles. Int J Environ Res Public Health 15:280

Rieux AD, Ragnarsson EGE, Gullberg E, Préat V, Schneider YJ, Artursson P (2005) Transport of nanoparticles across an in vitro model of the human intestinal follicle associated epithelium. Eur J Pharm Sci 25:455–465

Rocha-Santos T, Duarte AC (2015) A critical overview of the analytical approaches to the occurrence, the fate and the behavior of microplastics in the environment. TrAC Trends Anal Chem 65:47–53 Rochman CM (2018) Microplastics research – from sink to source. Science 360:28–29

Rochman CM, Hoh E, Hentschel BT, Kaye S (2013) Long-term field measurement of sorption of organic contaminants to five types of plastic pellets: implications for plastic marine debris. Environ Sci Technol 47:1646–1654

Rochman CM, Hentschel BT, Teh SJ (2014) Long-term sorption of metals is similar among plastic types: implications for plastic debris in aquatic environments. PLoS One 9:e85433

Rose D, Webber M (2019) Characterization of microplastics in the surface waters of Kingston Harbour. Sci Total Environ 664:753–760

Ruiz-Orejón LF, Sardá R, Ramis-Pujol J (2018) Now, you see me: high concentrations of floating plastic debris in the coastal waters of the Balearic Islands (Spain). Mar Pollut Bull 133:636–646

Rujnic-Sokele M, Pilipovic A (2017) Challenges and opportunities of biodegradable plastics: a mini review. Waste Manag Res 35:132–140

Ryberg MW, Hauschild MZ, Wang F, Averous-Monnery S, Laurent A (2019) Global environmental losses of plastics across their value chains. Resour Conserv Recycl 151:104459

Saeed T, Al-Jandal N, Al-Mutairi A, Taqi H (2020) Microplastics in Kuwait marine environment: results of first survey. Mar Pollut Bull 152:110880

Saliu F, Montano S, Garavaglia MG, Lasagni M, Seveso D, Galli P (2018) Microplastic and charred microplastic in the Faafu atoll, Maldives. Mar Pollut Bull 136:464–471

Savoca S, Capillo G, Mancuso M, Bottari T, Crupi R, Branca C, Spanò N (2019) Microplastics occurrence in the Tyrrhenian waters and in the gastrointestinal tract of two congener species of seabreams. Environ Toxicol Pharmacol 67:35–41

Schönlau C, Karlsson TM, Rotander A, Nilsson H, Engwall M, van Bavel B, Kärrman A (2020) Microplastics in sea-surface waters surrounding Sweden sampled by manta trawl and in-situ pump. Mar Pollut Bull 153:111019

Schwabl P, Köppel S, Königshofer P, Bucsics T, Trauner M, Reiberger T, Liebmann B (2019) Detection of various microplastics in human stool: a prospective case series. Ann Intern Med 171:453–457

Scopetani C, Cincinelli A, Martellini T, Lombardini E, Ciofini A, Fortunati A, Pasquali V, Ciattini S, Ugolini A (2018) Ingested microplastic as a two-way transporter for PBDEs in Talitrus saltator. Environ Res 167:411–417

Setala O, Norkko J, Lehtiniemi M (2016) Feeding type affects microplastic ingestion in a coastal invertebrate community. Mar Pollut Bull 102:95–101

Stock V, Böhmert L, Lisicki E, Block R, Cara-Carmona J, Pack LK, Selb R, Lichtenstein D, Voss L, Henderson CJ, Zabinsky E, Sieg H, Braeuning A, Lampen A (2019) Uptake and effects of orally ingested polystyrene microplastic particles in vitro and in vivo. Arch Toxicol 93:1817–1833

Sun X, Liang J, Zhu M, Zhao Y, Zhang B (2018) Microplastics in seawater and zooplankton from the Yellow Sea. Environ Pollut 242:585–595

Sutton R, Mason SA, Stanek SK, Willis-Norton E, Wren IF, Box C (2016) Microplastic contamination in the San Francisco Bay, California, USA. Mar Pollut Bull 109:230–235

Tan X, Yu X, Cai L, Wang J, Peng J (2019) Microplastics and associated PAHs in surface water from the Feilaixia Reservoir in the Beijiang River, China. Chemosphere 221:834–840

Tanaka K, Takada H, Yamashita R, Mizukawa K, Fukuwaka MA, Watanuki Y (2013) Accumulation of plastic-derived chemicals in tissues of seabirds ingesting marine plastics. Mar Pollut Bull 69:219–222
Teuten EL, Rowland SJ, Galloway TS, Thompson RC (2007) Potential for plastics to transport hydrophobic contaminants. Environ Sci Technol 41:7759–7764
Todd G, Wohlers D, Citra M (2003) Agency for toxic substances and disease registry. Agency for Toxic Substances and Disease Registry, Atlanta, GA
Tosetto L, Brown C, Williamson JE (2016) Microplastics on beaches: ingestion and behavioral consequences for beachhoppers. Mar Biol 163:199
Tourinho PS, Koci V, Loureiro S, Gestel CAM (2019) Partitioning of chemical contaminants to microplastics: sorption mechanisms, environmental distribution and effects on toxicity and bioaccumulation. Environ Pollut 252:1246–1256
Tsang YY, Mak CW, Liebich C, Lam SW, Sze ET, Chan KM (2017) Microplastic pollution in the marine waters and sediments of Hong Kong. Mar Pollut Bull 115:20–28
Tsiota P, Karkanorachaki K, Syranidou E, Franchini M, Kalogerakis N (2018) Microbial degradation of HDPE secondary microplastics: preliminary results. Proceedings of the international conference on microplastic pollution in the Mediterranean Sea, pp 181–188
Vajargah MF, Imanpoor MR, Shabani A, Hedayati A, Faggio C (2019) Effect of long-term exposure of silver nanoparticles on growth indices, hematological and biochemical parameters and gonad histology of male goldfish (Carassius auratus Gibelio). Microsc Res Tech 82:1224–1230
Vega RL, Epel D (2004) Stress-induced apoptosis in sea urchin embryogenesis. Mar Environ Res 58:799–802
Vethaak AD, Leslie HA (2016) Plastic debris is a human health issue. Environ Sci Technol 50: 6825–6826
Viršek MK, Lovšin MN, Koren Š, Kržan A, Peterlin M (2017) Microplastics as a vector for the transport of the bacterial fish pathogen species Aeromonas salmonicida. Mar Pollut Bull 125: 301–309
Wang F, Shih KM, Li XY (2015) The partition behavior of perfluorooctanesulfonate (PFOS) and perfluorooctanesulfonamide (FOSA) on microplastics. Chemosphere 119:841–847
Wang YL, Lee YH, Chiu IJ, Lin YF, Chiu HW (2020) Potent impact of plastic nanomaterials and micromaterials on the food chain and human health. Int J Mol Sci 21:1727
Windsor FM, Tilley RM, Tyler CR, Ormerod SJ (2019) Microplastic ingestion by riverine macroinvertebrates. Sci Total Environ 646:68–74
Wong SL, Nyakuma BB, Wong KY, Lee CT, Lee TH, Lee CH (2020) Microplastics and nanoplastics in global food webs: a bibliometric analysis (2009–2019). Mar Pollut Bull 158: 111432
Wright SL, Kelly FJ (2017) Plastic and human health: a micro issue? Environ Sci Technol 51:6634–6647
Wright SL, Rowe D, Thompson RC, Galloway TS (2013a) Microplastic ingestion decreases energy reserves in marine worms. Curr Biol 23:R1031–R1033
Wright SL, Thompson RC, Galloway TS (2013b) The physical impacts of microplastics on marine organisms: a review. Environ Pollut 178:483–492
Wu C, Zhang K, Huang X, Liu J (2016) Sorption of pharmaceuticals and personal care products to PE debris. Environ Sci Pollut Res Int 23:8819–8826
Yang D, Shi H, Li L, Li J, Jabeen K, Kolandhasamy P (2015a) Microplastic pollution in table salts from China. Environ Sci Technol 49:13622–13627
Yang Y, Yang J, Wu WM, Zhao J, Song Y, Gao L, Yang R, Jiang L (2015b) Biodegradation and mineralization of polystyrene by plastic-eating mealworms: part 1. Chemical and physical characterization and isotopic tests. Environ Sci Technol 49:12080–12086
Yu X, Ladewig S, Bao S, Toline CA, Whitmire S, Chow AT (2018) Occurrence and distribution of microplastics at selected coastal sites along the southeastern United States. Sci Total Environ 613-614:298–305

Yu F, Yang C, Zhu Z, Bai X, Ma J (2019a) Adsorption behavior of organic pollutants and metals on micro/nanoplastics in the aquatic environment. Sci Total Environ 694:133643

Yu F, Sun Y, Yang M, Ma J (2019b) Adsorption mechanism and effect of moisture contents on ciprofloxacin removal by three-dimensional porous graphene hydrogel. J Hazard Mater 374: 195–202

Yuan J, Ma J, Sun Y, Zhou T, Zhao Y, Yu F (2020) Microbial degradation and other environmental aspects of microplastics/plastics. Sci Total Environ 715:136968

Zhang Z, Chen Y (2020) Effects of microplastics on wastewater and sewage sludge treatment and their removal: a review. Chem Eng J 382:122955

Zhang W, Zhang S, Wang J, Wang Y, Mu J, Wang P, Ma D (2017) Microplastic pollution in the surface waters of the Bohai Sea, China. Environ Pollut 231:541–548

Zhang H, Fei Y, Wang H, Chen Y, Huang S, Yu B, Wang J, Tong Y, Wen D, Zhou B, He X, Xia X, Luo Y (2020) Interaction of microplastics and organic pollutants: quantification, environmental fates, and ecological consequences. In: He D, Luo Y (eds) Microplastics in terrestrial environments – emerging contaminants and major challenges, The handbook of environmental chemistry. Springer, Heidelberg, pp 1–24

Zhao S, Zhu L, Wang T, Li D (2014) Suspended microplastics in the surface water of the Yangtze estuary system, China: first observations on occurrence, distribution. Mar Pollut Bull 86:562–568

Zheng Y, Li J, Cao W, Liu X, Jiang F, Ding J, Sun C (2019) Distribution characteristics of microplastics in the seawater and sediment: a case study in Jiaozhou Bay, China. Sci Total Environ 674:27–35

Zhu D, Chen Q, An X, Yang X, Christie P, Ke X, Wu L, Zhu Y (2018) Exposure of soil collembolans to microplastics perturbs their gut microbiota and alters their isotopic composition. Soil Biol Biochem 116:302–310

Zhu J, Zhang Q, Li Y, Tan S, Kang Z, Yu X, Shi H (2019) Microplastic pollution in the Maowei Sea, a typical mariculture bay of China. Sci Total Environ 658:62–68

Ziajahromi S, Kumar A, Neale PA, Leusch FD (2019) Effects of PE microplastics on the acute toxicity of a synthetic pyrethroid to midge larvae (Chironomus tepperi) in synthetic and river water. Sci Total Environ 671:971–975

Zuo LZ, Li HX, Lin L, Sun YX, Diao ZH, Liu S, Zhang ZY, Xu XR (2019) Sorption and desorption of phenanthrene on biodegradable poly (butylene adipate co-terephtalate) microplastics. Chemosphere 215:25–32

Microplastic to Accumulate Microbial Pathogens in the Terrestrial Environment

20

Hetvi Kanojia, Harsh Purohit, Madhvi Joshi, Jignesh H. Kamdar, and Joydeep Chakraborty

Abstract

Plastics have gradually become one of the most ubiquitous household materials in the modern world. Although already infamously known to pollute the environment due to their persistence, plastics are gradually raising further concerns as and when they result in the release of microplastics in both aquatic and terrestrial environments. The hydrophobic surfaces of these microplastics act as a suitable cradle for generation and colonization of microbial biofilms. The current chapter brings an insight into the various sources of these microplastics, the major constituents of the so-called "plastisphere," and how the pathogenic members of these plastispheres pose a threat to mankind, as well as various factors influencing the growth of microbes on microplastics.

Keywords

Plastic pollution · Microplastics · Microbiome · Pathogens · Terrestrial ecosystem

20.1 Introduction

Plastics have transformed our lives since the advent of synthetic resins in the early 1900s (Cole et al. 2011; Sivan 2011). Its usage has risen globally in the past few decades (Nielsen et al. 2020; Shen et al. 2020; Yadav et al. 2020). This is mainly due

H. Kanojia · H. Purohit · M. Joshi · J. H. Kamdar
Department of Microbiology, School of Science, RK University, Rajkot, Gujarat, India

J. Chakraborty (✉)
Department of Biological Sciences, PD Patel Insitute of Applied Sciences, Charotar University of Science and Technology, Changa, Anand, Gujarat, India
e-mail: joydeepchakraborty.as@charusat.ac.in

M. Shahnawaz et al. (eds.), *Microplastic Pollution*,
https://doi.org/10.1007/978-981-99-8357-5_20

to plastic's extraordinary properties, versatility, and low cost (Debrah et al. 2021; Razeghi et al. 2021; Bucknall 2020; Klemeš et al. 2020). Human occupants depend greatly on the use of plastic products on a daily basis. In 2010, alone, it is predicted that eight million metric tons (Mt) of plastic trash from terrestrial sources made its way into the ocean (Geyer et al. 2017; Jambeck et al. 2015). Plastic production has globally increased from 1.5 to 335 million tons within a span of less than seven decades (1950–2016). Plastic trash is found throughout the world because synthetic polymers are frequently used in everyday life and persist for quite a long in the nature (Loganathan and Kizhakedathil 2022). Plastic overuse has not only resulted in the release of potentially hazardous chemicals in the environment, but the global carbon dioxide emission has also increased as a result of plastic production (PlasticsEurope 2018). Unfortunately, despite the advancement of novel re-usage strategies, around 40% of global plastic trash is disposed in landfills and undergoes very slow biodeterioration (Zhang et al. 2019).

Research on microplastics (MPs) dates back to the late 1990s (or early 2000s), when microscopic plastic particles were discovered for the first time in the Great Pacific Garbage Patch (Moore 2011). Research gradually expanded to freshwater bodies (Hu et al. 2020) and, more recently, to terrestrial habitats (Corradini et al. 2021). Based on what has been revealed so far, apart from being a silent threat to the terrestrial world, MPs also cause significant harm to marine life forms as well. They are also regarded as potential threat to food safety and agricultural sustainability. In recent years, MPs have been traced to transfer from terrestrial agricultural origins to the human food chain and into the animal bodies, which definitely will have devastating effects on ecosystems and human health (Rillig 2012; Duis and Coors 2016). However, the role of MPs in soil and terrestrial habitats and a complete understanding of associated risks is yet to be explored (Zhu et al. 2019; Choi et al. 2021). For example, MPs may act as carriers of chemical pollutants like heavy metals and organic contaminants from terrestrial to aquatic habitat (Mai et al. 2018). Monitoring MPs in different local environments, such as soil and sediments, is quite challenging (Campanale et al. 2022).

Plastic in the environment may undergo wear and tear, leading to the formation of smaller particles such as macroplastics (2 cm), mesoplastics (5 mm to 2 cm), MPs (5 mm), and nanoplastics (1 mm). MPs are eventually produced when disposable plastic products are discarded without being recycled (Steinmetz et al. 2016). They primarily come from two types of sources, viz. primary and secondary (Su et al. 2019). Primary sources comprise personal care products (such as shampoo and cosmetics), industrial abrasives (especially tire ware), and raw material, process sub-products, and usually generate particles smaller than 5 mm in size (Yurtsever 2019; Chia et al. 2021). Secondary sources include larger plastic materials that were originally larger in size (e.g., virgin plastics in the ecosystem), but were eventually broken down to smaller size, less than 5.0 mm due to physical, biological, and chemical weathering (Shahnawaz et al. 2019; Naik et al. 2020). The concentration of MPs in soil has been found to be increasing significantly over the past few years (Zhang et al. 2020a, b). From the terrestrial environment, these MPs may eventually end up in water or be transported to the subsoil thus, reaching the groundwater system. This poses an even greater danger as it can harm various terrestrial life forms, including both prokaryotes and eukaryotes (Uheida et al. 2021; Wanner 2021). MPs can have a negative impact on soil biota, which can affect the food web and human health (Kim et al. 2021).

Microplastics are ubiquitous pollutants and, interestingly, can be colonized with a wide range of microbes many of which can be potentially pathogenic in nature. Because they are small enough to be taken up by microbes, they eventually end up accumulating in the food chain. Additionally, microplastics have the capability to absorb contaminants onto their surfaces, thereby enriching themselves. Consequently, MPs introduce a new array of issues and pose a threat to the health of soil biota. Therefore, it is important to assess the presence of MPs in soil environments.

20.2 Sources of Microplastic in Terrestrial Environment

In the environment, plastic is easy to develop from many sources due to its slow breakdown. MP resin pellets may pose a particularly serious health danger. In fact, reports of plastic contamination in both marine and terrestrial settings have been making headlines for years (Duis and Coors 2016). MPs are not limited to the ocean, as they have been found in abundance in both freshwater and atmospheric samples. However, only a few studies have led to what we know so far about MPs in freshwater and sediments (Wagner et al. 2014; Cera et al. 2020; Li et al. 2020; Yao et al. 2020). Multiple paths are depicted for the generation, transport, and modification of MPs across diverse environmental compartments (including freshwater, marine water, groundwater, sediments, and soils) (Frost et al. 2022). There are several entry points and potential sources for MPs in terrestrial ecosystems. They have the potential to build up in the soil and alter its physical, chemical, and biological characteristics in both positive and negative ways. Soil ecosystem health and biota may be endangered because MPs interact with several antibiotics, pesticides, heavy metals, and persistent organic pollutants (POPs), which are just a few examples of potentially dangerous compounds (Dissanayake et al. 2022).

Polystyrene, polyethylene, polyethylene terephthalate (PET), polypropylene, and polyamide are the most commonly used polymeric raw materials, which gradually leads to the generation of secondary MPs. However, once generated, these MPs are difficult to identify, handle, and characterize because of their microscopic particle size (nearly undetectable to the human eye), complicated chemical composition, cross-contamination, and inhomogeneous nature, all of which necessitate standardized experimental protocols. Most of the plastic trash in the water originates on land, but some come from the ocean as well (Andrady 2011; Mehlhart and Blepp 2012). It is mostly unknown nearly how much MPs contribute to the environmental pollution. Even though it is assumed that many plastic products break down to form micro- or nano-sized fragments upon exposure to various environmental factors such as light, pressure, and heat, this has not been experimentally demonstrated for most of the synthetic polymers under varying environmental conditions (Kallenbach et al. 2022). Particle release rates and the size distribution of the particles that are released are poorly known so far. MP pollution in agricultural and urban areas has been highlighted in this section because they are anticipated to have a significant role in the release or effect of MP particles into the environment (He and Luo 2020). For example, release routes explain actions or certain practices that result in the discharge of MPs into the atmosphere. These may not be major sources, such as the formation of tire-worn particles during vehicle usage. In contrast, there are major sources as well, like the land use of sewage sludge, which leads to the discharge of MPs into the environment after being compiled from a wide variety of sources. Industrial abrasives, especially those used in air blasting to remove metal paint and clean various types of engines, also generate microplastic particles. Drilling fluids are utilized in oil and gas development (Gregory 1996; Derraik 2002; Sundt et al. 2014; Essel et al. 2015). Some of the MPs found in abrasives are acrylic, polystyrene (PS), melamine, polyester (PES), and polyallyl diglycol carbonate (Eriksen et al. 2013). They represent a threat to the ecosystem if not employed in closed systems and disposed offed correctly (Sundt et al. 2014). It is not uncommon for incorrect handling to lead to the release into the environment of the raw ingredients used for the formation of plastic items. Granulates made from recycled plastics are a good example, as are factory wastes from plastic production (de Souza Machado et al. 2018). It is generally accepted that the most significant pathways for plastic materials to access the environment are through general littering, the disposal of plastic trash, and the loss from inadequate management of disposal sites and during the collection of garbage. In addition, MPs may be found in large quantities in synthetic materials (Wong et al. 2020).

20.3 The Plastisphere Microbiome

MPs have recently been detected in marine, freshwater, and terrestrial ecosystems as potential pollutants and pathogen carriers (Iqbal et al. 2022). Microorganisms, hydrophobic organic compounds, persistent organic pollutants, and heavy metals can all be carried on the surface of these particles, where they can have a deleterious effect on living things. It is well known that they are poisonous to both marine and land life. Cell membranes, tissues, and physiological processes undergo detrimental effects due to this (Iqbal et al. 2022). Gaining knowledge about the microbial existence of plastic waste is essential for predicting and evaluating its potential ecological impact on both the environment and human beings. The hydrophobic surface of plastic trash is an advantageous ecological niche, the so-called "plastisphere," for microbial colonization and biofilm production (Zettler et al. 2013). In general, the term "plastisphere" refers to the microbiological biofilms associated with polymers or MPs. Investigating the organisms within the plastisphere deepens our understanding of how MPs function as a microhabitat and conduit for potentially pathogenic microbes (Gkoutselis et al. 2021). Currently, marine systems account for more than 96% of MP research. As mentioned earlier, although it has been established that the majority of marine MP pollution originates from land-based sources, their presence in terrestrial ecosystems should not be disregarded. Furthermore, the construction industry, landfills, improper disposal of waste, synthetic turf, road paints, tires, agricultural plastic films, and personal care and cosmetic products, all contribute to the release of MPs into the environment. The aquatic system is critical for MP transport. Since the 1950s, plastic products have been widely employed in our daily lives. Because of its widespread use, mishandling, and environmental persistence, plastic garbage is recognized to exert a range of ecological pressures (Xu et al. 2020). The discoveries of the mycobiome at the core of the plastisphere provide further proof that the land-based plastisphere is an adequate ecological habitat for a diverse range of bacterial and fungal species, encompassing significant animal and plant pathogens (Gkoutselis et al. 2021). MPs function as specialized artificial microenvironments, enticing not only distinctive fungal communities but also specific human pathogens such as those belonging to Cryptococcal- and Phoma-like species. Consequently, MPs should be regarded not only as potential carriers of pathogenic organisms but also as their persistent reservoirs. With the escalating the quantity of plastic trash in terrestrial ecosystems worldwide, this interconnected relationship could have significant effects for the global epidemiology of diseases spanning multiple kingdoms and organisms. These persistent substances not only significantly improve adhesion, thereby acting as sources of disease, but they can also be carried across extensive distances by wind, currents, and waves, ultimately resulting in the introduction of foreign communities to specific areas (Casabianca et al. 2019).

Pathogenic fungi such as *Candida*, *Fusarium*, and *Rhodotorula* have been found on various plastic surfaces, especially those used as medical, industrial, and domestic appliances (Wirth and Goldani 2012; Vethaak and Leslie 2016; Neu et al. 2018). As a result, MPs may have a role in the formation and dissemination of fungal infections in soil habitats that receive a high concentration of plastic solid waste, such as residential gardens, agricultural land, landfills, and roadsides (Parthasarathy et al. 2019). Plastics release additives, accumulate environmental toxins, and provide a surface for the formation of biofilms that contain harmful and unfamiliar organisms. As the size of plastic fragments decreases, the number of species capable of consuming them also decreases. Particles as tiny as 20 nm can penetrate cell membranes, intensifying the risks associated with exposure. Feeding, metabolic processes, reproduction, and behavior can all suffer adverse effects due to this exposure (Hale et al. 2020).

Microplastics (MPs), like any other material introduced into aquatic ecosystems, experience rapid biofouling, where they become colonized by living organisms that accumulate gradually. These organisms create biofilms, which are complex microbial communities consisting of various bacteria, single-celled algae, diatoms, fungi, and protozoa. The growth of these microorganisms that make up the plastisphere on the surface of MPs, and, in some cases, even their penetration into the internal structure can vary depending on the polymer composition. Pathogenic bacteria can be found among both prokaryotic and eukaryotic species residing in plastic biofilms. Numerous studies have demonstrated that the surfaces of microplastics (MPs) are abundant with detrimental bacteria, such as *Vibrio* and *Pseudomonas*, and additionally act as a storage site for fecal indicator organisms like *E. coli* (Forero-López et al. 2022).

20.4 Environmental Factors Affecting Accumulation of Microbes

Microorganisms capable of forming biofilms on the surface of MPs play a vital role in supporting plastisphere communities (Bowley et al. 2021). The colonization of microbes on MPs is influenced by a range of factors, including the physical and chemical attributes of the MP (such as surface charge, hydrophobicity/hydrophilicity, roughness, and hardness), the physiological characteristics of the colonizing microbes (such as fimbriae, nanofibers, and hyphae), and environmental conditions (such as temperature, oxygen levels, light exposure, pH, nutrient availability, and salinity) (Hossain et al. 2019; Cai et al. 2019; Qiu et al. 2022). Geographical location has been found to impact the microbial community structure attached to the MPs. Different marine locations exhibit diverse microbial community patterns within their microbial biofilms (Xu et al. 2019). Additionally, biofilm formation decreases vertically with increasing water depth in natural coastal environments. Distinct plastisphere ecosystems may exist in different coastal sediments, while the accumulation of microbes on MPs has also been associated with the low temperature and high pH of the marine environment (Rodrigues et al. 2019). The water column plays a significant role in influencing microbial growth on MPs. The location of MPs within the water column affects the energy source available to the microbes. Primary photosynthetic producers live on floating MPs in the ocean's light layer, where they create photosynthetic products to survive. As the thickness of MPs increases, they sink to the opaque layer, depriving the microbes of the light required for photosynthesis as the water depth decreases. Exposure time is another crucial factor that affects the microbial community structures within biofilms on MPs. Longer exposure times allow for the diversification of microbial communities due to the hydrophobic nature of MP surfaces and nutrient deficiencies (Xu et al. 2019). Biofilm formation and metabolic processes contribute to the production of vitamins and extracellular polysaccharides on the surfaces of microplastics. This alteration in hydrophobicity of MP surfaces subsequently promotes microbial colonization, leading to the formation of biofilms.

20.5 Microplastics: A Possible Carrier of Multidrug-Resistant Pathogens in the Environment

Plastic particles are now ubiquitously present over various ecological systems worldwide, including air, terrestrial landscapes, freshwater bodies and marine environments, and within different organisms, including seafood frequently consumed by humans. Consequently, as a result, they have emerged as a substantial contributor to microplastic pollution (Hurley et al. 2018; Jambeck et al. 2015; van

Wijnen et al. 2019). MPs can serve as a new substrate for marine microorganisms, including fish and human pathogens, and the plastics themselves can act as a reservoir for genes associated with metal resistance and multidrug resistance. The investigation of MPs is a relatively recent area of study that has captured the attention of scientists due to its potential involvement in the propagation of antibiotic resistance (Zettler et al. 2013; McCormick et al. 2014; Dang and Lovell 2016). Nevertheless, there is an inadequate amount of research examining the potential hazards that microplastics pose to human health through their indirect effects, especially regarding the safety of seafood and global food security (Bank et al. 2020; Bowley et al. 2021; Guo et al. 2020; Parthasarathy et al. 2019; Radisic et al. 2020).

The concept of the "plastisphere" is a crucial aspect of understanding the relationship between microplastics and antibiotic resistance. The initial publication that introduced this concept highlighted that the microbial communities attached with plastic waste were distinct from those in the surrounding seawater, characterized by their diversity and the presence of both producers and consumers, such as heterotrophs, autotrophs, symbionts, and predators. In the open ocean, the surfaces of these plastic particles provided a new substrate and/or natural environment (Amaral-Zettler et al. 2015; Bowley et al. 2021; Oberbeckmann et al. 2018; Wright et al. 2020; Zettler et al. 2013). Even though microplastic surfaces start with no charge and demonstrate hydrophobic properties upon entering water as newly produced synthetic materials from land-based systems, they quickly become inhabited by microbial biofilms (Bowley et al. 2021; Wright et al. 2020; Zettler et al. 2013). Due to the long residence time of MPs in their surroundings, their potential for long-range transport, and the possible consequences of pathogen and disease transfer, the development of the plastisphere concept was crucial in addressing fundamental questions about the overall direct and indirect implications of microplastic pollution (Bowley et al. 2021). Among all microbes, vibrios have been predominantly found in colonizing microplastic surfaces. While not all vibrios are harmful, they exhibit a preference for lower salinity conditions typically found in coastal and estuarine locations, highlighting the significance of the plastisphere in terms of the distribution, abundance, fate, and transportation of pathogenic bacteria such as *Vibrio* species (Bowley et al. 2021; Zettler et al. 2013; Zhang et al. 2020a, b). Overall, various factors including ocean currents, sources, dynamics of microplastic transport, trophic transfer and complexity of the food web, horizontal gene transfer, buoyancy and sinking characteristics of MPs, variation and uptake by various seafood species, and human exposures collectively contribute to the identified risks and processes associated with microplastics and microbial interactions.

The interaction between plastics and microbes is a growing concern, yet its implications for ecosystems and human health remain poorly understood. While the plastisphere communities encompass archaea, bacteria, unicellular, and oligocellular eukaryotes, metagenomic analyses have revealed that microplastics promote the growth of distinct microbial communities compared with those in the surrounding environment (Zettler et al. 2013; Oberbeckmann et al. 2018; Delacuvellerie et al. 2019). Furthermore, these communities' composition and succession vary depending on factors such as the type of polymer, geographical location, and time of year (De Tender et al. 2017; Kirstein et al. 2019). Consequently, microplastics serve as microhabitats with high selectivity and flexibility, leading to ecosystem-level consequences such as the proliferation of antibiotic resistance through the concentration of specific lineages or alterations to microbial nitrogen and carbon cycles (Arias-Andres and Rojas-Jimenez 2018; Hartmann et al. 2019). A recurring theme in this context is the role of microplastics as reservoirs and vectors for invasive and harmful microorganisms (Vethaak and Leslie 2016; Zettler et al. 2013; Kirstein et al. 2019). Consequently, microplastics' function as reservoirs and vectors for invasive bacteria is a recurring theme in relevant research.

Most highly evolved fungi, which do not rely on an aqueous environment for their proliferation, can generate significantly more biomass in certain soils compared with prokaryotes. As a result, there is a theoretical possibility that these fungi could systematically assimilate plastic waste deposited in soils (Haglund et al. 2002; Harms et al. 2011; Pramila and Ramesh 2011). Fungal isolates have been discovered in plastic waste and have shown evidence of being among the initial organisms to colonize floating microplastics, forming part of polymicrobial biofilms similar to those found on plastic surfaces in households. Plastics used in medical, industrial, and domestic equipment have been found to harbor pathogenic fungal species such as *Candida*, *Fusarium*, and *Rhodotorula*. Soil habitats that receive a significant amount of plastic waste, such as backyards, roadsides, farmland, and landfills (Pramila and Ramesh 2011), are particularly susceptible to the accumulation and spread of fungal infections, with microplastics potentially playing a role in this process (Wirth and Goldani 2012). However, research on the impacts of microplastics has primarily focused on aquatic systems and isolated regions, with little consideration given to the nearby human environment.

20.6 Conclusion and Future Prospective

Owing to its environmental persistence and associated detrimental effects, plastic pollution has already become a global matter of concern over the last few decades. The spread of pathogens, on the other hand, is an issue mankind has always been struggling with. The emergence of multi-drug-resistant pathogens has further worsened the situation. Recent studies revealing how MPs can be responsible for the spread of pathogenic microorganisms should be taken as a warning for potential MP-mediated disease outbreaks in future and measures must be designed and implemented to minimize the dumping of plastics in the environment, which is one of the primary sources of MP generation.

Acknowledgments The authors would like to acknowledge the Faculty of Doctoral Studies and Research, RK University, and the managment of Charotar University of Science and Technology for their kind support during the preparation of the chapter. The authors are also thankful to the editors and reviewers for their feedback and suggestions in modifying the contents of the chapter.

References

Amaral-Zettler LA, Zettler ER, Slikas B, Boyd GD, Melvin DW, Morrall CE, Proskurowski G, Mincer TJ (2015) The biogeography of the plastisphere: implications for policy. Front Ecol Environ 13:541–546

Andrady AL (2011) Microplastics in the marine environment. Mar Pollut Bull 62:1596–1605

Arias-Andres MKU, Rojas-Jimenez K (2018) Microplastic pollution increases gene exchange in aquatic ecosystems. Environ Pollut 237:253–261

Bank MS, Ok YS, Swarzenski PW (2020) Microplastic's role in antibiotic resistance. Science 369: 1315–1315

Bowley J, Baker-Austin C, Porter A, Hartnell R, Lewis C (2021) Oceanic hitchhikers–assessing pathogen risks from marine microplastic. Trends Microbiol 29:107–116

Bucknall DG (2020) Plastics as a materials system in a circular economy. Philos Trans A Math Phys Eng Sci 378:20190268

Cai L, Wu D, Xia J, Shic H, Kim H (2019) Influence of physicochemical surface properties on the adhesion of bacteria onto four types of plastics. Sci Total Environ 671:1101–1107

Campanale C, Galafassi S, Savino I, Massarelli C, Ancona V, Volta P, Uricchio VF (2022) Microplastics pollution in the terrestrial environments: poorly known diffuse sources and implications for plants. Sci Total Environ 805:150431

Casabianca S, Capellacci S, Giacobbe MG, Dell'Aversano C, Tartaglione L, Varriale F et al (2019) Plastic-associated harmful microalgal assemblages in marine environment. Environ Pollut 244: 617–626

Cera A, Cesarini G, Scalici M (2020) Microplastics in freshwater: what is the news from the world? Diversity 12:276

Chia RW, Lee JY, Kim H, Jang J (2021) Microplastic pollution in soil and groundwater: a review. Environ Chem Lett 19:4211–4224

Choi YR, Kim YN, Yoon JH, Dickinson N, Kim KH (2021) Plastic contamination of forest, urban, and agricultural soils: a case study of Yeoju City in the Republic of Korea. J Soils Sediments 21: 1962–1973

Cole M, Lindeque P, Halsband C, Galloway TS (2011) Microplastics as contaminants in the marine environment: a review. Mar Pollut Bull 62:2588–2597

Corradini F, Casado F, Leiva V, Huerta-Lwanga E, Geissen V (2021) Microplastics occurrence and frequency in soils under different land uses on a regional scale. Sci Total Environ 752:141917

Dang H, Lovell CR (2016) Microbial surface colonization and biofilm development in marine environments. Microbiol Mol Biol Rev 80:91–138

de Souza Machado AA, Kloas W, Zarfl C, Hempel S, Rillig MC (2018) Microplastics as an emerging threat to terrestrial ecosystems. Glob Chang Biol 24:1405–1416

De Tender C, Devriese LI, Haegeman A, Maes S, Vangeyte J, Cattrijsse A et al (2017) Temporal dynamics of bacterial and fungal colonization on plastic debris in the North Sea. Environ Sci Technol 51:7350–7360

Debrah JK, Vidal DG, Dinis MA (2021) Innovative use of plastic for a clean and sustainable environmental management: learning cases from Ghana, Africa. Urban Sci 5:12–23

Delacuvellerie A, Cyriaque V, Gobert S, Benali S, Wattiez R (2019) The plastisphere in marine ecosystem hosts potential specific microbial degraders including Alcanivorax borkumensis as a key player for the low-density polyethylene degradation. J Hazard Mater 380:120899

Derraik JG (2002) The pollution of the marine environment by plastic debris: a review. Mar Pollut Bull 44:842–852

Dissanayake PD, Kim S, Sarkar B, Oleszczuk P, Sang MK, Haque MN et al (2022) Effects of microplastics on the terrestrial environment: a critical review. Environ Res 209:112734

Duis K, Coors A (2016) Microplastics in the aquatic and terrestrial environment: sources (with a specific focus on personal care products), fate and effects. Environ Sci Eur 28:1–25

Eriksen M, Mason S, Wilson S, Box C, Zellers A, Edwards W et al (2013) Microplastic pollution in the surface waters of the Laurentian Great Lakes. Mar Pollut Bull 77:177–182

Essel R, Engel L, Carus M, Ahrens RH (2015) Sources of microplastics relevant to marine protection in Germany. Texte. 64:1219–1226
Forero-López AD, Brugnoni LI, Abasto B, Rimondino GN, Lassalle VL, Ardusso MG et al (2022) Plastisphere on microplastics: in situ assays in an estuarine environment. J Hazard Mater 440: 129737
Frost H, Bond T, Sizmur T, Felipe-Sotelo M (2022) A review of microplastic fibres: generation, transport, and vectors for metal(loid)s in terrestrial environments. Environ Sci Processes Impacts 24:504–524
Geyer R, Jambeck JR, Law KL (2017) Production, use, and fate of all plastics ever made. Sci Adv 3: e1700782
Gkoutselis G, Rohrbach S, Harjes J, Obst M, Brachmann A, Horn MA, Rambold G (2021) Microplastics accumulate fungal pathogens in terrestrial ecosystems. Sci Rep 11:13214
Gregory MR (1996) Plastic 'scrubbers' in hand cleansers: a further (and minor) source for marine pollution identified. Mar Pollut Bull 32:867–871
Guo XP, Sun XL, Chen YR, Hou L, Liu M, Yang Y (2020) Antibiotic resistance genes in biofilms on plastic wastes in an estuarine environment. Sci Total Environ 745:140916
Haglund C, Levin L, Forchiassin F, Lopez M, Viale A (2002) Degradation of environmental pollutants by Trametes trogii. Rev Arg Microbiol 34:157–162
Hale RC, Seeley ME, La Guardia MJ, Mai L, Zeng EY (2020) A global perspective on microplastics. J Geophys Res Oceans 125:e2018JC014719
Harms H, Schlosser D, Wick LY (2011) Untapped potential: exploiting fungi in bioremediation of hazardous chemicals. Nat Rev Microbiol 9:177–192
Hartmann NB, Huffer T, Thompson RC, Hassellöv M, Verschoor A, Daugaard AE et al (2019) Are we speaking the same language? Recommendations for a definition and categorization framework for plastic debris. Environ Sci Technol 53:1039–1047
He D, Luo Y (2020) Microplastics in terrestrial environments. In: Emerging contaminants and major challenges. Springer
Hossain MR, Jiang M, Wei Q, Leff LG (2019) Microplastic surface properties affect bacterial colonization in freshwater. J Basic Microbiol 59:54–61
Hu L, He D, Shi H (2020) Microplastics in inland small waterbodies. In: He D, Luo Y (eds) Microplastics in terrestrial environments: emerging contaminants and major challenges. Springer, pp 93–110
Hurley R, Woodward J, Rothwell JJ (2018) Microplastic contamination of river beds significantly reduced by catchment-wide flooding. Nat Geosci 11:251–257
Iqbal R, Khan MT, Bilal H, Aslam MM, Khan IA, Raja S et al (2022) Microplastics as vectors of environmental contaminants: interactions in the natural ecosystems. Hum Ecol Risk Assess 28: 1022–1042
Jambeck JR, Geyer R, Wilcox C, Siegler TR, Perryman M, Andrady A et al (2015) Plastic waste inputs from land into the ocean. Science 347:768–771
Kallenbach EM, Rødland ES, Buenaventura NT, Hurley R (2022) Microplastics in terrestrial and freshwater environments. In: Bennett ER (ed) Microplastic in the environment: pattern and process, pp 87–130
Kim YN, Yoon JH, Kim KHJ (2021) Microplastic contamination in soil environment—a review. Soil Sci Annu 71:300–308
Kirstein IV, Wichels A, Gullans E, Krohne G, Gerdts G (2019) The plastisphere–uncovering tightly attached plastic "specific" microorganisms. PLoS One 14:e0215859
Klemeš JJ, Fan YV, Jiang P (2020) Plastics: friends or foes? The circularity and plastic waste footprint. Energy Sources A Recovery Util Environ Eff 43:1549–1565
Li C, Busquets R, Campos LC (2020) Assessment of microplastics in freshwater systems: a review. Sci Total Environ 707:135578
Loganathan Y, Kizhakedathil MPJ (2022) A review on microplastics—an indelible ubiquitous pollutant. Biointerface Res Appl Chem 13:126

Mai L, Bao LJ, Wong CS, Zeng EY (2018) Microplastics in the terrestrial environment. In: Microplastic contamination in aquatic environments. Elsevier, pp 365–378
McCormick A, Hoellein TJ, Mason SA, Schluep J, Kelly JJ (2014) Microplastic is an abundant and distinct microbial habitat in an urban river. Environ Sci Technol 48:11863–11871
Mehlhart G, Blepp M (2012) Study on land-sourced litter (LSL) in the marine environment: review of sources and literature. In: The context of the initiative of the declaration of the global plastics associations for solutions on marine litter. Öko-Institut eV, Darmstadt/Freiburg
Moore C (2011) Plastic ocean: how a sea captain's chance discovery launched a determined quest to save the oceans. Penguin
Naik RA, Rowles LS, Hossain AI, Yen M, Aldossary RM, Apul OG, Conkle J, Saleh NB (2020) Microplastic particle versus fiber generation during photo-transformation in simulated seawater. Sci Total Environ 736:139690
Neu L, Bänziger C, Proctor CR, Zhang Y, Liu WT, Hammes F (2018) Ugly ducklings—the dark side of plastic materials in contact with potable water. npj Biofilms Microbiomes 4:7
Nielsen TD, Hasselbalch J, Holmberg K, Stripple J (2020) Politics and the plastic crisis: a review throughout the plastic life cycle. Wiley Interdiscp Reviews Energy Environ 9:1–18
Oberbeckmann S, Kreikemeyer B, Labrenz M (2018) Environmental factors support the formation of specific bacterial assemblages on microplastics. Front Microbiol 8:2709
Parthasarathy A, Tyler AC, Hoffman MJ, Savka MA, Hudson AO (2019) Is plastic pollution in aquatic and terrestrial environments a driver for the transmission of pathogens and the evolution of antibiotic resistance? Environ Sci Technol 53:1744–1745
PlasticsEurope (2018) Plastics—the facts 2018: an analysis of European plastics production. Demand and waste data
Pramila R, Ramesh KV (2011) Biodegradation of low density polyethylene (LDPE) by fungi isolated from municipal landfill area. J Microbiol Biotechnol Res 1:131–136
Qiu X, Qi Z, Ouyang Z, Liu P, Guo X (2022) Interactions between microplastics and microorganisms in the environment: modes of action and influencing factors. Gondwana Res 108:102–119
Radisic V, Nimje PS, Bienfait AM, Marathe NP (2020) Marine plastics from Norwegian west coast carry potentially virulent fish pathogens and opportunistic human pathogens harboring new variants of antibiotic resistance genes. Microorganisms 8:1200
Razeghi N, Hamidian AH, Wu C, Zhang Y, Yang M (2021) Microplastic sampling techniques in freshwaters and sediments: a review. Environ Chem Lett 19:4225–4252
Rillig MC (2012) Microplastic in terrestrial ecosystems and the soil? Environ Sci Technol 46:6453–6454
Rodrigues JP, Duarte AC, Santos-Echeandía J, Rocha-Santos T (2019) Significance of interactions between microplastics and POPs in the marine environment: a critical overview. Trends Anal Chem 111:252–260
Shahnawaz M, Sangale MK, Ade AB (2019) Microplastics. In: Bioremediation technology for plastic waste. Springer, Singapore, pp 11–19
Shen M, Huang W, Chen M, Song B, Zeng G, Zhang Y (2020) (Micro)plastic crisis: un-ignorable contribution to global greenhouse gas emissions and climate change. J Clean Prod 254:120138
Sivan A (2011) New perspectives in plastic biodegradation. Curr Opin Biotechnol 22:422–426
Steinmetz Z, Wollmann C, Schaefer M, Buchmann C, David J, Tröger J et al (2016) Plastic mulching in agriculture. Trading short-term agronomic benefits for long-term soil degradation? Sci Total Environ 550:690–705
Su Y, Zhang Z, Wu D, Zhan L, Shi H, Xie B (2019) Occurrence of microplastics in landfill systems and their fate with landfill age. Water Res 164:114968
Sundt P, Schulze PE, Syversen F (2014) Sources of microplastic-pollution to the marine environment. Mepex for the Norwegian Environment Agency, pp 86:20
Uheida A, Mejía HG, Abdel-Rehim M, Hamd W, Dutta J (2021) Visible light photocatalytic degradation of polypropylene microplastics in a continuous water flow system. J Hazard Mater 406:124299

van Wijnen J, Ragas AM, Kroeze C (2019) Modelling global river export of microplastics to the marine environment: sources and future trends. Sci Total Environ 673:392–401

Vethaak AD, Leslie HA (2016) Plastic debris is a human health issue. Environ Sci Technol 50: 6825–6826

Wagner M, Scherer C, Alvarez-Muñoz D, Brennholt N, Bourrain X, Buchinger S et al (2014) Microplastics in freshwater ecosystems: what we know and what we need to know. Environ Sci Europe 26:1–9

Wanner P (2021) Plastic in agricultural soils–a global risk for groundwater systems and drinking water supplies?—a review. Chemosphere 264:128453

Wirth F, Goldani LZ (2012) Epidemiology of Rhodotorula: an emerging pathogen. Interdisciplinary perspectives on infectious diseases 2012, 1

Wong JKH, Lee KK, Tang KHD, Yap PS (2020) Microplastics in the freshwater and terrestrial environments: prevalence, fates, impacts and sustainable solutions. Sci Total Environ 719: 137512

Wright RJ, Erni-Cassola G, Zadjelovic V, Latva M, Christie-Oleza JA (2020) Marine plastic debris: a new surface for microbial colonization. Environ Sci Technol 54:11657–11672

Xu X, Wang S, Gao F, Li J, Zheng L, Sun C, He C, Wang Z, Qu L (2019) Marine microplastic-associated bacterial community succession in response to geography, exposure time, and plastic type in China's coastal seawaters. Mar Pollut Bull 145:278–286

Xu C, Zhang B, Gu C, Shen C, Yin S, Aamir M, Li F (2020) Are we underestimating the sources of microplastic pollution in terrestrial environment? J Hazard Mater 400:123228

Yadav V, Sherly MA, Ranjan P, Tinoco RO, Boldrin A, Damgaard A, Laurent A (2020) Framework for quantifying environmental losses of plastics from landfills. Resour Conserv Recy 161: 104914

Yao L, Hui L, Yang Z, Chen X, Xiao A (2020) Freshwater microplastics pollution: detecting and visualizing emerging trends based on Citespace II. Chemosphere 245:125627

Yurtsever M (2019) Glitters as a source of primary microplastics: an approach to environmental responsibility and ethics. J Agric Environ Ethics 32:459–478

Zettler ER, Mincer TJ, Amaral-Zettler LA (2013) Life in the "plastisphere": microbial communities on plastic marine debris. Environ Sci Technol 47:7137–7146

Zhang S, Wang J, Liu X, Qu F, Wang X, Wang X et al (2019) Microplastics in the environment: a review of analytical methods, distribution, and biological effects. TrAC Trends Anal Chem 111: 62–72

Zhang Y, Kang S, Allen S, Allen D, Gao T, Sillanpää M (2020a) Atmospheric microplastics: a review on current status and perspectives. Earth Sci Rev 203:103118

Zhang Y, Lu J, Wu J, Wang J, Luo Y (2020b) Potential risks of microplastics combined with superbugs: enrichment of antibiotic resistant bacteria on the surface of microplastics in mariculture system. Ecotoxicol Environ Saf 187:109852

Zhu F, Zhu C, Wang C, Gu C (2019) Occurrence and ecological impacts of microplastics in soil systems: a review. Bull Environ Contam Toxicol 102:741–749

Impact of Microplastics on the Physiology of Benthos

21

Mubashar Hussain, Dominique Blache, and Shane K. Maloney

Abstract

Microplastic pollution presents a great menace to the biodiversity occupying aquatic ecosystems by accumulating in benthic environments and adversely affecting biocenoses globally. Microplastics are globally responsible for being a major source of plastics in freshwater and marine environments. Benthos, the bottom-dwelling communities, are likely to combat extreme pressure from microplastics. The content of microplastic fragments revealed the taxa-specific accumulations usually in the digestive tract in benthic organisms. Microplastic interferes with benthos by reducing feeding capacity, lowering energy reserves, slow growth, low reproduction, body deformations, abnormal behavior, affecting lipid metabolism, and interference with swimming. The gastropods and the bivalves exhibit the presence of microplastics, and the filter feeders display the highest quantity of microplastics. Microplastics contribute adversely to the growth and development of benthos. Some adverse effects include reduced growth, inflammation, decreased acetylcholinesterase activity, decreased swimming speed, and intestinal injury. For example, growth inhibition in the sandworm (Annelida), inhibition in the feeding and growth of urchins (Echinodermata), and neurotoxicity in mussels (Mollusca) are caused by microplastics. Microplastics absorb potentially toxic pollutants and contribute to their transfer across the food chain consequently inflicting indirect effects on the benthos. Microplastics have an established adverse impact on the physiology

M. Hussain (✉)
University of Gujrat, Gujrat, Punjab, Pakistan

University of Western Australia, Perth, WA, Australia
e-mail: dr.mubashar@uog.edu.pk

D. Blache · S. K. Maloney
University of Western Australia, Perth, WA, Australia

M. Shahnawaz et al. (eds.), *Microplastic Pollution*,
https://doi.org/10.1007/978-981-99-8357-5_21

of benthos in marine and freshwater ecosystems, which ultimately have ecological consequences. Microplastic management strategies include minimizing the microplastic input into marine ecosystems, restricting single-use plastics, reusing, and recycling plastics, and microbeads. The legislation to curb plastic production and usage, social awareness campaigns, and research initiatives to restrict the entry of microplastics into aquatic ecosystems could save benthos from this adversity.

Keywords

Microplastic pollution · Biodiversity · Benthic environments · Biocenoses · Microplastic impacts

21.1 Introduction

Microplastic contamination is a major environmental issue nowadays (Alfaro-Núñez et al. 2021), and omnipresent plastic waste is also found in marine environments (Guzzetti et al. 2018). These are not individual entities, but they constitute a variety of additives and polymers that can absorb pollutants, nutrients, and other substances (Guzzetti et al. 2018). Microplastics specifically refer to a size of <5 mm of plastic particles (Duis and Coors 2016; Taylor 2019). These tiny particles are hazardous and unfriendly due to the time taken for degradation (Chamas et al. 2020). In aquatic ecosystems, various biological, physical, and chemical components are known to affect the ecological services directly (Reiss et al. 2015). The entry of microplastics in marine and freshwater ecosystems around the globe has raised concerns regarding their long-term effects on ecosystem sustainability and biodiversity (Reid et al. 2019). Various factors including depth, salinity, water temperature, latitude, nature of substrate, competition, predation, and pollution affect the abundance and variety of benthos (Haegerbaeumer et al. 2019; Manokaran et al. 2022). Organic debris and plankton are the major food sources in benthos (Kress 2019; Taylor 2019), and increased levels of plastics in both aquatic and terrestrial ecosystems interfere with the life of benthos (Bellasi et al. 2020). Recent research studies have reported microplastic pollution from isolated and remote areas of the world including Antarctica (Sfriso et al. 2020). The improper disposal of plastic waste has resulted in increased accumulation and dispersion of plastics in the natural environment (Alfaro-Núñez et al. 2021; Teuten et al. 2009). Intense industrialization and overpopulation have resulted in the introduction of new hazards and new compounds in various ecosystems around the world including Antarctica (Stark et al. 2019). In Antarctica, tourism and scientific activities have caused various types of pollution, i.e., microplastics (Cincinelli et al. 2017; Isobe et al. 2017; Munari et al. 2017), organic enrichment (Conlan et al. 2004), metals (Guerra et al. 2011), and chemical contamination by hydrocarbons (Lenihan and Oliver 1995). In addition, localized contamination has been observed in the marine and terrestrial habitats near abandoned or current scientific bases (Bargagli 2008). In this chapter, we have discussed different aspects of plastics on the physiology of benthos such as clams, shrimp-like crustaceans, mussels, oysters, and worms. Environmental hazards of microplastics include bioaccumulation, desorption and toxicity of pollutants, leaching, physical effects, transport of invasive species, and toxicity of additives and monomers (Kärrman et al. 2016).

21.2 Benthic Communities

Benthic organisms can be either categorized as epiflora or epifauna (living on the substrate) and infauna (living inside the substrate by burrowing in the sediment). Organisms living in the benthic region might be sessile (living attached to the rocks or any substrate) or mobile (move freely on or inside the substrate) (Steller et al. 2003). The categorization of aquatic epifauna can be done by sieve mesh size and whether the organisms are mobile or sessile. Freshwater macrophytes are the mobile fauna inhabited by oligochaete worms, insect larvae, macrofaunal gastropods, and meiofaunal nematodes. Benthos occupies all habitats from the intertidal zone to abyssal plains (Taylor 2019). Macroalgae, corals, mussels, sea stars, barnacles, seagrasses, sea urchins, and macroalgae are examples of benthic organisms (Kress 2019; Taylor 2019). Polychaete worms, meiofaunal nematodes, macrofaunal gastropods, harpacticoid copepods, and isopod and amphipod crustaceans inhabit the marine and estuarine macrophytes (Taylor 2019). Increased distance from the shore and increased water depth result in decreased diversity and abundance of benthic communities (Zinger et al. 2011).

21.3 Types of Microplastics

Based on the sources, microplastics are classified as primary and secondary microplastics that are produced from commercial cosmetics and also include the microfibers shed from fishing nets and clothing. Secondary microplastics originate when large plastics degrade under the influence of environmental factors, i.e., ocean waves and solar radiations (Browne 2015). Regardless of their sources, microplastics are harmful to the environment as they cannot degrade into harmless molecules in a short duration of time (Campanale et al. 2020).

21.4 Microplastics in Aquatic Environment

In an aquatic environment, microplastics are consumed by marine biota, i.e., planktons, corals, invertebrates, whales, and fishes, hence becoming a part of the food chain (Sharma and Chatterjee 2017). Human health and ecosystem sustainability are at risk due to increasing levels of microplastics in aquatic ecosystems (Rubin and Zucker 2022). The entry of microplastics into aquatic environments poses a serious threat to the biodiversity and ecosystem (Vivekanand et al. 2021). Microplastics from coastal, domestic, and industrial activities enter into the aquatic ecosystem and are found distributed in various regions of the marine ecosystem (depending upon the shape, density, and size) and finally settle in the benthic zone (Sharma and Chatterjee 2017).

21.5 Ingestion of Microplastics by Benthos

Microplastic particles are readily available for ingestion to almost all biota in the benthic zone due to their smaller size (do Sul and Costa 2014). Crustaceans and planktons in the pelagic zone encounter low-density microplastics, whereas benthic biota (mollusks, tubifex and polychaete worms, and amphipods) are exposed to high-density microplastics (De Sá et al. 2018). Microbeads and microfibers are a common source for microplastics and have been reported in marine ecosystems for last three decades (Kärrman et al. 2016). The benthic marine biota ingests microplastics in the form of microbeads and microfibers (Chatterjee and Sharma 2019). The ingestion of microfibers has been found in six benthic invertebrate species (Piarulli et al. 2020). Ingestion of microplastics depends on the particle concentration, mode of feeding, and encounter rate (Setälä et al. 2016). Studies on the species-specific (such as bivalves and crustaceans) ingestion of microplastics showed that exposure of relatively low concentrations (10 μm microbeads) resulted in the ingestion of microbeads by all animals. Significantly higher levels of plastic beads were found in some bivalves such as *Mytilus trossulus* and *Macoma balthica* (Setälä et al. 2016). It was also observed that free-swimming crustaceans have demonstrated ingestion of more beads than benthic communities (Kärrman et al. 2016). Microplastic fragments mistakenly ingested as food particles by corals, lobsters, fish, sea urchins, and zooplanktons are transferred to higher trophic levels. The microplastics feeding in corals occur by ingestion of microplastics, retention of these fragments, and finally digestion (Lusher et al. 2015). Microplastics affect the physiology of corals by their retention in mesenterial tissue, and this results in reduced feeding capacity and lowered energy reserves (Reichert et al. 2018).

21.6 Residency of Microplastics in Benthos

Microplastic beads (0.5 μm) have been reported in crab tissues and hemolymph (Farrell and Nelson 2013). Mussels have shown the uptake of plastics (microspheres of 3 and 10 μm) into the gut and circulatory system (Bouwmeester et al. 2015; Browne et al. 2008). Bivalve species may have lysosomal perturbations and oxidative stress due to intracellular uptake of microplastics (Canesi et al. 2012), but ingestion and translocation of microplastics by larvae of the brine shrimp have shown adverse effects on feeding, physiology, and behavior (Bergami et al. 2016). The size and shape determine the time of plastics to stay inside an organism between ingestion and excretion. The smaller particle sizes tend to take longer time before excreted (Browne et al. 2008). Additionally, studies have reported that the residence time in organism body is specific (Farrell and Nelson 2013). For example, residence time in mussels (over 48 days) (Browne et al. 2008), crabs (less than 21 days) (Farrell and Nelson 2013), and lobsters over 24 h (Murray and Cowie 2011).

21.7 Effects of Microplastics on Benthos

The uptake of microplastics by the benthic marine biota impacts their development, growth, and reproductive system (Gao et al. 2022). Due to small size (<5 μm), a wide variety of organisms ingest the microplastics causing health issues, posing a threat to marine biota and marine ecosystem (Guzzetti et al. 2018). Extreme microplastic pressure has been confronted by the marine biota in the benthic region (Gao et al. 2022). Microplastics have the capability to accumulate in specific tissues of benthos resulting in reproductive toxicity and genotoxicity in benthic biota (Gao et al. 2022).

21.7.1 Health Hazard to Benthos

The health hazards associated with microplastics are not only because of their ingestion, but they also possess the properties to concentrate the pollutants in seawater, hence transferring them across the food chains (Guzzetti et al. 2018). Moreover, microplastics can also affect the ecosystem services and various ecological processes. Plastic debris has been considered emerging contaminant posing a great threat to marine biota across the globe (Guzzetti et al. 2018). The availability of microplastics for marine organisms is a major risk because they mimic food, hence obstructing and negatively affecting the functioning of the digestive system (Gall and Thompson 2015).

21.7.2 Impact of Microplastics on Feeding

Microplastics have been found to negatively influence the planktons, which play a very important role in marine ecosystem. Microplastics penetrate the cell walls of planktons and result in decreased chlorophyll absorption (Nerland et al. 2014). Heterotrophic planktons consume microplastics by phagocytosis, and these plastic fragments are retained in their tissues (Laist 1987). Zooplanktons are the primary consumers in marine food chains and play an important role in marine ecosystem. A wide range of feeding mechanisms have been observed in zooplanktons, and they mostly rely upon chemo-mechano receptors for capturing prey (DeMott 1988). Greater chances of interaction between zooplanktons and microplastics exist due to increased concentration of microplastics in marine water and the same dimensions of both (>333 μm) (Cole et al. 2013). Experimental evidence indicates the consumption of latex beads by zooplanktons after exposure to microplastics (Li et al. n.d.). Another research study reported the tendency of zooplanktons to ingest polystyrene beads with 1.7–30.6 μm dimensions. Another incidence of microplastic (7.3 μm) ingestion was reported in *Centropages typicus*, a copepod, resulting in the loss of feeding capability, hence negatively affecting their health (Cole et al. 2013).

21.7.3 Impact of Microplastics on Growth and Development

The exposure to polyhydroxybutyrate and polymethyl methacrylate microplastics results in decreased growth rate in *Gammarus fossarum* (Straub et al. 2017). Reduced reproduction and growth process has been observed due to polyethylene microplastic ingestion by *Hyalella azteca in the benthic region* (Au et al. 2015). A decreased feeding ability and weight loss have been observed in *Arenicola marina (marine lugworm) due to microplastic ingestion* (Besseling et al. 2013).

21.7.4 Impact of Microplastic on Behavior

Studies have shown that no effects of microplastic fragments on *G. fossarum* were reported (Blarer and Burkhardt-Holm 2016). Feeding behavior of *L. variegatus*, *Tubifex* spp., *A. aquaticus*, *G. pulex*, H. azteca, or S. corneum showed no alterations in the feeding behavior when exposed to microplastic fragments (Redondo-Hasselerharm et al. 2018).

21.7.4.1 Cellular Response

The uptake of microplastics adversely affects their growth, development, and reproductive system. In benthic organisms, these microplastics accumulate in gills and intestines resulting in toxic effects on tissues, which results in genotoxicity and reproductive toxicity (Gao et al. 2022). For example, the impact of microplastics on the cellular response such as reactive oxygen species production, gene expression, and enzyme activity at various concentrations on the mussel (D. polymorpha) was measured using various biomarker (Magni et al. 2018). Microplastics have the potential to cause respiratory inefficiencies, and rate of respiration in marine bivalves is affected by microplastics (Mkuye et al. 2022).

21.7.4.2 Impact of Microplastic on Reproduction

Microplastics interfere with the reproduction of benthic communities such as egg production, fecundity, fertility, growth rate, and population size. For example, a study reported that microplastic *P. crassirostris* showed more sensitivity to population size as compared to egg production (Heindler et al. 2017). *B. manjavacas* demonstrated significantly reduced population growth rates when encountered with microplastic spheres. Plastic beads showed a significant effect on the fecundity of *T. japonicus* (Lee et al. 2013). Plastic spheres affect ferritization rates of *P. lividus* significantly (Martínez-Gómez et al. 2017) and low numbers of oocytes of *C. gigas* (Sussarellu et al. 2016). The studies have reported negative effects of microplastics on *C. helgolandicus* (Cole et al. 2015), *T. gratilla* (Kaposi et al. 2014), *P. lividus*

(Martínez-Gómez et al. 2017), *M. galloprovincialis* and *M. edulis* (Paul-Pont et al. 2016), and *C. robusta* (Messinetti et al. 2018).

21.8 Conclusion

Microplastics despite their numerous valuable uses have shown several ecological and health consequences on marine ecosystem. Microplastics have now been reported to form a new marine microbial habitat called the "plastisphere." Microplastics now are considered a great menace to marine biodiversity and food chains. Marine life (mussels, worms, fish, and even plankton) ingests microplastics. Microplastics are not only reported to have deleterious effects directly but are also known to concentrate toxic pollutants (pesticides). Microplastics entering oceans in the form of microbeads interfere with the physiology of benthos when ingested resulting in poor growth and development. Ecologically, microplastics could create hurdles for benthos to move and reach for the food particles. The studies have reported that microplastics pose threats to the reproduction of benthic communities.

References

Alfaro-Núñez A, Astorga D, Cáceres-Farías L, Bastidas L, Soto Villegas C, Macay KC, Christensen JH (2021) Microplastic pollution in seawater and marine organisms across the tropical eastern Pacific and Galápagos. Sci Rep 11(1):1–8

Au SY, Bruce TF, Bridges WC, Klaine SJ (2015) Responses of Hyalella azteca to acute and chronic microplastic exposures. Environ Toxicol Chem 34(11):2564–2572

Bargagli R (2008) Environmental contamination in Antarctic ecosystems. Sci Total Environ 400(1–3):212–226

Bellasi A, Binda G, Pozzi A, Galafassi S, Volta P, Bettinetti R (2020) Microplastic contamination in freshwater environments: a review, focusing on interactions with sediments and benthic organisms. Environments 7(4):30

Bergami E, Bocci E, Vannuccini ML, Monopoli M, Salvati A, Dawson KA, Corsi I (2016) Nano-sized polystyrene affects feeding, behavior and physiology of brine shrimp Artemia franciscana larvae. Ecotoxicol Environ Saf 123:18–25

Besseling E, Wegner A, Foekema EM, Van Den Heuvel-Greve MJ, Koelmans AA (2013) Effects of microplastic on fitness and PCB bioaccumulation by the lugworm Arenicola marina (L.). Environ Sci Technol 47(1):593–600

Blarer P, Burkhardt-Holm P (2016) Microplastics affect assimilation efficiency in the freshwater amphipod Gammarus fossarum. Environ Sci Pollut Res 23:23522–23532

Bouwmeester H, Hollman PC, Peters RJ (2015) Potential health impact of environmentally released micro-and nanoplastics in the human food production chain: experiences from nanotoxicology. Environ Sci Technol 49(15):8932–8947

Browne MA (2015) Sources and pathways of microplastics to habitats. In: Marine anthropogenic litter, pp 229–244

Browne MA, Dissanayake A, Galloway TS, Lowe DM, Thompson RC (2008) Ingested microscopic plastic translocates to the circulatory system of the mussel, Mytilus edulis (L.). Environ Sci Technol 42(13):5026–5031

Campanale C, Stock F, Massarelli C, Kochleus C, Bagnuolo G, Reifferscheid G, Uricchio VF (2020) Microplastics and their possible sources: the example of Ofanto river in Southeast Italy. Environ Pollut 258:113284

Canesi L, Ciacci C, Fabbri R, Marcomini A, Pojana G, Gallo G (2012) Bivalve molluscs as a unique target group for nanoparticle toxicity. Mar Environ Res 76:16–21

Chamas A, Moon H, Zheng J, Qiu Y, Tabassum T, Jang JH et al (2020) Degradation rates of plastics in the environment. ACS Sustain Chem Eng 8(9):3494–3511

Chatterjee S, Sharma S (2019) Microplastics in our oceans and marine health. Field actions science reports. J Field Actions (Special Issue 19), 54–61

Cincinelli A, Scopetani C, Chelazzi D, Lombardini E, Martellini T, Katsoyiannis A et al (2017) Microplastic in the surface waters of the Ross Sea (Antarctica): occurrence, distribution and characterization by FTIR. Chemosphere 175:391–400

Cole M, Lindeque P, Fileman E, Halsband C, Goodhead R, Moger J, Galloway TS (2013) Microplastic ingestion by zooplankton. Environ Sci Technol 47(12):6646–6655

Cole M, Lindeque P, Fileman E, Halsband C, Galloway TS (2015) The impact of polystyrene microplastics on feeding, function and fecundity in the marine copepod Calanus helgolandicus. Environ Sci Technol 49(2):1130–1137

Conlan KE, Kim SL, Lenihan HS, Oliver JS (2004) Benthic changes during 10 years of organic enrichment by McMurdo Station, Antarctica. Mar Pollut Bull 49(1–2):43–60

De Sá LC, Oliveira M, Ribeiro F, Rocha TL, Futter MN (2018) Studies of the effects of microplastics on aquatic organisms: what do we know and where should we focus our efforts in the future? Sci Total Environ 645:1029–1039

DeMott WR (1988) Discrimination between algae and detritus by freshwater and marine zooplankton. Bull Mar Sci 43(3):486–499

do Sul JAI, Costa MF (2014) The present and future of microplastic pollution in the marine environment. Environ Pollut 185:352–364

Duis K, Coors A (2016) Microplastics in the aquatic and terrestrial environment: sources (with a specific focus on personal care products), fate and effects. Environ Sci Eur 28(1):1–25

Farrell P, Nelson K (2013) Trophic level transfer of microplastic: Mytilus edulis (L.) to Carcinus maenas (L.). Environ Pollut 177:1–3

Gall SC, Thompson RC (2015) The impact of debris on marine life. Mar Pollut Bull 92(1–2): 170–179

Gao D, Liu X, Junaid M, Liao H, Chen G, Wu Y, Wang J (2022) Toxicological impacts of micro (nano) plastics in the benthic environment. Sci Total Environ 836:155620

Guerra R, Fetter E, Ceschim LM, Martins CC (2011) Trace metals in sediment cores from deception and Penguin islands (South Shetland Islands, Antarctica). Mar Pollut Bull 62(11):2571–2575

Guzzetti E, Sureda A, Tejada S, Faggio C (2018) Microplastic in marine organism: environmental and toxicological effects. Environ Toxicol Pharmacol 64:164–171

Haegerbaeumer A, Mueller M-T, Fueser H, Traunspurger W (2019) Impacts of micro-and nano-sized plastic particles on benthic invertebrates: a literature review and gap analysis. Front Environ Sci 7:17

Heindler FM, Alajmi F, Huerlimann R, Zeng C, Newman SJ, Vamvounis G, van Herwerden L (2017) Toxic effects of polyethylene terephthalate microparticles and Di (2-ethylhexyl) phthalate on the calanoid copepod, Parvocalanus crassirostris. Ecotoxicol Environ Saf 141:298–305

Isobe A, Uchiyama-Matsumoto K, Uchida K, Tokai T (2017) Microplastics in the southern ocean. Mar Pollut Bull 114(1):623–626

Kaposi KL, Mos B, Kelaher BP, Dworjanyn SA (2014) Ingestion of microplastic has limited impact on a marine larva. Environ Sci Technol 48(3):1638–1645

Kärrman A, Schönlau C, Engwall M (2016) Exposure and effects of microplastics on wildlife: a review of existing data. Retrieved from Örebro University, Sweden

Kress N (2019) Chapter 3. Seawater quality for desalination plants. In: Marine impacts of seawater desalination, pp 35–52

Laist DW (1987) Overview of the biological effects of lost and discarded plastic debris in the marine environment. Mar Pollut Bull 18(6):319–326

Lee K-W, Shim WJ, Kwon OY, Kang J-H (2013) Size-dependent effects of micro polystyrene particles in the marine copepod Tigriopus japonicus. Environ Sci Technol 47(19):11278–11283

Lenihan HS, Oliver JS (1995) Anthropogenic and natural disturbances to marine benthic communities in Antarctica. Ecol Appl:311–326

Li S, Li F, Liua Q, Suzukib Y (n.d.) Processes and impacts. Management, 16, 17

Lusher AL, Hernandez-Milian G, O'Brien J, Berrow S, O'Connor I, Officer R (2015) Microplastic and macroplastic ingestion by a deep diving, oceanic cetacean: the True's beaked whale Mesoplodon mirus. Environ Pollut 199:185–191

Magni S, Gagné F, André C, Della Torre C, Auclair J, Hanana H et al (2018) Evaluation of uptake and chronic toxicity of virgin polystyrene microbeads in freshwater zebra mussel Dreissena polymorpha (Mollusca: Bivalvia). Sci Total Environ 631:778–788

Manokaran S, Joydas TV, Khan A (2022) Physico-chemical factors regulating marine benthos structure and function. In: Ecology and biodiversity of benthos. Elsevier, pp 209–250

Martínez-Gómez C, León VM, Calles S, Gomáriz-Olcina M, Vethaak AD (2017) The adverse effects of virgin microplastics on the fertilization and larval development of sea urchins. Mar Environ Res 130:69–76

Messinetti S, Mercurio S, Parolini M, Sugni M, Pennati R (2018) Effects of polystyrene microplastics on early stages of two marine invertebrates with different feeding strategies. Environ Pollut 237:1080–1087

Mkuye R, Gong S, Zhao L, Masanja F, Ndandala C, Bubelwa E et al (2022) Effects of microplastics on physiological performance of marine bivalves, potential impacts, and enlightening the future based on a comparative study. Sci Total Environ 838:155933

Munari C, Infantini V, Scoponi M, Rastelli E, Corinaldesi C, Mistri M (2017) Microplastics in the sediments of terra nova bay (ross sea, Antarctica). Mar Pollut Bull 122(1–2):161–165

Murray F, Cowie PR (2011) Plastic contamination in the decapod crustacean Nephrops norvegicus (Linnaeus, 1758). Mar Pollut Bull 62(6):1207–1217

Nerland IL, Halsband C, Allan I, Thomas KV (2014) Microplastics in marine environments: occurrence, distribution and effects

Paul-Pont I, Lacroix C, Fernández CG, Hégaret H, Lambert C, Le Goïc N et al (2016) Exposure of marine mussels Mytilus spp. to polystyrene microplastics: toxicity and influence on fluoranthene bioaccumulation. Environ Pollut 216:724–737

Piarulli S, Vanhove B, Comandini P, Scapinello S, Moens T, Vrielinck H et al (2020) Do different habits affect microplastics contents in organisms? A trait-based analysis on salt marsh species. Mar Pollut Bull 153:110983

Redondo-Hasselerharm PE, Falahudin D, Peeters ET, Koelmans AA (2018) Microplastic effect thresholds for freshwater benthic macroinvertebrates. Environ Sci Technol 52(4):2278–2286

Reichert J, Schellenberg J, Schubert P, Wilke T (2018) Responses of reef building corals to microplastic exposure. Environ Pollut 237:955–960

Reid AJ, Carlson AK, Creed IF, Eliason EJ, Gell PA, Johnson PT et al (2019) Emerging threats and persistent conservation challenges for freshwater biodiversity. Biol Rev 94(3):849–873

Reiss H, Birchenough S, Borja A, Buhl-Mortensen L, Craeymeersch J, Dannheim J et al (2015) Benthos distribution modelling and its relevance for marine ecosystem management. ICES J Mar Sci 72(2):297–315

Rubin AE, Zucker I (2022) Interactions of microplastics and organic compounds in aquatic environments: a case study of augmented joint toxicity. Chemosphere 289:133212

Setälä O, Norkko J, Lehtiniemi M (2016) Feeding type affects microplastic ingestion in a coastal invertebrate community. Mar Pollut Bull 102(1):95–101

Sfriso AA, Tomio Y, Rosso B, Gambaro A, Sfriso A, Corami F et al (2020) Microplastic accumulation in benthic invertebrates in Terra Nova Bay (Ross Sea, Antarctica). Environ Int 137:105587

Sharma S, Chatterjee S (2017) Microplastic pollution, a threat to marine ecosystem and human health: a short review. Environ Sci Pollut Res 24(27):21530–21547

Stark JS, Raymond T, Deppeler SL, Morrison AK (2019) Antarctic seas. In: World seas: an environmental evaluation. Elsevier, pp 1–44

Steller DL, Riosmena-Rodríguez R, Foster M, Roberts C (2003) Rhodolith bed diversity in the Gulf of California: the importance of rhodolith structure and consequences of disturbance. Aquat Conserv Mar Freshwat Ecosyst 13(S1):S5–S20

Straub S, Hirsch PE, Burkhardt-Holm P (2017) Biodegradable and petroleum-based microplastics do not differ in their ingestion and excretion but in their biological effects in a freshwater invertebrate Gammarus fossarum. Int J Environ Res Public Health 14(7):774

Sussarellu R, Suquet M, Thomas Y, Lambert C, Fabioux C, Pernet MEJ et al (2016) Oyster reproduction is affected by exposure to polystyrene microplastics. Proc Natl Acad Sci U S A 113(9):2430–2435

Taylor RB (2019) Epiflora and epifauna

Teuten EL, Saquing JM, Knappe DR, Barlaz MA, Jonsson S, Björn A et al (2009) Transport and release of chemicals from plastics to the environment and to wildlife. Philos Trans R Soc B Biol Sci 364(1526):2027–2045

Vivekanand AC, Mohapatra S, Tyagi VK (2021) Microplastics in aquatic environment: challenges and perspectives. Chemosphere 282:131151

Zinger L, Amaral-Zettler LA, Fuhrman JA, Horner-Devine MC, Huse SM, Welch DBM et al (2011) Global patterns of bacterial beta-diversity in seafloor and seawater ecosystems. PLoS One 6(9): e24570

Microplastics and Microorganisms: A Evolutionary Symbiosis in Terrestrial Environment

22

Rajalakshmi Sridharan and K. Veena Gayathri

Abstract

Plastics, an inevitable, unavoidable synthetic resource, have become part of every individual. The application of plastics has gained importance and increased in demand due to its cost-effectiveness and resistance to biotic and abiotic factors. Prolonged use of plastics increased the dispersion of waste irrespective of the ecosystem. Then, the threat started, which became a nightmare for every individual on the planet. The inevitability of plastic has made it difficult to avoid the usage or relay or an effective alternative. The current chapter focuses on the accumulation of plastic wastes (irrespective of the nature of plastics), the treatment processes, and specifically on microplastics. The facts and research propose that the interaction between microplastics and microbes dates back to the discovery of plastics. The high surface-to-volume ratio of microplastics attracts pollutants and microbes to coexist. The microbes, in turn, gained resistance to the plastic toxicity and fed on it as a carbon source for survival. This pattern paved a way for the discovery of biological degradation of plastics/microplastics. The symbiosis will exist for years to come until the availability of plastics and their derivatives, which provides researchers a platform to discover methods to eliminate plastic wastes in the environment. Microplastics, due to their smaller size, have entered the food chain, which is a major cause of ill effects in living organisms. Hence, understanding the symbiotic relationship and utilizing it to eradicate plastic wastes will reduce the environmental toxic impact in the forthcoming life.

R. Sridharan · K. Veena Gayathri (✉)
Department of Biotechnology, Stella Maris College (Autonomous), Affiliated to University of Madras, Chennai, Tamil Nadu, India
e-mail: veenagayathri@stellamariscollege.edu.in

M. Shahnawaz et al. (eds.), *Microplastic Pollution*,
https://doi.org/10.1007/978-981-99-8357-5_22

Keywords

Plastics · Microplastics · Biofilm · Microorganisms · Biodegradation

22.1 Introduction

Revolutions, of which the plastic revolution is one, help the advancement of science and the well-being of the world. Plastic industries originated in the 1900s, which eased the usage with increased production at a low cost. Plastic production increased multifold in a decade with widespread applications (Amelia et al. 2021). Plastics are known to be derived from fossil fuels and renewable resources. The survey states that nearly 70% of plastic waste is dumped into landfills or spread into the environment (Geyer 2020). Despite several applications of plastic in various sectors, the release of plastics into the environment leads to the accumulation of plastic debris, which makes it a global concern (Li et al. 2018b). The release of plastic debris into terrestrial and aquatic ecosystems mostly includes plastic bottles, covers, wrappers, and other forms of plastics (Blettler et al. 2017; Li et al. 2018a). As plastics originate from petroleum, it is resistant to natural degradation and decomposition. The aftermath of plastic usage is and will be faced in future as the consumption of plastics directly or indirectly by ingestion leads to the destabilization of the food chain. Plastic waste ingestion leads to several consequences and to name a few—decreased appetite, decrease in body mass, negative impact on feeding and fitness, and increased fatality (Farrell and Nelson 2013; Wright et al. 2013; Besseling et al. 2013; Lee et al. 2013). Despite the hazardous characteristics of plastics, the advantages of plastics outweigh them. Plastics that have been mechanically damaged experience cleavage of the polymer, which weathers the polymeric chain. According to their sizes, the created tiny particles are referred to as microplastics (MPs) and nanoplastics (NPs). According to Frias and Nash (2019), microplastics can range in size from 1 to 5 mm and can have regular or irregular shapes that come from primary or secondary sources. According to the sizes, it was categorized by Blettler et al. (2017). They claimed that the particles ranged in size from >2 cm for macroplastics to 5 cm for mesoplastics to 5 cm for nanoplastics. The morphology-based diversity of microplastics is shown in Fig. 22.1.

Plastic disintegration leading to MP formation is known to enter into terrestrial and aquatic ecosystems. The marine environments are reported to contain 140 particles/m^3 of MP (Thushari and Senevirathna 2020). Research on MPs resulted in another negative impact of it on the environment. The MPs are known to attract diverse hazardous pollutants such as heavy metals and persistent organic pollutants (POPs) (Kwon et al. 2017). Light-colored MPs are reported to contain low molecular weight PAHs (Fisner et al. 2017). PCBs and phenanthrene are also reported to be present in high-density MPs (Fries and Zarfl 2012). MPs are considered a threat to the biosphere, while their aging causes unrepairable aftermath such as increased surface area, high generation of reactive oxygen species, increased polarity, and roughness (Fries and Zarfl 2012; Fotopoulou and Karapanagioti 2012).

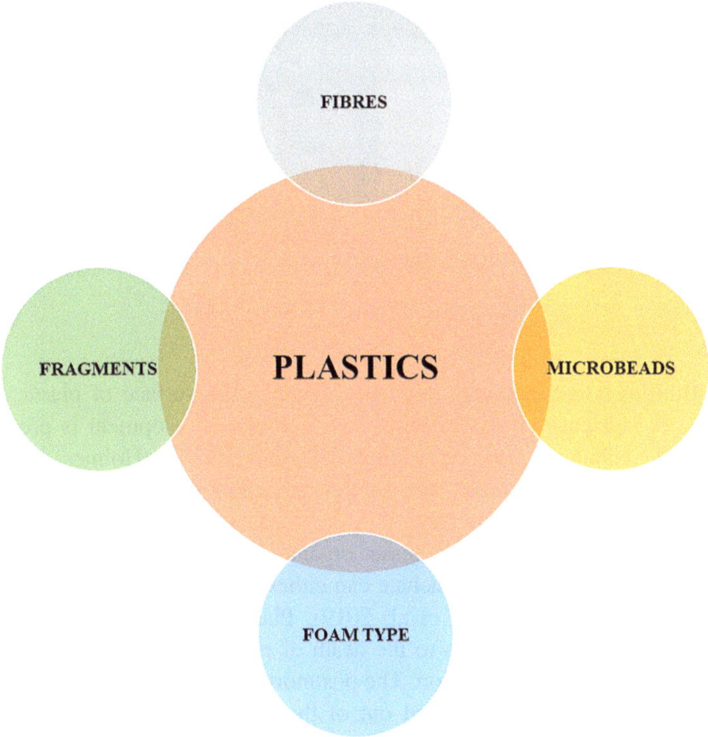

Fig. 22.1 Types of microplastics based on morphology

These characteristics further increase the sorption of pollutants (Hüffer et al. 2018). Crystallinity, stability, the method of manufacture, polarity, and the functional groups contained in MPs are all variables that affect the sorption of contaminants on MPs (Rochman et al. 2013). Hydrophobic MPs are known to adsorb hydrophobic pollutants like PAHs, while hydrophilic MPs are reported to adsorb hydrophilic antibiotics (Rochman et al. 2013; Li et al. 2018a, b). Organic matter (dissolved and particulate) alters the sorption capacity of MPs (Besseling et al. 2017). The acidic pH is said to decrease the sorption capacity of HDPE to metals such as Pb, Co, Ni, and Cd (Holmes et al. 2012).

22.2 Plastics and Microbes

Microbes, the key to sustainability in the biosphere, have its prime importance in several applications. They inhabit diverse substrates and hosts. The interactions between diverse pollutants and microbes are well reported, while the existence of microbes on plastic surfaces is still being explored. Although the connection between bacteria and plastics has long existed, it was not recognized until the

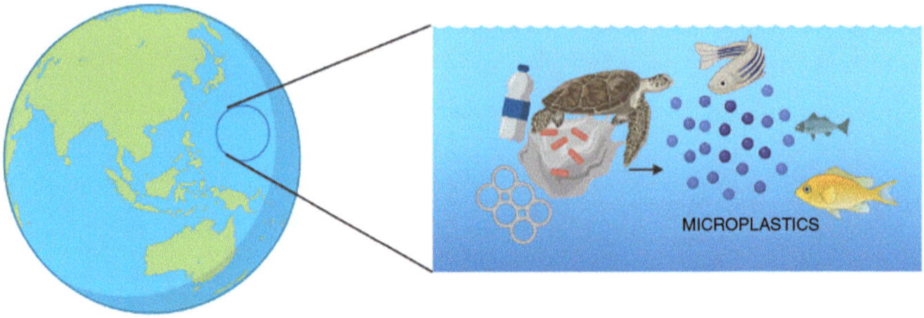

Fig. 22.2 Formation of MPS in marine environment

1970s. Biofilm is created when microbes colonize the surface of plastic trash or materials. A nutrient-rich environment for microbial development is provided by plastics with enhanced sorption capacity to contaminants (Holmes et al. 2012; Galgani et al. 2019). According to research by Dussud et al. (2018), Aestuariicella hydrocarbonica and Marinobacter sp. are the main bacteria that colonize polymers. Due to deterioration, plastics release organic material into the environment. Depending on its toxicity, the leachate can either promote or prevent the development of microorganisms (Tetu et al. 2019). Plastic pollution inhibits microbial diversity. Plastic pollution leads to the death of marine organisms majorly due to plastic contamination and ingestion. The postmortem report of the marine organism was brutal as plastic wastes spilled out of the gut during the process. Research suggests that the plastics tend to lodge onto the gastrovascular cavity of the organisms irrespective of the trophic level to which they belong (Wilcox et al. 2015). MP ingestion compromises host immunity. After a week of MP exposure, presence of MPs in the gut lumen, intestinal lumen, and other parts was observed. Reactive oxygen species (ROS) produced by plastics causes genotoxicity by breaking DNA strands. MPs are also reported to trigger immunological and inflammatory responses (Deng et al. 2017). Plasticizers used in the preparation process have an intense negative impact on ingested organisms. Dibutyl phthalates are reported to delay gonad development in fishes, while bisphenol A reduces sperm count in males and delays or inhibits ovulation in female fishes (Lahnsteiner et al. 2005). The ability of microbes to adsorb onto the MP surface increases the dispersion and survival of microbes. Microbes shield MPs from environmental stresses, while they get benefitted by horizontal gene transfer.

Plastics entering into the terrestrial ecosystem embed in the soil, therefore inhibiting soil–water interaction and soil microbial community (Drenovsky et al. 2004). Kirstein et al. (2016) stated the microbes found in the plastisphere as "dangerous hitchhikers" as these microbes are pathogenic, antibiotic-resistant, and toxic. Algal bloom caused by *Alexandrium* sp., *Ostreopsis* sp., *Coolia* sp., and *Ceratium* sp. forms on a floating plastic surface of which cysts and vegetative cells are toxic. It is reported in the Mediterranean Sea and Atlantic Ocean (Zettler et al. 2013). Figure 22.2 shows the formation of MPs by withering caused by microbes.

Table 22.1 Overview of the treatment process used globally to reduce plastic waste accumulation

Method	Merit	Demerit	References
Pyrolysis	It helps to faster commercialization of products	High capital cost	Evode et al. (2021); Awasthi and Li (2017)
Bioremediation	Sustainable remediation of post-consumer plastics	These techniques are harmful to the environment; therefore, a huge investment is required to employ the degradation of synthetic compounds by enzymes	Evode et al. (2021)
Landfills	Low-cost waste disposal method	Carbon dioxide emission	Ghasemi and Mohd. Yusuff (2016)
Incineration	Eliminates harmful germs and chemicals	Risk to human health, and living or working near an incineration facility can have social, economic, and psychological effects	Ghasemi and Mohd. Yusuff (2016)
Recycling	Energy conservation	Limited to a few cycles using the minor fraction of the purest and cleanest waste stream	Lange (2021)
Gasification	Energy conserved is higher than the energy generated by incineration	Tar formation and agglomeration	Antelava et al. (2021)

Plastic wastes are treated in various ways, which are tabulated in Table 22.1.

Pyrolysis is the commonly used method as it results in the production of liquid fuel while anaerobic degradation of plastics was an ancient biological method. These methods have their own pros and cons as the leachate from the landfill might contaminate the groundwater. While the ashes produced by the pyrolysis are still a problem, the mechanical method of plastic degradation reduces the volume of plastics, but it is not an effective process (Voss et al. 2022). Although the process of degradation is effectively studied, UNEP and NOAA organizations encourage the production of biodegradable plastics (Sheveealy et al. 2012) and about 1,00,000 bioplastics were produced in the year 2017 The polymeric blends are used to retain the characteristics of the synthetic plastics as bioplastics are not flexible as synthetic plastics. The blends are selected based on the application of the bioplastic (Narancic et al. 2018).

22.3 Plastic Degradation by Microorganisms

According to reports, the output of synthetic plastic was 400 metric tons in 2020 and was projected to reach 600 metric tons by 2050 (Cornwall 2021). Single-use plastics such as cups, covers, bottles, and masks made up the majority of the plastics manufactured, which led to trash creation. No matter the kind or branching of the polymeric chain, the properties of plastics are the same. The main element that helps plastics survive for years is their resistance. During the production of plastic, over 850 metric tons of carbon dioxide are created annually. Despite the creation of plastic that is biodegradable and compostable, it is urgent to remove the plastic garbage that has accumulated in the environment. Many pieces of prior work revealed the negative consequences that plastics made from fossil fuels have on biological things. Microorganisms are the only community that can survive the hazardous effects of plastics and their byproducts (Purohit et al. 2020). The enzymes released by bacteria, which break the link and result in depolymerization, help in the deterioration of plastics. This happens when hydrocarbons are converted into oligomers or monomers by hydrolysis or oxidation. Because of their tiny size, the cleavage compounds are transported to the cytosol, where assimilation takes place. As a result of the absorbed particles being used as a carbon source (Pathak and Navneet. 2017), biomass and carbon dioxide are created (Fig. 22.3).

The decomposition of PET is facilitated or improved by the enzymes known as PETase and MHETase, which turn the polymers (polymer) into a monomer, following the detection of the PET-degrading bacterium Ideonella sakaiensis. The homogeneity of PETase to hydrolase enzymes was shown by crystallographic analysis,

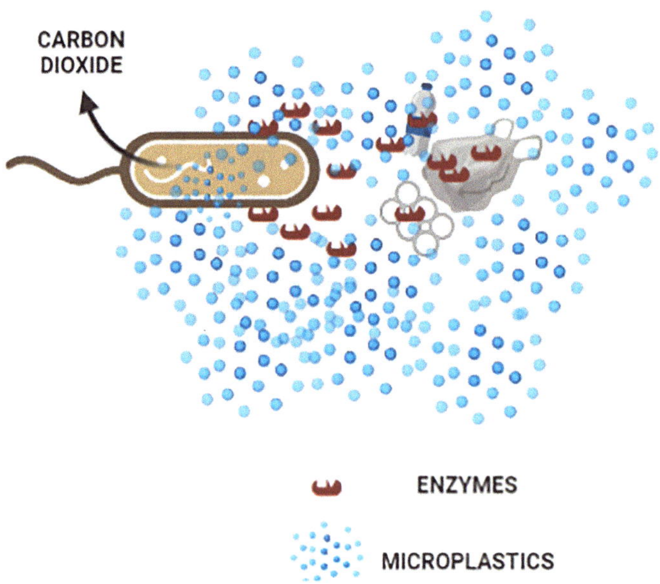

Fig. 22.3 Biodegradation of plastic wastes

which was used to characterize these enzymes (Yoshida et al. 2016; Lee et al. 2017; Joo et al. 2018). It was discovered that the enzymes responsible for degrading plastic had the hydrolase fold. Palm et al. (2019) conducted a thorough study of the MHETase active site and substrate binding. The degradation of plastics by bacterial enzymes is hardly reported, and there are very few studies indicating the molecular determination of the enzymes involved in plastic degradation.

Some microbes involved in the degradation of plastics are as follows: *P. aeruginosa* in PE degradation (Mouafo Tamnou et al. 2021), *Aspergillus* sp. consortium (DSouza et al. 2021), *Acinetobacter* sp., *Arthrobacter* sp., *Bacillus* sp., *Brevibacillus* sp., *Flavobacterium* sp., *Micrococcus* sp., *Paenibacillus* sp., *Pseudomonas* sp., *Rhodococcus* sp., *Staphylococcus* sp., *Streptomyces* sp., *Penicillium* sp., *Fusarium* sp., and *Phanerochaete* sp. The degradation of PE carried out by the soil burial method is also reported in PE degradation (Restrepo-Flórez et al. 2014). Winogradsky column is a soil ecosystem developed by Sergi Winogradsky to study the ecology of microbes and its interactions with the environment. The column was used to study the microbial community in specific areas and identify the biogeochemical cycle of the particular environment. Babcsányi et al. (2017) reported the biogeochemical gradient of the polluted wetland using the Winogradsky column. The prediction of the sulfur cycle used the Winogradsky column. There is no research extensively on the application of the same in pollutant degradation, which might affect/alter the biogeochemical cycle and microbial community. Thus, it helps us to gain knowledge in understanding the soil ecology of the polluted environment. PETase, cutinase, and lipase are commonly reported enzymes in plastic degradation (Tokiwa et al. 2009). Enzymes hydrolytically cleave the C–H backbone and form metabolites. Plastics such as polyethylene (PE), polyethylene terephthalate (PET), polyurethane (PU), and other types are reported to undergo slower degradation. The hydrophobic cleft in the hydrolase enzyme in the active site accommodates the hydrophobic groups in the plastic polymer and forms monomers or dimers (Wilkes and Aristilde 2017).

Plastic wastes are discarded/treated/recycled after manual separation of the wastes. There are studies that promote the pretreatment of plastic wastes using UV radiation to weaken the backbone. This leads to weathering of plastics into fragments, which could be easily utilized by microbes, but the pretreatment might consume time, which extends the duration of degradation. Thus, opting for the plastic-resistant microbes/plastic-consuming microbes in the removal is a wise alternative. Dey and Michael (2021) reported the degradation of unpretreated LDPE using *Stenotrophomonas* sp. and *Achromobacter* sp. These strains were isolated from dumpsites, which have gained resistance due to prolonged exposure. Gajendiran et al. (2016) reported degradation of LDPE using *Aspergillus* sp. isolated from landfill soil. Even though some bacterial strains are reported as efficient plastic degraders, none of the research focused on the utilization of the entire microbiota in the process of degradation. Microbes reported in LDPE degradation include *Pseudomonas* sp., *Rahnella* sp., *Staphylococcus* sp., *Streptomyces* sp., *Delftia* sp., *Nocardia* sp., *Acinetobacter* sp., and *Arthrobacter* sp. (Kumar Sen and Raut 2015). In India, there are hardly any reports on the microbial community and the ecology using the Winogradsky column. Sridharan et al. (2021) reported the degradation of low-density polyethylene (LDPE) using the Winogradsky column. They also reported the changes in the column and its effects on the degradation of LDPE. There are still gaps left that are yet to be studied. Sridharan et al. (2021) reported the degradation of LDPE using the Winogradsky column and characterized it for changes in mechanical strength, morphological changes using atomic force microscopy (AFM), and weight loss percentage. The isolation of specific and novel microbes producing enzymes that are involved in LDPE degradation might provide a breakthrough as these microbes are native to particular areas and can be utilized to the fullest in the process of degradation. The hydrophobic enzymes produced by the microbes attach on the hydrophobic surface of the plastics and aid in the degradation. The extracellular enzymes produced by fungi are highly reported to be efficient (Das and Kumar 2015). According to research using crystallization techniques, enzyme hydroxylase from the cutinase group features substrate-binding domains that are known to speed up the breakdown of plastic (Mohanan et al. 2020).

22.4 Microplastics and Microbes

Plastics can be converted into oligomers or monomers to create microplastics. According to reports, soil contains roughly 20 times more microplastics than aquatic environments. The widespread presence of microplastics in soils in countries such as China, Australia, and Europe is a result of the way that agriculture is practiced, which includes applying sludge and biofertilizer to agricultural land (Ya et al. 2021). The microbiota, enzyme activity, and soil properties are all changed by the prevalence of microplastics. Microplastics are known to have a large surface area, be hydrophobic, and absorb organic contaminants when compared to conventional plastics (Zhang et al. 2022; Wang et al. 2022). The use of biochar in agriculture to increase soil fertility demonstrates unequivocally that microplastics and biochar coexist (Mak-Mensah et al. 2021).

Microplastics' hydrophobic properties are said to promote the development of biofilms. By employing plastic additives as carbon sources, microbial metabolism is helped (McCormick et al. 2016). According to Zettler et al. (2013), the plastisphere is made up of heterotrophs, autotrophs, and symbionts, which together make up the biofilm-forming community. According to Harrison et al. (2018), the growth of biofilms in aquatic settings is influenced by factors such as light intensity, temperature, salinity, oxygen availability, and pressure. The investigations also claim that the presence of chemicals including phthalates, heavy metals, and bisphenol A inhibits the development of bacteria. The natural organic matter facilitates the interaction between contaminants and microplastics. Swedish hydrobiologists used the word "Neuston" to describe the bacteria that live on the surface of a body as a film in the aquatic environment (Naumann 1919). According to the characterization of bacterioneuston in earlier studies, there was a reduction in bacterial diversity in the jumper (Franklin et al. 2005). The electrostatic force initially facilitates bacterial contact with microplastics. The size, content, and surface characteristics of microplastics affect whether or not bacteria will adhere to them (Tuson and Weibel 2013). According to Debeljak et al. (2017), *Aeromonas, Rhodococcus, Pseudomonas, Enterobacter, Halomonas, Mycobacterium, Photobacterium,* and *Shigella* have been shown to biofilm on the surface of microplastics. SARS-CoV-2 viral attachment to plastic surfaces has been shown, and non-ionic forces are thought to be involved in the virus binding to polyethylene (Gassilloud et al. 2007). Although the route of viral entrance into the cells is altered, the viral metabolism does not change upon attachment to microplastics (Gassilloud and Gantzer 2005). According to the findings (Donlan 2002), any component produced by deceased bacteria binds to the microplastics.

22.5 Conclusion

The chapter provides a summary of plastics, their significance, and the effects of continued use. The variables driving the production of biofilms on plastic surfaces in general and on microplastics in particular are emphasized. While much of the literature reports on microplastics and their negative consequences, this chapter informs the reader about an overview of plastic and microplastic degradation that is facilitated by biofilm development. The sort of bacteria that create biofilm affect the nature of the microplastics. The environment in which biofilm grows influences its rate and type. As a result, microplastics are more attractive to bacteria than macroplastics. The possibility of pathogenic biofilm development may be decreased in the future by the elimination of microplastics from the environment.

References

Amelia TS, May WM, Khalik AW, Ong MC, Shao YT, Pan H-J, Bhubalan K (2021) Marine microplastics as vectors of major ocean pollutants and its hazards to the marine ecosystem and humans. Prog Earth Planet Sci 8:12. https://doi.org/10.1186/s40645-020-00405-4

Antelava A, Jablonska N, Constantinou A, Manos G, Salaudeen SA, Dutta A, Al-Salem SM (2021) Energy potential of plastic waste valorization: a short comparative assessment of pyrolysis versus gasification. Energy Fuels 35(5):3558–3571

Awasthi AK, Li J (2017) Management of electrical and electronic waste: a comparative evaluation of China and India. Renew Sust Energ Rev 76:434–447

Babcsányi I, Meite F, Imfeld G (2017) Biogeochemical gradients and microbial communities in Winogradsky columns established with polluted wetland sediments. FEMS Microbiol Ecol. 93(8). https://doi.org/10.1093/femsec/fix089

Besseling E, Wegner A, Foekema EM, van den Heuvel-Greve MJ, Koelmans AA (2013) Effects of microplastic on fitness and PCB bioaccumulation by the Lugworm Arenicola Marina (L.). Environ Sci Technol 47(1):593–600

Besseling E, Foekema EM, van den Heuvel-Greve MJ, Koelmans AA (2017) The effect of microplastic on the uptake of chemicals by the lugworm *Arenicola Marina* (L.) under environmentally relevant exposure conditions. Environ Sci Technol 51:8795. https://doi.org/10.1021/acs.est.7b02286

Blettler MCM, Ulla MA, Rabuffetti AP, Garello N (2017) Plastic pollution in freshwater ecosystems: macro-, Meso-, and microplastic debris in a floodplain Lake. Environ Monit Assess 189:581. https://doi.org/10.1007/s10661-017-6305-8
Cornwall W (2021) The plastic eaters. Science 373(6550):36–39
Das MP, Kumar S (2015) An approach to low-density polyethylene biodegradation by Bacillus amyloliquefaciens. 3 Biotech 5(1):81–86
Debeljak P, Pinto M, Proietti M, Reisser J, Ferrari FF, Abbas B, van Loosdrecht MCM, Slat B, Herndl GJ (2017) Extracting DNA from ocean microplastics: a method comparison study. Anal Methods 9(9):1521–1526
Deng Y, Zhang Y, Lemos B, Ren H (2017) Tissue accumulation of microplastics in mice and biomarker responses suggest widespread health risks of exposure. Sci Rep 7:46687
Dey T, Michael M (2021) Plastic possibilities: contrasting the uses of plastic 'waste' in India. Anthropol Today 37(3):11–15
Donlan RM (2002) Biofilms: microbial life on surfaces. Emerg Infect Dis 8(9):881–890
Drenovsky RE, Vo D, Graham KJ, Scow KM (2004) Soil water content and organic carbon availability are major determinants of soil microbial community composition. Microb Ecol 48(3):424–430
DSouza GC, Sheriff RS, Ullanat V, Shrikrishna A, Joshi AV, Hiremath L, Entoori K (2021) Fungal biodegradation of low-density polyethylene using consortium of Aspergillus species under controlled conditions. Heliyon 7(5):e07008
Dussud C, Hudec C, George M, Fabre P, Higgs P, Bruzaud S, Delort A-M et al (2018) Colonization of non-biodegradable and biodegradable plastics by marine microorganisms. Front Microbiol 9: 1571. https://doi.org/10.3389/fmicb.2018.01571
Evode N, Qamar SA, Bilal M, Barceló D, Iqbal HMN (2021) Plastic waste and its management strategies for environmental sustainability. Case Stud Chem Environ Eng 4(100142)
Farrell P, Nelson K (2013) Trophic level transfer of microplastic: Mytilus Edulis (L.) to Carcinus Maenas (L.). Environ Pollut 177:1–3
Fisner M, Majer A, Taniguchi S, Bícego M, Turra A, Gorman D (2017) Colour Spectrum and resin-type determine the concentration and composition of polycyclic aromatic hydrocarbons (PAHs) in plastic pellets. Mar Pollut Bull 122(1–2):323–330
Fotopoulou KN, Karapanagioti HK (2012) Surface properties of beached plastic pellets. Mar Environ Res 81:70–77
Franklin MP, McDonald IR, Bourne DG, Owens NJP, Upstill-Goddard RC, Colin Murrell J (2005) Bacterial diversity in the Bacterioneuston (sea surface microlayer): the Bacterioneuston through the looking glass. Environ Microbiol 7(5):723–736
Frias JPGL, Nash R (2019) Microplastics: finding a consensus on the definition. Mar Pollut Bull 138:145–147
Fries E, Zarfl C (2012) Sorption of polycyclic aromatic hydrocarbons (PAHs) to low and high density polyethylene (PE). Environ Sci Pollut Res Int 19(4):1296–1304
Gajendiran A, Krishnamoorthy S, Abraham J (2016) Microbial degradation of low-density polyethylene (LDPE) by Aspergillus clavatus strain JASK1 isolated from landfill soil. 3 Biotech 6(1):52
Galgani L, Tsapakis M, Pitta P, Tsiola A, Tzempelikou E, Kalantzi I, Esposito C, et al. (2019) Microplastics increase the marine production of particulate forms of organic matter. Environ Res Lett ERL [Web Site] 14(12):124085
Gassilloud B, Gantzer C (2005) Adhesion-aggregation and inactivation of poliovirus 1 in groundwater stored in a hydrophobic container. Appl Environ Microbiol 71(2):912–920
Gassilloud B, Huguet L, Maul A, Gantzer C (2007) Development of a viral concentration method for bottled water stored in hydrophobic support. J Virol Methods 142(1–2):98–104
Geyer R (2020) Production, use, and fate of synthetic polymers. In: Plastic waste and recycling. https://doi.org/10.1016/b978-0-12-817880-5.00002-5
Ghasemi MK, Mohd. Yusuff R (2016) Advantages and disadvantages of healthcare waste treatment and disposal alternatives: Malaysian scenario. Pol J Environ Stud 25(1):17–25

Harrison JP, Hoellein TJ, Sapp M, Tagg AS, Ju-Nam Y, Ojeda JJ (2018) Microplastic-associated biofilms: a comparison of freshwater and marine environments. In: Freshwater microplastics: emerging environmental contaminants?, pp 181–201

Holmes LA, Turner A, Thompson RC (2012) Adsorption of trace metals to plastic resin pellets in the marine environment. Environ Pollut 160(1):42–48

Hüffer T, Weniger A-K, Hofmann T (2018) Sorption of organic compounds by aged polystyrene microplastic particles. Environ Pollut 236:218–225

Joo S, Cho IJ, Seo H, Son HF, Sagong H-Y, Shin TJ, Choi SY, Lee SY, Kim K-J (2018) Structural insight into molecular mechanism of poly(ethylene terephthalate) degradation. Nat Commun 9(1):1–12

Kirstein IV, Kirmizi S, Wichels A, Garin-Fernandez A, Erler R, Löder M, Gerdts G (2016) Dangerous hitchhikers? Evidence for potentially pathogenic Vibrio Spp. on microplastic particles. Mar Environ Res 120:1. https://doi.org/10.1016/j.marenvres.2016.07.004

Kumar Sen S, Raut S (2015) Microbial degradation of low density polyethylene (LDPE): a review. J Environ Chem Eng 3(1):462–473

Kwon J-H, Chang S, Hong SH, Shim WJ (2017) Microplastics as a vector of hydrophobic contaminants: importance of hydrophobic additives. Integr Environ Assess Manag 13(3): 494–499

Lahnsteiner F, Berger B, Kletzl M, Weismann T (2005) Effect of bisphenol a on maturation and quality of semen and eggs in the Brown trout, Salmo Trutta F. Fario. Aquat Toxicol 75(3): 213–224

Lange J-P (2021) Managing plastic waste—sorting, recycling, disposal, and product redesign. ACS Sustain Chem Eng 9(47):15722–15738

Lee K-W, Shim WJ, Kwon OY, Kang J-H (2013) Size-dependent effects of micro polystyrene particles in the marine copepod Tigriopus Japonicus. Environ Sci Technol 47(19):11278–11283

Lee U, Han J, Wang M (2017) Evaluation of landfill gas emissions from municipal solid waste landfills for the life-cycle analysis of waste-to-energy pathways. J Clean Prod 166:335–342

Li J, Liu H, Paul Chen J (2018a) Microplastics in freshwater systems: a review on occurrence, environmental effects, and methods for microplastics detection. Water Res 137:362. https://doi.org/10.1016/j.watres.2017.12.056

Li X, Chen L, Mei Q, Dong B, Dai X, Ding G, Zeng EY (2018b) Microplastics in sewage sludge from the wastewater treatment plants in China. Water Res 142:75. https://doi.org/10.1016/j.watres.2018.05.034

Mak-Mensah E, Obour PB, Essel E, Wang Q, Ahiakpa JK (2021) Influence of plastic film mulch with biochar application on crop yield, evapotranspiration, and water use efficiency in northern China: a meta-analysis. PeerJ 9:e10967

McCormick AR, Hoellein TJ, London MG, Hittie J, Scott JW, Kelly JJ (2016) Microplastic in surface waters of urban rivers: concentration, sources, and associated bacterial assemblages. Ecosphere 7(11):e01556

Mohanan N, Montazer Z, Sharma PK, Levin DB (2020) Microbial and enzymatic degradation of synthetic plastics. Front Microbiol 11:580709

Mouafo Tamnou EB, Arfao AT, Nougang ME, Metsopkeng CS, Noah Ewoti OV, Moungang LM, Nana PA et al (2021) Biodegradation of polyethylene by the bacterium Pseudomonas aeruginosa in acidic aquatic microcosm and effect of the environmental temperature. Environ Challenges 3:100056

Narancic T, Verstichel S, Chaganti SR, Morales-Gamez L, Kenny ST, De Wilde B, Padamati RB, O'Connor KE (2018) Biodegradable plastic blends create new possibilities for end-of-life management of plastics but they are not a panacea for plastic pollution. Environ Sci Technol 52(18):10441–10452

Naumann E (1919) Some aspects of the ecology of the limnoplankton, with special reference to the phytoplankton. [Translation from: Svensk Botanisk Tidskrift] 13(2):129–163. https://aquadocs.org/handle/1834/22386

Palm GJ, Reisky L, Böttcher D, Müller H, Michels EAP, Walczak MC, Berndt L, Weiss MS, Bornscheuer UT, Weber G (2019) Structure of the plastic-degrading Ideonella Sakaiensis MHETase bound to a substrate. Nat Commun 10(1):1717

Pathak VM, Navneet. (2017) Review on the current status of polymer degradation: a microbial approach. Bioresources and Bioprocessing 4(1):1–31

Purohit J, Chattopadhyay A, Teli B (2020) Metagenomic exploration of plastic degrading microbes for biotechnological application. Curr Genomics 21(4):253–270

Restrepo-Flórez J-M, Bassi A, Thompson MR (2014) Microbial degradation and deterioration of polyethylene—a review. Int Biodeterior Biodegradation 88:83–90

Rochman CM, Hoh E, Hentschel BT, Kaye S (2013) Long-term field measurement of sorption of organic contaminants to five types of plastic pellets: implications for plastic marine debris. Environ Sci Technol 47(3):1646–1654

Sheveaiy S, Courtney K, Parks JE (2012) The Honolulu strategy : a global framework for prevention and management of marine debris. https://repository.library.noaa.gov/view/noaa/10789

Sridharan R, Vetriselvan M, Krishnaswamy VG, Sagaya Jansi R, Rishin H, Thirumal Kumar D, George Priya Doss C (2021) Integrated approach in LDPE degradation—an application using Winogradsky column, computational modeling, and pathway prediction. J Hazard Mater 412: 125336

Tetu SG, Sarker I, Schrameyer V, Pickford R, Elbourne LDH, Moore LR, Paulsen IT (2019) Plastic leachates impair growth and oxygen production in Prochlorococcus, the ocean's most abundant photosynthetic bacteria. Commun Biol 2:184. https://doi.org/10.1038/s42003-019-0410-x

Thushari GGN, Senevirathna JDM (2020) Plastic pollution in the marine environment. Heliyon 6(8):e04709

Tokiwa Y, Calabia BP, Ugwu CU, Aiba S (2009) Biodegradability of plastics. Int J Mol Sci 10(9): 3722–3742

Tuson HH, Weibel DB (2013) Bacteria-surface interactions. Soft Matter 9(18):4368–4380

Voss R, Lee RP, Fröhling M (2022) Chemical recycling of plastic waste: comparative evaluation of environmental and economic performances of gasification- and incineration-based treatment for lightweight packaging waste. Circ Econ Sustain 2(4):1369–1398

Wang K, Han T, Chen X, Rushimisha IE, Liu Y, Yang S, Miao X, Li X, Weng L, Li Y (2022) Insights into behavior and mechanism of tetracycline adsorption on virgin and soil-exposed microplastics. J Hazard Mater 440:129770

Wilcox C, Van Sebille E, Hardesty BD (2015) Threat of plastic pollution to seabirds is global, pervasive, and increasing. Proc Natl Acad Sci U S A 112(38):11899–11904

Wilkes RA, Aristilde L (2017) Degradation and metabolism of synthetic plastics and associated products by Pseudomonas sp.: capabilities and challenges. J Appl Microbiol 123(3):582–593

Wright SL, Rowe D, Thompson RC, Galloway TS (2013) Microplastic ingestion decreases energy reserves in marine worms. Curr Biol 23(23):R1031–R1033

Ya H, Jiang B, Xing Y, Zhang T, Lv M, Wang X (2021) Recent advances on ecological effects of microplastics on soil environment. Sci Total Environ 798:149338

Yoshida S, Hiraga K, Takehana T, Taniguchi I, Yamaji H, Maeda Y, Toyohara K, Miyamoto K, Kimura Y, Oda K (2016) A bacterium that degrades and assimilates poly(ethylene terephthalate). Science 351(6278):1196–1199

Zettler ER, Mincer TJ, Amaral-Zettler LA (2013) Life in the 'plastisphere': microbial communities on plastic marine debris. Environ Sci Technol 47(13):7137–7146

Zhang S, Li Y, Chen X, Jiang X, Li J, Yang L, Yin X, Zhang X (2022) Occurrence and distribution of microplastics in organic fertilizers in China. Sci Total Environ 844:157061

Insights of the Toxic Impact of Microplastic on Flora: Recent Trends and Challenges

23

Neeraja P. Dhole, Mudasir A. Dar, Rohit R. Kulkarni, Abhishek G. Chaudhari, Mohd. Shahnawaz, and Radhakrishna S. Pandit

Abstract

Microplastics generated from the fragmentation of plastic polymers constitute one of the emergent pollutants invading both terrestrial and aquatic environments. Microplastics pose serious consequences to life including plants. While the impact of microplastics on soil properties and microbial communities is discussed frequently, research detailing the consequences of microplastics on plants is still at its nascent stage. Recent research has demonstrated that microplastics can impact the plant growth positively and negatively in a variety of ways such as they can impact the water uptake by blocking soil pores; moreover, the impacts could be direct and indirect. Similar to other pollutants, microplastics impact the plant growth and development by exerting oxidative stress. The inhibiting mechanisms include obstructing pores or light, breaking down roots mechanically, preventing gene expression, releasing additives, and altering soil characteristics, microbial communities, and soil animals. It further impacts the bioavailability of other contaminants. Since plants exhibit different growth

[$] Authors contributed equally and are all first authors.

N. P. Dhole · R. R. Kulkarni · A. G. Chaudhari · R. S. Pandit (✉)
Department of Zoology, Savitribai Phule Pune University, Pune, India
e-mail: rspandit@unipune.ac.in

M. A. Dar
Biofuels Institute, School of the Environment and Safety Engineering, Jiangsu University, Zhenjiang, Jiangsu Province, China

M. Shahnawaz
Department of Botany, University of Ladakh, Kargil Campus, Khumbathang, Kargil, Ladakh UT, India

Present Address: Department of Botany, Shri Pandit Prem Nath Dogra Government Degree College Samba, Samba, Jammu and Kashmir UT, India

M. Shahnawaz et al. (eds.), *Microplastic Pollution*,
https://doi.org/10.1007/978-981-99-8357-5_23

patterns, the influence of microplastics varies from species to species depending on plastic type, particle size, available concentration, and uptake of polymer. Finally, the pros and cons of current research to understand the molecular effects of microplastics on aquatic and terrestrial plants are also highlighted.

Keywords

Plastic pollution · Uptake, toxicity · Microplastics, oxidative stress

23.1 Introduction

Plastics have become enormously prevalent in day-to-day lives, with a significant increase in their annual production (Verla et al. 2019). By 2050, the production of plastics will touch a staggering 33 billion tons (Rochman et al. 2013). Unfortunately, along with this surge in production, the microplastic amount of environmental plastic waste is also projected to rise, reaching approximately 67.8 million metric tons by this time (Ebere et al. 2019). According to the report, each year a significant volume of plastic, ranging from 4.8 to 12.7 million tons, is released into aquatic environments (Jambeck et al. 2015). Large plastic objects break down into smaller particles through the combined effects of mechanical abrasion and environmental factors such as ultraviolet (UV) radiation, heat, chemical reactions, and biological processes (Law and Thompson 2014). Within these fragmented particles, microplastics are presumed to be the most prevalent forms of plastic that contaminate the environment. Global concern over microplastic pollution and its environmental impact has received a great deal of attention. The migration of microplastics throughout the environment is significantly influenced by the substantial adsorption capacity of their surfaces (Mao et al. 2022). Detection of microplastics in water is easier and observed compared with their presence in soil. He et al. (2018) reported experiments for identifying microplastics in soil recently, while the process of developing effective methods for soil analysis is ongoing and poses significant challenges, especially when compared to the testing presence of microplastics in water. Microplastics present in soil have the potential to alter soil properties, which can have both favorable and adverse outcomes on plants and food quality. The rate at which microplastics decompose in soil remains unknown, and their continued presence and accumulation are presumed to pose a significant environmental risk, as stated by Rillig (2012). Considering the crucial soil's function in protecting biodiversity, facilitating nutrient cycling, and supporting food production, it becomes of utmost importance to understand the impact of microplastics on terrestrial environments, particularly soil ecosystems. Unfortunately, the study of microplastics in soil has been overlooked in previous years, making it imperative to address this knowledge gap and comprehensively investigate their impacts on soil environments (Zhu et al. 2019). The specific effects of microplastics on crop yields can be attributed to changes in soil biological and physicochemical properties as noted by Mateos-Cárdenas et al. (2021). These effects include changes in soil structure, nutrient immobilization, soil microbial community, and interactions with root symbionts (Rillig et al. 2019). However, in general, our understanding of the specific impacts of micro- and nanoplastics on plants remains limited and unknown.

When it comes to plants, there are two primary concerns: the ability of plants to absorb and accumulate microplastics and the impact of microplastics on plant growth and the quality of food produced. Unfortunately, there is a scarcity of information on these topics, due to the challenges associated with identifying microplastics within plant tissues and the insufficient attention given to studying the effects of microplastics on crops (Zhu et al. 2019). Additionally, plants can grow in areas contaminated with plastic pollution if they receive adequate sunlight. This increases the likelihood of plants accumulating microplastics, as large plastic debris undergoes natural weathering and sunlight-assisted degradation, eventually breaking down into smaller microplastics (Ebere et al. 2019). Plants encounter microplastics through various pathways, including direct contact with microplastics that have settled on plant surfaces, exposure through contaminated water or irrigation, and ingestion via uptake through the rhizosphere within the plastic-soil matrix. The main sources of microplastics and nanoplastics can be attributed to plastic particles originating from the textile industry, plastic manufacturing industry, cosmetics, and scrubbing agents. These are considered primary sources of microplastics and nanoplastics. On the other hand, fragmented plastic particles generated through natural weathering processes are referred to as secondary microplastics and nanoplastics (Tang et al. 2019). Despite the advantages associated with common agricultural practices such as mulching, which aims to reduce weed control costs and conserve water resources (Lira-Saldivar 2017), there is a hidden downside related to the polypropylene coating used in mulch materials. This coating can release plastic residues into the soil (Liu et al. 2014; Kader et al. 2017). Furthermore, the likelihood of crops being exposed to microplastics significantly increases when sewage sludge and organic manure are utilized as fertilizers (Watteau et al. 2018). Therefore, it is crucial to conduct investigations into the potential impact of various plastic polymers on different crop species. The current chapter aimed to provide a systematic review of types of microplastics, their pollution in the environment, and effects on plants.

23.2 Uptake of MPs in Plants

The presence of microplastics can hinder water uptake through temporary mechanical blockage of seed capsule pore openings, which can delay relative seed germination, root, and shoot growth. Also, the accumulation of microplastics close to the root hair slows its growth (Bosker et al. 2019). Due to their large size and high molecular weight, MPs are thought to be unable to be absorbed by plants because they cannot pass through the cellulose-rich plant cell walls, but when they are reduced to their nanoforms, reports have indicated their uptake (Wang et al. 2022a; Li et al. 2020a). MPs have been shown to penetrate seeds, roots, stems, leaves, fruits, and plant cells, but to a limited extent and depending on size and species (Dietz and Herth 2011). Microplastics clogging the pores may prevent water absorption and lower the rate of germination of plant seeds (Zhang et al. 2021). The presence of micro- and nanoplastics in edible fruits and vegetables has recently come under scrutiny, with emphasis placed on the significant risk to human health (Conti et al. 2020).

23.3 Plant Uptake and Translocation of Microplastics

Plastic particles are particularly absorbed on root hairs, where there has been evidence of MP and nanoparticle (NP) uptake in plants (Azeem et al. 2021). NPs are so small that they can directly enter plant tissue. The ability of NPs to move from leaves to nearby leaves stems, and roots has been demonstrated (Karny et al. 2018). According to Sun et al. (2021), stomatal uptake is one potential route for NPs to enter the leaves before traveling to the vasculature. Through transpiration, the vascular system move plastic particles from the root to the shoot. Confocal images showed that the vascular system of the stem was where plastic luminescence signals were primarily found. MPs and NPs can enter the vasculature that transports water by passing through tiny extracellular channels (Schwab et al. 2020). According to Li et al. (2020b), plastic particles as small as 200 nm and 2 μm were found in the xylem sap and leaves of wheat and lettuce. This revealed that the transpiration stream-driven vascular system was responsible for the root-to-shoot translocation of micro (nano)plastics. In one study, rice was exposed to polystyrene NPs (20 nm), and a significant distribution of PS was discovered in the intercellular spaces of the roots (Zhou et al. 2021a). Furthermore, after being exposed for 72 h, nano-polystyrene (50 nm) was discovered in various locations of the onion's primary roots, demonstrating that it can pierce several biological barriers before entering root cells. Confocal microscopy with 3D laser scanning made it clear that PS NPs are taken up by the root tips, while the fluorescent signal of PS NPs in leaves was blocked by the autofluorescence of plant tissues. Similarly, PS NPs that were absorbed in the xylem of the root and shoot were also found by SEM. Through the xylem pathway, PS NPs (100 nm) could be transported from root to shoot and then into the grain (Azeem et al. 2021).

The first in vitro report of plant uptake of micro(nano)plastics was made by Bandmann et al. (2012), who observed the endocytosis of fluorescent PS NPs (20 and 40 nm) by tobacco BY-2 cells. The results of Li et al. (2020a) showed that plastic particles enter the epidermal tissue of primary and secondary roots of wheat, are stimulated through the pericycle, and are then transported into the xylem. These particles can move to the plant's aerial portion inside the central cylinder via the xylem (Giorgetti et al. 2020). Inflicting damage on the roots caused by MPs reduced phenanthrene uptake in soybean roots and leaves. Reduced phenanthrene uptake in soybean roots was caused by the inhibition of soybean root activity by micron-size MPs and the consequent decline in the relative abundance of microbial in rhizosphere soil (Xu et al. 2021). Lian et al. (2021) research indicates that PS NPs attach to the plant's stomata, pierce the phloem, and reach the lettuce plant's roots. In a study by Bandmann et al. (2012) on tobacco (*Nicotiana tabacum*), plant cells showed that NPs can penetrate plant cell walls. Tobacco plants did not uptake 100-nanometer nanopolystyrene beads, but 20- to 40-nanometer beads were. The maize plant's stems and roots both had high concentrations of micro-fluorescent PS NPs, and fluorescence microscope imaging analysis also shows that PS NPs are frequently found in and near the stem's vascular system. According to these findings, NPs travel through the vascular bundle from the leaves to the stems and then to the roots (Azeem et al. 2021). Following a 21-day exposure period, PS nanospheres (100 nm) were found in the root and shoot xylems of wheat plants in a hydroponic study. This finding further supports the vital role the vascular system plays in the upward transport of micro(nano)plastics in terrestrial plants (Lian et al. 2020). The primary mode of NP uptake by plants is thought to be the crack entry mode. PS NPs were demonstrated to penetrate the cell wall and enter the cytoplasm in a study of wheat callus with cell walls. However, plants' cell wall pores (1.5–5 nm) are much smaller than NPs' particle sizes (Chesson et al. 1997). The understanding of how plants take up NPs has advanced significantly in a recent study involving wheat and lettuce. In this study, lateral root caps and apical meristems of wheat and lettuce roots were found to contain NPs in the 200 nm range. They postulated that root cap mucus retention encouraged microbeads to penetrate the cell wall. Fluorescence was primarily seen in the cell wall and intercellular zone of lettuce plants, indicating that 200 nm PS NPs entered the cortical zone through the space between epidermal cells but were unable to pass through the endothelial layer of the continuous area of the Casparian zone. In the cracks of the lateral root area (50–100 mm from the apex), strong PS luminescence signals were seen at the same time. The fact that the lateral root enters the endothelial layer and cortex suggests that these cracks are the primary entry points for the endothelial layer (Li et al. 2020b). The hypothesis for the uptake, according to Zhu et al. (2019), is that small-sized microplastics can bypass plant cell wall and membrane barriers. This hypothesis can be studied and monitored using fluorescent microbeads.

Early research suggested that plants would absorb NPs and then store them in their epidermal cells. However, in recent years, there has been a significant advancement in the ability to detect the absorption and migration of MPs and NPs in plants thanks to the development of granular plastic analysis technology (Wu et al. 2021). A recent study discovered that the xylem sap of wheat and lettuce contains 0.2 mm PS NPs and 2.0 mm PS-MPs, indicating that both submicron and micron plastic beads migrate from root to shoot through transpiration. The fluorescence intensity at the junction of the primary and secondary roots of wheat, which is thought to be the microbead entry channel, is also stronger under high transpiration conditions than it is under low transpiration conditions, suggesting that an increase in transpiration rate facilitates the absorption of beads by plants (Li et al. 2020b). Rice roots exposed to $HgCl_2$ as an aquaporin inhibitor saw a significant decrease in the accumulation of PS nanoparticles in root tissues. A potential explanation for the reduced uptake of PS particles by rice roots was the reduced transpiration flow brought on by the inhibition of aquaporins in root cells (Zhou et al. 2021a).

It is also thought that roots can absorb NPs through apoplast transport. The apoplastic pathway may be the primary transport pathway in plant roots for particles with relatively larger dimensions (e.g., a few hundred nanometers to several micrometers) (Li et al. 2020b; Sun et al. 2020). NPs may be trapped by mucus secreted by plant roots, gather on the root surface, and migrate because of plant transpiration and root pressure, according to one theory for the mechanism of apoplast transport. Both nano- and micro-sized PS can be absorbed by rice roots in the hydroponically grown rice experiment and then transferred to the above-ground portion. These microbeads primarily congregate on the root cortex tissue's cell wall, suggesting that apoplast transport may be the primary mode of absorption and the primary driver of PS translocation in rice tissue. (Sun et al. 2020; González-Morales et al. 2020; Lu et al. 2008; Schwab et al. 2020).

Another potential method for root cells to absorb NPs is symplast transport. It is a way for NPs to enter cells through endocytosis (Wu et al. 2021). In one experiment, in vitro-cultured BY-2 cells quickly internalized fluorescent plastic particles that were 20 nm in size (Bandmann et al. 2012). The midpoint of the plasmodesmata, which connect cells, has a diameter of 50–60 nm (Robards 1975). The symplast pathway may therefore allow exceedingly small plastic particles (such as those measuring less than 60 nm) to enter root cells and travel to the root xylem. The endocytosis of NPs by root cells of intact plants has not been studied, and this route is only based on cells that have been isolated and cultured independently (Hua et al. 2023). There is currently truly little known about how plants accumulate MP, and it is unclear what the actual uptake mechanism is. Therefore, more research is needed to better understand MP accumulation and distribution in plants.

23.4 Stimulatory Effects of Microplastic on Plants

Despite being classified as pollutants, microplastics promoted the growth of higher plants. Some plastics, which are an organic polymer, include nutrients such as nitrogen, the main building block of protein. Some microplastics may enhance soil structure, increase the availability of water and nutrients, and accelerate plant growth. For instance, polyester (PES) fibers can increase water-holding capacity and prolong soil moisture saturation (de Souza Machado et al. 2019). Microplastics have the potential to change the microbial population of soil and boost the activity of soil enzymes involved in nutrient cycling. Microplastics are thought to be effective vectors for environmental pollutants, such as heavy metals, hydrophobic organic contaminants, pharmaceuticals, and personal care products (Hartmann et al. 2017). This is because of their small particle size, large specific surface area, and strong hydrophobicity. Microplastics may therefore have an impact on the movement and bioavailability of environmental pollutants, which in turn may have an indirect impact on plants. Ren et al. (2021) concluded that the various PS-bead sizes had an impact on various plant growth parameters and may have altered plant growth and photosynthetic parameters by altering microbial metabolism through the influence of the correlation between microbes.

In soil exposed to microplastics such as polypropylene and low-density PE, for instance, the activities of urease and phosphatase were increased, which may increase the availability of N and P and their uptake by plants (Liu et al. 2017; Fei et al. 2020). According to Lozano and Rillig (2020), the addition of polyester (PES) microfibers (0.4%, w/w) increased the shoot mass of *Hieracium* by 85% during a drought and *Calamagrostis* by 66% when conditions were moist. This was attributed to microfibers, which led to improved aeration, decreased soil bulk density, and improved root penetration. According to de Souza Machado et al. (2019), PES with an average length of 5 mm significantly increased the soil microbes' colonization of *A. fstulosum* roots. On plant photosynthetic indicators and chlorophyll contents, MPs and NPs have both favorable and detrimental concentration-dependent effects. According to Pignattelli et al. (2020), polyethylene(PE), polypropylene (PP), and polyvinyl chloride (PVC) in soils can promote the synthesis of chlorophyll in *L. sativum* leaves. Foamy PE, fibrous PES, foamy polyurethane (PU), and fragmented polyethylene terephthalate (PET) in soils can also boost *D. carota* biomass (Lozano et al. 2021). In hydroponic experiments, Dong et al. (2020) discovered that polytetrafluoroethylene (PTFE) and polystyrene (PS) microplastics decreased rice seedlings' uptake of arsenic (As)(III). Three explanations were given for this phenomenon: microplastics' competition with as for sorption sites on the surface of roots, their direct adsorption of As, and their inhibition of root activity. According to Zong et al. (2021), PS could lessen the toxicity of heavy metals and prevent the accumulation of Copper (Cu) and cadmium (Cd) in wheat seedlings.

23.5 Negative Impacts of Microplastics on Plants

There are different ways by which microplastics can affect plants either directly or indirectly. Exposure to microplastics and nanoplastics has been shown to alter plant growth; however, the precise effects vary depending on the type of plant and the plastic. In hydroponically grown wheat (*Triticum aestivum*) species, the elongation of wheat roots and stems was dramatically reduced by PS-MPs at a concentration of 200 mg L^{-1} (Liao et al. 2019). In a pot experiment on *Triticum aestivum* done by Qi et al. (2018), polystyrene microplastic inhibited growth of above-ground and below-ground parts of the wheat plant during both vegetative and reproductive growth.

23.6 Biochemical and Oxidative Stress by Microplastics to Plants

The impact of microplastic pollution on plants is diverse (Rillig et al. 2019). Although the presence of microplastics in the environment can impact plants in a variety of ways, plant species appear to respond inadequately to this pollution. While most plant species showed reduced biomass (root and/or shoot) in the presence of microplastics (Urbina et al. 2020; Wu et al. 2020a; Pignattelli et al. 2021; Dong et al. 2021), others remain unaffected (Meng et al. 2021) or even show increased growth (de Souza Machado et al. 2019; Lozano and Rillig 2020). The preliminary findings indicate that microfibers can reduce the bulk density of soils (de Souza Machado et al. 2018), which in turn affects the penetration of root, soil aeration, and root development (Zimmermann and Kardos 1961). Microplastics can also disrupt plant metabolic pathways by affecting the number of amino acids and/or sugar alcohols (Lian et al. 2021; Zeb et al. 2022). Furthermore, toxic chemicals such as bisphenol A (vary with plastic materials) might leach out of the polymer matrix, thereby affecting plant fitness (Boots et al. 2019; Pflugmacher et al. 2020). In addition to these direct consequences, microplastics have an indirect impact on plant performance by altering soil properties and microbial compositions. It can alter both the physical and chemical properties of the soil (Wang et al. 2022b), which may further alter the microbial community (Huang et al. 2019; Yi et al. 2021; Han et al. 2022) and soil fauna (Zhu et al. 2018; Boots et al. 2019), thereby affecting the overall interactions of the plants in a habitat. The photosynthetic apparatus in plants may also suffer under stressful conditions, thus in turn impacting the pigment production and tetra pyrrole pathway (Dalal and Tripathy 2012).

The mechanism of microplastic uptake by the plants is elusive (Yang et al. 2017); however, it is a well-known fact that nano-sized particles could enter plant via roots and exert damage to cell membrane, intracellular molecules, and/or oxidative stress (Navarro et al. 2008). Plastic films of 2, 5, and 10 mm size have been demonstrated to generate channels for water circulation, resulting in higher water evaporation (Wan et al. 2019). This might cause soil dryness, which could pose oxidative stress and consequently deleterious effect on plant performance. Microplastics can cause cytotoxicity by influencing gene expression (Sun et al. 2020; Wu et al. 2022) or even damage DNA (i.e., genotoxicity) through direct absorption of tiny particles or indirectly through increased oxidative stress (Jiang et al. 2019; Giorgetti et al. 2020; Wang et al. 2022a).

23.6.1 Antioxidant System of Plants

Like other pollutants, microplastics affect the plant growth and development by exerting oxidative stress (Table 23.1), leading to the formation of reaction oxygen species (ROS; Wang et al. 2022c). Under normal conditions, plants maintain the balance between ROS production and scavenging reactions (Hasanuzzaman et al. 2020). Generally, ROS act as signaling molecules or secondary messengers that regulate the physiological activities of the plants for proper survival. The imbalance of the ROS levels leads to the disruption of the plant growth causing cellular damage and apoptosis (Raja et al. 2017). The four main types of reactive oxygen are singlet oxygen ($^1O_2$), hydroxyl radicals (OH), superoxide anion (O_2), and hydrogen peroxide (H_2O_2). These molecules vary in terms of origin, reactivity, lifespan, and biological effect. The major ROS, known as the superoxide anion, are produced by the membrane-bound enzyme, NADPH oxidase (also known as the respiratory burst oxidase isoform), which is part of the electron transport chain in mitochondria and chloroplasts (Ekner-Grzyb et al. 2022). Since this ROS is extremely reactive, it can easily be transformed into H_2O_2, which is only moderately reactive (Demidchik 2015; Fichman and Mittler 2020; Farnese et al. 2016). Hydrogen peroxide (H_2O_2)

Table 23.1 List of the plants along with their oxidative responses to polystyrene microplastic contaminant

Plant species	Common name	Particle size (nm)	Concentration tested	Treatment days	Effects detected	Reference(s)
Lactuca sativa	Lettuce	80	1, 5, 10 mg/L	7	High H_2O_2 and SO, SOD, CAT, ASA, and GSH; low polyphenol and flavonoids	Gao et al. (2022)
Oryza sativa	Rice	50	50 mg/L	7	Altered POD, SOD, and CAT; low TBRAS by PS-COOH; high TBARS by PS-NH2	Wang et al. (2022d)
Lemna minor	Duckweed	100	2–50 μg/L	3	High H_2O_2, SO, TBARS	Xiao et al. (2022)
Oryza sativa	Rice	19	10, 50, and 100 mg/L	4	Low H_2O_2 and TBARS in roots	Spanò et al. (2022)
Lactuca sativa	Lettuce	100	0.1 and 1 mg/L	30	Reduced antioxidant activity	Lian et al. (2021)
Cucumis sativus	Cucumber	100, 300, 500, 700	50 mg/L	14 and 65	Modified lipid peroxidation and vitamin C levels	Li et al. (2021a, b)
Arabidopsis thaliana	Arabidopsis	55, 71	300 and 1000 mg/L	7, 10 and 49	High H_2O_2 and SO; modified gene expression	Sun et al. (2020)
Triticum aestivum	Wheat	87	10 mg/L	21	High lipid peroxidation	Lian et al. (2020)
Allium cepa	Onion	50	10, 100, and 1000 mg/L	3	H_2O_2 accumulation; lipid peroxidation	Giorgetti et al. (2020)
Vicia faba	Castor bean	100	10, 50, and 100 mg/L	2	Lipid peroxidation, high activity of ROS enzymes	Jiang et al. (2019)

PS polystyrene, *TBARS* thiobarbituric acid reactive substances, *ROS* reactive oxygen species, *ASA* ascorbic acid, *CAT* catalase, *POD* peroxidases, *SOD* superoxide dismutase, *SO* superoxide anion, H_2O_2 hydrogen peroxide has a comparatively long lifetime (>1 ms) and specific transporters called aquaporins (Rodrigues et al. 2017; Bienert et al. 2007). Consequently, it can enter several cellular compartments and react with proteins. The least stable ROS is the hydroxyl radical, whose half-life is measured in nanoseconds (ns). Hydroxyl radicals are highly reactive and cause extensive structural damage. Singlet oxygen is produced within chloroplasts through the process of energy transfer from three-dimensional chlorophyll to oxygen. This highly reactive form of oxygen, known for its exceptional reactivity, plays a significant role in causing oxidative damage to the photosystems within chloroplasts (Das and Roychoudhury 2014; Dumanović et al. 2021).

The ROS species are countered by both enzymatic and non-enzymatic antioxidants (Noctor et al. 2016). Peroxidases (POX), superoxide dismutase (SOD), and catalase (CAT) are the primary antioxidant enzymes. Ascorbic acid (ASC), tocopherol, carotenoids, glutathione (GSH), and phenolic chemicals are a few examples of the non-enzymatic antioxidants (Dumanović et al. 2021). The ability of ROS molecules to affect protein structure and function is mediated by the oxidation of cysteine residues, resulting in the formation of disulfide bridges. Mitogen-activated protein kinase (MAPK), glyceraldehyde-3-phosphate dehydrogenase (GAPDH), and H_2O_2 detecting HPCA1 are some proteins whose activity is controlled by cysteine oxidation (Mittler 2017; Castro et al. 2021). Certain proteins, such as glutaredoxins (GRX) and thioredoxins (TRX), have a propensity to undergo oxidation of cysteine residues, which enables them to function as transmitters of redox signals. These proteins can act as redox switches in other enzymes also by mediating the oxidation and reduction states of cysteine residues (Martı et al. 2020). In addition to impacting protein functions, reactive oxygen species (ROS) have the ability to modify signaling pathways by interacting with various elements such as calcium ions (Ca_{2+}), reactive sulfur and nitrogen species, plant hormones, mitogen-triggering protein kinases, and transcription factors (Castro et al. 2021; Farnese et al. 2016). Given that signaling pathways are interconnected within signaling hubs, it is plausible that NADPH oxidase (RBOH) serves as a site for integrating and coordinating these signaling processes. This membrane-bonding enzyme can be controlled by changes in calcium levels, direct phosphorylation, and phosphatidic acid binding. Evidence shows that the activity of NADPH oxidase is affected by S-nitrosation and presulfidation can have an impact on function, thus linking its pathways with reactive nitrogen and sulfur molecules. The activation of NADPH oxidase leads to the production of more O_2, which is then converted into the primary ROS molecule, H_2O_2.

23.6.2 Antioxidant Responses of Plants to Microplastics

23.6.2.1 Oxidative Stress by Microplastics and Its Enzymatic Response

In plants, oxidative stress has been recognized as one of the main responses to abiotic stress caused by microplastics (Wang et al. 2022c). Many studies have stated that the effect of plastic polymers on H_2O_2 levels can change over time in plants

(Pignattelli et al. 2020). However, the low concentrations of the pollutant do not show measurable antioxidative responses. Lian et al. (2020) observed that the low levels of the polystyrene nano-plastics do not impact the CAT activity and dry biomass of the wheat (*Triticum aestivum*). Similarly, Gong et al. (2021) stated that plants show species-specific responses to polystyrene nanoparticles. The authors inferred that the higher doses/concentrations (50 and 100 compared to 10 mg/L) of the plastic nanoparticles depict measurable SOD, POD, and CAT activities (Zhou et al. 2021b). Other investigations (Wu et al. 2020b; Shi et al. 2022) have shown that CAT activity increased at lower microplastic concentrations but showed marked reductions at higher doses. The size of the microplastic particles is also known to impact the CAT activities (Jiang et al. 2019). Although CAT is essential for ROS detoxification and remains functional during stress (Mittler 2002), scavenging of H_2O_2 is not solely maintained by CAT. The expression of other antioxidant enzymes may be higher if H_2O_2 generation occurs other than the peroxisomes. Additionally, non-enzymatic antioxidants or factors also contribute to the tolerance of plants to microplastics (Raja et al. 2017). The PS (100 nm) enhanced MDA enzyme activity in faba bean (*Vicia faba*) by 37% at 100 mg L^{-1}, while it lowered the activity by 39% at lower concentrations of 50 mg L^{-1} (Jiang et al. 2019). Further, the short exposure did not show any effect on MDA activity in maize. Based on the available literature, it is plausible to state that high levels of oxidative stress cause damage to the tertiary structure of antioxidant enzymes, thereby reducing their activities (Wang et al. 2022c).

The majority of the studies have concluded that ROS is built up immediately after a brief exposure to nano-plastics. When onion seedlings were exposed to PS NP for a duration of 72 h at a concentration of 1000 mg/L, a notable rise in H_2O_2 production was observed. Conversely, exposure to lower concentrations, as low as 10 and 50 mg/L, showed no discernible impact. These observations were supported by the microscopic analysis that revealed concentration-dependent production of H_2O_2 in the roots of treated plants (Giorgetti et al. 2020). Another hypothesis is that plants may develop tolerance and resistance to oxidative stress by microplastics due to long-time exposures (Gao et al. 2019).

Some studies have revealed that oxidative stress depends on the surface charge of nano-plastics. A rise in H_2O_2 and O_2 levels was seen in Arabidopsis in response to positively charged PS-NH2 NP than the negatively charged PS-SO_3H particles (Sun et al. 2020). Similar inferences have been reported in dandelion and duckweed, where ROS was significantly increased by both the positively and negatively charged NP. However, in the case of PS-NH2, the ROS production was more pronounced (Gao et al. 2022; Xiao et al. 2022). Similarly, PS-NH2 exhibited a stronger effect on lipid peroxidation even at lower concentrations than the PS-SO3H in duckweed (Xiao et al. 2022). In fact, PS NP enhanced the activity of APX, POX, and catalase (CAT) in the shoots, while in shoots, the activities of guaiacol peroxidase (POX), ascorbate peroxidase (APX), and superoxide dismutase (SOD) were induced (Spanò et al. 2022). Further, investigations suggested that activation of peroxidases (POX), SOD, and CAT is induced by PS NP in the roots of rice plants (Wang et al. 2022d; Zhou et al. 2021a). In dandelion, SOD and CAT activities are stimulated by PS NP, with pronounced induction in the presence of positively charged particles than neutrally or negatively charged particles (Gao et al. 2022). However, polyphenol content, including flavonoids that constitute bioactive substances with antioxidant activity, was decreased as a result of exposure to PS NP (Gao et al. 2022). In natural environments, plants face combined stress exerted by several factors. In a recent study, the additive effect of Fe NP and PS NP on lettuce demonstrated that the nanoparticles agglomerate in heterogeneous mixtures. The combined exposure to PS NP and Fe NP caused the release and accumulation of Fe in plants. Additionally, it was noted that treated plants experience increased lipid peroxidation and progressed cell damage (Gong et al. 2022).

23.6.2.2 Non-Enzyme Responses of Plants to Microplastics

Apart from enzymes, plants also respond to oxidative stress through various macromolecules. Microplastics cause oxidative stress through the production of H_2O_2 in plants (Pignattelli et al. 2020). The production of H_2O_2 is mediated through the consumption of oxygen (O_2) during respiration. These ROS molecule (H_2O_2) plays a dual role. At low concentrations, it acts as signaling molecules, while at higher concentrations it leads to oxidative bursts causing cell death (Quan et al. 2008). Pignattelli et al. (2021) observed that *L. sativum* plants with C+ and PE + PVC+ treated plants showed higher production of H_2O_2 and glutathione (GSH), whereas the PVC+ treated plants produced high amount of ascorbic acid (AsA). The H_2O_2 level of the shoots was higher in case of the PE+ and PVC+ treatments than in roots when exposed for a short period. However, the production of AsA at foliar levels was higher for short exposure while recorded highest in roots for longer exposures (Pignattelli et al. 2021). Ascorbate and glutathione (GSH) constitute the key players of the detoxification system of plants (Quan et al. 2008). The antioxidative GSH (low-molecular-weight thiolic antioxidant) participates in several physiological functions of the detoxification system. Notably, it performs the production of ascorbate from dehydroascorbate (DHA) for H_2O_2 metabolism and the reduction of superoxide (Noctor et al. 2018). Similar to H_2O_2, ascorbic acid also serves dual function of cofactor for peroxidases and can directly eliminate ROS in plants through ascorbate/glutathione pathway (Noctor et al. 2018).

Lipid peroxidation is frequently used as a marker for measuring oxidative stress. It is often estimated by analyzing the products such as malondialdehyde (MDA) and thiobarbituric acid reactive substances (TBARS). Independent research on seedlings of broad bean and onion plants exposed for a brief period to PS NP showed concentration-dependent increases in lipid peroxidation. Application of the highest doses (100 mg/L for broad bean and 1000 mg/L for onion) led to an increase in lipid peroxidation, whereas lower concentrations (10 mg/L) resulted in a decrease in TBARS/MDA levels (Giorgetti et al. 2020; Jiang et al. 2019). In cucumber, the lipid peroxidation occurred in a size-dependent manner. After 14 days of treatment with PS NPs, the MDA levels could be measured in 300 and 500 nm only with no effect for particles of 100 and 700 nm size (Li et al. 2021a, b). In another study, rice seedlings showed that the effects also vary with the charge of the nano-plastics. However, in roots, lipid peroxidation was increased neutrally, negatively, as well as positively charged nano-plastics with the most prominent effect detected in neutral PS nano-plastics. Conversely, positively charged plastic particles caused accumulation of TBARS in the shoots, whereas neutrally charged NP had little effect on lipid peroxidation (Wang et al. 2022c).

23.7 Conclusion

The findings of the previous studies imply that microplastics indeed affect the growth and development of the plants, but recent research has demonstrated that microplastics might hinder greater plant growth both directly and indirectly. The inhibiting mechanisms include obstructing pores or light, breaking down roots mechanically, preventing gene expression, releasing additives, and altering soil characteristics, microbial communities, and soil animals, and it further impacts the bioavailability of other contaminants. Since plants show different patterns of growth, their response to pollutants varies substantially. The antioxidative response of the plants to microplastics varies from species to species depending on the type of plastic polymer encountered as well as uptake and concentration of the polymer available in the surrounding environment. Despite the fact, current research on oxidative stress in plants by microplastics varies depending on the techniques used, microplastic type, its size, and concentration, besides exposure time, which makes it challenging to draw standard conclusions. Nonetheless, studies on the effect of microplastics on the plants in field settings are lacking due to technical and associated challenges. The analysis of the available literature indicates that there are still many knowledge gaps that can elucidate the exact molecular impact of the plastic particles on plants, particularly in aquatic habitats.

Acknowledgements A research fellowship was provided to NPD by the Council of Scientific and Industrial Research (CSIR), Government of India. This work was supported by the National Natural Science Foundation of China (32250410285). RSP is thankful to Savitribai Phule Pune University (SPPU), Pune.

References

Azeem I, Adeel M, Ahmad MA et al (2021) Uptake and accumulation of nano/microplastics in plants: a critical review. Nanomaterials (Basel) 11:2935. https://doi.org/10.3390/nano11112935

Bandmann V, Müller JD, Köhler T, Homann U (2012) Uptake of fluorescent nano beads into BY2-cells involves clathrin-dependent and clathrin-independent endocytosis. FEBS Lett 586: 3626–3632. https://doi.org/10.1016/j.febslet.2012.08.008

Bienert GP, Møller ALB, Kristiansen KA, Schulz A, Møller IM, Schjoerring JK et al (2007) Specific aquaporins facilitate the diffusion of hydrogen peroxide across membranes. J Biol Chem 282:1183–1192. https://doi.org/10.1074/jbc.M603761200

Boots B, Russell CW, Green DS (2019) Effects of microplastics in soil ecosystems: above and below ground. Environ Sci Technol 53:11496–11506. https://doi.org/10.1021/acs.est.9b03304

Bosker T, Bouwman LJ, Brun NR et al (2019) Microplastics accumulate on pores in seed capsule and delay germination and root growth of the terrestrial vascular plant Lepidium sativum. Chemosphere 226:774–781. https://doi.org/10.1016/j.chemosphere.2019.03.163

Castro B, Citterico M, Kimura S, Stevens DM, Wrzaczek M, Coaker G (2021) Stress-induced reactive oxygen species compartmentalization, perception and signalling. Nat Plants 7:403–412. https://doi.org/10.1038/s41477-021-00887-0

Chesson A, Gardner PT, Wood TJ (1997) Cell wall porosity and available surface area of wheat straw and wheat grain fractions. J Sci Food Agric 75:289–295. https://doi.org/10.1002/(SICI)1097-0010(199711)75:3<289::AID-JSFA879>3.0.CO;2-R

Conti G, Ferrante M, Banni M et al (2020) Micro- and nano-plastics in edible fruit and vegetables. The first diet risks assessment for the general population. Environ Res 187:109677. https://doi.org/10.1016/j.envres.2020.109677

Dalal VK, Tripathy BC (2012) Modulation of chlorophyll biosynthesis by water stress in rice seedlings during chloroplast biogenesis. Plant Cell Environ 9:1685–1703

Das K, Roychoudhury A (2014) Reactive oxygen species (ROS) and response of antioxidants as ROS-scavengers during environmental stress in plants. Front Environ Sci 2:53. https://doi.org/10.3389/fenvs.2014.00053

de Souza Machado AA, Kloas W, Zarfl C et al (2018) Microplastics as an emerging threat to terrestrial ecosystems. Glob Chang Biol 24:1405–1416. https://doi.org/10.1111/gcb.14020

de Souza Machado AA, Lau CW, Kloas W et al (2019) Microplastics can change soil properties and affect plant performance. Environ Sci Technol 53:6044–6052. https://doi.org/10.1021/acs.est.9b01339

Demidchik V (2015) Mechanisms of oxidative stress in plants: from classical chemistry to cell biology. Environ Exp Bot 109:212–228. https://doi.org/10.1016/j.envexpbot.2014.06.021

Dietz K-J, Herth S (2011) Plant nanotoxicology. Trends Plant Sci 16:582–589. https://doi.org/10.1016/j.tplants.2011.08.003

Dong Y, Gao M, Song Z, Qiu W (2020) Microplastic particles increase arsenic toxicity to rice seedlings. Environ Pollut 259:113892. https://doi.org/10.1016/j.envpol.2019.113892

Dong Y, Gao M, Qiu W, Song Z (2021) Uptake of microplastics by carrots in presence of as (iii): combined toxic effects. J Hazard Mater 411:125055. https://doi.org/10.1016/j.jhazmat.2021.125055

Dumanović J, Nepovimova E, Natić M, Kuča K, Jaćević V (2021) The significance of reactive oxygen species and antioxidant defense system in plants: a concise overview. Front Plant Sci 11:552969. https://doi.org/10.3389/fpls.2020.552969

Ebere E, Wirnkor V, Evelyn Ngozi V (2019) Uptake of microplastics by plant: a reason to worry or to be happy?

Ekner-Grzyb A, Duka A, Grzyb T, Lopes I, Chmielowska-Bąk J (2022) Plants oxidative response to nanoplastic. Front Plant Sci 13:1027608. https://doi.org/10.3389/fpls.2022.1027608

Farnese FS, Menezes-Silva PE, Gusman GS, Oliveira JA (2016) When bad guys become good ones: the key role of reactive oxygen species and nitric oxide in the plant responses to abiotic stress. Front Plant Sci 7:471. https://doi.org/10.3389/fpls.2016.00471

Fei Y, Huang S, Zhang H et al (2020) Response of soil enzyme activities and bacterial communities to the accumulation of microplastics in an acid cropped soil. Sci Total Environ 707:135634. https://doi.org/10.1016/j.scitotenv.2019.135634

Fichman Y, Mittler R (2020) Rapid systemic signaling during abiotic and biotic stresses: is the ROS wave master of all trades? Plant J 102:887–896. https://doi.org/10.1111/tpj.14685

Gao M, Liu Y, Song Z (2019) Effects of polyethylene microplastic on the phytotoxicity of di-n-butyl phthalate in lettuce (Lactuca sativa L. var. Ramosa Hort). Chemosphere 237:124482. https://doi.org/10.1016/j.chemosphere.2019.124482

Gao M, Bai L, Li X, Wang S, Song Z (2022) Effects of polystyrene nanoplastics on lead toxicity in dandelion seedlings. Environ Pollut 306:119349. https://doi.org/10.1016/j.envpol.2022.119349

Giorgetti L, Spanò C, Muccifora S et al (2020) Exploring the interaction between polystyrene nanoplastics and allium cepa during germination: internalization in root cells, induction of toxicity and oxidative stress. Plant Physiol Biochem 149:170–177. https://doi.org/10.1016/j.plaphy.2020.02.014

Gong W, Zhang W, Jiang M et al (2021) Species-dependent response of food crops to polystyrene nanoplastics and microplastics. Sci Total Environ 796:148750. https://doi.org/10.1016/j.scitotenv.2021.148750

Gong D, Bai X, Weng Y, Kang M, Huang Y, Li F et al (2022) Phytotoxicity of binary nanoparticles and humic acid on *Lactuca sativa* L. Environ Sci Process Impacts 24:586–597. https://doi.org/10.1039/d2em00014h

González-Morales S, Parera CA, Juárez-Maldonado A, et al (2020) The ecology of nanomaterials in agroecosystems. In: Nanomaterials for agriculture and forestry applications, pp 313–355. https://doi.org/10.1016/b978-0-12-817852-2.00013-5

Han L, Chen L, Li D et al (2022) Influence of polyethylene terephthalate microplastic and biochar co-existence on paddy soil bacterial community structure and greenhouse gas emission. Environ Pollut 292:118386. https://doi.org/10.1016/j.envpol.2021.118386

Hartmann NB, Rist S, Bodin J et al (2017) Microplastics as vectors for environmental contaminants: exploring sorption, desorption, and transfer to biota. Integr Environ Assess Manag 13:488–493. https://doi.org/10.1002/ieam.1904

Hasanuzzaman M, Bhuyan MHM, Zulfiqar F et al (2020) Reactive oxygen species and antioxidant defense in plants under abiotic stress: revisiting the crucial role of a universal defense regulator. Antioxidants 9:681. https://doi.org/10.3390/antiox9080681

He D, Luo Y, Lu S et al (2018) Microplastics in soils: analytical methods, pollution characteristics and ecological risks. TrAC Trends Anal Chem 109:163–172. https://doi.org/10.1016/j.trac.2018.10.006

Hua Z, Ma S, Ouyang Z et al (2023) The review of nanoplastics in plants: detection, analysis, uptake, migration and risk. TrAC Trends Anal Chem 158:116889. https://doi.org/10.1016/j.trac.2022.116889

Huang Y, Zhao Y, Wang J et al (2019) LDPE microplastic films alter microbial community composition and enzymatic activities in soil. Environ Pollut 254:112983. https://doi.org/10.1016/j.envpol.2019.112983

Jambeck JR, Geyer R, Wilcox C et al (2015) Plastic waste inputs from land into the ocean. Science 347:768–771. https://doi.org/10.1126/science.1260352

Jiang X, Chen H, Liao Y, Ye Z, Li M, Klobučar G (2019) Ecotoxicity and genotoxicity of polystyrene microplastics on higher plant vicia faba. Environ Pollut 250:831–838. https://doi.org/10.1016/j.envpol.2019.04.055

Kader MA, Senge M, Mojid MA, Ito K (2017) Recent advances in mulching materials and methods for modifying soil environment. Soil Tillage Res 168:155–166. https://doi.org/10.1016/j.still.2017.01.001

Karny A, Zinger A, Kajal A et al (2018) Therapeutic nanoparticles penetrate leaves and deliver nutrients to agricultural crops. Sci Rep 8:7589. https://doi.org/10.1038/s41598-018-25197-y

Law KL, Thompson RC (2014) Oceans. Microplastics in the seas. Science 345:144–145. https://doi.org/10.1126/science.1254065

Li H, Zhang L, Lu H et al (2020a) Macro-/nanoporous al-doped zno/cellulose composites based on tunable cellulose fiber sizes for enhancing photocatalytic properties. Carbohydr Polym 250:116873. https://doi.org/10.1016/j.carbpol.2020.116873

Li L, Luo Y, Li R et al (2020b) Effective uptake of submicrometre plastics by crop plants via a crack-entry mode. Nat Sustain 3:929–937. https://doi.org/10.1038/s41893-020-0567-9

Li Z, Li W, Li X, Ma Z (2021a) Toxicity assessment of cationic microplastics on the seed germination, growth, and development of wheat. Environ Sci Pollut Res 28(9):10753–10762

Li Z, Li Q, Li R, Zhou J, Wang G (2021b) The distribution and impact of polystyrene nanoplastics on cucumber plants. Environ Sci Pollut Res 28:16042–16053. https://doi.org/10.1007/s11356-020-11702-2

Lian J, Wu J, Xiong H et al (2020) Impact of polystyrene nanoplastics (psnps) on seed germination and seedling growth of wheat (Triticum aestivum L.). J Hazard Mater 385:121620. https://doi.org/10.1016/j.jhazmat.2019.121620

Lian J, Liu W, Meng L et al (2021) Foliar-applied polystyrene nanoplastics (psnps) reduce the growth and nutritional quality of lettuce (Lactuca sativa L.). Environ Pollut 280:116978. https://doi.org/10.1016/j.envpol.2021.116978

Liao Y-C, Nazygul J, Li M, et al (2019) [Effects of microplastics on the growth, physiology, and biochemical characteristics of wheat (Triticum aestivum)]. Huan Jing Ke Xue 40:4661–4667. https://doi.org/10.13227/j.hjkx.201903113

Lira-Saldivar RH (2017) Gas exchange, yield and fruit quality of Cucurbita pepo cultivated with zeolite and plastic mulch, pp 123–139. https://doi.org/10.12871/00021857201723

Liu EK, He WQ, Yan CR (2014) 'White revolution' to 'white pollution'—agricultural plastic film mulch in China. Environ Res Lett 9:091001

Liu H, Yang X, Liu G et al (2017) Response of soil dissolved organic matter to microplastic addition in Chinese loess soil. Chemosphere 185:907–917. https://doi.org/10.1016/j.chemosphere.2017.07.064

Lozano YM, Rillig MC (2020) Effects of microplastic fibers and drought on plant communities. Environ Sci Technol 54:6166–6173. https://doi.org/10.1021/acs.est.0c01051

Lozano YM, Lehnert T, Linck LT et al (2021) Microplastic shape, polymer type, and concentration affect soil properties and plant biomass. Front Plant Sci 12:616645. https://doi.org/10.3389/fpls.2021.616645

Lu L-L, Tian S-K, Yang X-E et al (2008) Enhanced root-to-shoot translocation of cadmium in the hyperaccumulating ecotype of Sedum alfredii. J Exp Bot 59:3203–3213. https://doi.org/10.1093/jxb/ern174

Mao X, Xu Y, Cheng Z et al (2022) The impact of microplastic pollution on ecological environment: a review. Front Biosci (Landmark Ed) 27:46

Martı MC, Jiménez A, Sevilla F (2020) Thioredoxin network in plant mitochondria: cysteine s-posttranslational modifications and stress conditions. Front Plant Sci 11:571288. https://doi.org/10.3389/fpls.2020.571288

Mateos-Cárdenas A, van Pelt FNAM, O'Halloran J, Jansen MAK (2021) Adsorption, uptake and toxicity of micro- and nanoplastics: effects on terrestrial plants and aquatic macrophytes. Environ Pollut 284:117183. https://doi.org/10.1016/j.envpol.2021.117183

Meng F, Yang X, Riksen M et al (2021) Response of common bean (phaseolus vulgaris L.) growth to soil contaminated with microplastics. Sci Total Environ 755:142516. https://doi.org/10.1016/j.scitotenv.2020.142516

Mittler R (2002) Oxidative stress, antioxidants and stress tolerance. Trends Plant Sci 7:405–410. https://doi.org/10.1016/s1360-1385(02)02312-9

Mittler R (2017) ROS are good. Trends Plant Sci 22:11–19. https://doi.org/10.1016/j.tplants.2016.08.002

Navarro E, Baun A, Behra R et al (2008) Environmental behavior and ecotoxicity of engineered nanoparticles to algae, plants, and fungi. Ecotoxicology 17:372–386. https://doi.org/10.1007/s10646-008-0214-0

Noctor G, Mhamdi A, Foyer CH (2016) Oxidative stress and antioxidative systems: recipes for successful data collection and interpretation. Plant Cell Environ 39:1140–1160. https://doi.org/10.1111/pce.12726

Noctor G, Reichheld J-P, Foyer CH (2018) Ros-related redox regulation and signaling in plants. Semin Cell Dev Biol 80:3–12. https://doi.org/10.1016/j.semcdb.2017.07.013

Pflugmacher S, Sulek A, Mader H et al (2020) The influence of new and artificial aged microplastic and leachates on the germination of Lepidium sativum L. Plants 9:339. https://doi.org/10.3390/plants9030339

Pignattelli S, Broccoli A, Renzi M (2020) Physiological responses of garden Cress (L. sativum) to different types of microplastics. Sci Total Environ 727:138609. https://doi.org/10.1016/j.scitotenv.2020.138609

Pignattelli S, Broccoli A, Piccardo M et al (2021) Short-term physiological and biometrical responses of Lepidium sativum seedlings exposed to pet-made microplastics and acid rain. Ecotoxicol Environ Saf 208:111718. https://doi.org/10.1016/j.ecoenv.2020.111718

Qi Y, Yang X, Pelaez AM et al (2018) Macro- and micro-plastics in soil-plant system: effects of plastic mulch film residues on wheat (Triticum aestivum) growth. Sci Total Environ 645:1048–1056. https://doi.org/10.1016/j.scitotenv.2018.07.229

Quan L-J, Zhang B, Shi W-W, Li H-Y (2008) Hydrogen peroxide in plants: a versatile molecule of the reactive oxygen species network. J Integr Plant Biol 50:2–18. https://doi.org/10.1111/j.1744-7909.2007.00599.x

Raja V, Majeed U, Kang H et al (2017) Abiotic stress: interplay between Ros, hormones and mapks. Environ Exp Bot 137:142–157. https://doi.org/10.1016/j.envexpbot.2017.02.010

Ren X, Tang J, Wang L, Liu Q (2021) Microplastics in soil-plant system: effects of nano/microplastics on plant photosynthesis, rhizosphere microbes and soil properties in soil with different residues. Plant Soil 462:561–576. https://doi.org/10.1007/s11104-021-04869-1

Rillig MC (2012) Microplastic in terrestrial ecosystems and the soil? Environ Sci Technol 46:6453–6454. https://doi.org/10.1021/es302011r

Rillig MC, Lehmann A, de Souza Machado AA, Yang G (2019) Microplastic effects on plants. New Phytol 223:1066–1070. https://doi.org/10.1111/nph.15794

Robards AW (1975) Plasmodesmata. Annu Rev Plant Physiol 26:13–29. https://doi.org/10.1146/annurev.pp.26.060175.000305

Rochman CM, Browne MA, Halpern BS et al (2013) Classify plastic waste as hazardous. Nature 494:169–171. https://doi.org/10.1038/494169a

Rodrigues O, Reshetnyak G, Grondin A, Saijo Y, Leonhardt N, Maurel C et al (2017) Aquaporins facilitate hydrogen peroxide entry into guard cells to mediate ABA- and pathogen-triggered stomatal closure. Proc Natl Acad Sci U S A 114:9200–9205. https://doi.org/10.1073/pnas.1704754114

Schwab F, Rothen-Rutishauser B, Petri-Fink A (2020) When plants and plastic interact. Nat Nanotechnol 15:729–730. https://doi.org/10.1038/s41565-020-0762-x

Shi R, Liu W, Lian Y et al (2022) Phytotoxicity of polystyrene, polyethylene and polypropylene microplastics on tomato (Lycopersicon esculentum L.). J Environ Manag 317:115441. https://doi.org/10.1016/j.jenvman.2022.115441

Spanò C, Muccifora S, Ruffini Castiglione M, Bellani L, Bottega S, Giorgetti L (2022) Polystyrene nanoplastics affect seed germination, cell biology and physiology of rice seedlings in-short term treatments: evidence of their internalization and translocation. Plant Physiol Biochem 172:158–166. https://doi.org/10.1016/j.plaphy.2022.01.012

Sun X-D, Yuan X-Z, Jia Y et al (2020) Differentially charged nanoplastics demonstrate distinct accumulation in Arabidopsis thaliana. Nat Nanotechnol 15:755–760. https://doi.org/10.1038/s41565-020-0707-4

Sun H, Lei C, Xu J, Li R (2021) Foliar uptake and leaf-to-root translocation of nanoplastics with different coating charge in maize plants. J Hazard Mater 416:125854. https://doi.org/10.1016/j.jhazmat.2021.125854

Tang C-C, Chen H-I, Brimblecombe P, Lee C-L (2019) Morphology and chemical properties of polypropylene pellets degraded in simulated terrestrial and marine environments. Mar Pollut Bull 149:110626. https://doi.org/10.1016/j.marpolbul.2019.110626

Urbina MA, Correa F, Aburto F, Ferrio JP (2020) Adsorption of polyethylene microbeads and physiological effects on hydroponic maize. Sci Total Environ 741:140216. https://doi.org/10.1016/j.scitotenv.2020.140216

Verla AW, Enyoh CE, Verla EN, Nwarnorh KO (2019) Microplastic–toxic chemical interaction: a review study on quantified levels, mechanism and implication. SN Appl Sci 1:1400. https://doi.org/10.1007/s42452-019-1352-0

Wan Y, Wu C, Xue Q, Hui X (2019) Effects of plastic contamination on water evaporation and desiccation cracking in soil. Sci Total Environ 654:576–582. https://doi.org/10.1016/j.scitotenv.2018.11.123

Wang W, Yuan W, Xu EG et al (2022a) Uptake, translocation, and biological impacts of micro (nano)plastics in terrestrial plants: progress and prospects. Environ Res 203:111867. https://doi.org/10.1016/j.envres.2021.111867
Wang F, Wang Q, Adams CA, Sun Y, Zhang S (2022b) Effects of microplastics on soil properties: current knowledge and future perspectives. J Hazard Mater 424:127531
Wang F, Feng X, Liu Y, Adams CA, Sun Y, Zhang S (2022c) Micro(nano)plastics and terrestrial plants: up-to-date knowledge on uptake, translocation, and phytotoxicity. Resour Conserv Recycl 185:106503
Wang J, Lu S, Guo L, Wang P, He C, Liu D et al (2022d) Effects of polystyrene nanoplastics with different functional groups on rice (Oryza sativa l.) seedlings: combined transcriptome, enzymology, and physiology. Sci Total Environ 834:155092. https://doi.org/10.1016/j.scitotenv.2022.155092
Watteau F, Dignac M-F, Bouchard A et al (2018) Microplastic detection in soil amended with municipal solid waste composts as revealed by transmission electronic microscopy and pyrolysis/GC/MS. Front Sustain Food Syst 2:81
Wu C, Li Y, Cui S, Zhou J, Zhao Y (2020a) Effects of polystyrene microplastics on the growth and physiological processes of maize seedlings. Ecotoxicol Environ Saf 190:110071. https://doi.org/10.1016/j.ecoenv.2019.110071
Wu X, Liu Y, Yin S, Xiao K, Xiong Q, Bian S, Liang S, Hou H, Hu J, Yang J (2020b) Metabolomics revealing the response of rice (Oryza sativa L.) exposed to polystyrene microplastics. Environ Pollut 266:115159
Wu X, Lu J, Du M et al (2021) Particulate plastics-plant interaction in soil and its implications: a review. Sci Total Environ 792:148337. https://doi.org/10.1016/j.scitotenv.2021.148337
Wu X, Hou H, Liu Y, Yin S, Bian S, Liang S, Wan C, Yuan S, Xiao K, Liu B, Hu J, Yang J (2022) Microplastics affect rice (Oryza sativa L.) quality by interfering metabolite accumulation and energy expenditure pathways: a field study. J Hazard Mater 422:126834. https://doi.org/10.1016/j.jhazmat.2021.126834
Xiao F, Feng L-J, Sun X-D, Wang Y, Wang Z-W, Zhu F-P et al (2022) Do polystyrene nanoplastics have similar effects on duckweed (Lemna minor L.) at environmentally relevant and observed-effect concentrations? Environ Sci Technol 56:4071–4079. https://doi.org/10.1021/acs.est.1c06595
Xu G, Liu Y, Yu Y (2021) Effects of polystyrene microplastics on uptake and toxicity of phenanthrene in soybean. Sci Total Environ 783:147016. https://doi.org/10.1016/j.scitotenv.2021.147016
Yang J, Cao WD, Rui YK (2017) Interactions between nanoparticles and plants: phytotoxicity and defense mechanisms. J Plant Interact 12:158–169
Yi M, Zhou S, Zhang L, Ding S (2021) The effects of three different microplastics on enzyme activities and microbial communities in soil. Water Environ Res 93:24–32
Zeb A, Liu W, Meng L, Lian J, Wang Q, Lian Y, Chen C, Wu J (2022) Effects of polyester microfibers (PMFs) and cadmium on lettuce (Lactuca sativa) and the rhizospheric microbial communities: a study involving physiobiochemical properties and metabolomic profiles. J Hazard Mater 424:127405
Zhang Q, Zhao M, Meng F et al (2021) Effect of polystyrene microplastics on rice seed germination and antioxidant enzyme activity. Toxics 9:179. https://doi.org/10.3390/toxics9080179
Zhou C-Q, Lu C-H, Mai L et al (2021a) Response of rice (Oryza sativa L.) roots to nanoplastic treatment at seedling stage. J Hazard Mater 401:123412. https://doi.org/10.1016/j.jhazmat.2020.123412
Zhou J, Wen Y, Marshall MR et al (2021b) Microplastics as an emerging threat to plant and soil health in agroecosystems. Sci Total Environ 787:147444. https://doi.org/10.1016/j.scitotenv.2021.147444
Zhu B-K, Fang Y-M, Zhu D, Christie P, Ke X, Zhu Y-G (2018) Exposure to nanoplastics disturbs the gut microbiome in the soil oligochaete Enchytraeus crypticus. Environ Pollut 239:408–415

Zhu F, Zhu C, Wang C, Gu C (2019) Occurrence and ecological impacts of microplastics in soil systems: A review. Bull Environ Contam Toxicol 102:741–749. https://doi.org/10.1007/s00128-019-02623-z

Zimmermann RP, Kardos LT (1961) Effect of bulk density on root growth. Soil Sci 91:280–288

Zong X, Zhang J, Zhu J et al (2021) Effects of polystyrene microplastic on uptake and toxicity of copper and cadmium in hydroponic wheat seedlings (Triticum aestivum L.). Ecotoxicol Environ Saf 217:112217. https://doi.org/10.1016/j.ecoenv.2021.112217

Toxicity of Microplastics on the Flora and Fauna of the Freshwater Ecosystem: An Overview

24

Abel Inobeme, John Tadso, Babatunde Oluwafemi Adetuyi, Charles Oluwaseun Adetunji, Oluwabukola Atinuke Popoola, Falana Yetunde Olaitan, Olalekan Akinbo, Mohd. Shahnawaz, Oluwafemi Adebayo Oyewole, Eniola K.I.T., Mohammed Bello Yerima, and Olatunji Matthew Kolawole

Abstract

Microplastics (MPs) are defined as minute plastic particles that typically exhibit dimensions that span from 1 μm to 5 mm. The aforementioned materials exhibit

A. Inobeme
Department of Chemistry, Edo University Iyamho, Auchi, Edo State, Nigeria

J. Tadso
Department of Chemistry, Ibrahim Badamasi Babangida University Lapai, Lapai, Niger State, Nigeria

B. O. Adetuyi
Biochemistry Unit, Department of Natural Sciences, Faculty of Pure and Applied Sciences, Precious Cornerstone University, Ibadan, Nigeria

C. O. Adetunji (✉)
Applied Microbiology, Biotechnology and Nanotechnology Laboratory, Department of Microbiology, Edo State University Iyamho, Auchi, Edo State, Nigeria
e-mail: adetunji.charles@edouniversity.edu.ng

O. A. Popoola · F. Y. Olaitan
Plant Improvement Unit, Agricultural Biotechnology Department, National Biotechnology Development Agency, Abuja, Nigeria

O. Akinbo
Centre of Excellence in Science, Technology, and Innovation, AUDA-NEPAD, Johannesburg, South Africa

M. Shahnawaz
Department of Botany, University of Ladakh, Kargil Campus, Khumbathang, Kargil, Ladakh UT, India

Present Address: Department of Botany, Shri Pandit Prem Nath Dogra Government Degree College Samba, Samba, Jammu and Kashmir UT, India

M. Shahnawaz et al. (eds.), *Microplastic Pollution*,
https://doi.org/10.1007/978-981-99-8357-5_24

insolubility in aqueous media and can be subdivided into primary particles, generated initially at diminutive dimensions, and secondary particles, formed as a consequence of the disintegration of larger plastic constituents. Microplastics (MPs) tend to accumulate in various ecosystems, including those of freshwater origin, and subsequently become subject to degradation. Given their omnipresence, freshwater ecosystems represent a paramount facet of human existence and remain persistently vulnerable to the presence of minute plastic particles. MP pollution is regarded as a major global concern in contemporary times due to its detrimental impacts on all facets of the freshwater ecosystem. The deleterious effects are primarily associated with the toxic nature of MPs, which pose a grave threat to the various components of the said environment. The present chapter elucidates upon the subject of freshwater contaminants of emerging concern, specifically with regard to MPs. The toxicity and mechanism of action of these contaminants are comprehensively discussed, along with their biodegradation mechanism. The repercussions of exposure of indigenous aquatic life forms, including autotrophic organisms and flora and fauna within the freshwater environment, are also examined in detail. The degree of toxicity inflicted by microplastics (MPs) is contingent upon the species exposed to MPs, as well as the manner of exposure.

Keywords

Freshwater · Toxicity · Aquatic autotrophic organisms · Microplastics

24.1 Introduction

The acceleration of productivity and deceleration of biotic decomposition of plastic within ecosystems have resulted in its accumulation in the environment, ultimately leading to deleterious consequences for aquatic life. The long-lasting presence of plastics in the marine environment has been observed to persist for several centuries and even millennia. As a result of mechanical and photochemical processes, the plastics undergo fragmentation leading to the production of significantly smaller particles, commonly referred to as microplastics (measuring less than 5 mm) or even

O. A. Oyewole
Department of Microbiology, Federal University of Technology, Minna, Nigeria

E. K.I.T.
Department of Biological Sciences, Joseph Ayo Babalola University, Ikeji Arakeji, Osun State, Nigeria

M. B. Yerima
Department of Microbiology, Sokoto State University, Sokoto, Sokoto State, Nigeria

O. M. Kolawole
Department of Microbiology Faculty of Life Sciences, University of Ilorin, Ilorin, Kwara State, Nigeria

nanoplastics (measuring less than 1 μm) (Espinosa et al. 2016). In the year 2016, as per the Gregorian calendar, it was observed and recorded.

Microplastics (MPs) represent a novel class of contaminants, encompassing minute plastic particles with dimensions smaller than 5 mm. The escalating concentration of pollutants has emerged as a significant environmental concern in light of their persistency, ubiquitous presence, and plausible aquatic habitat toxicity, as documented by Anderson et al. (2017). The situation poses a looming threat to biodiversity and could potentially have adverse implications for human health, according to Chen et al. (2020) findings.

Freshwater systems are considered to be highly complex systems in terms of their ability to transport and retain microplastics. These systems are capable of both receiving microplastics from terrestrial environments and serving as conduits for their transportation to marine environments. Additionally, these systems are capable of generating microplastics through the fragmentation of larger particles. Furthermore, they act as sinks for microplastics, retaining them within sediments. Furthermore, the notion of "freshwater" encompasses an extensive array of aqueous bodies, including rivers, streams, ditches, lakes, and ponds, each characterized by distinct physical and chemical properties.

Insufficient waste disposal may contribute to the introduction of larger plastic entities into the freshwater ecosystem, arising from an array of sources such as haphazard dumping or leakage from landfills, and subsequent transportation by means of surface runoff or atmospheric breezes. Furthermore, freshwater systems experience substantial direct discharge of microplastics, in addition to macroplastics. The input of agricultural plastics or sewage sludge-derived fibers and microbeads can occur as a consequence of agricultural drainage and runoff stemming from agricultural activities. Stormwater drainage and urban runoff typically lack proper filtration and treatment and may be contaminated with microplastics originating from the deterioration of road paint as well as vehicular abrasion (Horton et al. 2017). Wastewater treatment plants have proven to be effective in removing microplastics; however, caution must be exercised as direct effluent input may still contain microplastics. Furthermore, when confronted with high flow conditions, combined sewage overflows (CSOs) are designed to discharge untreated sewage into nearby rivers in order to alleviate the burden on drainage systems, thereby discharging both micro- and macroplastic waste. According to research, while clusters of microplastic pollution may manifest in the vicinity of urban centers, the preponderance of microplastics appears to inundate bodies of water as a consequence of drainage structures. Therefore, it is imperative to carefully scrutinize diverse sources of microplastic pollutants, including combined sewer overflows (CSOs), stormwater systems, and discharge channels, which may be segregated from densely populated regions (Browne et al. 2011).

While the major emphasis in freshwater microplastic research is on rivers, the prevalence of microplastics in ponds and lakes has also been acknowledged (Free et al. 2014). Similar to rivers, standing waterbodies are susceptible to receiving inputs from land runoff and wind-dispersed debris. Nonetheless, owing to the enclosed nature of lakes, there is a higher likelihood of persistent accumulation of microplastics resulting from these inputs (Vaughan et al. 2017).

To thoroughly comprehend the ramifications of microplastic contamination in freshwater ecosystems, a comprehensive investigation and analysis of various aspects are imperative. The aforementioned elements may encompass various dimensions, ranging from environmental and ecological contingencies to socioeconomic and public health determinants. It should be noted that these aspects are not exhaustive in their nature. A comprehensive analysis of the aforementioned factors is likely to offer a heightened comprehension of the extensiveness and severity of the matter under consideration. The present discourse prevalent in academic literature concerns the comprehensive comprehension of microplastics, encompassing their origins, dispersion, classification, and consequences. The genesis of contaminant substances in freshwater ecosystems can typically be ascribed to the existence of synthetic textiles and personal care products, in addition to industrial raw materials. The principal factors that are known to contribute to the occurrence of microplastics in riverine habitats are commonly attributed to wastewater discharge, encompassing untreated industrial effluent and household wastewater disposal. The relatively small size of microplastics elicits a sense of increased concern, due to their potential to accumulate to a greater extent within various systems, as noted by Frias and Nash (2019). The present investigation elucidates that the composition of microplastics (MP) may experience modifications upon being exposed to aquatic pollutants, including metals and persistent, bioaccumulative, and toxic compounds. This phenomenon can be attributed to the relatively greater surface-to-volume ratio of MP. In their 2012 studies, Engler demonstrated a direct correlation between plastic waste and persistent bioaccumulative toxins (PBTs); numerous investigations have been carried out on polychlorinated biphenyls (PCBs) and dichlorodiphenyltrichloroethane (DDT), as well as on polycyclic aromatic hydrocarbons, as reported by Antunes et al. (2018). Due to variations in the spectrum of contaminants, there exist discernible differences in the contaminant composition profiles between freshwater and marine ecosystems. The issue of water pollution in stationary riverine systems such as ponds and lakes is a severe concern. The identified problem can be ascribed to the negligent conduct of local residents or to the ecologically disruptive interventions of the tourism industry. The conglomeration of disintegrable and non-disintegrable refuse intensifies this predicament. These life forms sustain their flotation within the uppermost stratum of the aqueous environment and later get entrapped in the deposits of the lake bed. Numerous investigations have conferred substantial evidence to verify the presence of microplastics within the drinking water system, thus establishing it as a matter of notable concern (Li et al. 2018). The study of river systems and watersheds contributes to the extant knowledge that enables individuals to comprehend the issue of microplastic accumulation within freshwater systems (Miller et al. 2017). The role of the river's flow is crucial in the process of removing plastic fragments from the environment. An important observation is that following precipitation events, a significant amount of microplastics can be identified in both fluvial and sedimentary settings (Lima et al. 2014). Over time, the freshwater system also plays a notable role in the dispersion of pollutants in marine ecosystems. This is due to the conflux of riverine and oceanic bodies which result in the amalgamation of minerals, sediments, soil, and pollutants. Hence, the import of freshwater systems as a potent contributor of microplastics to the marine ecosystem is not negligible. The non-uniform spatial distribution of microplastics in freshwater ecosystems is subject to the influence of point sources of wastewater discharge in close proximity.

The present study noted a higher occurrence of water contamination in riverine areas proximate to industrial zones relative to those located in residential localities. Wastewater treatment plants are typically established by various organizations, yet their level of efficiency in removing large plastic waste exceeds that of meso- and microplastic waste. This is due to the installation of various technologies that effectively remove large-sized particles; however, the specificity of these technologies in retaining microplastics remains uncertain (Mani et al. 2015). The effluent discharged from wastewater treatment facilities possesses hazardous compounds and micro- and nano-scale plastic particles that have the potential to enter the food chain and have a significant impact on the biotic ecosystem.

Amidst the numerous anthropogenic stressors which perpetuate in aquatic ecological systems, the accrual of plastic debris stands as a discernibly conspicuous yet inadequately investigated variable. Although plastics confer remarkable sociocultural advantages (Andrady and Neal 2009), the era of plastics brings with it certain detrimental consequences. The accumulation of plastics in natural habitats is a result of durability, unsustainable usage, and inappropriate waste management practices (Barnes et al. 2009). Plastics of diverse sizes and sources are ubiquitous in the marine environment and exert a deleterious impact on numerous species that become entangled in or ingest them, as noted by Gregory (2009). According to the prevailing environmental conditions, larger plastic objects tend to deteriorate into minuscule fragments referred to as microplastics (MP), measuring less than 5 mm in diameter. In addition to the production of primary microplastics, it is notable that secondary microplastics, which are degradation products, may also occur. As an illustration, microplastics (MPs) are deliberately employed as resin pellets—a primary constituent of plastic product manufacturing—or as a component within personal care items, such as exfoliating scrubs and shower gels.

The deterioration of MPs at a biological level predominantly occurs as a result of microbial activity, which encompasses a range of microorganisms such as bacteria, actinomycetes, algae, and fungi that are typically present within the surrounding ecosystem. The present study details a multi-step procedure involving the synthesis of extracellular enzymes, their attachment onto the surface of microplastics (MPs), the reduction of MPs into smaller structures, and ensuing mineralization leading to the production of oxidized metabolites. The aforementioned findings were brought to the fore in a recent investigation undertaken by Othman et al. (2021). The existence of MP pollutants within freshwater ecosystems constitutes a significant and concerning danger to the indigenous plant and animal life. As a result, notable biochemical and molecular modifications occur in the regulatory processes that govern homeostasis, encompassing genotoxicity, reactive oxygen species generation, hepatic dysfunction, and growth retardation, among other factors. Microplastic (MP) particles, due to their small size, have the potential to be easily incorporated into food sources and subsequently consumed or assimilated by various freshwater fauna. These faunae include fishes (Saad et al. 2022), bivalves (Pastorino et al. 2021), and caddisflies (Ehlers et al. 2019), among other species. Enyoh et al. (2020) have reported inhalation as an important route of exposure to microplastics (MPs) in the animal kingdom, alongside ingestion.

24.2 Toxicity of Microplastics in Flora

The proliferation of plastic waste in the environment is as a result of population growth and advancement in technology (Bosker et al. 2019). Microplastics can originate from different sources like through the cosmetics, large plastic items, synthetic clothing, wastewater treatment plants (WWTP), wind abrasion, tire wear, and the air–water interface (Apetogbor et al. 2023; O'Brien et al. 2023; Wu et al. 2023). Consequently, making MP to be ubiquitous, they have been found in the air, water, soil, and other environmental compartments (Apetogbor et al. 2023; O'Brien et al. 2023; Wu et al. 2023).

Research has suggested that the concentration of MPs in the soil is about 23 times higher than in the ocean (Khalid et al. 2023), primarily because of sources like plastic mulching, landfilling of plastic wastes, utilization of sludge, wastewater irrigation, and atmospheric deposition (Khalid et al. 2023). MPs have large surface area, so they can adsorb so many pollutants (Ashjar et al. 2023). When MPs are released into the soil through WWTP, not all plants have the necessary nets that can trap these particles due to their small size (Ashjar et al. 2023; Huang et al. 2023).

MPs due to their large surface area can serve as a carrier of several pollutants (Ashjar et al. 2023). When interacting with Cd in soil-plant systems, MPs have been observed to affect the behaviors and biological effects of Cd (Banaee et al. 2019). Huang et al. (2023) investigated the presence of polyethylene MP and Cd in soil, and their study revealed that the presence of MP increased the bioavailability Cd in soil, leading to higher plant Cd concentrations, but also resulted in hindering plant growth.

Studies have shown that MPs have a negative impact on the growth of plants (Zhou et al. 2023). For instance, MPs can reduce seed germination rates, inhibit root growth, and decrease plant biomass (Shorobi et al. 2023). Moreover, MPs can alter various cellular characteristics of plants. A study on tomato plants exposed to MPs had reduced activity of certain enzymes important for cell metabolism and growth (Liu et al. 2023).

In addition, MPs can interfere with the physiological processes in plants (Wang et al. 2023). A study by Bosker et al. (2019) on the effects MPs on *Lepidium sativum* revealed that plastic particles were present in the pores of the plant, inhibiting water uptake and causing delays in germination. However, there was no change in the chlorophyll content. Another study examined the effects of propylene MPs on tomato and cherry tomato plants and found that germination rate was not significantly affected. After 7 days, the root of the plants became elongated, and nutrient uptake was found to be dependent on the type of nutrient and plant species (Shorobi et al. 2023).

MPs in soils have been proven to reduce the inorganic nitrogen content enzyme in plants. MPS effect was also investigated in peanuts; Meng et al. reported that MPs inhibit nitrogen uptake and disrupt the nitrogen cycle in the plants. It additionally causes damage to the root introducing oxidative stress in the peanut plant (Liu et al. 2023; Meng et al. 2022). It was researched and found out that plants find it difficult to take up nutrients from soil because of some microbes that had interrupted the nutrient dynamics because of its blockage by MPs.

24.3 Microplastics in Fauna

They are pervasive in the environment due to the widespread use of plastics, which are known for their durability, low-cost, light weight nature (Zhou et al. 2023; Inobeme et al. 2023). MPs originate from different sources, such as the breakdown of larger plastics, personal care products, shedding of microfibers from clothing. Consequently, they can be found in various environmental compartments, given their extensive use (Apetogbor et al. 2023; Inobeme et al. 2023; Wu et al. 2023). Numerous studies have revealed that MPs can have detrimental effects on animals (Table 24.1). Moreover, MPs can contain hazardous chemicals, including flame retardants, plasticizers, bis phenol A which were added to plastics. They can leach into the animal's tissues potentially leading to toxic effects.

In marine fauna, such as fish, sea turtles, and marine mammals, ingestion of MPs has been linked to a range of negative effects such as reduced growth rates, reproductive failure, and death. Also, it has been pointed out that MPs can accumulate in the tissues of animals, leading to long-term health effects. MP toxicity was investigated on the gills of *Litopenaeus vannamei*; after 14 days of exposure to MPs, it was noted that the gills were deformed. Also, oxidative stress was induced by the expression of some genes, MPs caused the alteration in the metabolism of carbohydrates, lipids, and amino acids in the Shrimp.

In another study, polypropylene MPs were used to analyze *Daphnia similis*; the polypropylene MPs were obtained from polypropylene bags discovered in the river. The mobility of *D. similis* was seen to be hampered by the MPs, and the quantity of the enzyme acetylcholinesterase (AChE) also reduces the intensity and duration of the neurotransmitter activity.

A study examined the toxicity of polystyrene (PS) MPs on earthworms in soil by exposing them to 14 days. Results showed that the MPs did not influence the weight, survival rate, or gut microbiota diversity, but did cause disruptions in some metabolic processes and osmoregulation. Another research by examined the effects of PS MPs on mice over a period of 5 weeks and discovered that they resulted in body weight loss and changes in the makeup of intestinal microbiota that resulted in problems with lipid metabolism in the liver. Furthermore, it was shown that the presence of PS MPs led to fibrosis and death in the ovarian granulosa cells of female Wistar rats (An et al. 2022). These results underline the need for more study to fully comprehend MPs' possible negative impacts on various creatures and the potential harm they may cause.

24.4 Report on Toxicity on Flora and Fauna

24.4.1 Toxicity on Fauna

Microplastic pollution can also have detrimental effects on fauna, including various animal species (Fig. 24.1). The toxic impacts of microplastics on fauna can be attributed to several mechanisms: Ingestion: Many animals, both marine and terrestrial, can mistakenly ingest microplastics. Microplastics can resemble food particles or be associated with their natural prey. When animals consume microplastics, they can experience physical blockage or damage in their digestive system, leading to reduced feeding efficiency, nutrient absorption issues, and potential starvation (Ziani et al. 2023).

Accumulation and biomagnification: Microplastics can accumulate in the tissues of animals. They can be retained in the digestive tract or absorbed through the gut

Table 24.1 Common additives, uses, sources, and health effects to fauna

Sr. no.	Additives	Uses	Sources	Health effects	References
1.	BPA	1. Module of epoxy resins to coat the inner surfaces of cans 2. Helps to extend the shelf life of food and beverage products	Plastics, medical devices	Cardiac repolarization	Ma et al. (2023)
2.	Plasticizer	Used in the production of medical devices, furniture, floor tiles	Dismantling of electronic waste, usage of personal care products like shampoo, creams, soap, cosmetics. Industrial solvents, pesticides, soft plastics, toy, food container	Delay in wound healing	
3.	Brominated flame retardants	Used to reduce combustion, prevent fire outbreak, and also delay the spread of fire	Electronic equipment (televisions, monitor, computer, fax machine), animal-derived food	Unfavorable influences on the reproductive system, cardiovascular system, liver, and hypothalamic–pituitary–thyroid (HPT) axis	

into the circulatory system. Over time, the accumulation of microplastics can lead to chronic toxicity, impacting organ function and overall health. Moreover, microplastics can undergo biomagnification in food chains, where predator animals accumulate higher levels of microplastics by consuming prey that contains them (Miller et al. 2020).

Chemical toxicity: Microplastics can contain or adsorb various toxic chemicals from the surrounding environment. These chemicals may include persistent organic pollutants (POPs), heavy metals, and other harmful substances. When animals ingest microplastics laden with toxic chemicals, these substances can leach into their tissues and cause systemic toxic effects, such as organ damage, developmental abnormalities, endocrine disruption, and immune system suppression (Campanale et al. 2020; Menéndez-Pedriza and Jaumot 2020).

Physical injury and entanglement: Larger microplastic fragments, such as microfibers and small plastic debris, can cause physical harm to fauna. Animals can become entangled in microplastics, which can impair their movement, feeding, and escape abilities, leading to injuries, stress, or even death. Entanglement can occur in marine species like seals, turtles, and marine birds, as well as terrestrial animals like birds and mammals (Usman et al. 2020).

Disruption of behavior and reproduction: Microplastic exposure can influence animal behavior and reproductive processes. Research suggests that certain chemicals associated with microplastics can alter hormone levels and reproductive functions, leading to reduced fertility, impaired embryonic development, and disrupted mating behaviors. These impacts can have long-term consequences for the population dynamics and viability of affected species. It is important to note that the specific effects of microplastic toxicity on fauna can vary across different species, habitats, and exposure scenarios. Furthermore, the interactions between microplastics and fauna in ecosystems can be complex and can have cascading effects on food webs and ecosystem functioning. Ongoing research is necessary to deepen our understanding of the full extent and long-term consequences of microplastic toxicity on fauna and to develop strategies for mitigation and prevention (Dubey et al. 2022; Urli et al. 2023).

24.4.2 Toxicity on Flora

Microplastic toxicity on flora refers to the adverse effects of microplastic pollution on plants and plant ecosystems (Fig. 24.1). While the impacts of microplastics on marine environments have been extensively studied, research on their effects on terrestrial flora is relatively limited but growing. Some potential toxic effects of microplastics on flora:

Growth inhibition: Microplastics can interfere with the growth and development of plants. They can accumulate in the root zone, affecting root elongation, nutrient absorption, and water uptake. This can lead to stunted growth, reduced biomass production, and overall impairment of plant health (Azeem et al. 2021).

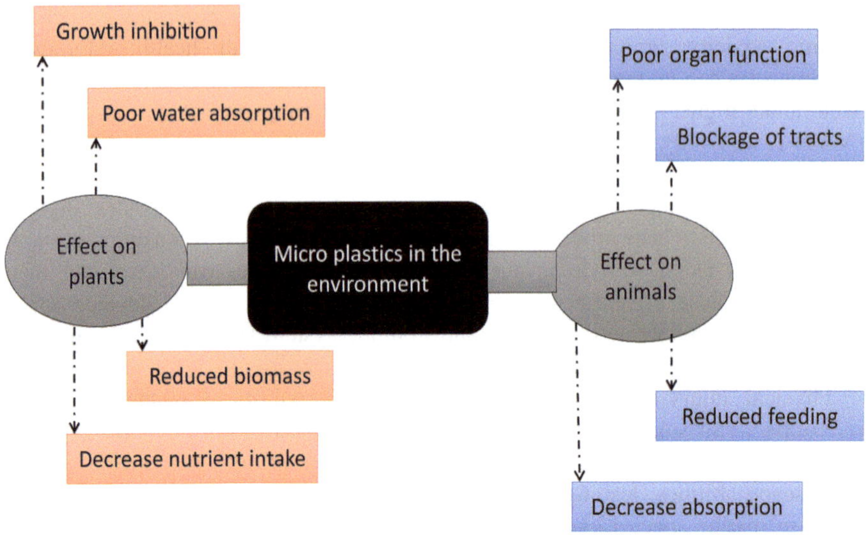

Fig. 24.1 Effect of microplastics in flora and fauna

Physiological disturbances: Microplastics can disrupt various physiological processes in plants. They can interfere with photosynthesis by blocking sunlight, reducing chlorophyll production, and inhibiting gas exchange through leaf surfaces. This can result in reduced energy production and nutrient assimilation, leading to weakened plants (Li et al. 2023).

Nutrient imbalances: Microplastics can alter the nutrient availability and cycling in soil, affecting nutrient uptake by plants. They can modify soil properties, such as water-holding capacity and nutrient retention, which can disrupt nutrient balance in plants. Imbalances in essential nutrients can impact plant growth, flowering, fruiting, and overall productivity (Inobeme et al. 2022; Prado et al. 2023).

Oxidative stress: The presence of microplastics can lead to the generation of reactive oxygen species (ROS) within plant tissues, causing cellular damage and impairing physiological processes. Oxidative stress can disrupt cellular metabolism, disturb hormone regulation, and contribute to plant senescence and cell death (Ekner-Grzyb et al. 2022).

Altered gene expression: Microplastics can affect gene expression patterns in plants. Studies have shown that exposure to microplastics can lead to changes in the expression of genes associated with stress responses, defense mechanisms, and growth regulation. These alterations in gene expression can have cascading effects on plant physiology and development (Roy and Dey 2022).

Disruption of symbiotic relationships: Microplastics can interfere with beneficial plant–microbe interactions, such as mycorrhizal associations. Mycorrhizal fungi play a crucial role in nutrient uptake and plant health. Microplastic pollution can disrupt the establishment and functioning of mycorrhizal symbiosis, impairing nutrient acquisition and overall plant performance. It is important to recognize that the effects of microplastic toxicity on flora can vary depending on factors such as plastic type, size, concentration, exposure duration, and plant species (Leifheit et al. 2021).

24.5 Conclusion

The toxicity of microplastics on the flora and fauna of freshwater ecosystems is a significant concern that has gained attention in recent years. This overview highlights the key findings and implications regarding the impacts of microplastics on the delicate balance of freshwater ecosystems. The accumulation of microplastics in freshwater environments poses a direct threat to the flora and fauna that inhabit these ecosystems. Studies have revealed that microplastics can adversely affect aquatic plants, algae, and macrophytes, leading to reduced growth, impaired photosynthesis, and altered nutrient uptake. These impacts can disrupt the ecological functioning of freshwater ecosystems and have cascading effects on higher trophic levels. Microplastics also pose significant risks to freshwater fauna. Aquatic organisms such as zooplankton, insects, mollusks, and fish can ingest microplastics, mistaking them for food. The ingestion of microplastics can lead to physical damage, internal injuries, and blockage of the digestive system, affecting feeding, reproduction, and overall fitness. Additionally, microplastics can act as carriers for other toxic pollutants, such as persistent organic pollutants (POPs) and heavy metals, exacerbating the toxicity and bioaccumulation potential in aquatic organisms. Furthermore, public awareness and education are vital in promoting responsible plastic use and fostering behavioral changes that contribute to the conservation of freshwater ecosystems. Collaborative efforts between scientists, policymakers, industries, and communities are necessary to develop comprehensive and sustainable solutions to mitigate the impacts of microplastics on flora and fauna in freshwater ecosystems. Overall, understanding the toxicity of microplastics on freshwater ecosystems is crucial for developing effective strategies to protect these valuable habitats and the organisms that rely on them. By addressing this issue, we can strive to safeguard the health and biodiversity of freshwater ecosystems, ensuring their sustainability for future generations. Microplastics are major formed from plastic bottles, containers, fibers, and textiles. Their deleterious effects on plants and animals in the environment make them serious issue of health concern in recent.

References

Anderson PJ, Warrack S, Langen V, Challis JK, Hanson ML, Rennie MD (2017) Microplastic contamination in Lake Winnipeg, Canada. Environ Pollut 225(2017):223–231

Andrady AL, Neal MA (2009) Applications and societal benefits of plastics. Philos Trans R Soc B Biol Sci 364:1977–1984

Antunes J, Frias J, Sobral P (2018) Microplastics on the Portuguese coast. Mar Pollut Bull 131:294–302

Apetogbor K, Pereao O, Sparks C, Opeolu B (2023) Spatio-temporal distribution of microplastics in water and sediment samples of the Plankenburg river, Western Cape, South Africa. Environ Pollut 323:121303

Ashjar N, Keshavarzi B, Moore F, Zarei M, Busquets R, Zebarjad SM, Mohammadi Z (2023) Microplastics (MPs) distribution in surface sediments of the Freidounkenar paddy wetland. Environ Pollut 317:120799

Azeem I, Adeel M, Ahmad MA, Shakoor N, Jiangcuo GD, Azeem K, Ishfaq M, Shakoor A, Ayaz M, Xu M, Rui Y (2021) Uptake and accumulation of nano/microplastics in plants: a critical review. Nanomaterials (Basel, Switzerland) 11(11):2935. https://doi.org/10.3390/nano11112935

Banaee M, Soltanian S, Sureda A, Gholamhosseini A, Haghi BN, Akhlaghi M, Derikvandy A (2019) Evaluation of single and combined effects of cadmium and micro-plastic particles on biochemical and immunological parameters of common carp (*Cyprinus carpio*). Chemosphere 236:124335

Barnes DK, Galgani F, Thompson RC, Barlaz M (2009) Accumulation and fragmentation of plastic debris in global environments. Philos Trans R Soc B Biol Sci 364:1985–1998

Bosker T, Bouwman LJ, Brun NR, Behrens P, Vijver MG (2019) Microplastics accumulate on pores in seed capsule and delay germination and root growth of the terrestrial vascular plant *Lepidium sativum*. Chemosphere 226:774–781

Browne MA, Crump P, Niven SJ, Teuten E, Tonkin A, Galloway T, Thompson R (2011) Accumulation of microplastic on shorelines worldwide: sources and sinks. Environ Sci Technol 45(21):9175–9179

Campanale C, Massarelli C, Savino I, Locaputo V, Uricchio VF (2020) A detailed review study on potential effects of microplastics and additives of concern on human health. Int J Environ Res Public Health 17(4):1212. https://doi.org/10.3390/ijerph17041212

Chen G, Feng Q, Wang J (2020) Mini-review of microplastics in the atmosphere and their risks to humans. Sci Total Environ 703:135504

Dubey I, Khan S, Kushwaha S (2022) Developmental and reproductive toxic effects of exposure to microplastics: a review of associated signaling pathways. Front Toxicol 4:901798. https://doi.org/10.3389/ftox.2022.901798

Ehlers SM, Manz W, Koop JH (2019) Microplastics of different characteristics are incorporated into the larval cases of the freshwater caddisfly *Lepidostoma basale*. Aquat Biol 28:67–77

Ekner-Grzyb A, Duka A, Grzyb T, Lopes I, Chmielowska-Bąk J (2022) Plants oxidative response to nanoplastic. Front Plant Sci 13:1027608. https://doi.org/10.3389/fpls.2022.1027608

Enyoh CE, Shafea L, Verla AW, Verla EN, Qingyue W, Chowdhury T, Paredes M (2020) Microplastics exposure routes and toxicity studies to ecosystems: an overview. Environ Anal Health Toxicol 35(1):e2020004

Espinosa C, Esteban MÁ, Cuesta A (2016) Microplastics in aquatic environments and their toxicological implications for fish. In: Larramendy M, Soloneski S (eds) Toxicology—new aspects to this scientific conundrum. IntechOpen, London, pp 113–145. https://doi.org/10.5772/64815

Free CM, Jensen OP, Mason SA, Eriksen M, Williamson NJ, Boldgiv B (2014) High-levels of microplastic pollution in a large, remote, mountain lake. Mar Pollut Bull 85(1):156–163

Frias JP, Nash R (2019) Microplastics: finding a consensus on the definition. Mar Pollut Bull 138:145–147

Gregory MR (2009) Environmental implications of plastic debris in marine settings-entanglement, ingestion, smothering, hangers-on, hitch-hiking and alien invasions. Philos Trans R Soc B Biol Sci 364:2013–2025

Horton AA, Svendsen C, Williams RJ, Spurgeon DJ, Lahive E (2017) Large microplastic particles in sediments of tributaries of the river Thames, UK—abundance, sources and methods for effective quantification. Mar Pollut Bull 114(1):218–226

Huang F, Hu J, Chen L, Wang Z, Sun S, Zhang W, Jiang H, Luo Y, Wang L, Zeng Y, Fang L (2023) Microplastics may increase the environmental risks of Cd via promoting Cd uptake by plants: a meta-analysis. J Hazard Mater 448:130887

Inobeme A, Nayak V, Mathew TJ, Okonkwo S, Ekwoba L, Ajai AI, Bernard E, Inobeme J, Agbugui MM, Singh KR (2022) Chemometric approach in environmental pollution analysis: a critical review. J Environ Manag 309:114653. https://doi.org/10.1016/j.jenvman.2022.114653

Inobeme A, Adetunji CO, Okonkwo S, Bamigboye MO, Ajai AI, Inobeme J, Olusanya J, Afoso E (2023) Fate and occurrence of microplastic and nanoplastic pollution in industrial wastewater. In: Shah MP (ed) Microbial degradation and detoxification of pollutants. Walter de Gruyter, GmbH, Berlin. https://doi.org/10.1515/9783110743623-013

Khalid N, Aqeel M, Noman A, Rizvi ZF (2023) Impact of plastic mulching as a major source of microplastics in agroecosystems. J Hazard Mater 445:130455

Leifheit EF, Lehmann A, Rillig MC (2021) Potential effects of microplastic on arbuscular mycorrhizal fungi. Front Plant Sci 12:626709. https://doi.org/10.3389/fpls.2021.626709

Li J, Green C, Reynolds A, Shi H, Rotchell JM (2018) Microplastics in mussels sampled from coastal waters and supermarkets in the United Kingdom. Environ Pollut 241:35–44

Li H, Song F, Song X, Zhu K, Lin Q, Zhang J, Ning G (2023) Single and composite damage mechanisms of soil polyethylene/polyvinyl chloride microplastics to the photosynthetic performance of soybean (*Glycine max* [L.] merr.). Front Plant Sci 13:1100291. https://doi.org/10.3389/fpls.2022.1100291

Lima AR, Costa MF, Barletta M (2014) Distribution patterns of microplastics within the plankton of a tropical estuary. Environ Res 132:146–155

Liu Y, Xu F, Ding L, Zhang G, Bai B, Han Y, Xiao L, Song Y, Li Y, Wan S, Li G (2023) Microplastics reduce nitrogen uptake in peanut plants by damaging root cells and impairing soil nitrogen cycling. J Hazard Mater 443:130384

Mani T, Hauk A, Walter U, Burkhardt-Holm P (2015) Microplastics profile along the Rhine River. Sci Rep 5(1):1–7

Menéndez-Pedriza A, Jaumot J (2020) Interaction of environmental pollutants with microplastics: a critical review of sorption factors. Bioaccumul Ecotoxicol Effects Toxics 8(2):40. https://doi.org/10.3390/toxics8020040

Meng F, Yang X, Riksen M, Geissen V (2022) Effect of different polymers of microplastics on soil organic carbon and nitrogen—a mesocosm experiment. Environ Res 204:111938

Miller ME, Kroon FJ, Motti CA (2017) Recovering microplastics from marine samples: a review of current practices. Mar Pollut Bull 123(1–2):6–18

Miller ME, Hamann M, Kroon FJ (2020) Bioaccumulation and biomagnification of microplastics in marine organisms: a review and meta-analysis of current data. PLoS One 15(10):e0240792. https://doi.org/10.1371/journal.pone.0240792

O'Brien S, Rauert C, Ribeiro F, Okoffo ED, Burrows SD, O'Brien JW, Wang X, Wright SL, Thomas KV (2023) There's something in the air: a review of sources, prevalence and behaviour of microplastics in the atmosphere. Sci Total Environ 874:162193

Othman AR, Hasan HA, Muhamad MH, Ismail NI, Abdullah SR (2021) Microbial degradation of microplastics by enzymatic processes: a review. Environ Chem Lett 19:3057–3073

Pastorino P, Prearo M, Anselmi S, Menconi V, Bertoli M, Dondo A, Pizzul E, Renzi M (2021) Use of the zebra mussel *Dreissena polymorpha* (*Mollusca, Bivalvia*) as a bioindicator of microplastics pollution in freshwater ecosystems: a case study from Lake Iseo (North Italy). Water 13(4):434

Prado J, Alvarenga P, Ribeiro H, Fangueiro D (2023) Nutrient potential leachability in a sandy soil amended with manure-based fertilisers. Agronomy 13(4):990. https://doi.org/10.3390/agronomy13040990

Roy T, Dey TK (2022) Microplastic/nanoplastic toxicity in plants: an imminent concern. Environ Monit Assess 195(1):27. https://doi.org/10.1007/s10661-022-10654-z

Saad D, Chauke P, Cukrowska E, Richards H, Nikiema J, Chimuka L, Tutu H (2022) First biomonitoring of microplastic pollution in the Vaal river using Carp fish (*Cyprinus carpio*) "as a bio-indicator". Sci Total Environ 836:155623

Shorobi FM, Vyavahare GD, Seok YJ, Park JH (2023) Effect of polypropylene microplastics on seed germination and nutrient uptake of tomato and cherry tomato plants. Chemosphere 329: 138679

Urli S, Corte Pause F, Crociati M, Baufeld A, Monaci M, Stradaioli G (2023) Impact of microplastics and nanoplastics on livestock health: an emerging risk for reproductive efficiency. Animals 13(7):1132. https://doi.org/10.3390/ani13071132

Usman S, Abdull Razis AF, Shaari K, Amal MN, Saad MZ, Mat Isa N, Nazarudin MF, Zulkifli SZ, Sutra J, Ibrahim MA (2020) Microplastics pollution as an invisible potential threat to food safety and security, policy challenges and the way forward. Int J Environ Res Public Health 17(24): 9591. https://doi.org/10.3390/ijerph17249591

Vaughan R, Turner SD, Rose NL (2017) Microplastics in the sediments of a UK urban lake. Environ Pollut 229:10–18

Wang Y, Bai J, Wen L, Wang W, Zhang L, Liu Z, Liu H (2023) Phytotoxicity of microplastics to the floating plant *Spirodela polyrhiza* (L.): plant functional traits and metabolomics. Environ Pollut 322:121199

Wu C, Song X, Wang D, Ma Y, Ren X, Hu H, Shan Y, Ma X, Cui J, Ma Y (2023) Tracking antibiotic resistance genes in microplastic-contaminated soil. Chemosphere 312:137235

Zhou J, Jia R, Brown RW, Yang Y, Zeng Z, Jones DL, Zang H (2023) The long-term uncertainty of biodegradable mulch film residues and associated microplastics pollution on plant-soil health. J Hazard Mater 442:130055

Ziani K, Ioniță-Mîndrican CB, Mititelu M, Neacşu SM, Negrei C, Moroşan E, Drăgănescu D, Preda OT (2023) Microplastics: a real global threat for environment and food safety: a state-of-the-art review. Nutrients 15(3):617. https://doi.org/10.3390/nu15030617

Photocatalytic Degradation of Microplastic in the Environment 25

Kousar Parveen, Aqsa Parvaiz, Mishal Subhan, and Afsar Khan

Abstract

Microplastic pollution is reported as an emerging environmental pollutant that needs a promising solution because of its uncontrolled occurrence, distribution, non-visibility, and long lifetime. Different methods were tested such as incineration, filtration, coagulation, adsorption, chemical oxidation, and biodegradation, but they did not give effective results. Either it requires additional chemicals and energy for processing or generates by-products or sludge that needs further treatments. Recently, photocatalytic degradation emerges as a promising solution for the degradation of microplastic. Different photocatalysts are reported for the degradation of low-density polyethylene, high-density polypropylene, and many others. Optimization of photocatalytic reaction is assessed for successful application of photocatalytic degradation of microplastic.

Keywords

Polypropylene · Polystyrene · Polyethylene · Photocatalytic degradation · Microplastic

K. Parveen (✉)
Department of Environmental Sciences, The Women University Multan, Multan, Pakistan
e-mail: kousar.parveen@wum.edu.pk

A. Parvaiz
Department of Biochemistry and Biotechnology, The Women University Multan, Multan, Pakistan

M. Subhan
Department of Microbiology and Molecular Genetics, The Women University Multan, Multan, Pakistan

A. Khan
School of Minerals Processing and Bioengineering, Central South University, Changsha, Hunan, China

M. Shahnawaz et al. (eds.), *Microplastic Pollution*,
https://doi.org/10.1007/978-981-99-8357-5_25

25.1 Introduction

Plastic products are commonly used in our daily life as food packaging material, carrying bags, stationery items, building fitting pipes, water bottles, containers, draining pipes, and electrical equipment, etc. It is frequently used because of its lightweight, low cost, and highly durable. The production of plastic has reached 390 million metric tons in 2021 (Statista Research Department 2023). Plastics are made up of different polymers such as polyethylene, polypropylene, polyvinyl chloride, polystyrene, and polyurethane (Usha et al. 2011). It is mostly synthesized for single use and thrown into the environment without concern. Only 20% of plastics are recycled and 80% are deposited into environmental medium (soil, water) for 100–150 years without decomposition (Yu et al. 2022). Plastic is of different sizes such as macro-plastic (>5 mm), meso-plastics (≤5 to >1 mm), microplastic (≤1 mm to >0.1 μm), and nano-plastics (≤0.1 μm) reported by Lambert et al. (2014).

Primary microplastic and secondary microplastic are types of microplastic. Primary microplastic is used for manufacturing synthetic clothes, pharmaceutical products, cosmetic products, abrasive components, fish nets, fishmeal, etc., while secondary microplastics are produced as a result of fragmentation. The adverse effect of microplastic pollution is also inevitable. Microplastic can release harmful substances into soil and alter the soil chemistry, soil physic, and biology that ultimately disturb nutrient cycling. It can act as a carrier for metals, dyes, and pesticides in the environment. It causes endocrine disruption, inflammation, respiratory, cardiovascular, and gastrointestinal diseases in humans (Yu et al. 2022). Keeping in view the effect of microplastic pollution, it is necessary to find out a promising solution to tackle such persistent pollutants. Different conventional methods such as filtration, coagulation, flocculation, sonication, and sedimentation are in practice. Unfortunately, these methods have some disadvantages such as being energy-intensive, incapable to remove small-size pollutants, releasing sludge as a secondary pollutant, and time-consuming, etc. Photocatalytic degradation emerges as a promising method for the treatment of microplastic pollution.

This chapter provides a basic understanding of the photocatalytic degradation of microplastic in the environment. It also gives detailed information about the degradation potential of some selected photocatalysts. The different analytical tools required to evaluate the degradation efficacy are also mentioned. The implications and limitations of photocatalytic degradation are also highlighted which helps to set the future direction for further improvement in the practical application of photocatalytic degradation.

25.2 Understanding of Photocatalytic Degradation in the Environment

Light source and photocatalyst are two basic components required for photocatalytic degradation. UV and visible irradiation are used as light sources while semiconducting materials are considered the best material as a photocatalyst because of their strong oxidizing capability, low toxicity, and ease of preparation (Elgohary et al. 2021). The selection of a light source depends on the band gap of the semiconducting material (Lany 2015). The wide band gap-bearing semiconductors such as gallium oxide, titanium oxide, zinc oxide, silicon dioxide become activated under UV light and transparent to visible light while narrow band gap-bearing semiconductors such as manganese oxide, iron oxide, and cobalt oxide become activated under visible light and transparent to UV light. When light fall on semiconducting material, the valence electron absorbs the light and becomes excited. The excited electron moves to the conduction band and creates an electron–hole (e^- and h^+) pair. These charge carriers initiate the redox reaction and break the complex structure of pollutants such as microplastics (Yousif and Haddad 2013; Elgohary et al. 2021).

The photocatalytic degradation of microplastic occurs in four main steps, i.e., initiation, propagation, branching, and termination (Hamd et al. 2022).

1. An abstraction of hydrogen and generation of alkyl free radicals as active species occurs in the initiation step after light absorption.
2. Peroxy and hydroperoxides are formed when free radicals interact with oxygen in the chain propagation step and lead to the chain reaction.
3. Oxidation makes the polymer surface brittle and continuous light exposure accelerates the formation of micro-cracks and breakdown the complex structure of the polymer into simpler compounds like aldehydes, ketones, and carbonyl groups in the branching step.
4. Complete mineralization occurs in the termination step, i.e., water and carbon dioxide.

Different factors such as temperature, pH, light types, light intensity, photocatalyst nature, surface area, and photocatalyst doping or surface functionalization greatly affect the photocatalytic degradation of microplastic.

The high surface area of the photocatalyst means more active sites which is suitable for better photocatalytic activity because it accommodates more pollutants. The surface area of the photocatalyst can be enhanced with surface modification. It influences the geometry of substrates and incorporates their physical and chemical properties into photocatalysts. Temperature augments the mechanisms of charge transfer means the generation of more reactive species (hydroxide and oxygen radicles). The pH also influences the initiation step where polymer radical formation took place. The selection of light depends on the energy band gap (wide and narrow) of the photocatalyst. An increase in light intensity increases the active radical production that facilitates the photocatalytic degradation process.

25.3 Reports on the Photocatalytic Degradation of Microplastic in the Environment

Different types of photocatalysts are designed using metal oxide, mixed metal oxide, doped or functionalized metal oxides, hybrids, or composite materials and polymers that showed different potentials toward different types of microplastics. Tofa et al. (2019a) synthesized zinc oxide nano-rod using the hydrothermal method and used it as a photocatalyst for the degradation of low-density polyethylene (LDPE) film, a prominent microplastic pollutant. The degradation study was performed under a 50 W dichroic halogen lamp at a 10 cm distance for 175 h. Upon exposure to light, the degradation process has started and the formation of new functional groups is identified (Fig. 25.1). Uheida et al. (2021) tested the potential of zinc oxide as a photocatalyst immobilized onto the glass for the degradation of polypropylene under visible light. The degradation of polypropylene is performed underwater to mimic real environmental conditions. The reduction in polypropylene size (up to 65%) is noticed via optical micrograph and particle size distribution analyzer as shown in Table 25.1.

Nabi et al. (2020) fabricated titanium oxide film using water, ethanol, Triton X-100 denoted as WT, ET, and TXT and evaluated the photodegradation potential against polystyrene (PS) under UV light irradiation. The effect of UV treatment on polystyrene for 6 and 12 h well explained the morphological changes as shown in Fig. 25.2. The film with Triton X-100 showed complete mineralization (98%) of polystyrene in 12 h. Titanium oxide in the film acts as an activation site for the absorption of UV irradiation and is super-activated for the degradation process.

Li et al. (2010) synthesized polypyrrole and titanium oxide nanocomposites using the sol-gel method and emulsion polymerization method. The efficiency of as-synthesized composite as a photocatalyst was tested to degrade the polyethylene plastic under sunlight. After continuous exposure to 240 h sunlight, a maximum of 35% weight loss of polyethylene is achieved. Furthermore, cavity formation is evidence of the evaporation of volatile organics from the polymer matrix.

Different research groups started to synthesize the different types of composites with other metals (titanium oxide, zinc oxide, etc.) or polymers (polystyrene, polyamide) to incorporate the physical and chemical properties that enhance the degradation efficiency.

Platinum-deposited zinc oxide nanocomposite is reported as an efficient photocatalyst for the degradation of low-density polyethylene (LDPE). The platinum helps to absorb more visible light and generated more active radicals. Fourier transform infrared spectroscopy confirmed the changes in the chemical structure such as the formation of carbonyl, vinyl, hydroperoxide, and peroxides depicting the degradation (Tofa et al. 2019a).

Ariza-Tarazona et al. (2019) synthesized nitrogen-doped titanium oxide from two sources, fresh blue mussels (protein-based) and urea for the degradation of high-density polyethylene. The comparative results showed that the photocatalyst synthesized from fresh blue mussels possesses an excellent photocatalytic degradation capacity with a mass loss of 1.85% in 16 h of under visible irradiation. The same

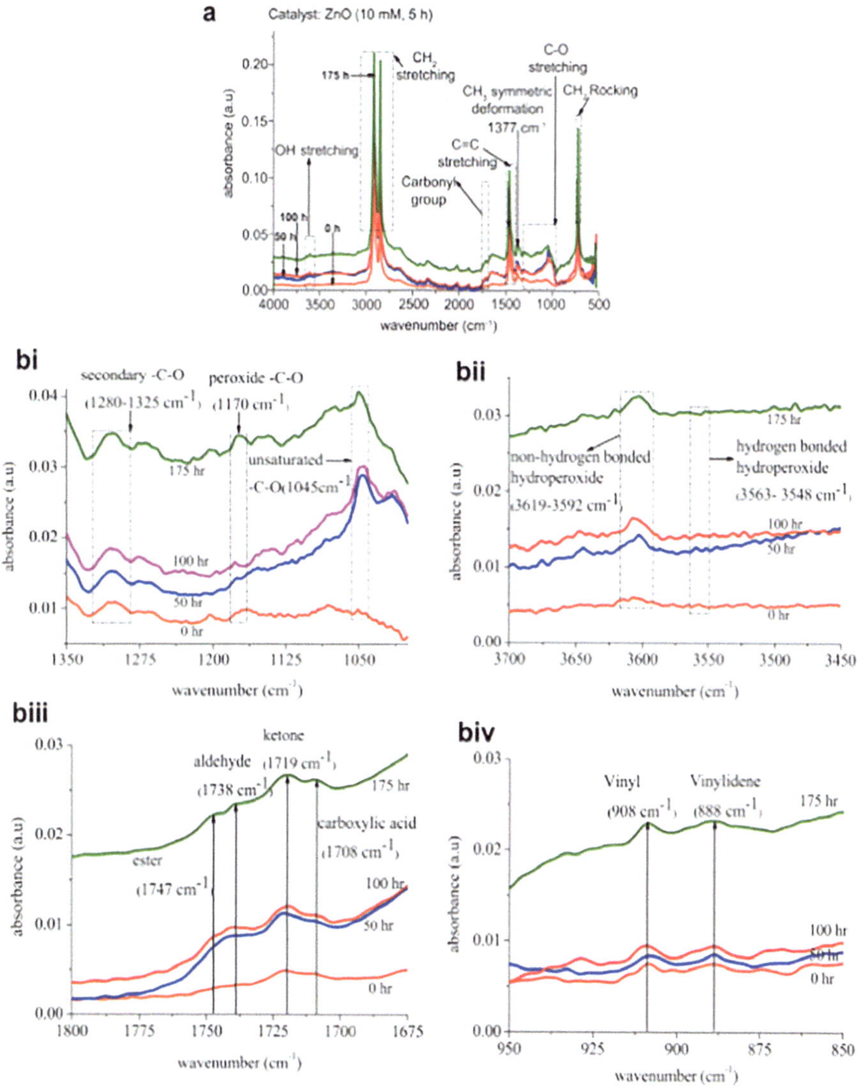

Fig. 25.1 FTIR spectra of polyethylene film (**a**), development of different functional, i.e., peroxides (**bi**), hydroperoxide (**bii**), carbonyl (**biii**), and iv unsaturated groups (**biv**) at different wavelengths (Credit: Tofa et al. 2019b)

research group attempted to evaluate the carbon and nitrogen-doped titanium oxide capacity for the degradation of HDPE under visible light by varying the different parameters (pH and temperature). The result showed that the low pH incorporates more hydrogen ion (H^+) into the system and favors HDPE degradation while low

Table 25.1 Particle size distribution and optical micrograph of polypropylene

<table>
<tr><th>Exposure Time (hours)</th><th>Particle size (μm)</th><th>Particle size distribution</th><th>Optical micrograph image</th></tr>
<tr><td>0
(as received)</td><td>154.8± 1.4</td><td>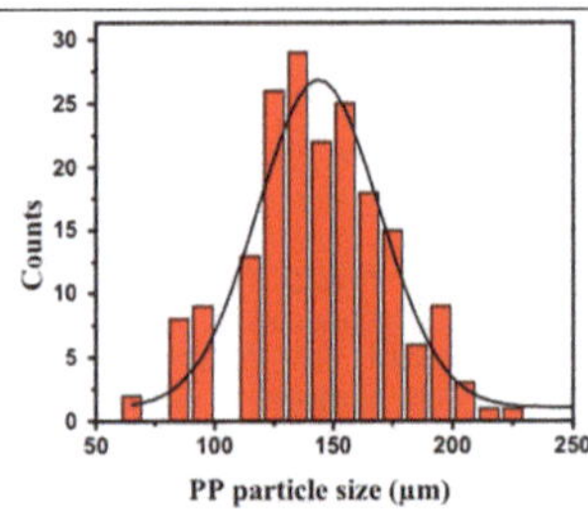
</td><td>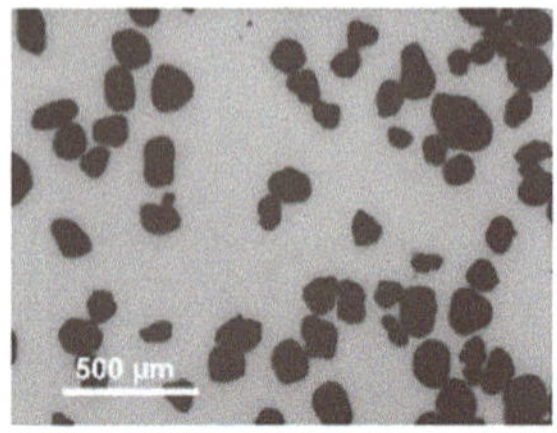
</td></tr>
<tr><td>72</td><td>123.3±
0.89</td><td>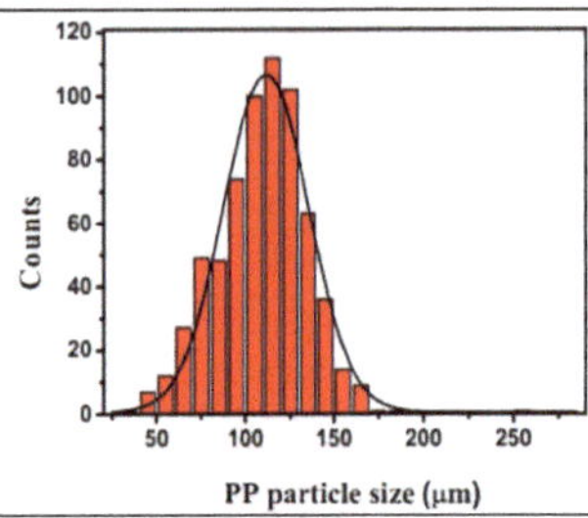
</td><td>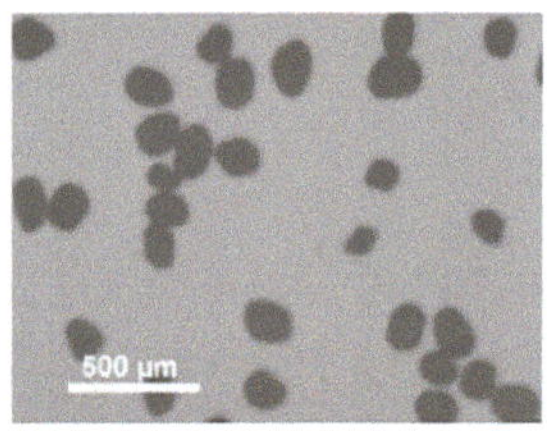
</td></tr>
<tr><td>240</td><td>111.5±
0.94</td><td>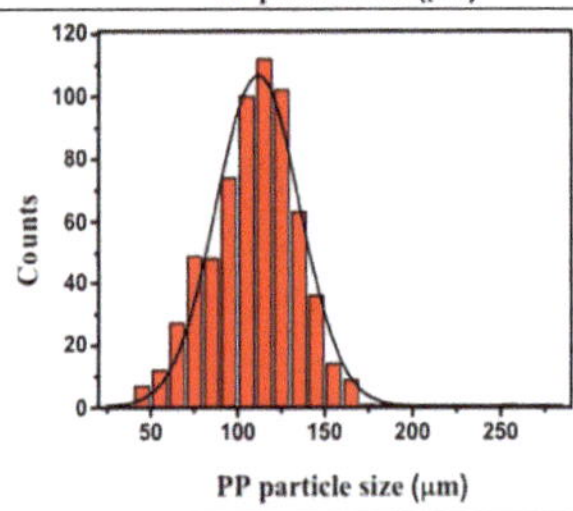
</td><td>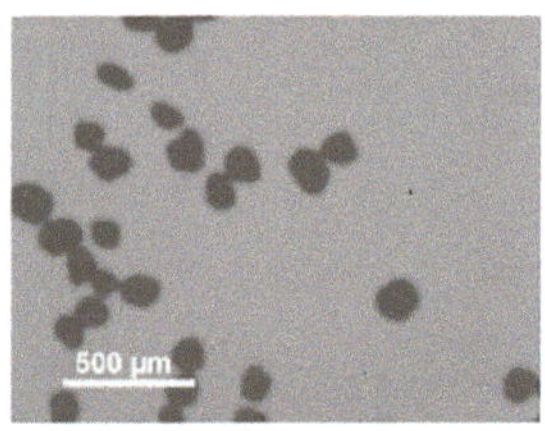
</td></tr>
<tr><td>456</td><td>108.2± 2.5</td><td>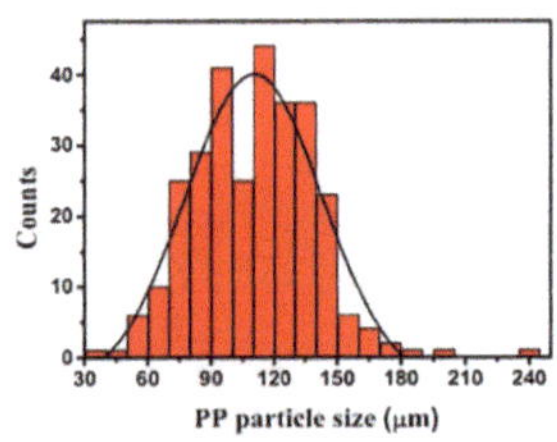
</td><td>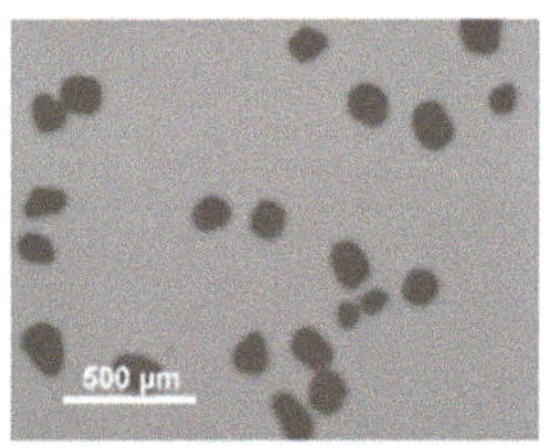
</td></tr>
</table>

(Credit: Uheida et al. 2021)

temperature initiated the cracking of microplastic, thus driving the degradation process faster (Ariza-Tarazona et al. 2020).

Jiang et al. (2021) synthesized hydroxy-rich bismuth nanosheet. The hydroxy group accelerates the charge transfer mechanism and showed 24 times more potential degradation capacity than bismuth nanosheet for microplastics. Ali et al. (2016) fabricated polyethylene films with different 1, 2, 7, and 10% titanium nanotubes as

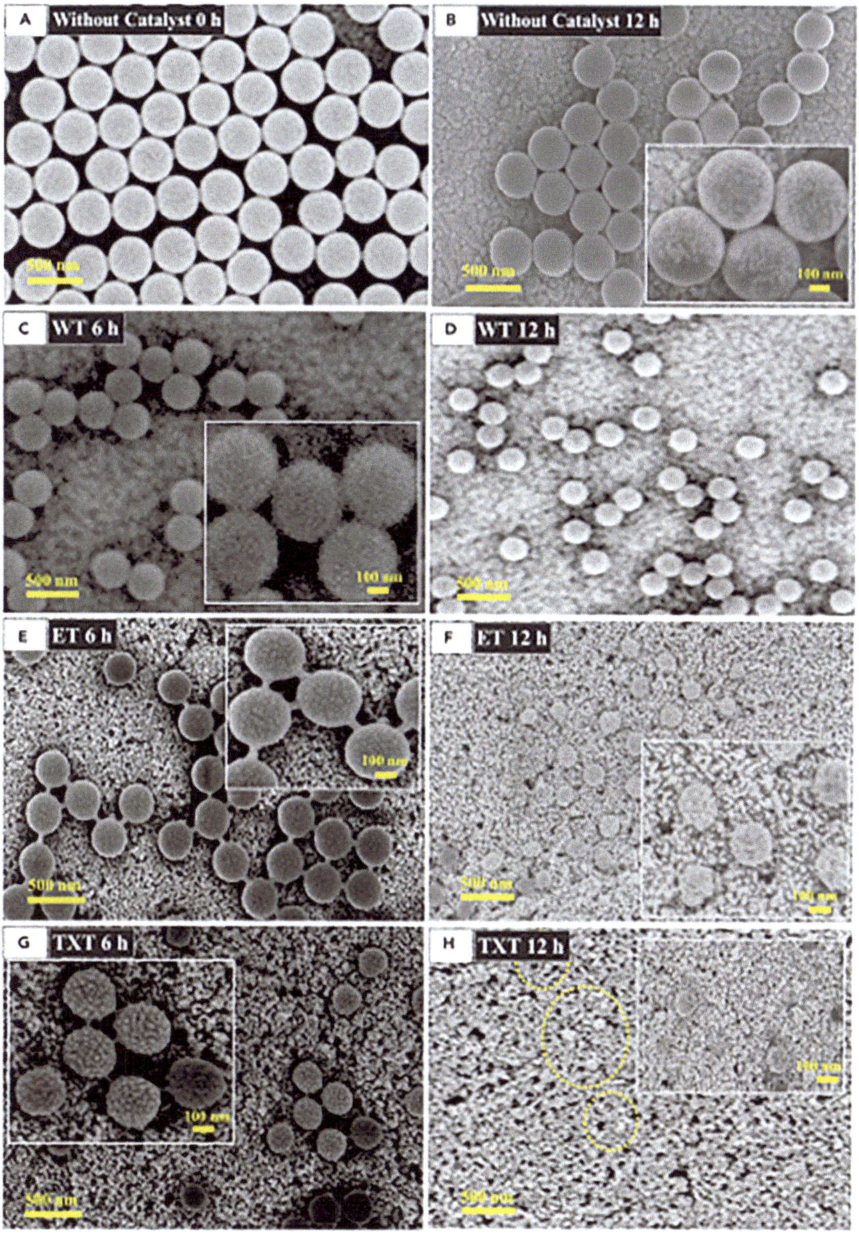

Fig. 25.2 FESEM of Polystyrene Spheres after different Irradiation under UV Lighttreatments (365 nm), A- Without catalyst before irradiation, B- Without catalyst after 12 h irradiation, C- Water-based TiO_2 (WT) after 6 h irradiation, D- Water-based TiO_2 (WT) after 12 h irradiation, E- Ethanol-based TiO_2 (ET) after 6 h irradiation, F- Ethanol-based TiO_2 (ET) after 12 h irradiation, G- Triton X-100 based TiO_2 (TXT) after 6 h irradiation, H-Triton X-100 based TiO_2 (TXT) after 12 h irradiation. (Credit: Nabi et al. 2020)

Table 25.2 Table shows the mass degradation rate

Catalyst	Mass of microplastic (mg) at time				
	0 h	1 h	2 h	3 h	4 h
Without catalyst	50	50	50	50	50
TiO_2P_{25}	50	40	32	96	22
3% Ag/TiO_2	50	30	24	19	16
3% Ag/TiO_2–1% RGO	50	28	18	14	12

(Credit: Fadli et al. 2021)

dopants. The maximum degradation of around 78% was achieved using 10% titanium nanotubes under visible light. Bandara et al. (2017) did a comparative study to assess the catalytic power of zirconium oxide and titanium oxide toward polyethylene and polypropylene degradation under sunlight. Different concentrations of zirconium and titanium oxide were tested and degradation results are measured via carbonyl index. The results showed maximum degradation by zirconium oxide at 11,000 mg/L on the considerable exposure of 20-h sunlight. Zirconium oxide contains a mesoporous structure and has a relatively high abundance of surface oxygen than titanium oxide which leads to maximum degradation potential. Nickel aluminate composite was tested for the degradation of polyethylene under visible light. FTIR data revealed 12.5% weight loss in 5 h (Venkataramana et al. 2021). Commercial titanium oxide powder is used as a photocatalyst for the degradation of PA66 micro-fibers under UVC irradiation. It showed 97% average mass loss in 48 h. The carbonyl index showed 25% oxidation in comparison to without photocatalyst and UVC light (Lee et al. 2020). Silver and graphene-doped titanium oxide is synthesized to control the recombination rate that facilitates the photodegradation of polyethylene under UV radiation (Fadli et al. 2021). The degradation experiment was conducted for 4 h, and mass loss was noted with 1-h intervals as shown in Table 25.2.

The as-synthesized photocatalyst showed a significant degradation potential of 76% compared to pure TiO_2. Another study showed that silver as a dopant reduced the band gap of titanium oxide from 3.19 to 3.02 eV and become active under visible light for the degradation of polyethylene. It showed 100% degradation after the completion of 120 min (Maulana and Ibadurrohman 2021). Multi-walled carbon nanotubes and titanium dioxide composites showed 35% degradation of polyethylene in 180 h (An et al. 2014). Dispersion of titanium oxide and ferric stearate in commercial polystyrene enhanced the degradation under UV light. It is noticed that the average molecular weight of composites has decreased by 63% (Fa et al. 2013). Similarly, titanium-doped polypropylene showed an increase in carbonyl index after UV irradiation that reflects the degradation process (Kamrannejad et al. 2014).

Different analytical techniques are used to study the effect of different treatments on the physical and chemical properties, and the evaluation of photocatalyst's efficiency. Optical and scanning electron microscopy is used to study the morphological changes before and after surface modification. Fourier transform infrared spectroscopy is used to identify the formation of different intermediates during the degradation process. Brunauer–Emmett–Teller is used to determine the photocatalyst's total surface area. Gas chromatography–mass spectrometry is used to identify the intermediate and end product after the completion of the degradation process that confirms the complete mineralization.

25.4 Implications and Future Directions

The adverse effects of microplastic pollution on the environment forced the scientific community to find out a promising solution to manage such notorious pollutants. The effort of the scientific community become successful to trace the different sources, distribution, pathways, and emerging threats to the different ecosystems and biotic communities. These research findings could help to create public awareness and encourage the stakeholders and policymakers to formulate policies and laws. These laws and policies extend the producer's responsibility to meet the international standard, design sustainable and eco-friendly plastic products as alternate solutions, and promote a circular economy concept. Along with this effort, different remediation methods are also needed to deal with microplastic pollution such as sedimentation, coagulation, flocculation, biodegradation, and filtration. The photocatalytic degradation method emerged as the eco-friendly method widely used for wastewater treatment, air purification, clean energy production, and antibacterial disinfection due to its better efficiency, less toxicity, and cost-effective technique. It used natural sunlight as a source of energy to initiate and process the degradation without the addition of any other chemical as initiator, or binder. It is a self-driven method and does not require an intense source of energy as fuel. The photocatalytic degradation of microplastic is at its early stage, and only a few materials were tested. There is still room to identify the optimal design and reaction conditions where these materials show better performance. There is also a need to scale up specific photocatalysts as a prototype for practical demonstration under real environmental conditions. The toxicological profile of photocatalysts is need to be documented for safe application in the environment. Understanding the degradation pattern helps to identify the intermediate products which can be reused as raw material. It is important to note that no single method can completely remediate microplastic pollution. Therefore, a combined approach is needed where different methods are integrated with each other for the effective removal of microplastic. The cost-effectiveness of the photocatalytic method is a critical factor for successful adoption.

There is also a need to make collaboration with other scientists and share data, new insights, and discoveries. Overall, the development of photocatalytic degradation for microplastic removal is an exciting area of research that shows promise for addressing one of the most important environmental concerns of our time.

25.5 Conclusion

Microplastic pollution is an emerging pollutant in the environment affecting all of us. This chapter provides inclusive information about sources, adverse effects, different types of microplastic, and potential photocatalysts. It also highlights the different factors affecting the degradation efficiency, mechanism, limitation, and implication. It also gives future direction for further research and development for the practical application of photocatalytic degradation in the treatment of microplastic pollution.

References

Ali SS, Qazi IA, Arshad M, Khan Z, Voice TC, Mehmood CT (2016) Photocatalytic degradation of low-density polyethylene (LDPE) films using titania nanotubes. Environ Nanotechnol Monitor Manag 5:44–53

An Y, Hou J, Liu Z, Peng B (2014) Enhanced solid-phase photocatalytic degradation of polyethylene by TiO_2–MWCNTs nanocomposites. Mater Chem Phys 148(1–2):387–394

Ariza-Tarazona MC, Villarreal-Chiu JF, Barbieri V, Siligardi C, Cedillo-González EI (2019) New strategy for microplastic degradation: green photocatalysis using a protein-based porous N-TiO_2 semiconductor. Ceram Int 45(7):9618–9624

Ariza-Tarazona MC, Villarreal-Chiu JF, Hernández-López JM, De la Rosa JR, Barbieri V, Siligardi C, Cedillo-González EI (2020) Microplastic pollution reduction by carbon and nitrogen-doped TiO_2: effect of pH and temperature in the photocatalytic degradation process. J Hazard Mater 395:122632

Bandara WL, de Silva RM, de Silva KN, Dahanayake D, Gunasekara S, Thanabalasingam K (2017) Is nano ZrO_2 a better photocatalyst than nano TiO_2 for the degradation of plastics? RSC Adv 7(73):46155–46163

Elgohary EA, Mohamed YM, El Nazer HA, Baaloudj O, Alyami MS, El Jery A, Assadi AA, Amrane A (2021) A review of the use of semiconductors as catalysts in the photocatalytic inactivation of microorganisms. Catalysts 11(12):1498

Fa W, Guo L, Wang J, Guo R, Zheng Z, Yang F (2013) Solid-phase photocatalytic degradation of polystyrene with TiO_2/Fe$(St)_3$ as a catalyst. J Appl Polym Sci 128(5):2618–2622

Fadli MH, Ibadurrohman M, Slamet S (2021) Microplastic pollutant degradation in water using modified TiO_2 photocatalyst under UV irradiation. In: IOP conference series: materials science and engineering, vol 1011. IOP Publishing, p 012055

Hamd W, Daher EA, Tofa TS, Dutta J (2022) Recent advances in photocatalytic removal of microplastics: mechanisms, kinetic degradation, and reactor design. Front Mar Sci 9:885614

Jiang R, Lu G, Yan Z, Liu J, Wu D, Wang Y (2021) Microplastic degradation by hydroxy-rich bismuth oxychloride. J Hazard Mater 405:124247

Kamrannejad MM, Hasanzadeh A, Nosoudi N, Mai L, Babaluo AA (2014) Photocatalytic degradation of polypropylene/TiO_2 nano-composites. Mater Res 17:1039–1046

Lambert S, Sinclair CJ, Boxall ABA (2014) Occurrence, degradation and effects of polymer-based materials in the environment. Rev Environ Contam Toxicol 227:1–53. https://doi.org/10.1007/978-3-319-01327-5_1

Lany S (2015) Semiconducting transition metal oxides. J Phys Condens Matter 27(28):283203

Lee JM, Busquets R, Choi IC, Lee SH, Kim JK, Campos LC (2020) Photocatalytic degradation of polyamide 66; evaluating the feasibility of photocatalysis as a microfibre-targeting technology. Water 12(12):3551

Li S, Xu S, He L, Xu F, Wang Y, Zhang L (2010) Photocatalytic degradation of polyethylene plastic with polypyrrole/TiO_2 nanocomposite as photocatalyst. Polym Plast Technol Eng 49(4): 400–406

Maulana DA, Ibadurrohman M (2021) Synthesis of nano-composite Ag/TiO_2 for polyethylene microplastic degradation applications. In: IOP conference series: materials science and engineering, vol 1011. IOP Publishing, p 012054

Nabi I, Li K, Cheng H, Wang T, Liu Y, Ajmal S, Yang Y, Feng Y, Zhang L (2020) Complete photocatalytic mineralization of microplastic on TiO_2 nanoparticle film. iScience 23(7):101326

Statista Research Department (2023) Global plastic production 1950–2021. https://www.statista.com/statistics/282732/global-production-of-plastics-since-1950/#statisticContainer. Accessed 11th May 2023

Tofa TS, Kunjali KL, Paul S, Dutta J (2019a) Visible light photocatalytic degradation of microplastic residues with zinc oxide nanorods. Environ Chem Lett 17:1341–1346

Tofa TS, Ye F, Kunjali KL, Dutta J (2019b) Enhanced visible light photodegradation of microplastic fragments with plasmonic platinum/zinc oxide nano-rod photocatalysts. Catalysts 9(10):819

Uheida A, Mejía HG, Abdel-Rehim M, Hamd W, Dutta J (2021) Visible light photocatalytic degradation of polypropylene microplastics in a continuous water flow system. J Hazard Mater 406:124299

Usha R, Sangeetha T, Palaniswamy M (2011) Screening of polyethylene degrading microorganisms from garbage soil. Libyan Agric Res Center J Int 2(4):200–204

Venkataramana C, Botsa SM, Shyamala P, Muralikrishna R (2021) Photocatalytic degradation of polyethylene plastics by $NiAl_2O_4$ spinels-synthesis and characterization. Chemosphere 265: 129021

Yousif E, Haddad R (2013) Photodegradation and photo stabilization of polymers, especially polystyrene. Springerplus 2(1):1–32

Yu H, Zhang Y, Tan W, Zhang Z (2022) Microplastics as an emerging environmental pollutant in agricultural soils: effects on ecosystems and human health. Front Environ Sci 10:217

Part V

Microbial Approaches to Degrade Microplastic Pollution

Biodegradation of Microplastic Using Bacterial Cultures

26

Lakshmi Mohan, Sony Elizabeth Manoj, Linta Sunil, Adarsh V. Raju, and K. S. Anandha Krishnan

Abstract

Plastics are large molecular-weight organic polymers, commonly derived from different petrochemicals. There is a considerable increase in the global demand for plastics because of the ease of processing into various products used for packaging, handling, and storage of materials. There are many uses for various plastics such as polyolefin, polyester, polyvinyl chloride (PVC), polystyrene (PS), polyamide, ethylene vinyl alcohol, polyethylene (PE), polycarbonate, polypropylene (PP), polyurethane (PU), and polyethylene terephthalate (PET) in daily living. Microplastics are tiny plastic particles defined as plastics less than 5 mm in diameter. They are of two types, the first being primary microplastics which are tiny particles intended for commercial use in cosmetic products, and microfibers shed from textiles and fishing nets. Secondary microplastics are particles generated by the breakdown of bigger plastic items exposed to abiotic factors such as the sun's rays and ocean waves. However, considering the rate of degradation or non-biodegradability of these organic polymers, they turn out to be a major environmental threat. They are highly stable and therefore do not enter into the degradation cycle of the biosphere easily. Degradation of plastics using physical and chemical methods may lead to innumerable environmental hazards. Microorganisms break down microplastics by producing intracellular or extracellular enzymes that turn long-chain polymers into oligomers and monomers that are then used by microbial cells. Some recent studies show that *Pseudomonas* sp. *Ideonella sakaiensis*, *Klebsiella pneumoniae*, *Bacillus* sp., *Paenibacillus* sp.,

L. Mohan (✉) · S. E. Manoj · L. Sunil · A. V. Raju · K. S. A. Krishnan
Department of Food Technology, Saintgits College of Engineering (Autonomous), Kottayam, Kerala, India
e-mail: lakshmi.mohan@saintgits.org

M. Shahnawaz et al. (eds.), *Microplastic Pollution*,
https://doi.org/10.1007/978-981-99-8357-5_26

Bacillus siamensis, Desulfotomaculum nigrificans, Alcaligenes sp. are some of the predominant bacterial species capable of degrading plastics.

Keywords

Plastics · Microplastics · Degradation · Bioremediation · Bacterial cultures

26.1 Introduction

Plastic is a versatile material that has many useful properties, such as strength, durability, and flexibility, making it perfect for a variety of uses, including the packing of solids and liquids. It is used in a variety of products, such as food packaging, containers, bottles, bags, and other household and industrial items (United States Environmental Protection Agency 2021).

However, the widespread use of plastic has also led to significant environmental concerns. Around 380 million tonnes of plastic are produced every year around the world; however, only a small percentage of this plastic is recycled or incinerated. The majority of plastic waste ends up in landfills, where it can take hundreds of years to decompose, or in the natural environment, where it can have harmful impacts on wildlife and ecosystems (United States Environmental Protection Agency 2021).

Plastic pollution is a major problem, and it is concerning to see how much plastic waste has accumulated in water bodies and other wild areas. Plastic waste has been found inside dead animals and seabirds, and it has also been detected in the air, soil, and water, where it can have harmful effects on the environment and human health (United Nations Environment Programme 2018).

To address these concerns, there is a growing movement to lessen the use of single-use plastic and increase efforts aimed at recycling and reuse of plastic products. This may involve the development of alternatives to plastic, such as biodegradable polymers or recycled materials, and the implementation of laws and rules to promote ethical plastic waste management and disposal.

Huge amounts of plastic are released into the atmosphere during the manufacture of plastics by industries, which harms ecosystems and wildlife. Polyethylene (PE), a type of plastic that is present in the ecosystem, can impair soil fertility and endanger animal life. Additionally, it may result in environmental pollution and landfill exhaustion. Finding suitable methods for disposing of plastic debris and reducing its negative effects on the environment is therefore necessary. The use of native microorganisms to break down plastic debris in polluted soils is one strategy that is currently being investigated. To create new methods for the biological degradation of plastic waste, it is crucial to conduct a study on the characterization of microorganisms that break down plastic (Bonhomme et al. 2003).

26.2 Everyday Plastics: Familiarity and Usage

Plastics are a group of synthetic materials made from polymers, which have become ubiquitous in our daily lives due to their versatility, durability, and low cost. Plastic products are used in almost every aspect of modern life, from packaging and transportation to electronics and medical equipment. Despite the negative impact of plastics on the environment, consumer behaviour towards plastic products has been slow to change. This is partly due to the convenience and affordability of plastic products, as well as a lack of awareness about the negative impact of plastic waste (Hopewell et al. 2009).

Polymerization, a chemical reaction in which monomers (small molecules) are joined together to form a polymer (a long chain of repeating monomer units) is of two main types: condensation polymerization, and addition polymerization (Jambeck et al. 2015).

Condensation polymerization is a process by which polymers are formed through the reaction of monomers that have at least two functional groups. These functional groups can be alcohols, amines, or carboxylic groups, and the reaction between them results in the formation of a polymer chain and the release of small molecules such as water or methanol (Sperling 2006). One of the main advantages of condensation polymerization is that it allows for the formation of polymers with a high molecular weight, as the monomers can react multiple times to form longer and longer chains. This process is also more efficient than addition polymerization, as it does not require the use of initiators or catalysts and can be carried out at lower temperatures (Carraher 2000). There are several different types of condensation polymers, including polyesters, polyamides, and polyurethanes. These polymers have a variety of applications, including the production of textiles, plastics, and coatings. Polyesters are a type of condensation polymer that is formed through the reaction of dicarboxylic acids and diols. They are known for their high strength and durability, as well as their resistance to moisture and chemicals. Polyesters are commonly used in the production of fabrics, films, and resins, and they can also be moulded into various shapes and forms (Hill and McGrail 2003). Polyamides, also known as nylons, are another type of condensation polymer that is formed through the reaction of diamines and dicarboxylic acids. They are known for their high strength and toughness, as well as their resistance to wear and tear. Polyamides are commonly used in the production of fibres, films, and moulded parts, and they are also used in the production of gears, bearings, and other mechanical components (Stevens 2009). Polyurethanes are a type of condensation polymer that is formed through the reaction of isocyanates and polyols. They are known for their excellent durability and flexibility, as well as their resistance to chemicals and UV radiation. Polyurethanes are commonly used in the production of foams, coatings, and elastomers, and they are also used in the production of adhesives, sealants, and binders (Lange and Wyser 2003). Another kind of polymerization, known as Addition polymerization or polyaddition is one where the polymer chain grows by the addition of monomer units without the formation of by-products. This type of polymerization typically involves unsaturated monomers with double bonds, such as alkenes. Some examples of addition polymers include polyethylene (PE), polypropylene (PP), and polyvinyl chloride (PVC) (Muralisankar and Arthanareeswari 2018). Polymer-based products known as plastics have been used in a variety of settings, including food packaging. Polyolefins (like polyethylene), polyesters (like polyethylene terephthalate), polystyrene, polyvinyl chloride (PVC), polyamide, polyvinylidene chloride and ethylene vinyl alcohol are some of the most popular plastics used in food packing (Odian 2004). Polyester material known as polyethylene terephthalate (PET) is frequently used in food containers, especially for beverages and mineral water. The interaction between carboxylic acid and alcohol produces a condensation polymer. PET is increasingly being used to create plastic cups for carbonated beverages. Polyesters used in food containers include polycarbonate and polyethylene naphthalate (PEN). The high strength and heat and chemical resistance of PEN make it ideal for use in vacuum-sealed food packaging, among other uses (Menczel and Prime 2009). Although plastics are convenient and useful materials, there are also concerns about their environmental impact. The production and disposal of plastics can contribute to greenhouse gas emissions, and plastic waste can accumulate in the environment, leading to pollution and harm to wildlife. As a result, there has been a push towards the development of more sustainable and biodegradable plastics, as well as efforts to reduce the use of single-use plastics (Groh et al. 2019).

Microplastics are tiny plastic particles measuring less than 5 mm in size. They can come from a variety of sources, including the breakdown of larger plastic items, microbeads found in personal care products, and industrial processes (Browne et al. 2011). Microplastics have been found in oceans, rivers, and even in the air we breathe, and have negative impacts on marine life and the environment. Microplastics are a growing environmental concern because they are not biodegradable and can persist in the natural environment for hundreds of years. They can be ingested by marine life and other organisms, potentially causing harm to their health and altering food chains.

Microplastics can be found in oceans, rivers, and other bodies of water, as well as on land. They can also be transported through the air and have been found in remote places such as the Arctic (Hahladakis et al. 2018).

Some common sources of microplastics include:

- Plastic bottles and bags breaking down into small particles
- Microbeads in personal care products such as toothpaste and exfoliating scrubs
- Synthetic clothing fibers shed during washing
- Tire and road wear particles
- Industrial processes such as fracking and oil extraction (Allen-Gil et al. 2020)

Since microplastics are so small, it's difficult to clean up and remove them from the environment. However, reducing the amount of plastic waste produced in the first place is the best way to prevent the build-up of microplastics. This can be done by using reusable bags and bottles, avoiding personal care products with microbeads, and properly disposing of waste (Geyer et al. 2017).

26.3 Plastics and Their Environmental Impact

Global plastic garbage distribution is correlated with human population density. The rise in the global population has led to an increase in the demand for plastics and plastic products. However, improper disposal of plastic waste can cause environmental pollution, leading to various negative impacts such as soil degradation, harm to wildlife and aquatic life, and the loss of the natural beauty of the environment (Sacharow and Griffin 1980).

Environmental impacts due to these plastic wastes have caused pollution of the soil, water, air, and wildlife as well. Plastic granules that are no larger than 5 mm are known as microplastics (Galloway 2015). Primary and secondary are the two groups of microplastics. Specifically added to goods like cosmetics, textiles, and packaging, primary microplastics are manufactured as microbeads, fibres, or pellets. Secondary microplastics are created from the degradation of larger plastic materials due to environmental factors such as UV radiation, wind, and wave action (Andrady 2015).

Microplastics are not biodegradable and can persist in the environment for hundreds of years. As a result, MPs accumulate in the environment and can cause harm to wildlife and ecosystems (Rochman 2018). MPs can be ingested by marine organisms, including plankton, fish, and shellfish, and can accumulate in the food chain. MPs have been found in a variety of seafood products, including fish, mussels, and oysters, which can pose a risk to human health (Wagner et al. 2014).

26.3.1 Land Pollution

Plastic products find widespread usage in both residential and workplace environments. Improper disposal and contamination of plastic products can pose a threat to the terrestrial ecosystem, which can further extend to the aquatic ecosystem (O'Brine and Thompson 2010). Plastics are abiotically and biotically degraded when they are dumped on the ground or landfilled. As a result, additives from the plastics can leach and ultimately percolate into the environment, contaminating the soil and water (Lee et al. 2001). Chlorinated plastics can release hazardous chemicals into the earth, which can then seep into nearby aquatic systems or underground water, polluting the environment.

26.3.2 Water Pollution

Recent research indicates that in 2012, approximately 165 million tonnes of plastic waste could be found in the world's seas, and an estimated eight million tonnes of plastic are being added to the oceans annually, leading to around 5 trillion plastic fragments floating on its surface (Andrady 2017). The breakdown of plastic in water can also cause water pollution by releasing harmful substances such as polystyrene and BPA (Stilwell et al. 1991).

Microplastics have been shown to concentrate and absorb contaminants found in seawater from various sources within the oceanic environment, harming aquatic life as well. About 80% of the waste discovered in the oceans is made of plastic, which is bad for aquatic life (Koelmans et al. 2016).

26.3.3 Air Pollution

During the decomposition of plastics dumped into landfills, carbon dioxide and methane will be released into the environment which affects the air quality and also the life of birds and other species (Hopewell et al. 2009).

When plastics and plastic products are burned, they emit carbon dioxide (CO_2) into the atmosphere. This gas can trap heat and hinder it from leaving the earth, which contributes to the problem of global warming (Zubris and Richards 2005). Environmental pollution is responsible for causing more than six million deaths annually due to related health issues. Air pollution is considered one of the most significant threats to public health. The practice of open burning of plastics and plastic goods results in the release of harmful pollutants, including heavy metals, dioxins, PCBs, and furans. These pollutants can negatively affect one's health, especially respiratory conditions when inhaled (Miller 2005).

26.3.4 Effect on the Life of Animals

Toxic substances from plastic waste and plastic products can harm animals, which could have adverse effects on human food sources. Animals are typically exposed to plastic waste through two primary means: ingestion and entanglement. Although entanglement is less frequent, ingestion is more prevalent. Various studies have demonstrated that more than 260 species of vertebrate and invertebrate animals have ingested or become entangled in plastics or plastic products, causing the death of over 400,000 marine mammals (Adane and Muleta 2011).

Similar issues emerge when it comes to fish and sea birds, both of which may confuse plastic waste for their natural prey. Fish can also mistake plastic waste for cuttlefish. Ingesting plastic wastes can physically harm birds' digestive systems and obstruct them, which reduces the ability of the system to digest food and ultimately results in death (Rathje et al. 1985). These are all the problems or effects on the environment brought on by the build-up of plastic waste in civilization.

26.4 Bioremediation

The process of bioremediation involves biologically converting organic wastes to a harmless form or quantities below the regulatory agencies' established limits. Microorganisms are well suited for the job of contaminant destruction because they have enzymes that allow them to use environmental toxins as sustenance.

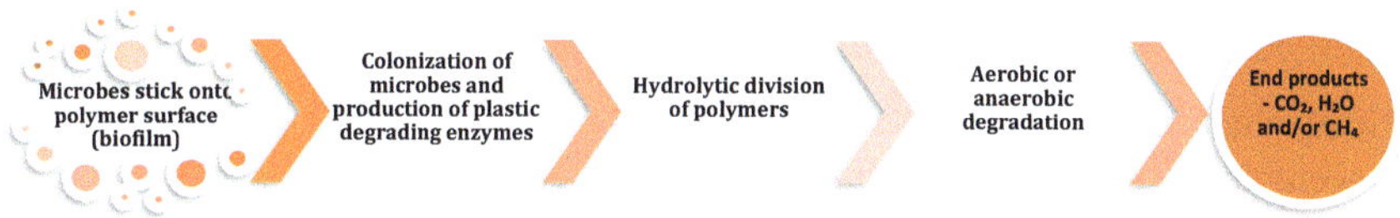

Fig. 26.1 Mechanism of bioremediation (schematic representation)

Several enzymes have an astonishingly wide range of substrates they can degrade because of their general and specific substrate affinities (Fig. 26.1). The implementation of bioremediation frequently involves altering environmental conditions to encourage microbial growth and deterioration because bioremediation is only effective in environments that promote microbial activity and growth (Kumar et al. 2011). The technology used in bioremediation is primarily founded on biodegradation. It refers to the full conversion of harmful or naturally occurring organic pollutants into substances that are safe for people, animals, plants, and aquatic life, such as carbon dioxide, water, and inorganic compounds (Jain and Bajpai 2012).

The microbes that degrade MPs most effectively have been discovered to be bacteria. It has been discovered that several bacterial species, involving *Pseudomonas*, *Bacillus*, and *Rhodococcus*, are capable of breaking down MPs. In a study conducted by Wu et al. (2019), it was found that Bacillus subtilis could break down 40% of polystyrene microplastics in 6 weeks. Another research by Lin et al. (2022) discovered that *Pseudomonas putida* could degrade 33% of polyethylene microplastics in just 30 days. However, the efficiency of bacterial degradation of microplastics is influenced by several factors, such as the plastic type, particle size, and bacterial species employed. Some plastics, like polyethylene terephthalate (PET), are more resistant to biodegradation than others (Guo and Zhang 2020).

There are numerous types of technologies for treatment or methods that fall under the heading of bioremediation procedures. The basic methods of bioremediation are bio-stimulation, augmentation, venting, attenuation, and piles (Chae and An 2018).

26.4.1 Biostimulation

To encourage the activity of local microorganisms, this type of strategy entails adding specific nutrients to the region (soil or groundwater) (Ang et al. 2005). Its primary objective is the promotion of nearby or locally existing bacterial and fungal communities. First, fertilizers, growth promoters, and trace nutrients are given out. Additional environmental needs, such as pH, temperature, and oxygen, can be supplied to hasten their metabolic rate and pathway (Adams et al. 2015).

26.4.2 Bioattenuation

Bioattenuation, also known as natural attenuation, is the process by which pollutant concentrations are eliminated from the atmosphere. It is accomplished through a variety of biological processes, such as anaerobic and aerobic biodegradation, plant and animal absorption, physical mechanisms like diffusion, dilution, and ion exchange, chemical reactions like complexation, and abiotic transformation (Mulligan and Yong 2004).

26.4.3 Bioaugmentation

"Bioaugmentation" is a technique that involves introducing genetically modified bacteria into a system to act as a bioremediator to effectively and rapidly eliminate complex contaminants (Mulligan and Yong 2004). Microorganisms are collected from the remediation site, cultured separately, genetically altered, and then released back onto the site (Niu et al. 2009).

26.4.4 Bioeventing

The process of bioventing entails the introduction of oxygen into the soil to encourage the growth of indigenous or introduced bacteria and fungi by supplying them with the oxygen they need to thrive (Lee et al. 2006).

26.4.5 Biopiles

"Biopiles" refer to a technique utilized for treating soil that has been excavated and contaminated with hydrocarbons that can be remediated aerobically. In this method, the air is introduced into the biopile system through a piping and pump system that can either supply air into the pile with positive pressure or draw air out of the pile with negative pressure (Delille et al. 2008).

26.5 Major Classes of Bacteria Involved in Bioremediation

Numerous bacterial species have been found to have the potential to degrade polymers. However, their activity towards polymers is found to be highly specific (Table 26.1).

Some of the predominant bacterial species capable of polymer degradation are discussed below:

Table 26.1 List of plastics and bacteria responsible for their bioremediation

Substrate	Bacterium	Enzymes involved in degradation	References
Different types of plastic bags	*Bacillus* sp. *BCBT21*	Carboxymethyl cellulase, chitinase, lipase, protease and xylanase	Dang et al. (2018)
HDPE and LDPE	*Bacillus cereus, Bacillus sphaericus*	Peroxidase	Sudhakar et al. (2008)
Polycaprolactone	*Alcaligenes faecalis*	Hydrolase	Fotopoulou and Karapanagioti (2017)
Polycaprolactone, polyesters	*Clostridium botulinum*	Esterase	Ghosh et al. (2013) and Perz et al. (2016)
Polyethylene	*Brevibacillus borstelensis*	Laccase	Hadad et al. (2005) and Ndahebwa Muhonja et al. (2018)
Polyethylene	*Pseudomonas aeruginosa*	Alkane monooxygenase, alkane hydrolases, rubredoxin reductase	Jeon and Kim (2015)
Polyethylene	*Rhodococcus ruber*	Laccase	Orr et al. (2004)
Polyethylene glycol	*Sphingomonas terrae*	PEG-dehydrogenase	Sugimoto et al. (2001)
Polyethylene terephthalate	*Bacillus licheniformis, Bacillus subtilis,* and *Thermobifida fusca*	Carboxylesterases	Barth et al. (2016)
Polyethylene terephthalate	*Ideonella sakaiensis*	Petase, mhetase	Austin et al. (2018) and Palm et al. (2019)
Polyethylene terephthalate	*Thermobifida fusca*	Hydrolase	Müller et al. (2005)
Polyurethane	*Pseudomonas chlororaphis*	Polyurethanes	Howard et al. (2007)
Polyurethane	*Pseudomonas protegens*	Lipase	Hung et al. (2016)
Polyvinyl chloride	*Pseudomonas putida*	Alkene monooxygenase	Danko et al. (2004)
Styrene	*Paraglaciecola agarilytica* and *Marinobacterium litorale*	Styrene monooxygenases	Pu et al. (2018)
UV-irradiated polyethylene	*Bacillus cereus*	Laccase	Sowmya et al. (2014)
UV-treated polyethylene films	*Rhodococcus ruber* C208	Laccase	Sivan et al. (2006)

26.5.1 *Ideonella sakaiensis*

A new species of the genus *Ideonella* called *Ideonella sakaiensis* is a Gram-negative, rod-shaped, aerobic non-spore-forming bacterium. It is also known as a bacterium that eats plastic. This bacterial strain can use PET as a major source of energy and carbon, as shown by a large number of scientific investigations. Scientists now think that the bacteria primarily adhere to the plastic via some type of small, arm-like appendages after conducting extensive research on bacterial metabolism. One exo-enzyme that generates a chemical intermediate is then released. The PET can be absorbed by the cell and catabolized for use in metabolism by a second enzyme once it has experienced significant degradation (Yang et al. 2016). The PET hydrolase, or PETase, produced by microbes breaks down PET to produce mono-(2-hydroxyethyl)-terephthalic acid (MHET), which is a heterodimer of terephthalic acid (TPA) and ethylene glycol (EG).

I. sakaiensis and many other bacteria can easily absorb and use products containing ethylene glycol, but they are unable to do so with TPA, so MHET hydrolase is utilized in its place (Dahal 2023).

26.5.2 *Bacillus subtilis*

Studies have found that *Bacillus subtilis*, a type of bacteria, can biodegrade various polymers, including polyethylene and polyurethane, which are commonly used in food packaging, by utilizing them as a source of energy. The microbial species produce surface-active compounds, known as biosurfactants, which accelerate the process of breakdown. When polyethylene films were prepared and inoculated with *Bacillus subtilis*, along with the addition of the bacteria's biosurfactant (surfactin), a substantial weight loss percentage was observed, indicating the effectiveness of this approach. The amphiphilic nature of the biosurfactant contributes to the adhesion of microorganisms to hydrophobic surfaces. Therefore, the presence of biosurfactants helps microorganisms adhere to the surface of PE films and utilize the polymer as a carbon source at a faster rate. It was also observed that PE films with lower thickness degraded more rapidly, leading to greater weight loss (Vimala and Mathew 2016).

26.5.3 *Pseudomonas putida*

At the General Electric Research and Development Center in 1971, Prof. Chakrabarty created a genetically modified Pseudomonas species, popularly known as a "superbug" or "oil-eating bacteria." He developed a methodology for genetic cross-linking via plasmid transfer to transfer the genes needed for oil breakdown, which led to the creation of a new stable bacterial species (now known as *Pseudomonas putida*) (Pandey and Arora 2020).

A bacterium called *Pseudomonas putida* is frequently discovered in water. It's a Gram-negative bacterium, like *E. coli*, and is frequently used in the environmental study because of its ability to break down plastics and various aromatic substances. *P. putida* can withstand and break down a multitude of hazardous and hard polymers compounds. Studies have demonstrated that the surface of plastic materials can change from a smooth to a rough texture with cracking as a result of compounds released by certain bacterial species, disrupting the intricate molecular structure of plastics (Saminathan et al. 2014).

26.5.4 *Alcaligenes faecalis*

Alcaligenes faecalis is a type of rod-shaped, Gram-negative bacteria that is typically found in the environment. As the macromolecules of the polymer degrade, small-molecular weight oligomers having oxygen groups like carboxylic, hydroperoxide, and carbonyl groups are created. *A. faecalis* can then use the process of biofragmentation to further reduce the molecular weight of polymers through enzymatic degradation, resulting in the production of alkyl, alkanes, and esters that have even lower molecular weights (Nag et al. 2021).

The hydrophobic properties of PET present a substantial barrier to microbial colonization of the polymer surface, inhibiting hydrolytic enzyme access and preventing effective adsorption to cause polymer degradation. *A. faecalis* may grow in colonies and create biofilms on a PET surface, proving that bacteria can use the material as a source of carbon for growth. However, additional research is required to confirm if bacteria are responsible for the degradation of PET (Atthoff and Hilborn 2007).

26.5.5 *Rhodococcus ruber*

Rhodococcus ruber is a Gram-positive, non-spore-forming, non-motile bacteria. *R. ruber* is capable of forming a biofilm out of polymers like polyethylene and polystyrene. It will colonise the surface and reproduce in a carbon-deficient environment. For *R. ruber*, the polymer serves as a substrate and a source of carbon, enhancing its metabolic activity while forming a biofilm. Through hydrophobic interactions and surfactant qualities connected to its cellular structure, it can stick to the surface. The bacterium continues to grow after forming a monolayer, forming multicellular structures that are durable and speed up the biodegradation of polymers like polyethylene and polystyrene (Mor and Sivan 2008).

26.6 Conclusions

According to the vast body of the literature on the subject, there is a great deal of room for developing new, more effective plastic-degrading agents for microbial bioremediation. The most economical, environmentally friendly, and generally accepted approach to solving the plastic waste crisis is microbial bioremediation.

It is encouraging to note that many microorganisms that break down natural polymers are also capable of breaking down synthetic polymers. However, the current photochemical and microbial degradation processes fall far short of the amount of microplastic generated daily. Thus, controlling plastic usage and disposal practices are crucial in reducing the burden on the environment. Unfortunately, the convenience and low cost of synthetic polymers make it challenging to persuade people to switch to biodegradable alternatives. Implementing regulations, awareness campaigns, and education among younger generations will be necessary to change this trend. The investigation of the interactions among different microorganisms engaged in the biodegradation of polymers in natural environments can aid in a better understanding of the process. With the increasing demand for biodegradable plastic alternatives, more attention and funding are required in this area of research. While numerous enzymes and microorganisms affect plastics, further studies are needed to identify and isolate the most effective organisms that can break down the most widely used plastics. Due to their heavy molecular weights and hydrophobic surfaces, establishing robust biofilms on polymers is challenging for bacteria.

Future studies should concentrate on molecular methods that can identify microbial groups that coordinate and speed up decomposition. It appears that both intelligent search algorithms for finding more effective enzymes by mining metagenome data and effective enzymes that can degrade plastic at contaminated sites will be successful. The discovery of sequencing, cloning, and genetic enrichment of catabolic pathways for pollutant breakdown were the main areas of attention in the past. We can forecast the metabolic abilities of microbes and their role in tackling this issue thanks to the latest developments in genomics and proteomics. It is essential to keep in mind that microplastics frequently include additives and impurities that may affect the optimization and standardization of the methods for analyzing microbial plastic biodegradation. The method of synthetic biology can create microorganisms that effectively break down plastic materials, and the byproducts that are produced could be used to make polymers or other products with added value. Another potential approach that could significantly reduce the ecological impact of plastic is the discovery and application of enzymes that can break down plastics to create biopolymers.

Acknowledgments The authors place on records their gratitude towards the editors, especially Dr. Mohd. Shahnawaz, for his insightful feedback and guidance, which helped to refine the structure and content of the work, and ensured its overall coherence and quality.

References

Adams GO, Fufeyin PT, Okoro SE, Ehinomen I (2015) Bioremediation, biostimulation and bioaugmentation: a review. Int J Environ Bioremediat Biodegrad 3(1):28–39

Adane L, Muleta D (2011) Survey on the usage of plastic bags, their disposal and adverse impacts on environment: a case study in Jimma City, Southwestern Ethiopia. J Toxicol Environ Health Sci 3:234–248

Allen-Gil G, Wabnitz CC, Beck MW et al (2020) From waste to wealth: unlocking economic benefits from a Sustainable Ocean Economy. World Resources Institute, Washington, DC

Andrady AL (2015) Microplastics in the marine environment. Mar Pollut Bull 97(1–2):122–123

Andrady AL (2017) The plastic in microplastics: a review. Mar Pollut Bull 119(1):12–22

Ang EL, Zhao H, Obbard JP (2005) Recent advances in the bioremediation of persistent organic pollutants via biomolecular engineering. Enzyme Microb Technol 37(5):487–496

Atthoff B, Hilborn J (2007) Protein adsorption onto polyester surfaces: is there a need for surface activation? J Biomed Mater Res B Appl Biomater 80B(1):121–130

Austin HP, Allen MD, Donohoe BS, Rorrer NA, Kearns FL, Silveira RL, Pollard BC, Dominick G, Duman R, El Omari K, Mykhaylyk V (2018) Characterization and engineering of a plastic-degrading aromatic polyesterase. Proc Natl Acad Sci 115(19):E4350–E4357

Barth M, Honak A, Oeser T, Wei R, Belisário-Ferrari MR, Then J, Schmidt J, Zimmermann W (2016) A dual enzyme system composed of a polyester hydrolase and a carboxylesterase enhances the biocatalytic degradation of polyethylene terephthalate films. Biotechnol J 11(8): 1082–1087

Bonhomme S, Cuer A, Delort A, Lemaire J, Sancelme M, Scott G (2003) Environmental biodegradation of polyethylene. Prog Polym Sci 81:441–452

Browne MA, Crump P, Niven SJ, Teuten E, Tonkin A, Galloway T, Thompson R (2011) Accumulation of microplastic on shorelines woldwide: sources and sinks. Environ Sci Technol 45(21):9175–9179

Carraher CE (2000) Introduction to polymer chemistry. CRC Press, Boca Raton

Chae Y, An YJ (2018) Current research trends on microbial degradation of plastics: an overview. J Environ Manag 217:56–73

Dahal P (2023) *Ideonella sakaiensis* (plastic-eating bacteria)—an overview microbe notes. https://microbenotes.com/ideonella-sakaiensis-plastic-eating-bacteria/. Accessed 15 July 2023

Dang TC, Nguyen DT, Thai H, Nguyen TC, Hien Tran TT, Le VH, Nguyen VH, Tran XB, Thao Pham TP, Nguyen TG, Nguyen QT (2018) Plastic degradation by thermophilic Bacillus sp. BCBT21 isolated from composting agricultural residual in Vietnam. Adv Nat Sci Nanosci Nanotechnol 9(1):015014

Danko AS, Luo M, Bagwell CE, Brigmon RL, Freedman DL (2004) Involvement of linear plasmids in aerobic biodegradation of vinyl chloride. Appl Environ Microbiol 70(10):6092–6097

Delille D, Duval A, Pelletier E (2008) Highly efficient pilot biopiles for on-site fertilization treatment of diesel oil-contaminated sub-Antarctic soil. Cold Reg Sci Technol 54(1):7–18

Fotopoulou KN, Karapanagioti HK (2017) Degradation of various plastics in the environment. In: Takada H, Karapanagioti HK (eds) Hazardous chemicals associated with plastics in the marine environment. Springer, Cham, pp 71–92

Galloway TS (2015) Micro-and nano-plastics and human health. In: Bergmann M, Gutow L, Klages M (eds) Marine anthropogenic litter. Springer, Cham, pp 343–366

Geyer R, Jambeck JR, Law KL (2017) Production, use, and fate of all plastics ever made. Sci Adv 3(7):e1700782

Ghosh SK, Pal S, Ray S (2013) Study of microbes having potentiality for biodegradation of plastics. Environ Sci Pollut Res 20:4339–4355

Groh KJ, Backhaus T, Carney-Almroth B, Geueke B, Inostroza PA, Lennquist A, Leslie HA, Maffini M, Slunge D, Trasande L, Warhurst AM (2019) Overview of known plastic packaging-associated chemicals and their hazards. Sci Total Environ 651:3253–3268

Guo H, Zhang Y (2020) Advances in microbial degradation of microplastics. Engineering 6(8): 817–824

Hadad D, Geresh S, Sivan A (2005) Biodegradation of polyethylene by the thermophilic bacterium *Brevibacillus borstelensis*. J Appl Microbiol 98(5):1093–1100

Hahladakis JN, Velis CA, Weber R, Iacovidou E, Purnell P (2018) An overview of chemical additives present in plastics: migration, release, fate and environmental impact during their use, disposal and recycling. J Hazard Mater 344:179–199

Hill DJ, McGrail PT (eds) (2003) Condensation polymers: synthesis, reactions, and applications. Springer Science & Business Media

Hopewell J, Dvorak R, Kosior E (2009) Plastics recycling: challenges and opportunities. Philos Trans R Soc B Biol Sci 364(1526):2115–2126

Howard GT, Mackie RI, Cann IK, Ohene-Adjei S, Aboudehen KS, Duos BG, Childers GW (2007) Effect of insertional mutations in the pueA and pueB genes encoding two polyurethanases in *Pseudomonas chlororaphis* contained within a gene cluster. J Appl Microbiol 103(6): 2074–2083

Hung CS, Zingarelli S, Nadeau LJ, Biffinger JC, Drake CA, Crouch AL, Barlow DE, Russell JN Jr, Crookes-Goodson WJ (2016) Carbon catabolite repression and Impranil polyurethane degradation in *Pseudomonas protegens* strain Pf-5. Appl Environ Microbiol 82(20):6080–6090

Jain PK, Bajpai V (2012) Biotechnology of bioremediation—a review. Int J Environ Sci 3:535–549

Jambeck JR, Geyer R, Wilcox C, Siegler TR, Perryman M, Andrady A, Narayan R, Law KL (2015) Plastic waste inputs from land into the ocean. Science 347(6223):768–771

Jeon HJ, Kim MN (2015) Functional analysis of alkane hydroxylase system derived from *Pseudomonas aeruginosa* E7 for low molecular weight polyethylene biodegradation. Int Biodeterior Biodegrad 103:141–146

Koelmans AA, Bakir A, Burton GA, Janssen CR (2016) Microplastic as a vector for chemicals in the aquatic environment: critical review and model-supported reinterpretation of empirical studies. Environ Sci Technol 50(7):3315–3326

Kumar A, Bisht BS, Joshi VD, Dhewa T (2011) Review on bioremediation of polluted environment: a management tool. Int J Environ Sci 1(6):1079–1093

Lange J, Wyser Y (2003) Recent innovations in barrier technologies for plastic packaging—a review. Packag Technol Sci Int J 16(4):149–158

Lee K, Tanabe S, Koh C (2001) Contamination of polychlorinated biphenyls (PCBs) in sediments from Kyeonggi Bay and nearby areas, Korea. Mar Pollut Bull 42:273–279

Lee TH, Byun IG, Kim YO, Hwang IS, Park TJ (2006) Monitoring biodegradation of diesel fuel in bioventing processes using in situ respiration rate. Water Sci Technol 53(4–5):263–272. https://doi.org/10.2166/wst.2006.131;2006

Lin Z, Jin T, Zou T, Xu L, Xi B, Xu D, He J, Xiong L, Tang C, Peng J, Zhou Y (2022) Current progress on plastic/microplastic degradation: fact influences and mechanism. Environ Pollut 304:119159

Menczel JD, Prime RB (eds) (2009) Thermal analysis of polymers: fundamentals and applications. Wiley, New York

Miller R (2005) The landscape for biopolymers in packaging. In: Miller-Klein associates report. Summary and full report available from the National non-Food Crops Centre. National Non-Food Crops Centre, Heslington

Mor R, Sivan A (2008) Biofilm formation and partial biodegradation of polystyrene by the actinomycete *Rhodococcus ruber*: biodegradation of polystyrene. Biodegradation 19:851–858

Müller RJ, Schrader H, Profe J, Dresler K, Deckwer WD (2005) DBM. Enzymatic degradation of poly(ethylene terephthalate): rapid hydrolyse using a hydrolase from *T. fusca*. Macromol Rapid Commun 26(17):1400–1405

Mulligan CN, Yong RN (2004) Natural attenuation of contaminated soils. Environ Int 30(4): 587–601

Muralisankar T, Arthanareeswari M (2018) Polyethylene terephthalate (PET)-a review on manufacturing, recycling process and environmental impacts. In: Sanghi R, Singh V (eds) Green chemistry for environmental remediation. Wiley, New York, pp 307–321

Nag M, Lahiri D, Dutta B, Jadav G, Ray RR (2021) Biodegradation of used polyethylene bags by a new marine strain of *Alcaligenes faecalis* LNDR-1. Environ Sci Pollut Res 28:41365–41379. https://doi.org/10.1007/s11356-021-13704-0

Ndahebwa Muhonja C, Magoma G, Imbuga M, Makonde HM (2018) Molecular characterization of low-density polyethene (LDPE) degrading bacteria and fungi from Dandora dumpsite, Nairobi, Kenya. Int J Microbiol 2018:1–10. https://doi.org/10.1155/2018/4167845

Niu GL, Zhang JJ, Zhao S, Liu H, Boon N, Zhou NY (2009) Bioaugmentation of a 4-chloronitrobenzene contaminated soil with *Pseudomonas putida* ZWL73. Environ Pollut 157(3):763–771. https://doi.org/10.1016/j.envpol.2008.11.024
O'Brine T, Thompson RC (2010) Degradation of plastic carrier bags in the marine environment. Mar Pollut Bull 60(12):2279–2283
Odian G (2004) Principles of polymerization. Wiley, New York
Palm GJ, Reisky L, Böttcher D, Müller H, Michels EA, Walczak MC, Berndt L, Weiss MS, Bornscheuer UT, Weber G (2019) Structure of the plastic-degrading *Ideonella sakaiensis* MHETase bound to a substrate. Nat Commun 10(1):1717
Pandey P, Arora NK (2020) Prof. Ananda Mohan Chakrabarty: the superbug superhero! Environ Sustain 3(3):333–335. https://doi.org/10.1007/s42398-020-00117-x
Perz V, Baumschlager A, Bleymaier K, Zitzenbacher S, Hromic A, Steinkellner G, Pairitsch A, Łyskowski A, Gruber K, Sinkel C, Küper U (2016) Hydrolysis of synthetic polyesters by *Clostridium botulinum* esterases. Biotechnol Bioeng 113(5):1024–1034
Pu W, Cui C, Guo C, Wu ZL (2018) Characterization of two styrene monooxygenases from marine microbes. Enzyme Microb Technol 112:29–34
Rathje WL, Reilly MD, Hughes WW (1985) Household garbage and the role of packaging—the United States/Mexico City household refuse comparison. Solid Waste Council of the Paper Industry Publishing, p 116
Rochman CM (2018) Microplastics research—from sink to source. Science 360(6384):28–29
Sacharow S, Griffin RC (1980) Principles of food packaging, 2nd edn. AVI Publishing Company, Westport
Saminathan P, Sripriya A, Nalini K, Sivakumar T, Thangapandian V (2014) Biodegradation of plastics by *Pseudomonas putida* isolated from garden soil samples. J Adv Bot Zool 1(3):34–38
Sivan A, Szanto M, Pavlov V (2006) Biofilm development of the polyethylene-degrading bacterium *Rhodococcus ruber*. Appl Microbiol Biotechnol 72:346–352
Sowmya HV, Ramalingappa MK, Thippeswamy B (2014) Biodegradation of polyethylene by *Bacillus cereus*. Adv Polym Sci Technol Int J 4(2):28–32
Sperling LH (2006) Introduction to physical polymer science. Wiley, New York
Stevens MP (2009) Polymer chemistry: an introduction. Oxford University Press, Oxford
Stilwell EJ, Canty RC, Kopf PW, Montrone AM (1991) Packaging for the environment: a partnership for progress. American Management Assn. (Amacom) Publishing, New York
Sudhakar M, Doble M, Murthy PS, Venkatesan R (2008) Marine microbe-mediated biodegradation of low-and high-density polyethylenes. Int Biodeterior Biodegrad 61(3):203–213
Sugimoto M, Tanabe M, Hataya M, Enokibara S, Duine JA, Kawai F (2001) The first step in polyethylene glycol degradation by sphingomonads proceeds via a flavoprotein alcohol dehydrogenase containing flavin adenine dinucleotide. J Bacteriol 183(22):6694–6698
United Nations Environment Programme (2018) Single-use plastics: a roadmap for sustainability. https://wedocs.unep.org/bitstream/handle/20.500.11822/25496/singleUsePlastic_sustainability.pdf?sequence=1&isAllowed=y. Accessed 18 June 2023
United States Environmental Protection Agency (2021) Plastics: material-specific data. https://www.epa.gov/facts-and-figures-about-materials-waste-and-recycling/plastics-material-specific-data. Accessed 18 June 2023
Vimala PP, Mathew L (2016) Biodegradation of polyethylene using *Bacillus subtilis*. Proced Technol 24:232–239
Wagner M, Scherer C, Alvarez-Muñoz D, Brennholt N, Bourrain X, Buchinger S, Fries E, Grosbois C, Klasmeier J, Marti T, Rodriguez-Mozaz S (2014) Microplastics in freshwater ecosystems: what we know and what we need to know. Environ Sci Eur 26(1):1–9
Wu X, Pan J, Li M, Li Y, Bartlam M, Wang Y (2019) Selective enrichment of bacterial pathogens by microplastic biofilm. Water Res 165:114979

Yang Y, Yang J, Jiang L (2016) Comment on "A bacterium that degrades and assimilates poly (ethylene terephthalate)". Science 353(6301):759

Zubris KA, Richards BK (2005) Synthetic fibers as an indicator of land application of sludge. Environ Pollut 138(2):201–211

Orr, I.G., et al., Colonization, biofilm formation and biodegradation of polyethylene by a strain of Rhodococcus ruber. 2004. 65(1): p. 97-104.

Mycodegradation of Microplastics

27

Nilam A. Salunkhe-Patil, Pranjali Mahamuni-Badiger,
Hrishikesh Khodade, and Maruti J. Dhanavade

Abstract

Microplastic pollution has gained much more attention in the research field as it has created a greater threat to biota worldwide. A major source of these microplastic materials is derived from mismanaged wastes in the environment. Current approaches like recycling and incineration approaches do not provide a sustainable solution to tackle the huge amount of microplastic pollution. There is an urgent need to develop innovative approaches for their disposal. Fungal-based biodegradation (mycoremediation) of microplastic is primed focused since last decades. Fungi are diverse eukaryotic groups that can degrade microplastic by using it as a food source under the influence of temperature and pH. Fungi are paling crucial role in degradation by securing varieties of enzymes like cutinase, lipases, proteases, cellulolytic enzymes, and also pro-oxidant ions. The action of these enzymes on microplastic and enhance the hydrophilicity and in turn convert

Nilam Ashok Salunkhe-Patil and Pranjali Mahamuni-Badiger contributed equally with all other contributors.

N. A. Salunkhe-Patil
Department of Microbiology, Bharati Vidyapeeth's Dr. Patangrao Kadam Mahavidyalaya, Sangli, Sangli, Maharashtra, India

P. Mahamuni-Badiger
Department of Microbiology, Rayat Shikshan Sanstha's, S.M. Joshi College, Pune, Maharashtra, India

H. Khodade
Department of Botany, Rayat Shikshan Sanstha's, S.M. Joshi College, Pune, Maharashtra, India

M. J. Dhanavade (✉)
Department of Microbiology, Bharati Vidyapeeth's Dr. Patangrao Kadam Mahavidyalaya, Sangli, Sangli, Maharashtra, India
e-mail: maruti.dhanavade@bharatividyapeeth.edu.in

M. Shahnawaz et al. (eds.), *Microplastic Pollution*,
https://doi.org/10.1007/978-981-99-8357-5_27

complex molecule into simple molecule which is an low molecular weight compound. Even though mycodegradation can be a promising tool for microplastic removal, it is highly demanding to finalize different parameters and determination of optimum conditions for successful degradation of plastics. This chapter summarizes different sources of microplastics along with their types. Furthermore, it focuses on different fungal isolates responsible for the biodegradation of microplastics, various modes of it, and different enzymes responsible for degradation. In addition, microplastic degradation is affected by varieties of biological and physiological factors. Therefore, different environmental factors like pH and temperature affecting microplastic degradation and degradation effects are discussed. Further, enhancing factors of microplastic degradation are briefly discussed. Finally, we have focused on future research directions and prospective for controlling microplastic pollution by mycodegradation. This chapter encompasses summary of the mycodegradation of microplastic degradation and a bunch of information for future innovative research to control microplastic pollution worldwide.

Keywords

Microplastic · Mycoremediation · Fungi · Mycodegradation · Biodegradation

27.1 Introduction

Recently, in the whole world, the focus of researchers shifted on the microplastic pollution which is increasing astonishingly. Microplastics are fragments of plastics ranging less than 5 mm in diameter. Microplastics have innumerable unique properties with low cost and having highest demand for plastic-based materials. It is considered an obligatory thing in the modern lifestyle. Plastic-based materials have gained popularity with their use in many industrial sectors such as food preservation, electronics, medical applications, construction, etc. Plastic is the third most abundant human-made material after steel and concrete and is easily disposed of in the environment without any prior treatment. Fragments, fibers, beads, and different flakes with pellets are considered the sources of microplastics. Microplastics are available everywhere in nature such as the atmosphere, soil, freshwater, and marine water. These are also investigated in surface water sources. Additionally, the ingestion of microplastics is reported in marine animals such as invertebrates and fish (GESMAP 2015). The variable forms of plastic-based materials are to be considered as a key source of inland pollution and crucially it was reported that plastic material degradation has issues in the soil environment (Klemchuk 1990). Though the emphasis on microplastic pollution has gained attention in the aquatic ecosystem, many researchers are working on the bio-degradation of microplastics in terrestrial environments (Wagner et al. 2014).

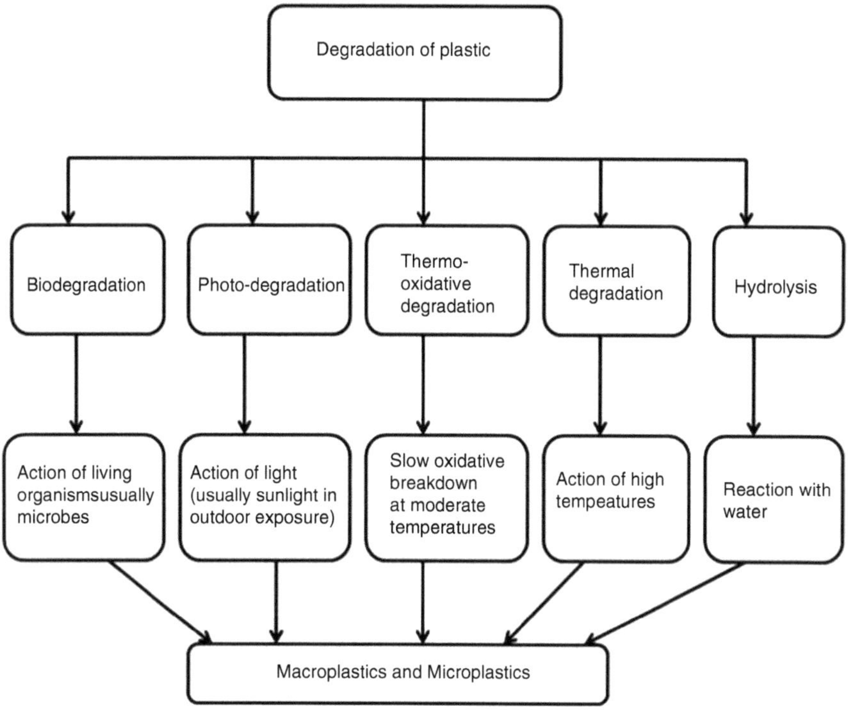

Fig. 27.1 Environmental degradation methods under different conditions (Sharma and Chatterjee 2017)

The losses of microplastic pellets such as industrial abrasives, printing powders, personal care products, etc. are the preliminary microplastic sources (Klemchuk 1990). In a terrestrial ecosystem, the mechanical breakdown (by wind and UV radiation) of plastic materials into small-sized pieces takes place (UNEP Year Book 2016; NEA 2014). This is called as physical weathering of plastic material (UNEP Year Book 2013). This process produces a secondary type of microplastics (UNEP Year Book 2013). Any type of additional breakdown of secondary microplastic polymer is not enough to completely convert it into monomers such as methane, carbon dioxide, ammonia, hydrogen, water, and other inorganic compounds.

Currently, methods employed for microplastic degradation are mechanical degradation, landfilling, recycling, and biological degradation (Fig. 27.1). Due to tiny microplastic particles, they are not easily recycled. Biological degradation of environmental pollutants is an environment-friendly microplastic degradation process (Patil and Kurhekar 2020). The use of potential microbes like bacteria and fungi has become an effective way for microplastic degradation to overcome the pollution problem (Yu et al. 2022). Especially many fungal species (*Aspergillus* and *Fusarium* species) reported by researchers as potent degraders of microplastics (Yu et al. 2022) which is called as mycodegradation. The fungal species showed better ability to degrade microplastic's than bacteria with great efficacy (Yu et al. 2022).

The following are the reasons enlisted:

1. Fungi produce significantly more enzymes than bacteria in terms of numbers. Fungi can firmly adhere to microplastic surfaces with the help of mycelium and they can easily penetrate the interior of microplastic particles (Li et al. 2020).
2. Due to penetration, the surface hydrophobicity of microplastic particles could reduce by endorsing the chemical bonds formation such as carboxyl, carbonyl, and ester groups in the microplastics (Li et al. 2020).
3. To achieve microplastics degradation, microorganisms take at least 21 days; however, it is a too slow process as compared to fungi (Du et al. 2021).

27.2 Different Sources of Microplastic Pollution

Microplastics can be produced from manufacturing processes or the fragmentation of larger plastic materials. Figure 27.2 represents different sources of microplastic pollution. Primary sources include the use of microbeads in medical products, facial cleansers, daily cleaning products, toothpaste, and cosmetic goods. Whereas, the fragments or particles of plastics from a fishing net, plastic garbage, plastic furniture, plastic film, etc. are included in secondary microplastics (Sun et al. 2022). Both types of microplastics are found ubiquitously in both terrestrial as well as in marine environment (Sadri and Thompson 2014). Importantly, there are known microplastic sources are noticed and characterized into three forms. First, different types of fibers from textile industries, e. g., synthetic textiles. Second, the larger pieces of

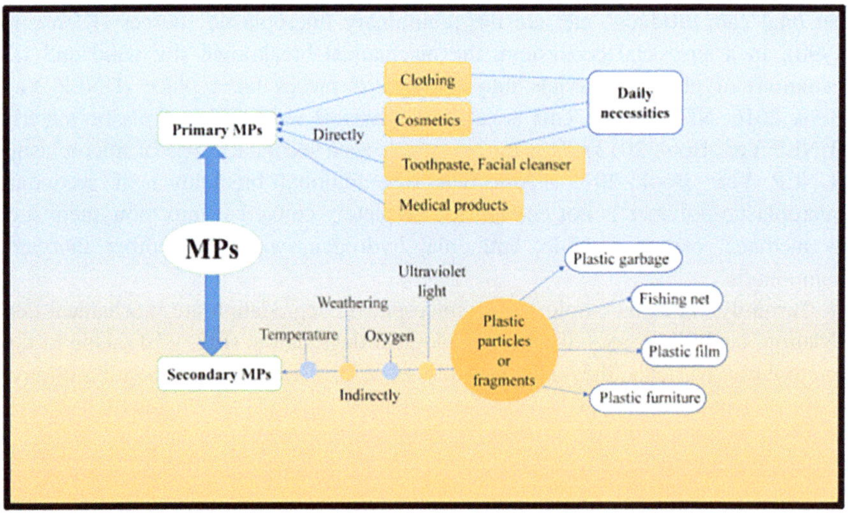

Fig. 27.2 Different microplastic sources in the environment (Sun et al. 2022)

plastics-based materials become small-sized particles, i.e., microplastics by fragmentation process (Browne 2015). Third, the natural weathering of ropes, tires, fishing nets, plastics in varnishes, carpets, and paints is an vital microplastics source in environment.

27.2.1 Fibers

Washing of clothes is the process through which a variety of small-sized fibers from polyester or nylon entered the atmosphere (Browne et al. 2015). Till today, the complete removal of microplastics is not studied from sewage treatment plants, ultimately fibers retain in the sewage effluent. As a result, fibers will be entered into natural water streams such as lakes, rivers, seas, etc., and also in terrestrial ecosystems. Fibers are highly abundant type of microplastics present in the sediments and marine wildlife (Thompson 2015).

27.2.2 Plastic Debris

An opportunistic or accidental entry from waste treatment systems bigger-sized particles of plastics-based materials can be entered into the ocean. Sometimes plastic debris can enter the marine environment and has originated from inland sources and traveled via wind or rivers. Either based on weight or number, the larger plastic debris can be studied. Eriksen et al. (2014) reported, around 891 visual surveys regarding the larger plastic debris detected on the ocean surfaces of the world. Out of all, 20% of matters are generated from fishing process, 58% as non-fishing materials, and 22% are grouped as miscellaneous. However, the fishing gear accounting about 70% of the total weight which is the most dominant form of litter observed. 10% from non-fishing gear like items from plastic films or bags, 18% from pieces of foamed polystyrene and plastic bottles. To date, there is limited data reported on suspended and sunk types of plastic debris at sea bottom (Eriksen et al. 2014).

27.2.3 Microbeads

Microbeads are spheres of plastics especially used in cosmetics, industrial blasting, and, daily household cleaning products. In toothpaste, it can be used as an exfoliant. The microbead emission from facial exfoliation alone was reported as 16–86 tonnes per annum in the UK (Napper et al. 2015).

27.3 Types of Microplastics

Plastics are polymers. These are characterized based on their buoyancy, toxicity, and degradability. Mainly plastics are made up of enlisted types of polymers: polypropylene (PP), polystyrene (PS), polyethylene (PE), polyurethane (PUR), polyvinyl chloride (PVC), and polyethylene terephthalate (PET) (GESMAP 2015) (Fig. 27.3). Usually, pigments are present in plastic materials. Sometimes flame retardants and stabilizers are also used in the plastic material-making process, which is used for the property change of end products. The addition of these types of additives reduced the speed of degradation of plastic debris or microplastics. Ultimately, it will leach out into the environment.

27.4 Hazardous Effect of Microplastic Pollution

Inevitably, the animal body serves as a receiver point for all types of environmental pollutants including microplastics. Microplastics can enter an animal's body in direct or indirect ways. Especially bio filtering animals such as Mollusca, zooplanktons, and ascidia are prone to the microplastics accumulation (Smith et al. 2018). Researcher Kwon et al. (2020) reported that an average fish carries around 35–40 MPs in their gastrointestinal tract. Microplastics are also detected in the terrestrial food chain: chicken feces (105 particles/g) and in the chicken stomach (5.1 particles/g) (Van den Berg et al. 2020).

1. Microplastics can be inhaled by a human being from the air, and also ingested through food and water. Microplastics are found in honey (0.10 particles/g), sugar (0.44 particles/g), alcohol (32.27 particles/g), salt (0.11 particles/g), bottled water (94.37 particles/g), tap water (4.23 particles/g), and in inhaled air (9.80 particles/m^3) (Cox et al. 2019). Around 50% of microplastics in food items are composed

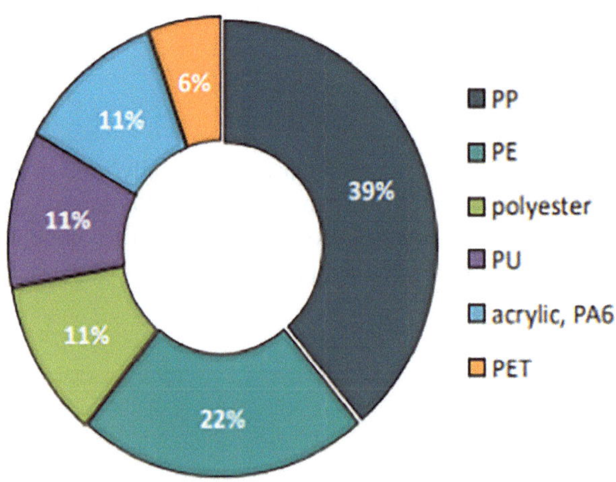

Fig. 27.3 Polymeric composition of microplastics from wastewater treatment plants

of PP, PS, PE, and PET. These are fibers with small filaments. As compared to spherical particles of microplastics the fibers have toxic effects at lower concentrations or doses (Kwon et al. 2020). Microplastics less than 5 μm diameter has toxic effects than bigger (100 μm) sized plastic particles in fishes (Barboza et al. 2020). It is evidenced that the change in motor activity and feeding behavior was observed in adult fish. Maternal transfer of microplastics to the offspring has been noticed, and ultimately it negatively affects fertility (Sucharitakul et al. 2021). Biological and histological changes were observed in the tissues of fish because of accumulated microplastic particles (Wang et al. 2019). Figure 27.4 depicts the details of all toxic effects in fish observed due to microplastic exposure.

The toxic effect of microplastic exposure in terrestrial mammals is studied and reported by around 30 research articles listed in Scopus and PubMed. During the study, the strain and species of animal under study are to be important.

Exposure time, exposure length, type of doses, and routes of microplastic administration are the variable that should be considered during the study. When MPs entered mice body via orally it will exert a harmful effect on the gastrointestinal tract (Jin et al. 2019). Several reports depicted that accumulated microplastics (diameter 0.5–50 μm) causes damage to the epithelium in the colon of mice (Lu et al. 2018). The hazardous effects of microplastics in mice is shown in Fig. 27.5. Whereas, the detrimental effects of microplastics on human health is explained in Fig. 27.6. It was demonstrated by several research studies that, due to microplastic exposure to the human body many changes were observed morphologically such as fat deposition in tissues, alteration in hormone levels, and difficulty in breathing and intestinal blocking. From a behavioral point of view, reduced vigor and loss of energy are noticed. Also, it affects mobility. Microplastics affect badly on the reproductive system also as it delays ovulation, cause infertility, and effects the secretion of reproductive hormones in humans. There are some toxicological impact on marine animals.

27.5 Biodegradation of Microplastic

The production of microplastics can be controlled using conventional techniques. Like other waste, plastic waste is kept out of the ecosystem by being buried in landfills (He et al. 2019). But due to their refractory qualities, which cover a lot of areas and prevent their relocation for a long time, discarded plastic waste is still kept in the same place for a long period, which affects the sustainable use of land (Natesan et al. 2021). In addition, the groundwater was contaminated due to the emission of plastic waste (Chia et al. 2021). As already indicated the breakdown of large plastics increases the bioavailability of micro-scale plastics and poses a serious risk to the atmosphere. The microplastics are having small size and inert qualities; hence, these are hard to eliminate from the aquatic settings (Imran et al. 2019).

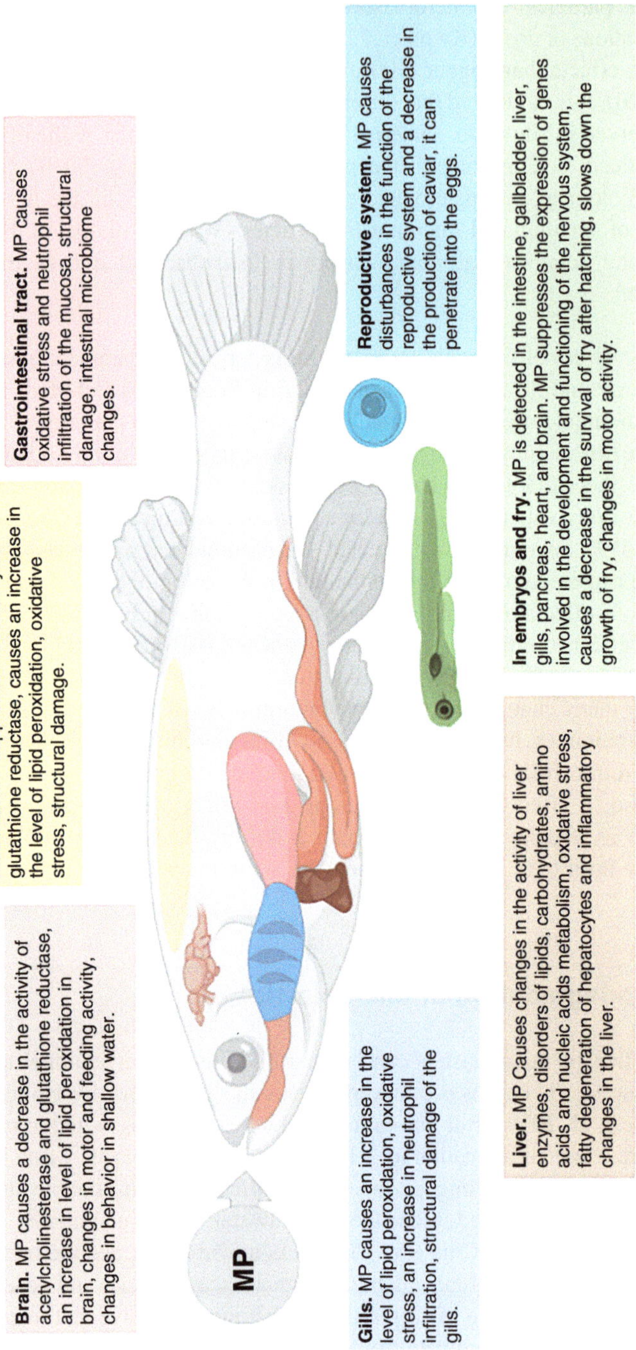

Fig. 27.4 Harmful effects of microplastics in fish (Zolotova et al. 2022)

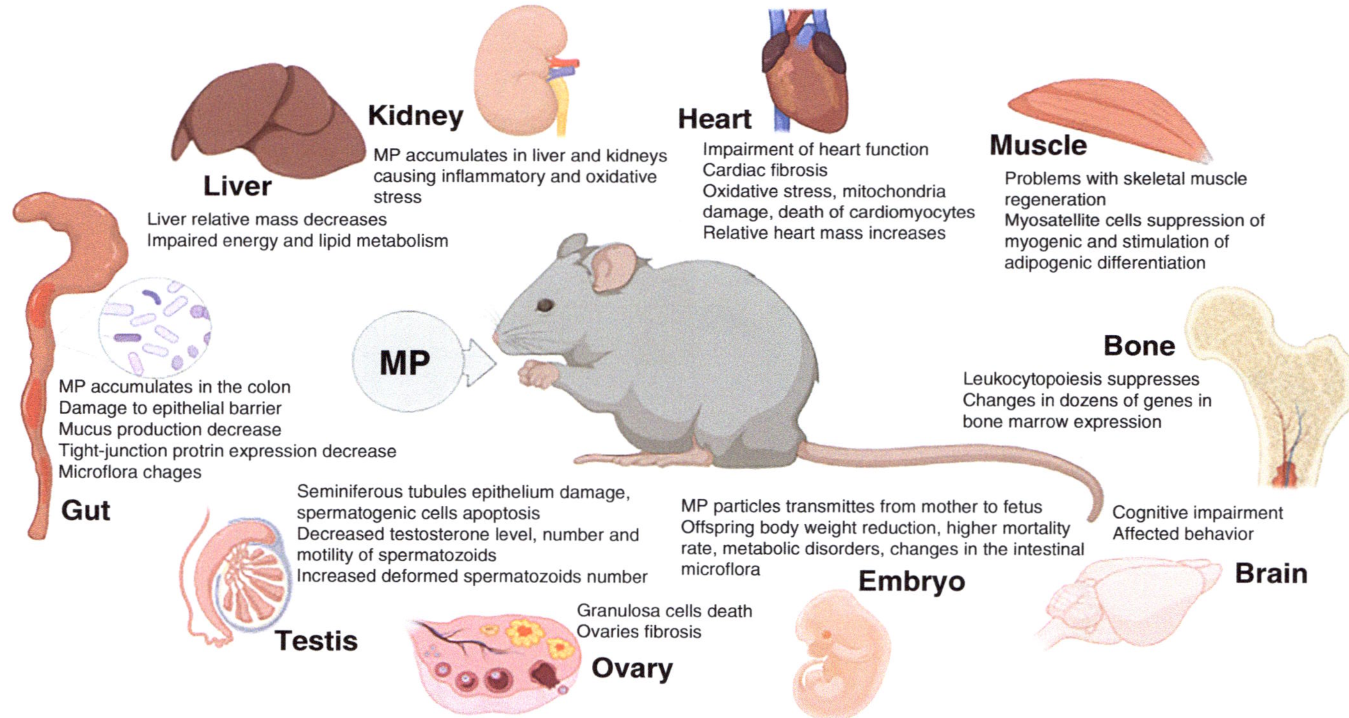

Fig. 27.5 The harmful properties of microplastics in mice (Zolotova et al. 2022)

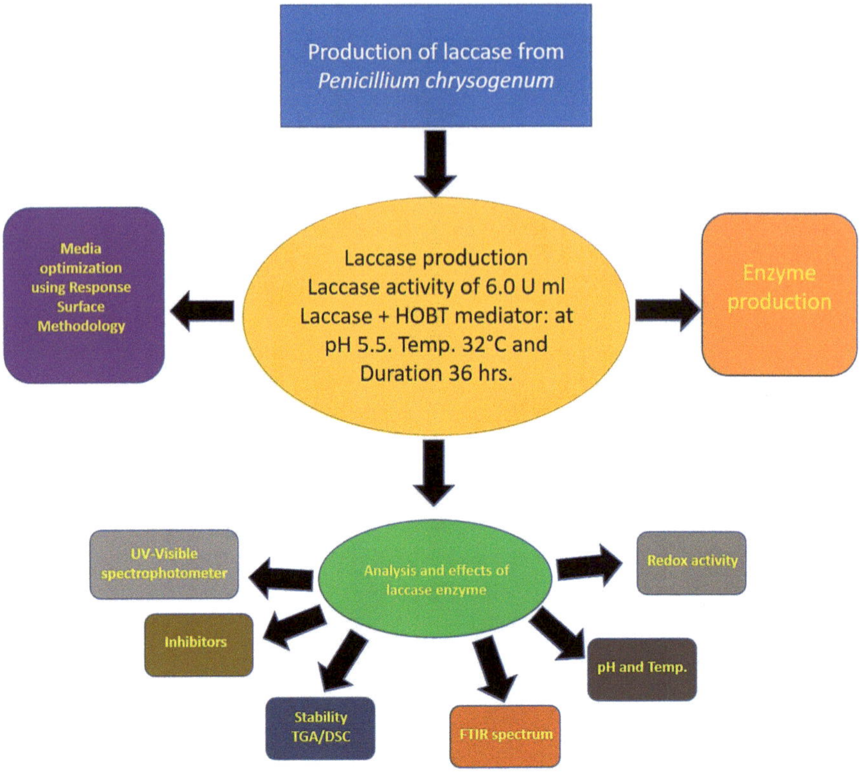

Fig. 27.6 Production, optimization and characterization of laccase enzyme

Specific plastic-degrading enzymes accelerate bio-fragmentation, acting as more effective catalysts for the depolymerization reaction. Bio-fracture oligomers are used after bacterial fragmentation (aggregation) (Liu et al. 2017). Finally, bacteria and fungi use several metabolic processes to extract all the energy from the oligomers and release the oxidized waste into the environment (mineralization). Mineralized byproducts (water, molecular nitrogen, methane, and carbon dioxide) can then re-enter the atmosphere (Ramezani et al. 2021).

Most microplastics in wastewater treatment plants (WWTPs), especially large microplastic particles, can be eliminated by different methods of treatment (Cristaldi et al. 2020). Degradation of microplastics is now being proposed by means of method like advanced oxidation processes (AOPs) as well as biological degradation (Ricardo et al. 2021; Zurier and Goddard 2021). The chemical bonds of polymer microplastics can be broken by these two methods to make smaller molecules, that are further transformed into valuable chemicals or completely converted to CO_2 and H_2O (Ricardo et al. 2021). The microplastic polymer is then broken down by secreted enzymes into low (600 Da) molecular weight oligomers that may or may not be particularly suited to the plastic substrate (biodegradation) (Zurier and

Goddard 2021). Since this step in the biodegradation process is rate-limiting, it has received much scientific attention. Plastic must be colonized by microorganisms before it can biodegrade. Plastic waste in the marine environment quickly becomes contaminated with both organic and inorganic substances (Derraik 2002). Pollutants that affect ecosystems are converted into byproducts that are cycled by ecosystems during biodegradation. There are sequential stages of plastic biodegradation because microbial community development has taken place on the surface of the polymer. The plastic matrix (biodegradation) is weakened by macromolecular changes, either physical (expansion of biofilms pushing pits and fissures) or chemical (secreted acids degrade the plastic structure) (Dionisi et al. 2005).

Due to the variety of plastic types and the substrate specificity of bio-enzymes, oxidoreductases engaged in redox processes are also essential for microplastic breakdown (Sánchez 2020). Polyolefin and other polyesters made from petroleum (like PET) have a reduced breakdown efficiency in bacterial processes. In recent investigations, several intermediate products and particular enzymes were extensively studied. For instance, *Bacillus cereus* and *P. putida* can alter physical properties and produce certain intermediate products, including aldehydes and ketones (Chandra et al. 2020).

Fungi, in addition to bacteria, can adhere to and consume microplastics. The hydrophobicity of microplastics can be reduced by fungi by encouraging the synthesis of various chemical bonds amongst various groups such as carboxyl, carbonyl and ester present in microplastics (Harms et al. 2011). Fungi are widely distributed and diverse group of microorganisms having wide reproductive capacity. Fungi can also promote oxidation and the movement of chemicals. Currently, the degradation of microplastics by fungi under various conditions is being studied and some progress has been made (Hyde et al. 1998). Yeasts, filamentous fungi, and dimorphic fungi are morphologically classified as one of the important members of kingdom fungi of included in eukaryotic organism domain. These organisms may be obligate, opportunistic, or saprotrophs (decomposing dead matter) (decomposers, sympathizers, or pathogens). Fungi can be found in a variety of habitats, and few members of them have progressed to adjust and thrive in harsh terrestrial and marine surroundings (Lee et al. 2019; Bianco and Perrotta 2015). Although there are few anaerobic fungi found in freshwater lakes, landfills, and in the rumen of herbivores, however mostly aerobic fungi are seeming to be present in these environments. Due to their filamentous structure, fungi can locate and grow in a places where other microorganisms can be rarely found, in search of nutrients (Lee et al. 2019). By stimulating important chemical processes, numerous environmental contaminants are successfully decomposed and mineralized by filamentous fungi. As heterotrophic organisms, fungi obtain their food by absorbing nutrients from the environment (Bianco and Perrotta 2015). They exocytose out of their hyphae to release some digestive enzymes which cleave macro/organic compounds into smaller organic molecules so that these molecules can be reabsorbed. If the substrate is mineralized, in aerobic condition the enzymes release CO_2 and H_2O (whereas in anaerobic condition CH_4 is released). To cope with a wide range of complex molecules, some of which are harmful and even polluting, fungi have developed many surprising tricks. These techniques comprise a prevailing enzymatic system, adsorption capacity, and natural biosurfactants which allow them to use polymers like plastics as a carbon and electron sources, supplementing them with energy sources and cellular components (Purohit et al. 2018). Recently the interest to use fungi in the microplastic biodegradation as well as in different applications has increased. In near future screening and research will be highly demanded to confirm the degradation of functional fungi working on microplastics which is not easily degradable and it is highly resistant to processes such as corrosion and oxidation (Yuan et al. 2020).

To be strong and resistant to biological and inorganic degradation, non-biodegradable polymers are produced. However, a significant problem arises in the durability of these polymers once they are released into the environment. The PE, PP, PS, PU, and PVC are non-biodegradable microplastics commonly found in marine environments. These polymers can be degraded by bacteria in part because they contain C–C chains only or ether bonds that connects its monomeric units. These polymers are generally hard to degrade (Kyrikou and Briassoulis 2007).

27.6 Modes of Mycoremediation

Generally, there are two categories of mycoremediation like in situ and ex situ. While restoration is carried out on the contaminated site in the first instance, it is excavated and transported to another location in the second (Purohit et al. 2018).

27.6.1 In-Situ Remediation

At the contaminated location, in situ cleanup is carried out. It is an approach that works well and is economical.

27.6.1.1 Biosparging

The biosparging is an example of in-situ remediation technique which is also called as sparging. In this technique local microbes are used to degrade organic materials in the saturated zone. To increase the biological activity of the local microorganisms in the saturated zone, sparging entails infusing air (or oxygen) and nutrients (if necessary) into the saturated zone. Biosparging can minimize petroleum chemicals that are present in the capillary fringe, adsorbed to the soil under the water table, and dissolved in groundwater (Williams 2002).

27.6.1.2 Bioventing

Using existing soil microorganisms which requires oxygen, the method of "bioventing" encourages the in-situ biodegradation of pollutants in soil. Low airflow rates are used in bioventing to deliver just enough oxygen to support microbial activity in the vadose zone. The most typical method of supplying oxygen to soil with residual pollution is direct air injection. As vapors flow slowly through biologically active soil, volatile chemicals are biodegraded in addition to the destruction of adsorbed fuel residuals (Lee and Swindoll 1993).

27.6.1.3 Bioaugmentation

Treatment of polluted soil or water involves the introduction of a collection of indigenous microbial strains. Activated sludge bioreactors are often used in municipal wastewater treatment facilities (Tyagi et al. 2011).

27.6.2 Ex-Situ Remediation

It entails digging out contaminated dirt from the ground. Ex situ cleanup is costly and harms the ecosystem when pollutants are moved from one place to another.

27.6.2.1 Land Farming

This is straightforward method where polluted soil is removed, put on a bed which was already prepared, and repeatedly tilled so that the pollutants are broken down. This procedure was carried out to promote the aerobic breakdown of pollutants by local biodegradative bacteria (Lin et al. 2010).

27.6.2.2 Composting

This method involves adding organic materials to polluted soil to encourage the rich development of microbes. There is requirement of high temperature for composting procedure (Cristorean et al. 2016).

27.6.2.3 Biopiling

Excavated soils are combined with some additives of soil. This mixture was kept on a treatment area to carryout bioremediation process with proper aeration. This is a full-scale method. Carbon dioxide and water are left as the only pollutants (Beškoski et al. 2011).

27.6.2.4 Fungi Used in Mycoremediation

Because of their distinct physical, physiological, and genetic characteristics, fungi are unusual creatures. They may colonize soil, water, and other natural settings, making them omnipresent and contributing to the balance of the ecosystem. As a result of environmental adaptability, fungi have acquired special bioremediation abilities. By producing enzymes like proteases, amylases and lipases which utilizes substrates like proteins, starches and lipids. Few fungal species can almost destroy any natural organic substance. Different molecules such as cellulose, hemicellulose and pectin can be used as carbon sources by a small number of species. The primary degraders of naturally occurring, complex polymers that are microbially resistant are successfully degraded by fungi. These polymers includes as lignin, keratin, and chitin, are under assault (Anastasi et al. 2013).

Several examples of useful mycoremediation viz. Ligninolytic Fungal Degradation, White Rot Fungi (WRF), Brown Rot Fungi, Soft Rot Fungi, Soil Fungal

Biosorption, and Mycorrhizal Fungal Degradation (Anastasi et al. 2013; Periasamy et al. 2019).

For their energy demands, microorganisms, including fungi, may biochemically break down, absorb, and metabolite complex organic molecules, xenobiotics, and refractory chemicals. The biodegradation of complicated and polluting polymers is now being improved and promoted through research into a variety of species and distinct methods. In contaminated areas, for instance, the introduction of genetically modified bacteria with catabolic genes that are transmitted through plasmids in to the native bacterial population, enhancing their potential for xenobiotic breakdown (Periasamy et al. 2019).

27.7 Different Enzymes Participating in Mycoremediation

Hydrolases and oxidoreductases are the two primary groups of enzymes involved in the biodegradation of plastic. Since these enzymes are used in both industrial and natural processes, much research has been done on them. These enzymes are playing important role to degrade lignocellulose, as well as it is involved in fungal pathogenicity, and the hydrolysis of triacylglycerol into fatty acids in both nature and industry. These enzymes are used in bioremediation procedures as well as in the food and textile sectors (Hou and Majumder 2021).

27.7.1 Laccases

In the plant species, laccases were originally identified in 1883. Numerous species of Basidiomycetes, Ascomycetes, and Deuteromycetes fungi were found to contain laccases (Nunes and Kunamneni 2018). The capacity of white-rot fungus to make laccases to break down lignin is the group that has been investigated the most. These laccases are multicopper monomeric glycoproteins that are a subclass of the blue copper oxidases of enzymes. When employing free enzymes or crude enzyme extracts to biotransform a variety of contaminants, laccases have demonstrated significant promise. However, the complex composition, high salt content, and low pH values of polluted media that impact protein stability, recovery, and recycling limit their widespread utilization in bioremediation and water treatment. One of the key properties of these enzymes is their redox potential, which is directly correlated with the energy needed to remove an electron from the reducing substrate. Because they can oxidize substrates with high redox potential, laccases with a high redox potential, such as fungal laccases, are of particular importance in biotechnology (Witayakran 2008).

27.7.2 Peroxidases

Peroxidases are a class of oxidoreductases that include heme and catalyze the reduction of hydrogen peroxide as well as the oxidation of organic and inorganic molecules. Peroxidases are exclusively found in plants and fungi and are further divided into 3 subclasses such as Class I—Intracellular except animals, Class II—Extracellular Fungal peroxidases, and Class III—Extracellular Plant Peroxidases. Depending on what they employ as the reducing substrate, the primary fungus peroxidases are manganese peroxidases, lignin peroxidases, multifunctional peroxidases, and dye decolorizing peroxidases. Peroxidases' non-specificity and capacity to oxidize substrates with high redox potential are their key traits. These characteristics have made these enzymes useful in a wide range of applications. The production of biofuels, paper, waste management, and bioremediation of industrial contaminants including synthetic colors and polycyclic aromatic hydrocarbons all include peroxidases (Hofrichter et al. 2010).

27.7.3 Cutinases

Cutinases are extracellular serine esterases that belong to one bacterial subfamily, two fungal subfamilies, and cutinases. The fundamental structure of these subfamilies varies depending on whether they are prokaryotic or eukaryotic. Cutinases play a role in the development of fungal diseases in nature because they can break down ester bonds (Martínez and Maicas 2021). Due to their capacity to catalyze hydrolysis processes, esterifications, and transesterification, cutinases are also multifunctional enzymes with several commercial uses. Due to these characteristics, they can be used to modify synthetic fibers and degrade high molecular-weight polyesters (Nikolaivits et al. 2018).

27.7.4 Lipases

Extracellular triacylglycerol acyl hydrolases known as fungal lipases may hydrolyze ester bonds from insoluble monoglycerides, diglycerides and triglycerides substrates into free fatty acids and glycerol (Desnuelle and Savary 1963). Extracellular triacylglycerol acyl hydrolases known as fungal lipases can hydrolyze ester bonds from insoluble substrates of tri-, di-, and mono-glycerides into free fatty acids and glycerol. Lipases are thus engaged in the metabolism of lipids throughout procedures including digestion, absorption, and reconstitution. Fungal lipases have a variety of industrial uses, including in the manufacture of detergents, cosmetics, and medicines as well as in the food, textile, and manufacturing sectors. Additionally, lipases may break down polyurethane and fatty waste (Sarmah et al. 2018).

27.7.5 Cellulolytic Enzymes

Xylanolytic and cellulolytic processes are carried out by a class of hydrolytic enzymes known as cellulolytic enzymes. The majority of them are extracellular enzymes produced by a different fungus. Cellulases, hemicellulases, pectinases, chitinases, amylases, proteases, phytases, manages, and xylanases are examples of fungal cellulolytic enzymes (Bala and Singh 2019). These include *Trichoderma reesei, A. awamori, A. fumigatus, Trichoderma* sp., *Aspergillus niger, Sclerotium* (*Trichoderma* sp., *Penicillium* sp., *Aspergillus* sp., *Thermomonospora* sp.). Extra-cellular chemicals produced by brown rot fungus can enter the wood cell wall very deeply and take part in breakdown processes. In addition, they can penetrate into cell wall along micropores by breaking the glycosidic bonds of polysaccharides (Arantes and Goodell 2014). These compounds break the bond between the fibril structures and depolymerize polysaccharides of wood, that causes an minimal weight loss at the initial stages of decay. Fungal enzymes are capable of degrading complex materials which include;

(a) Chlorinated aromatic compounds degradation
(b) Dyes degradation
(c) Herbicide degradation
(d) Insecticide degradation
(e) PAH degradation
(f) Nitroaromatic degradation
(g) Hydrocarbon degradation
(h) Microplastic degradation

(Bala and Singh 2019); (Arantes and Goodell 2014)

27.8 Different Factors Affecting the Rate of Mycodegradation

Different environmental factors comprise microplastic composition, pH, temperature, moisture content, enzymes, and the type of microbial strain. These factors can cause changes in the mass, crystallinity, and color of plastics. To fully degrade plastics under specific circumstances, it is essential to collect plastics and enzymes/ microbes that consume plastic (Buyer et al. 2010).

27.8.1 pH

The pH of the soil is one of the most important elements impacting the microbial community there. Abiotic variables including the solubility of metals, the availability of nutrients, and the availability of carbon are all significantly influenced by pH. The biomass composition of bacteria and fungi, for example, may be influenced by soil pH in the soil. The fact that soil pH has a wide range of impacts on several factors makes it difficult to analyze those effects (Buyer et al. 2010). The ability to adapt to these widely varying environmental circumstances is essential for the survival of fungi. They can adapt to quick changes in their extracellular environment because of their capacity to react to a wide pH range, in particular. By placing a physiological restriction on microbial survival and growth, soil pH may have a direct impact on the makeup of the microbial community. If the pH of the soil is outside of a specific range, some microbial taxa may not be able to develop or survive. In general, acidic environment favor fungi growth (Zhalnina et al. 2015).

27.8.2 Temperature

The growth, germination, and reproduction of fungi are all significantly influenced by temperature. Most fungi generally reach their optimum growth rates at a temperature of around 25 °C; however, this might vary based on other environmental parameters including pH levels. The ideal temperature varies depending on the species. In addition to affecting fungus behavior, temperature also has an impact on the chemical and physical makeup of microplastics as well as fungi's capacity to biodegrade them. By employing the fungi species, the basic biodegradation efficiency may be increased. We must compare the impact of temperature on the biodegradation of hydrocarbons to the enzymes participating in the degradation pathways (44) (Doohan et al. 2003). Laccases, Peroxidase, Cutinases, and Lipases are the primary enzymes involved in the biodegradation of hydrocarbons. The activity of these degradative enzymes decreases at extremely high and low temperatures, which is why they need the ideal temperature. At particular temperature ranges that are ideal for the processes that produce enzymes, microorganisms that biodegrade hydrocarbons are very successful. Temperatures between 20 and 35 °C are ideal for the biodegradation of hydrocarbons by many microorganisms and offer the best biodegradation values. Lower temperatures resulted in low enzyme activity and altered microplastics' physical characteristics. Because of this, the hydrocarbon is resistant to microbial biodegradation (Alekseyeva et al. 2022).

27.9 Advantages and Challenges Associated with Mycodegradation

27.9.1 Advantages

Mycodegradation is the process through which potent fungal strains can degrade plastics by secreting enzymes. The biological degradation of plastics is an environmentally friendly process. This can be applied for the removal of contaminants effectively. Mycodegradation can be applied for a large microplastics contaminated area. After the decomposition of MPs, they can be used for another purpose (Fig. 27.7).

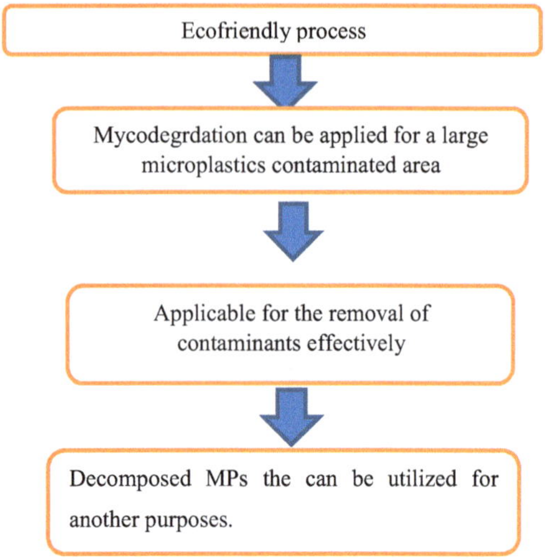

Fig. 27.7 Advantages of mycodegradation of microplastic pollution

27.9.2 Challenges Associated with Mycodegradation

The biological degradation of microplastics has many limitations and challenges. At first, the period of treatment is should be properly studied. In the large-scale application of potential fungal or bacterial strains for the microplastic pollution, there should be an availability of biomass of microbes. The data should be available regarding the maximum microplastics concentration at the site. Nutrient availability or supply for the growth of fungal species during the microplastics degradation process is a challenging task. There is a requirement of expertise having training in degradation treatment process in detail (Fig. 27.8).

27.10 Conclusion and Future Prospects

27.10.1 Conclusion

Worldwide microplastic pollution is becomes a serious issue as MPs are present everywhere in nature (e.g., oceans, soil, and fresh water). The current understanding and knowledge of different sources. Microplastics are reported and identified in many samples like soil, water, and food. The distribution of MPs and their dangerous effects on both aquatic animals and humans are described here. The MPs bring threatened toxic effects to the aquatic environment and also has an adverse effect on human beings. Microplastics have long shelf life as they can stay a long time in the environment. Mycodegradation is an eco-friendly path for the complete reduction of microplastics from the environment. The biological degradation of microplastics using potential fungal strains is an effective way of microplastic degradation as compared to the use of potent bacteria. There is a need for better understanding and more focused research on screening and characterization of novel fungal species

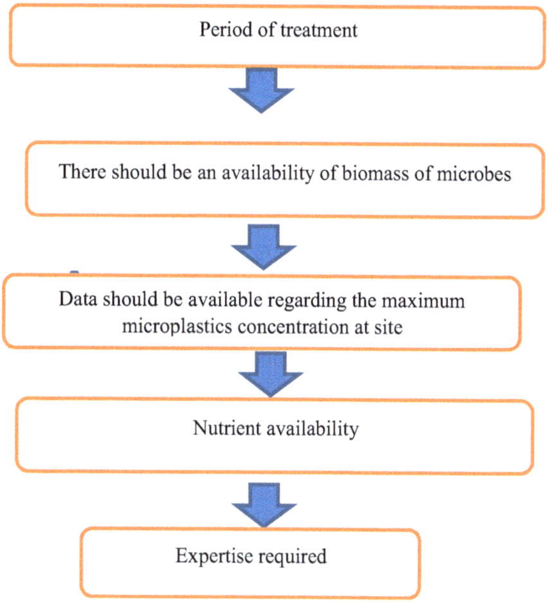

Fig. 27.8 Challenges associated with mycodegradation

having the potential of complete conversion of microplastics polymers into monomers naturally.

27.10.2 Future Prospects

1. Research should be explored on the study of the detailed mechanism of toxicity of microplastics.
2. Still the exact value of microplastics' entry into the terrestrial environment is not yet properly studied well. There is a need of the development of rapid and proper analyzing methods of microplastics.
3. In the current trend, need for study on novel and highly active fungal as well as bacterial strains having high degradation ability.
4. Research work should be focused on the development of effective consortia of potent microbes (either bacteria or fungi) proved as a weapon in biological degradation of MPs.

References

Alekseyeva KS, Herndl GJ, Baltar F (2022) Extracellular enzymatic activities of oceanic pelagic fungal strains and the influence of temperature. J Fungi 8(6):571. https://doi.org/10.3390/jof8060571

Anastasi A, Tigini V, Varese GC (2013) The bioremediation potential of different ecophysiological groups of fungi. Springer, Cham, pp 29–49. https://doi.org/10.1007/978-3-642-33811-3_2

Arantes V, Goodell B (2014) Current understanding of brown-rot fungal biodegradation mechanisms: a review. ACS Symp Ser 1158:3–21. https://doi.org/10.1021/bk-2014-1158.ch001
Bala A, Singh B (2019) Cellulolytic and xylanolytic enzymes of thermophiles for the production of renewable biofuels. Renew Energy 136:1231–1244. https://doi.org/10.1016/j.renene.2018.09.100
Barboza LGA, Lopes C, Oliveira P, Bessa F, Otero V, Henriques B, Guilhermino L (2020) Microplastics in wild fish from North East Atlantic Ocean and its potential for causing neurotoxic effects, lipid oxidative damage, and human health risks associated with ingestion exposure. Sci Total Environ 717:134625
Beškoski VP, Gojgić-Cvijović G, Milić J, Ilić M, Miletić S, Šolević T, Vrvić MM (2011) Ex situ bioremediation of a soil contaminated by mazut (heavy residual fuel oil)—a field experiment. Chemosphere 83(1):34–40. https://doi.org/10.1016/j.chemosphere.2011.01.020
Bianco L, Perrotta G (2015) Methodologies and perspectives of proteomics applied to filamentous fungi: from sample preparation to secretome analysis. Int J Mol Sci 16(3):5803–5829. https://doi.org/10.3390/ijms16035803
Browne MA (2015) Sources and pathways of microplastics to habitats. In: Marine anthropogenic litter. Springer, Cham, pp 229–244
Buyer JS, Teasdale JR, Roberts DP, Zasada IA, Maul JE (2010) Factors affecting soil microbial community structure in tomato cropping systems. Soil Biol Biochem 42(5):831–841. https://doi.org/10.1016/j.soilbio.2010.01.020
Chandra P, Enespa, Singh DP (2020) Microplastic degradation by bacteria in aquatic ecosystem. In: Microorganisms for sustainable environment and health. Wiley. https://doi.org/10.1016/B978-0-12-819001-2.00022-X
Chia RW, Lee JY, Kim H, Jang J (2021) Microplastic pollution in soil and groundwater: a review. Environ Chem Lett 19(6):4211–4224. https://doi.org/10.1007/s10311-021-01297-6
Cox KD, Covernton GA, Davies HL, Dower JF, Juanes F, Dudas SE (2019) Human consumption of microplastics. Environ Sci Technol 53(12):7068–7074
Cristaldi A, Fiore M, Zuccarello P, Conti GO, Grasso A, Nicolosi I, Copat C, Ferrante M (2020) Efficiency of wastewater treatment plants (WWTPs) for microplastic removal: a systematic review. Int J Environ Res Public Health 17(21):1–23. https://doi.org/10.3390/ijerph17218014
Cristorean C, Micle V, Sur IM (2016) A critical analysis of ex-situ bioremediation technologies of hydrocarbon polluted soils. ECOTERRA J Environ Res Prot 13(1):17–29. www.ecoterra-online.ro
Derraik JGB (2002) The pollution of the marine environment by plastic debris: a review. Mar Pollut Bull 44(9):842–852. https://doi.org/10.1016/S0025-326X(02)00220-5
Desnuelle P, Savary P (1963) Specificities of lipases. J Lipid Res 4(4):369–384. https://doi.org/10.1016/s0022-2275(20)40278-0
Dionisi D, Beccari M, Gregorio SD, Majone M, Papini MP, Vallini G (2005) Storage of biodegradable polymers by an enriched microbial community in a sequencing batch reactor operated at high organic load rate. J Chem Technol Biotechnol 80(11):1306–1318. https://doi.org/10.1002/jctb.1331
Doohan FM, Brennan J, Cooke BM (2003) Influence of climatic factors on *Fusarium* species pathogenic to cereals. Eur J Plant Pathol 109(7):755–768. https://doi.org/10.1023/A:1026090626994
Du H, Xie Y, Wang J (2021) Environmental impacts of microplastics on fishery products: an overview, vol 98. Gondwana Res, p 63
Eriksen M, Lebreton LC, Carson HS, Thiel M, Moore CJ, Borerro JC, Reisser J (2014) Plastic pollution in the world's oceans: more than 5 trillion plastic pieces weighing over 250,000 tons afloat at sea. PLoS One 9(12):e111913
GESMAP (2015) Sources, fate and effects of microplastics in the marine environment: a global assessment joint group of experts on the scientific aspects of marine environmental protection. 90:96

Harms H, Schlosser D, Wick LY (2011) Untapped potential: exploiting fungi in bioremediation of hazardous chemicals. Nat Rev Microbiol 9(3):177–192. https://doi.org/10.1038/nrmicro2519

He P, Chen L, Shao L, Zhang H, Lü F (2019) Municipal solid waste (MSW) landfill: a source of microplastics? Evidence of microplastics in landfill leachate. Water Res 159:38–45. https://doi.org/10.1016/j.watres.2019.04.060

Hofrichter M, Ullrich R, Pecyna MJ, Liers C, Lundell T (2010) New and classic families of secreted fungal heme peroxidases. Appl Microbiol Biotechnol 87(3):871–897. https://doi.org/10.1007/s00253-010-2633-0

Hou L, Majumder ELW (2021) Potential for and distribution of enzymatic biodegradation of polystyrene by environmental microorganisms. Materials 14(3):1–20. https://doi.org/10.3390/ma14030503

Hyde KD, Jones EBG, Leaño E, Pointing SB, Poonyth AD, Vrijmoed LLP (1998) Role of fungi in marine ecosystems. Biodivers Conserv 7(9):1147–1161. https://doi.org/10.1023/A:1008823515157

Imran M, Das KR, Naik MM (2019) Co-selection of multi-antibiotic resistance in bacterial pathogens in metal and microplastic contaminated environments: an emerging health threat. Chemosphere 215:846–857. https://doi.org/10.1016/j.chemosphere.2018.10.114

Jin Y, Lu L, Tu W, Luo T, Fu Z (2019) Impacts of polystyrene microplastic on the gut barrier, microbiota, and metabolism of mice. Sci Total Environ 649:308–317

Klemchuk PP (1990) Degradable plastics: a critical review. Polym Degrad Stab 27(2):183–202

Kwon OY, Kang JH, Hong SH, Shim WJ (2020) Spatial distribution of microplastic in the surface waters along the coast of Korea. Mar Pollut Bull 155:110729

Kyrikou I, Briassoulis D (2007) Biodegradation of agricultural plastic films: a critical review. J Polym Environ 15(2):125–150. https://doi.org/10.1007/s10924-007-0053-8

Lee MD, Swindoll CM (1993) Bioventing for in situ remediation. Hydrol Sci J/Journal Des Sciences Hydrologiques 38(4):273–282. https://doi.org/10.1080/02626669309492674

Lee MR, Powell JR, Oberle B, Cornwell WK, Lyons M, Rigg JL, Zanne AE (2019) Good neighbors aplenty: fungal endophytes rarely exhibit competitive exclusion patterns across a span of woody habitats. Ecology 100(9):e02790. https://doi.org/10.1002/ecy.2790

Li C, Busquets R, Campos LC (2020) Assessment of microplastics in freshwater systems: a review. Sci Total Environ 707:13557

Lin TC, Pan PT, Cheng SS (2010) Ex situ bioremediation of oil-contaminated soil. J Hazard Mater 176(1–3):27–34. https://doi.org/10.1016/j.jhazmat.2009.10.080

Liu H, Yang X, Liu G, Liang C, Xue S, Chen H, Ritsema CJ, Geissen V (2017) Response of soil dissolved organic matter to microplastic addition in Chinese loess soil. Chemosphere 185:907–917. https://doi.org/10.1016/j.chemosphere.2017.07.064

Lu K, Qiao R, An H, Zhang Y (2018) Influence of microplastics on the accumulation and chronic toxic effects of cadmium in zebrafish (*Danio rerio*). Chemosphere 202:514–520

Martínez A, Maicas S (2021) Cutinases: characteristics and insights in industrial production. Catalysts 11(10):1194. https://doi.org/10.3390/catal11101194

Napper IE, Bakir A, Rowland SJ, Thompson RC (2015) Characterisation, quantity and sorptive properties of microplastics extracted from cosmetics. Mar Pollut Bull 99(1–2):178–185

Natesan U, Vaikunth R, Kumar P, Ruthra R, Srinivasalu S (2021) Spatial distribution of microplastic concentration around landfill sites and its potential risk on groundwater. Chemosphere 277:130263. https://doi.org/10.1016/j.chemosphere.2021.130263

NEA (2014) Sources of microplastics to the marine environment. Norwegian Environmental Agency

Nikolaivits E, Kanelli M, Dimarogona M, Topakas E (2018) A middle-aged enzyme still in its prime: recent advances in the field of cutinases. Catalysts 8(12):612. https://doi.org/10.3390/catal8120612

Nunes CS, Kunamneni A (2018) Laccases-properties and applications. In: Enzymes in human and animal nutrition: principles and perspectives, vol 4, pp 133–161. https://doi.org/10.1016/B978-0-12-805419-2.00007-1

Patil NS, Kurhekar JV (2020) Optimization of protease production by *Bacillus isronensis* strain KD3 isolated from dairy industry effluent. Nat Environ Pollut Technol 19(3):1257
Periasamy D, Mani S, Ambikapathi R (2019) White rot fungi and their enzymes for the treatment of industrial dye effluents, vol 3. Springer, Cham. https://doi.org/10.1007/978-3-030-25506-0_4
Purohit J, Chattopadhyay A, Biswas MK, Singh NK (2018) Mycoremediation of agricultural soil: bioprospection for sustainable development. Springer, pp 91–120. https://doi.org/10.1007/978-3-319-77386-5_4
Ramezani M, Ghahari S, Ghahari S (2021) Microbial rejuvenation of polluted environment, vol 3. Springer
Ricardo IA, Alberto EA, Silva Júnior AH, Macuvele DLP, Padoin N, Soares C, Gracher Riella H, Starling MCVM, Trovó AG (2021) A critical review on microplastics, interaction with organic and inorganic pollutants, impacts and effectiveness of advanced oxidation processes applied for their removal from aqueous matrices. Chem Eng J 424:130282. https://doi.org/10.1016/j.cej.2021.130282
Sadri SS, Thompson RC (2014) On the quantity and composition of floating plastic debris entering and leaving the Tamar Estuary, Southwest England. Mar Pollut Bull 81(1):55–60
Sánchez C (2020) Fungal potential for the degradation of petroleum-based polymers: an overview of macro- and microplastics biodegradation. Biotechnol Adv 40:107501. https://doi.org/10.1016/j.biotechadv.2019.107501
Sarmah N, Revathi D, Sheelu G, Yamuna Rani K, Sridhar S, Mehtab V, Sumana C (2018) Recent advances on sources and industrial applications of lipases. Biotechnol Prog 34(1):5–28. https://doi.org/10.1002/btpr.2581
Sharma S, Chatterjee S (2017) Microplastic pollution, a threat to marine ecosystem and human health: a short review. Environ Sci Pollut Res 24(27):21530–21547
Smith M, Love DC, Rochman CM, Neff RA (2018) Microplastics in seafood and the implications for human health. Curr Environ Health Rep 5(3):375–386
Sucharitakul P, Pitt KA, Welsh DT (2021) Trophic transfer of microbeads to jellyfish and the importance of aging microbeads for microplastic experiments. Mar Pollut Bull 172:112867
Sun Q, Li J, Wang C, Chen A, You Y, Yang S, Li Y (2022) Research progress on distribution, sources, identification, toxicity, and biodegradation of microplastics in the ocean, freshwater, and soil environment. Front Environ Sci Eng 16(1):1–14
Thompson RC (2015) Microplastics in the marine environment: sources, consequences and solutions. In: Marine anthropogenic litter. Springer, Cham, pp 185–200
Tyagi M, da Fonseca MMR, de Carvalho CCCR (2011) Bioaugmentation and biostimulation strategies to improve the effectiveness of bioremediation processes. Biodegradation 22(2): 231–241. https://doi.org/10.1007/s10532-010-9394-4
UNEP Year Book (2013) Emerging issues in our global environment. United Nation Environmental Program, DEW/1565/NA, p 78
UNEP Year Book (2016) Emerging issues in our global environment. United Nation Environmental Program, DEW/1565/NA, p 78
Van den Berg P, Huerta-Lwanga E, Corradini F, Geissen V (2020) Sewage sludge application as a vehicle for microplastics in eastern Spanish agricultural soils. Environ Pollut 261:114198
Wagner M, Scherer C, Alvarez-Muñoz D, Brennholt N, Bourrain X, Buchinger S, Reifferscheid G (2014) Microplastics in freshwater ecosystems: what we know and what we need to know. Environ Sci Eur 26(1):1–9
Wang J, Liu X, Li Y, Powell T, Wang X, Wang G, Zhang P (2019) Microplastics as contaminants in the soil environment: a mini-review. Sci Total Environ 691:848–857
Williams J (2002) Bioremediation of contaminated soils: a comparison of in situ and ex situ techniques
Witayakran S (2008) Laccase in organic synthesis and its applications. https://smartech.gatech.edu/handle/1853/26665
Yu Y, Luan Y, Dai W (2022) Time evolution of protein corona formed by polystyrene nanoplastics and urease. Int J Biol Macromol 218:72–81

Yuan J, Ma J, Sun Y, Zhou T, Zhao Y, Yu F (2020) Microbial degradation and other environmental aspects of microplastics/plastics. Sci Total Environ 715:136968. https://doi.org/10.1016/j.scitotenv.2020.136968

Zhalnina K, Dias R, de Quadros PD, Davis-Richardson A, Camargo FAO, Clark IM, McGrath SP, Hirsch PR, Triplett EW (2015) Soil pH determines microbial diversity and composition in the park grass experiment. Microb Ecol 69(2):395–406. https://doi.org/10.1007/s00248-014-0530-2

Zolotova N, Kosyreva A, Dzhalilova D, Fokichev N, Makarova O (2022) Harmful effects of the microplastic pollution on animal health: a literature review. PeerJ 10:e13503

Zurier HS, Goddard JM (2021) Biodegradation of microplastics in food and agriculture. Curr Opin Food Sci 37:37–44. https://doi.org/10.1016/j.cofs.2020.09.001

Reports of Actinomycetes to Degrade Microplastic

28

Ria Mukhopadhyay and Koushik Biswas

Abstract

Microplastics, originating from a variety of polymers, range in size from 1 μm to 5 mm. Because of their tiny size, reduced density, structure, and endurance in the surroundings, microplastics spread throughout the world via ocean currents. Microplastics are microscopic sized particles used in cosmetics, homes and industrial cleaning products, and have accumulated in aquatic as well as terrestrial ecosystem. Microplastics can be consumed by numerous marine animals by mistaking them as phytoplankton and these will accumulate in the food chain causing serious health problems. Thus, the degradation of these microplastics is inevitable and it can be done in two ways—abiotic and biotic. The abiotic degradation of microplastics can be done by thermal, mechanical, and chemical means, while in biotic degradation, microbial enzymes work to breakdown polymers. Several reports suggest the role of fungi and bacteria in degrading the microplastics but actinomycetes, a Gram-positive bacteria present ubiquitously help in the biodegradation process of microplastics. A little amount of actinomycetes can form a layer of biofilm around the polymer and aids in its degradation viz. *Rhodococcus ruber. Rhodococcus rhodochrous* degrades the microplastics after getting treatment with abiotic factors. Actinomyces in combination with bacteria changes the physical, chemical structure of the microplastics polluting the environment which leads to its degradation. Microplastic deterioration in the environment is triggered with a pre-treatment of nitric acid before applying on the actinomycetes *Microbacterium paraoxydans*. Numerous

R. Mukhopadhyay
School of Agriculture, Swami Vivekananda University, Barrackpore, West Bengal, India

K. Biswas (✉)
Department of Agriculture, Brainware University, Kolkata, West Bengal, India
e-mail: dkb.ag@brainwareuniversity.ac.in

M. Shahnawaz et al. (eds.), *Microplastic Pollution*,
https://doi.org/10.1007/978-981-99-8357-5_28

actinomycetes break down the micron sized polymers by secreting such enzymes which has high degrading capability. Some novel marine-derived actinomycetes like *Streptomyces gougerotii, Micromonospora matsumotoense,* and *Nocardiopsis prasina* degrade different microplastics and use them as a source of carbon to produce biodegradable plastics.

Keywords

Microplastics · Degradation · Actinomycetes · Biotic degradation · Marine-derived actinomycetes

28.1 Introduction

The emergence of plastic represents a significant advancement in the realm of materials. The benefits of plastics in terms of cost, effectiveness, and adaptability have been increasingly clear as industrial technology has advanced. Today, a vast variety of plastic items are utilized often in daily life. However, only 20% of plastics out of total are reused and the remaining 80% eventually end up in the ecosystem of the land, rivers, and oceans (Trevor 2020). The plastic waste that accumulates in the environment is gradually broken down into smaller pieces under chemical, physical, or biological force, producing microplastic (MP) particles that are less than 5 mm in size (Thompson et al. 2004).

Marine plastic contamination has a detrimental impact on the economy, society, and environment. Aquatic animals can swallow plastic debris floating in the water, leading to digestive and nutritional issues (Oliveira et al. 2020). The types and levels of MPs contamination varied between lakes and rivers. PE, PP, and fibres are identified as the three main forms of MPs in fresh waterways, alike to the marine environment. Furthermore, variations in MPs in freshwaters linked to human activity were discovered. Due to the activity of people and companies that were close to freshwater sources, MPs pollution levels were high (Rezaniaa et al. 2018). When it comes to MPs pollution in biota, the bulk of consumed MPs is collected in biota's intestine. Due to their strong hydrophobic characteristics, small particle size, huge surface area, chemical stability, and capacity to carry environmental pollutants like heavy metals and antibiotics, MPs have the potential to concentrate, migrate, and disperse in the environment. According to Alimi et al. (2018) and Kumar et al. (2020), MP pollution is now a distinct category of environmental issue that must be addressed by the entire world. After entering the soil environment, MPs scatter throughout the soil due to biological disruptions dry and wet cycles, and soil management practices (O'Connor et al. 2019) which alters the physical and chemical characteristics of soils. Additionally, MPs have an impact on soil microorganisms, microbial activity, community structures, and microbial diversity (Hou et al. 2021). Additionally, MPs emit pollutants and have the capacity to absorb a variety of hazardous compounds, which worsen soil contamination and alter soil characteristics. Additionally, plastics have been found across the food chain levels, which is also causing complications in human health.

It is a great challenge to develop novel approaches and solutions to the issue of ongoing microplastic contamination. Apart from abiotic and chemical degradation, bioremediation is a possible method to attain the microplastic degradation, and it is vital to look for new viable, time-saving, and cost-effective techniques. Thus, evolving microorganisms, notably from marine habitats, can be studied successfully, for the focussed identification of microorganisms, specialised for plastic biodegradation and developing new, inexpensive degradation methods (Oliveira et al. 2020). MPs can be broken down by various microorganisms such as fungi, bacteria, actinomycetes, etc. Here, we report how actinomycetes degrade the microplastics which are a major concern to the environment and animals.

28.1.1 What are Microplastics?

Plastics are used extremely because of their exceptional rheological and mechanical qualities, which give high resistance, stability and durability, chemical inertness, malleability, low water permeability, light weight, and low cost. These factors contribute to the widespread production and use of plastics (Hopewell et al. 2009). However, the present practices of manufacturing and absorption by the mankind have adverse impacts on the environment (Savelli et al. 2017). Polypropylene (PP), which accounts for 21.1% of the global market, polyethylene (PE), which includes low-density polyethylene (LDPE), 19.9%, high-density polyethylene (HDPE), 16.1%, polyvinyl chloride (PVC), which accounts for 11.8%, polyethylene terephthalate (PET), accounts for 10.2%, polyurethane (PU), which accounts for 8.4%, and polystyrene (PS) 7.8%, are the common polymers that predominate worldwide (Geyer et al. 2017).

Plastics can be dangerous, especially when they are reduced to tiny particles called microplastics, which is a major environmental problem. Microplastics (fragments <5 mm) and macroplastics (particles >5 mm) are the important categories of aquatic wastes. Microplastics are small plastic particles, whose size varies from 1 μ to 5 mm, made from a variety of polymers. Microplastics are made from diverse molecules, have different molecular structures, and come from products with a variety of applications, such as tyres, textiles, and packaging. Microplastics are particles with a variety of sizes, shapes, and colors, which makes them distinct from pesticides and other chemical pollutants. Microplastics, as we refer to the varied group of contaminants that make up microplastic particles, are not just "microplastics" (Rochman et al. 2019). Microplastics are carried throughout the world by ocean currents due to their tiny size, structure, and persistence in the environment (Desforges et al. 2014). In addition, microplastics include plastic material of microscopic size, such as abrasives used in household, and industrial raw materials used in the manufacture of plastics, and plastic fibres and particles resulting due to the disintegration of larger plastic products. These particles can be found in a variety of ecosystems, such as freshwater, marine, and terrestrial ones.

They possess large molecular weight, are hydrophobic, and are extremely resistant to biodegradation. As a result, microplastics are regarded as emergent contaminants (Zhang et al. 2019).

Microplastics are categorized into two forms-primary and secondary and are derived from several sources. Microplastics are plastic particles smaller than 5 mm in size, as opposed to secondary microplastics, which are produced when macroplastics break down. The main microplastics are plastic particles used in various industry processes, plastic microbeads used in ingredients for cosmetics, and plastic powders used in molding. When products are used, such as clothes, and paints, or when they are improperly disposed of, secondary microplastics can build or the materials can deteriorate due to physical, chemical, and biological environmental conditions. Particles size lesser than 5 mm are produced as a result of the heat and photo oxidation reactions that impair the plastic characteristics (Correia et al. 2020). The marine environment contains microplastics in variable amounts along the water column, with varied polymers and physical properties. The sort of microplastic that predominates in surface waters—whether it is shards, pellets, styrofoam, or fibres—also varies greatly. The primary final locations for microplastics in the aquatic environment are deposition sites, including beaches (Turra et al. 2014). Microplastics can pose ecological risks after being dumped on a sand beach because they alter heat transfer, resulting in slow heating process and low maximum sedimentation temperatures than sediments without microplastics. This can affect the time it takes for eggs to incubate and hatch as well as the sex determination of several animals based on temperature. Moreover, microplastics have a hazardous impact on man as well as aquatic ecosystems, either by ingestion (into the food chain), hazardous chemical leaching, or the buildup of organic contaminants. The colonisation of microplastics by bacteria and algae can also result in the emergence of novel microbial species and habitats. The surrounding landscapes and new communities diverge greatly (Zhang et al. 2019). Microplastics can be linked to serious pollution in environment as they can be detected also in remote areas and are dispersed from their manmade pollution sources. The concentration of microplastics in seawater can exceed 1 lakh particles/m^3, despite the fact that they still make up a very modest portion of the total amount of plastics in the ocean (Correia et al. 2020). This growing concern is due to the fact that even when plastic discharges cease, macroplastic debris in the ocean will continue to degrade and persist for many years.

28.2 Different Forms of Microplastics

28.2.1 Various Microplastic Products

Microplastics as a class of contaminants are typically categorised as either primary or secondary and arise from a wide range of product categories. The size requirement for primary microplastics is less than 5 mm. Tiny pieces of plastic known as secondary microplastics are produced accidentally when larger plastic goods are broken up and fragmented by biological, physical, and chemical processes.

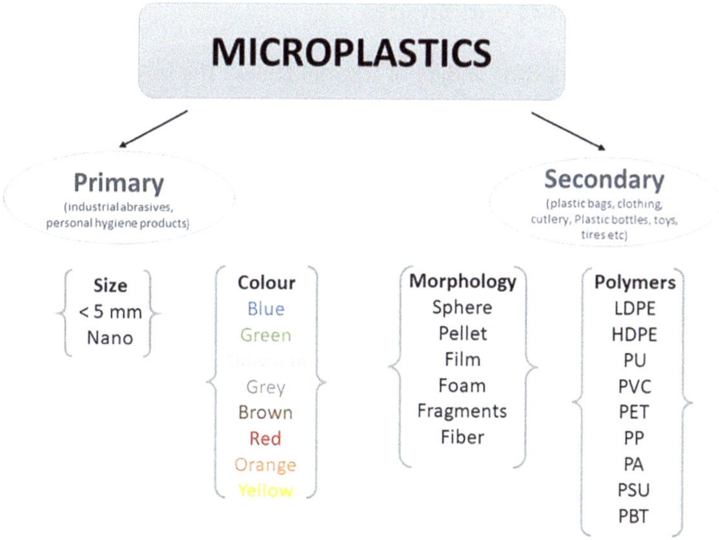

Fig. 28.1 Microplastics and their diverse forms, sizes, colour, morphology, and polymers

Secondary microplastics can develop during product use (such as the microfibers that come off of textiles after washing or wear particles from tyres) or when they are exposed in the environment (by fragmentation) (Gewert et al. 2015). Many plastic items, such as furniture, clothing, food packaging, agricultural products, building materials, and agricultural materials, can produce microplastics as a by-product (Fig. 28.1).

28.2.2 Microplastics of Different Sizes

According to the National Oceanographic and Atmospheric Association, they are often described as any plastic particle smaller than 5 mm in one dimension. Some advocate using the metric system to categorise sizes (e.g., 1–999 mm are microplastics). However, there is typically no lower bound, and this definition often includes nanosized polymers (0.1 mm). Several researchers report discrete sizes and size ranges in different ways. The diameters of the sieves used by researchers to size-fractionate samples allows them to be classified in a number of ways (Rochman et al. 2019). The ability to detect and identify ever-tinier microparticles is prompting researchers to broaden the scope of their analytical methods. The sizes of microplastics documented in the literature are becoming increasingly diverse, including a larger range that is bound by the lowest limit defined by sampling or analytical processes.

28.2.3 Various Colors and Shapes

Microplastics come in a wide variety of sizes and hues. The source of a microplastic can be determined by categorizing it based on its shape (Helm 2017). According to structure or morphology, researchers typically utilize between four and seven different categories, such as fibre, fibre bundle, fragment, sphere (or bead), pellet, film, and foam (Fig. 28.1). We are aware that specific shapes are often shed from distinct items, such as, pellets are usually linked with feeds obtained from industry, spheres may be microbeads from cosmetics or industrial leftovers, and foam comes from expanded polystyrene. Fibers and fibre bundles are frequently shed from clothing, upholstery, and carpet.

28.2.4 Different Types of Polymers

Microplastics are made up of a variety of different polymer kinds. All plastic polymers are made up of repeating monomers, which serve as the substance's structural core. The fundamentally dissimilar backbone structures of the various polymer types dictate the physical and chemical characteristics of plastics. Polypropylene, low-density polyethylene (LDPE), high-density polyethylene (HDPE), polyvinyl chloride (PVC), polyurethane, polyethylene terephthalate (PET), and polystyrene are the different polymers that are produced and mostly used (Plastics Europe 2017). In order to meet the needs of plastics' numerous applications, this diversity of polymers is required. For instance, PET is utilized in place of LDPE since it is too frail to be used in water bottles. Moreover, LDPE is frequently used in film, food packaging, and single-use shopping bags.

28.3 Impact of Microplastics on the Environment and Human Health

28.3.1 Effects of Microplastic on Soil

In order to evaluate the toxicological characteristics of microplastics (MPs), an assessment of MPs' effects on the soil environment is necessary. Multiple forms of MPs have been found in the soil of various nations across the world, including Australia, China, Chile, and Switzerland, with high concentrations of 67.5 g/kg (Zhou et al. 2020). The existence of MPs poses a concern to the soil ecology and ecosystems due of their lengthy life spans, persistence, and potentially harmful consequences. MPs obtained from soil have negative effects on the soil's ability to store water and cause the degradation of the soil's structure and infiltration capacity for irrigation and rainwater. The presence of microplastics in the environment drastically alters the structure of the soil, particularly its porosity. MPs are mostly found in agricultural soil due to air deposition, biosolids (sewage sludge), municipal waste (compost), leftover plastic shards from mulching, and plastic coat fertiliser.

A favorable temperature and moisture environment in the presence of microplastics can boost plant root exudation and promote root growth. Microplastics in the soil on a large scale reduce infiltration from irrigation and rainwater, having an undesirable effect on the soil's ability to retain water and potentially leading to oxygen deficiency. Additionally, these pollutants hinder evaporation, which can raise soil moisture levels. According to Siwek et al. (2015), by increasing the proportion of stable water aggregates in soil, MPs may eventually alter the physical characteristics and structure of the soil. The physical properties of soil, including its bulk density, pore structure, and capacity to store water, change when it is contaminated with MPs. According to Wan et al. (2019), MPs had an impact on soil shrinkage, cracking, and water evaporation. Additionally, increased water loss from the surface of the soil may have an effect on the leaching of pollutants into deeper soil layers. MPs of High-density polyethylene (HDPE) and polypropylene (PP) were found to have considerable adverse effects on the diversity of soil bacteria. It was observed that these MPs reduced bacterial population through altering the composition of bacterial communities. Additionally, some authors have demonstrated how MPs negatively affect soil microbial activity, organic carbon (C), nitrogen (N), and nutrient cycling. Conversely, by acting as a medium, MPs can increase the bioavailability of zinc and the likelihood that earthworms will come into touch with it; though, nothing is known regarding the potential risk to earthworms.

The presence of plastic waste in the soil has a significant impact on the soil's enzymatic activity, which reflect soil fertility. MPs had a significant impact on the catalase and urease activity. Due to the presence of MPs in the soil, the activities of urease, dehydrogenase, and FDase were decreased. Furthermore, it decreased the enzymatic activities involved in the pathways of soil like N-(leucine aminopeptidase), P-(alkaline-phosphatase), and C-(beta glucosidase and cellobiohydrolase) (Awet et al. 2018). Additionally, it was confirmed that a separate microbiota was discovered on the MPs' surfaces, which can alter the fundamental ecological functions and biogeochemical processes in the soil environment (Fig. 28.2).

According to Zhou et al. (2020), the MPs present in the soil have a detrimental effect on soil microbial activity. As MPs absorb harmful contaminants like toxic metal ions and hydrophobic organic pollutants on their surface, their accumulation in soil environments disrupt microbial activity. MPs may offer soil microorganisms an ecological habitat. MPs have the ability to enhance particular microbial populations and influence how plants and microbes can interact, creating microbial hotspots on MP surfaces (Zang et al. 2020) (Fig. 28.2). On the outermost layer of the MPs, colonies of microorganisms such as *Vibrio* spp. can be seen.

28.3.2 Impact of MPs on Plant

MPs have the potential to have a wide-ranging toxic effect on a variety of biochemical and physiological processes in plants, including delaying in seed germination, restricting plant growth and development, altering root characters, lowering

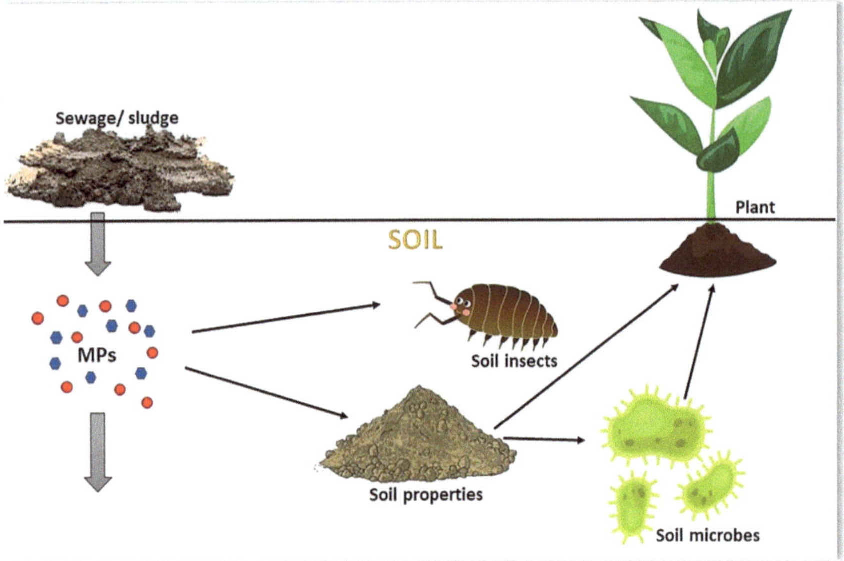

Fig. 28.2 Effect of MPs on soil

biomass, decreasing fruit yield, disrupting photosynthesis, leading to oxidative injury, and causing genotoxicity. MPs have the potential to bind to plant roots and alter their characters, which makes it more difficult for the plants to absorb water and nutrients. According to Jiang et al. (2019), MPs having smaller size are more detrimental to plants. MPs may enter the plant's body and injure it by altering the status of intracellular molecules and cell membranes and producing oxidative stress. MPs can interfere with plants' ability to absorb other compounds including Fe, Mn, Cu, and Zn.

The characteristics, variety, and the biological community of soil all have a significant role in how well the plant's function. MPs modify the physicochemical properties of the soil and the microbial populations, and these modifications may have a subtly negative effect on plants by changing the environment for plant growth, and the availability of nutrients. According to Wan et al. (2019), for instance, MPs considerably speed up the rate of soil water evaporation, which may cause soil dryness and perhaps harm plant performance. MPs may decrease the fertility of the soil and result in loss of nutrients in plants. Additionally, MPs may decrease the diversity of soil microbes or the quantity of rhizosphere fungal symbionts, which may then result in a reduction in plant diversity.

28.3.3 Aquatic Environment Pollution by MPs

MPs pollution is regarded as a severe problem for the aquatic environment and marine biota because of interactions with pollutants such as various types of heavy metals and hydrophobic pollutants. MPs contamination of marine habitats, including sediments, open waterways, and organisms, has been documented. According to Clark et al. (2016), it is yet unknown how MPs are dispersed and what will happen to them over time in the marine environment. It relies on a number of variables, including their chemical composition, density relative to sea water, ease of weatherability, compounds used in their manufacture, polymer nature, ecological effects, and fragmentation ability. MPs, which include microfibers, make up the bulk of the plastic found in the seas. For example, the average MP abundance in the Atlantic and Mediterranean seas, with recoverable depths varying between 1176 to 4843 m, is up to 1 MP per 25 cm^3. La Daana et al. (2018) found that MPs with a predominance of polyester and fibrous materials occur in the sub-surface waters (8–4369 m) in the Arctic central basin. Sea beaches are a source of extremely broken up plastic waste which carries MPs to the shore and then to oceans. Along the continental coast of Galapagos and eastern pacific, 240 samples of 16 different species of squid, shrimps, and fish which all are used for human consumption were gathered. All the marine creatures and water samples contained microplastic particles, and mainly they were tiny microplastics, within sizes of 150–500 m (Alfaro-Núñez et al. 2021).

MPs are extensively dispersed in the freshwaters, and since a few years ago, more focus has been given to both their existence and how it affects the ecosystem. They hover in the top layer of water, linger there, and eventually sink into the lake's sediments. Despite being more stable than water, sediments move through the earth at a slower rate compared to those floating in the water's surface (Rezaniaa et al. 2018). Therefore, it can be said that MPs pollution levels in sediments and the distance from the source of contamination are directly related. Lately, it has been noted that MPs are also polluting the drinking water.

Many aquatic creatures such as fish, turtles, and whales ingest the microplastics floating in the ocean or sea water which aggregate in their digestive system. There are substantial physical and toxicological effects on marine organisms when MPs are swallowed by them. Through consumption by marine species, MPs go into the food chain. When people eat animals from contaminated habitats, it puts their health and that of predators in danger (Rezaniaa et al. 2018).

28.3.4 Effect of MPs on Human Health

MPs move from one trophic level to another trophic level until they reach the top of the human food chain through direct absorption (by polluted food and beverages especially in plastic bottles) or indirect absorption (by polluted edible plants and animals) (Mercogliano et al. 2020). A significant number of MPs have been found in the daily diet in a variety of foods and drinks, including drinking water, beer, honey, and sugar. After ingesting MPs, infections, parasite vectors, chemical toxicity, and particle toxicity may all have a negative impact on a person's health. MPs ingested by humans has the potential to damage the variety and composition of the intestinal microbiota and result in diseases of the intestinal microbial population. It can also disrupt hormone balance and impact normal gastrointestinal functioning.

Agricultural soil microplastics are a serious threat to mankind because they get into the body through eating, inhaling, and skin contact and cause conditions including asthma, obesity, digestive issues, and cardiovascular issues (Yu et al. 2022).

28.4 Degradation of Microplastics

Microplastics, one of the most destructive agents of twenty first century, is constantly assimilating in our body due to our daily lifestyle habit and it eventually have the potential to bring about death of aquatic organisms when being ingested. Thus, it is an urgent need to find out a viable amendment technique that can mitigate the risk of health issues arises due to microplastics. Due to the rise in environmental issues in now a day, the implication of physical and chemical methods to break down plastic garbage is not to be the alternative measures to reduce plastic pollution. The biodegradation of plastic materials using bacteria and fungi has gained popularity recently. The fact of plastic biodegradation is more imperative because of their efficacious, affordability, friendly towards environment, and viability (Mukherjee and Chatterjee 2014). Numerous scientific research has shown that some fungi and bacteria have the ability to break down microplastics. This is done by releasing an enzyme called polyhydroxyalkanoate depolymerases, which speeds up the biodegradation process of microplastics (Anani and Adetunji 2021). Some reports confirmed that most of the recalcitrant microplastics can be digested partially by a number of microbes and some of them are *Ideonella sakaiensis*, having the potential to efficiently degrade PET (polyethylene terephthalate) (Joo et al. 2018). Since 1970s, certain strains of bacteria from the genera Penicillium, Brevibacterium, Aspergillus, Moraxella, Nocardia, Streptococcus, Streptomyces, Pseudomonas, Staphylococcus, and Bacillus have been utilized to break down plastic wastes (Jones et al. 1974). The genus contains microorganisms that can break down PET and LDPE, including Arenibacter and Marinobacter. The authors suggest *Alcanivorax borkumensis* as the most promising choice for decomposing plastic wastes produced from petroleum sources. Their active potentials are due to their capacity to develop dense biofilms on low-density polyethylene (LDPE) (Delacuvellerie et al. 2019).

In 2016, Divyalakshmi and Subhashini identified and screened a few microorganisms from various soil habitats to determine their ability to degrade polyethylene in MSM (mineral salt medium). Polyethylene experienced severe weight depreciation (13.6%) after 30 days of incubation at 37 °C. The Fourier-Transform Infrared (FTIR) examination results showed chemical alterations in the polyethylene characteristics as a result of the microorganisms' degrading activities. Among the 12 screened strains and the 32 isolated strains examined, the 16S rRNA micro-biome analysis performed on the Fourier Transform revealed that strain KX344032 (*Staphylococcus arlettae*) was the most active microbe. In conclusion, the scientists proposed that, in order to speed up the plastic decomposition, vigor genes isolation from *Staphylococcus arlettae* might play a significant role in advanced recombinant DNA biotechnology. It is well accepted among people that plastic and polythene wastes that are carelessly dumped in the environment constitute a major ecological issue. However, microbes have also been at the forefront of managing the environment sustainably by degrading these wastes (Divyalakshmi and Subhashini 2016).

In another examination, various microorganisms and fungi found during the screening in rubbish soil having their ability to degrade plastics and polyethylene. The population of actinomycetes, fungi, and heterotrophic microbial counts in polythene and plastics ranged from 72.54×10^4 to 64.75×10^4, 44.32×10^2 to 35.62×10^2, and 62.71×10^4 to 56.52×10^4, respectively, according to the result observed after 2, 4, and 6 months of incubation periods. The breakdown of polymers was discovered to be caused by the actinomycetes, fungi, and bacterial species *Streptomyces, Aspergillus flavus, Aspergillus nidulans, Staphylococcus* sp., *Bacillus* sp., and *Pseudomonas*. After a 6-month incubation period, the results of the bacteria's breakdown efficiency showed that the Pseudomonas strain had broken down 28.42% of plastics and 37.09% of polythene. Although the fungi strain Streptomyces degraded 35.78% of plastics and 46.16% of polythene wastes which established the conclusion that Streptomyces had a greater capacity for degrading plastics and polythene when corelated to different fungi and bacterial strains (Usha et al. 2011).

A review stated that plastic decomposition by microbes in marine habitat is happening under very cold temperature. The experts said that the majority of human waste found in the earth's crust is composed of man-made plastics. As a result of this occurrence, the impact of plastic on the marine ecosystem may cause the water resources, energy, and food to become unstable. Hence, plastic pollution recently has redirected to remarkable changes at global level. However, the problem infuses that there may be several methods that have potential in reducing the impact of plastic on the environment (Urbanek et al. 2018). However, recent research on plastic waste degradation claiming its major occurrence in cold environments. The advantages of isolating particular bacteria used in the eradication and breakdown of aquatic plastics were highlighted. It was determined that there should be a significant need for study on microorganisms, plastic debris, the arctic marine ecology, and human effect on the environment since the buildup of bio consortia of bacteria are still little known. This is because the plant and future earth are solely dependent on us. Without it, the ecosystem may become depleted and bacteria may not be able to adapt to the environment's permanent degradation by plastic garbage.

From a colonized and degradation prone area of polyethylene, a specific type of microorganisms have been identified, tested, and screened from coastal marine saltwater. A sharp inclement of non-weathered polyethylene (5.8%) and weathered polyethylene (3.7%) as compared to the controls (0.6%) was observed as a result of sequencing of 16S rRNA protein sequence after 2 days incubation period in the coastal water. After 9-day incubation period, a reduction in the relative abundance of the matured biofilms (0.3%) was observed. *Aestuariibacter* and *Roseobacter* are two closely related genera that considerably enhanced the early stage of colonisation with percentage averages of 1.8 and 26.9 in the consortia, respectively. This could be due to the existence of short chains of oxidation as a result of weathering process.

The findings of their research showed the pace at which early stage hydrocarbon-degrading bacteria colonised and improved consortia (Erni-Cassola et al. 2019).

Vague et al. (2019) reported that there has been a substantial increase in plastic wastes in landfills and coastal seas. They discovered few positive strains of Bacillus (two) and Pseudomonas (three) among the clusters involved in breakdown of PET in the littoral environment. According to the authors, when the positive bacterial consortia were placed next to the control (*E. coli*) during the incubation period under ultraviolet radiation and where carbon was the main source of wastes, there was an elevated tendency for the decomposition of LDPE, HDPE, and PET. However, the spectroscopic data showed that the bacteria and related biofilms caused molecular vibrations and surface degradation of the PET. This was also showing consistent results with the SEM (scanning electron microscope), which displayed an induced early stage of plastic breakdown. According to the authors' research, some Pseudomonas strains can break down LDPE quickly. In order to unlock latent potentials for future PET and related plastics chain breakdown, they concluded by recommending additional advanced lipase positive bacteria screening.

Brunner et al. (2018) identified several fungus strains from floating waste generated from the shoreline in an aquatic body and examined their capacity to degrade plastic garbage. According to them, these fungal strains can break down complex cellulose lignin, including C-polymers. Over 100 plastic-degrading fungi were isolated and examined, one genus Oomycota, was found to be a very effective degradation agent. Four fungal strains were identified in the findings of the biodegradation test on polyurethane and polyethylene, among which three were saprophytes, and one of them (*Leptosphaeria* sp.) was a plant pathogen (*Penicillium griseofulvum, Xepiculopsis graminea*, and *Cladosporium cladosporioides*).

28.5 Actinomycetes: A Potential Bio-Tool for Micro-Plastics Degradation

Actinomycetes, the microorganisms, well known for their ability to produce antibiotic, and numerous studies have demonstrated their potential for generating and degrading enzymes under in vitro condition. Actinomycetes are typically considered to be most active during the late stages of the breakdown of plants and other materials, playing a significant role in the degradation of polyester (Tokiwa and Pranamuda 2001). Recently, some information confirmed that actinomycetes can degrade polymer at high temperature regime (Kleeberg et al. 1998; Jarerat and Tokiwa 2001; Calabia and Tokiwa 2004). In order to study the distribution and population of microorganisms that break down polyesters at a temperature of 50 °C, the authors tried to extract the thermophilic actinomycetes from various regions and tested their polymer-degrading ability by using the clear zone methods.

The thermophilic actinomycetes, having the ability to degrade the polyesters PCL, PHB, and PES, were isolated from a variety of environments. Actinomycetes that degrade PCL are more common than those that degrade PHB or PES, and soil from the ground, gardens, and hot springs has the highest concentration of polyester degraders. The genus Actinomadura, Microbispora, Streptomyces, Thermoactinomyces, and Saccharomonospora were used to identify a variety of 31 isolates of actinomycetes that are having the depredation potential against PHB-, PCL-, and PES. As per the facts shared by the authors, the above report depicting the first review of thermophilic actinomycetes that degrade polyester from Taiwan and the first explanation of the breakdown of polyester in the genera Actinomadura and Saccharomonospora (Tseng et al. 2007).

Low-density polyethylene, or LDPE, is frequently used in plastics as a packing material. The continuous accumulation of plastic in the environment could pose a hazard to both the ecosystem and humanity. In order to prevent the accumulation of plastic and free the environment from plastic, microbes are segregated from the waste soil in the Andhra Pradesh and Telangana regions in India. To verify the degrading activity, these microorganisms are screened by following inhibition zone technique using polythene powder. As a result of microbial attachment to the polymer granules, isolated organisms are able to biodegrade the granules over time, causing physical and structural changes (Deepika and Jaya 2015). In another research work on actinomycete, 83 isolates were earlier isolated from an oil-contaminated soil sample at a naval dockyard in Visakhapatnam for the manufacture of biosurfactants and were examined in the lab for testing the degradation capacity on low density polyethylene (LDPE). Twenty of them showed significantly good characteristics using the viability indicator 2,3,5-triphenyl tetrazolium chloride (TTC) and they were able to flourish in a mineral salt medium supplemented with LDPE, the only source of carbon. On mineral salts media with emulsified polyethylene (carbon source), the NDYS-4 isolate produced the best results, which generates pink colour constituents namely triphenylformazan (TPF) within 5–7 days. Based on the findings, it is possible that enzyme activity and cell-surface hydrophobicity are both involved in this breakdown. The PE-degrading actinomycete selected for this investigation shown a variable degree of capacities to break down polyethylene and other polymers, and it can be used to clean up polyethylene-contaminated areas also (Duddu et al. 2015).

In a notable report on actinomycetes, hydrolase enzyme isolated from bacteria *Thermobifida fusca* under the group of actinomycete, can successfully depolymerize PET and which is typically considered to be "non-biodegradable". After incubation at 55.8 °C, erosion rates of 8–17 mm per week were observed. PET was not degraded by lipases from *Pseudomonas* sp. or *Candida antarctica* in similar circumstances. On the enzymatic attack on PET, PBT, and PHB, the effects of crystallinity, melting point, and glass transition temperature are discussed (Müller et al. 2005). A group of researchers conducted a study to find and isolate microorganisms in agricultural soils that might break down plastics. Soil samples were inoculated onto nutrient rich potato dextrose and starch–casein agar, in that order, for the isolation of bacteria, fungi, and actinomycetes. The most efficient degraders of PE plastic bags among the bacteria, actinomycetes, and fungi were the un-sequenced bacterial and actinomycete isolates, namely B1, A3, Aspergillus sp. (F7), B1 and A3 isolates. The retrospective study clarifies the knowledge and necessity for the bio-prospecting of agricultural soils, water bodies, and landfills that include plastic wastes, which may result in the identification of more effective microbial species with the capacity to digest plastics (Nakei et al. 2022).

28.6 Conclusions

Plastics have a significant impact on modern society and have elevated the standard of living due to their widespread use in carrying different products and packaging of food. Plastics, however, have a low biodegradability and persistent in the environment and have grown to be a significant cause of pollution. Microplastics which are petroleum based product and their use in daily life is becoming a matter of great concern to the plant and human health. Due to their small size, it can easily enter into every ecosystem and finally into the living system and finally show their adverse and toxic effects. Regarding the technologies now in use for managing plastic waste, the breakdown of plastics employing advantageous soil microorganisms has lately come to the front light because of their capacity to break down a variety of plastics, including polyethylene (PE) polymers. The degradation of PE plastics by various genera and species, including *Bacillus cereus* (bacteria), *Streptomyces werraensis* (actinomycete), *Aspergillus*, and *Phoma* sp. (fungi) and many other microbial species were identified and functionally characterized based on their degradation potential of plastics in the mentioned studies. Some required more research because they could not currently be identified. Some of the identified isolates, particularly the highly effective, have the potential to be used environmentally (in landfills) or industrially (in fermenters) to break down waste plastics. Actinomycetes in particular produce hydrolytic enzymes that let high molecular weight substances break down into simpler ones, enabling them to grow on various polymers. For instance, the bio-degradation of polymers has been linked to the genera Rhodococcus and Streptomyces. The degree of bio-degradation is often determined by changes in surface morphology and chemical composition, as well as dry weight loss. Thus, the identification of more effective microbial species with the capacity to digest plastics could result from the bio-prospecting of agricultural soils, aquatic bodies, and landfills that include plastic waste. As a result of these ecological threats, our future generations face a significant challenge. We must find a solution that ensure the environmental sustainability, so bioremediation with the help of available microbial community with biotechnological intervention could be a potential tool for the degradation of microplastics, a major global concern.

References

Alfaro-Núñez A, Astorga D, Cáceres-Farías L, Bastidas L, SotoVillegas C, Macay K, Christensen JH (2021) Microplastic pollution in seawater and marine organisms across the Tropical Eastern Pacific and Galápagos. Sci Rep 11:6424.10.1038/s41598-021-85939-3

Alimi OS, Farner Budarz J, Hernandez LM, Tufenkji N (2018) Microplastics and nanoplastics in aquatic environments: aggregation, deposition, and enhanced contaminant transport. Environ Sci Technol 52(4):1704–1724. https://doi.org/10.1021/acs.est.7b05559

Anani OA, Adetunji CO (2021) Bioremediation of polythene and plastics using beneficial microorganisms. In: Microorganisms for sustainability, vol 27. Springer. https://doi.org/10.1007/978-981-15-7459-7

Awet TT, Kohl Y, Meier F, Straskraba S, Grün AL, Ruf T et al (2018) Effects of polystyrene nanoparticles on the microbiota and functional diversity of enzymes in soil. Environ Sci Eur 30(1):11. https://doi.org/10.1186/s12302-018-0140-6

Brunner I, Fischer M, Rüthi J, Stierli B, Frey B (2018) Ability of fungi isolated from plastic debris floating in the shoreline of a lake to degrade plastics. PLoS One 13(8):e0202047. https://doi.org/10.1371/journal.pone.0202047

Calabia BP, Tokiwa Y (2004) Microbial degradation of poly(D-3-hydroxybutyrate) by a new thermophilic Streptomyces isolate. Biotechnol Lett 26:15–19

Clark JR, Cole M, Lindeque PK, Fileman E, Blackford J, Lewis C, Galloway TS (2016) Marine microplastic debris: a targeted plan for understanding and quantifying interactions with marine life. Front Ecol Environ 14(6):317–324. https://doi.org/10.1002/fee.1297

Correia J, João P, Lopes I, Duarte AC, Rocha-Santos T (2020) Environmental exposure to microplastics: an overview on possible human health effects. Sci Total Environ 702:134455. https://doi.org/10.1016/j.scitotenv.2019.134455

Daana L, Gårdfeldt KK, Lyashevska K, Hassellöv O, Thompson M, O'Connor RC (2018) Microplastics in sub-surface waters of the Arctic Central Basin. Mar Pollut Bull 130:8–18

Deepika S, Jaya MR (2015) Biodegradation of low density polyethylene by microorganisms from garbage soil. J Exp Biol Agric Sci 3:1–5

Delacuvellerie A, Cyriaque V, Gobert S, Benali S, Wattiez R (2019) The plastisphere in marine ecosystem hosts potential specific microbial degraders including *Alcanivorax borkumensis* as a key player for the low-density polyethylene degradation. J Hazard Mater 380(15):120899. https://doi.org/10.1016/j.jhazmat.2019.120899

Desforges JPW, Galbraith M, Dangerfield N, Ross PS (2014) Widespread distribution of microplastics in subsurface seawater in the NE Pacific Ocean. Mar Pollut Bull 79:94–99. https://doi.org/10.1016/j.marpolbul.2013.12.035

Divyalakshmi S, Subhashini A (2016) Screening and isolation of polyethylene degrading bacteria from various soil environments. IOSR J Environ Sci Toxicol Food Technol 10(12):01–07

Duddu MK, Tripura KL, Guntuku G, Divya DS (2015) Biodegradation of low density polyethylene (LDPE) by a new biosurfactant-producing thermophilic *Streptomyces coelicoflavus* NBRC 15399T. Afr J Biotechnol 14(4):327–340

Erni-Cassola G, Wright RJ, Gibson MI, Christie-Oleza JA (2019) Early colonization of weathered polyethylene by distinct bacteria in marine coastal seawater. Microb Ecol 79:1–10. https://doi.org/10.1007/s00248-019-01424-5

Gewert B, Plassmann MM, MacLeod M (2015) Pathways for degradation of plastic polymers floating in the marine environment. Environ Sci Process Impacts 17:1513–1521

Geyer R, Jambeck JR, Law KL (2017) Production, use, and fate of all plastics ever made. Sci Adv 3: e1700782. https://doi.org/10.1126/sciadv.1700782

Helm PA (2017) Improving microplastics source apportionment: a role for microplastic morphology and taxonomy? Anal Methods 9:1328–1331

Hopewell J, Dvorak R, Kosior E (2009) Plastics recycling: challenges and opportunities. Philos Trans R Soc 364:2115–2126

Hou J, Xu X, Yu H, Xi B, Tan W (2021) Comparing the long-term responses of soil microbial structures and diversities to polyethylene microplastics in different aggregate fractions. Environ Int 149:106398. https://doi.org/10.1016/j.envint.2021.106398

Jarerat A, Tokiwa Y (2001) Degradation of poly(tetramethylene succinate) by thermophilic actinomycetes. Biotechnol Lett 23:647–651

Jiang X, Chen H, Liao Y, Ye Z, Li M, Klobučar G (2019) Ecotoxicity and genotoxicity of polystyrene microplastics on higher plant *Vicia faba*. Environ Pollut 250:831–838. https://doi.org/10.1016/j.envpol.2019.04.055

Jones PH, Prasad D, Heskins M, Morgan MH, Guillet JE (1974) Biodegradability of photodegraded polymers I. Development of experimental procedures. Environ Sci Technol 8:919–923

Joo S, Cho IJ, Seo H, Son HF, Sagong HY, Shin TJ, Choi SY (2018) Structural insight into molecular mechanism of poly(ethylene terephthalate) degradation. Nat Commun 9:382

Kleeberg L, Hetz C, Kroppenstedt RM, Müller RJ, Deckwer WD (1998) Biodegradation of aliphatic–aromatic copolyesters by *Thermomonospora fusca* and other thermophilic compost isolates. Appl Environ Microbiol 64:1731–1735

Kumar M, Xiong X, He M, Tsang DCW, Gupta J, Khan E et al (2020) Microplastics as pollutants in agricultural soils. Environ Pollut 265:114980. https://doi.org/10.1016/j.envpol.2020.114980

Mercogliano R, Avio CG, Regoli F, Anastasio A, Colavita G, Santonicola S (2020) Occurrence of microplastics in commercial seafood under the perspective of the human food chain. A review. J Agric Food Chem 68(19):5296–5301. https://doi.org/10.1021/acs.jafc.0c01209

Mukherjee S, Chatterjee SA (2014) Comparative study of commercially available plastic carry bag biodegradation by microorganisms isolated from hydrocarbon effluent enriched soil. Int J Curr Microbiol Appl Sci 3(5):318–325

Müller RJ, Schrader H, Profe J, Dresler K, Deckwer WD (2005) Enzymatic degradation of poly (ethylene terephthalate): rapid hydrolyse using a hydrolase from *T. fusca*. Macromol Rapid Commun 26(17):1400–1405

Nakei MD, Misinzo G, Tindwa H, Semu E (2022) Degradation of polyethylene plastic bags and bottles using microorganisms isolated from soils of Morogoro Tanzania. Front Microbiol 13: 1077588. https://doi.org/10.3389/fmicb.2022.1077588

O'Connor D, Pan S, Shen Z, Song Y, Jin Y, Wu WM et al (2019) Microplastics undergo accelerated vertical migration in sand soil due to small size and wet-dry cycles. Environ Pollut 249:527–534. https://doi.org/10.1016/j.envpol.2019.03.092

Oliveira J, Belchior A, da Silva VD, Rotter A, Petrovski Ž, Almeida PL, Lourenço ND, Gaudêncio SP (2020) Marine environmental plastic pollution: mitigation by microorganism degradation and recycling valorization. Front Mar Sci 7:567126

Plastics Europe (2017) Plastics—The Facts 2017. PlasticsEurope, Brussels, Belgium. https://www.plasticseurope.org/application/files/5715/1717/4180/Plastics_the_facts_2017_FINAL_for_website_one_page.pdf. Accessed 12 Nov 2018

Rezaniaa S, Parka J, Md Dinb MF, Mat Taibb S, Talaiekhozanic A, Yadavd KK, Kamyab H (2018) Microplastics pollution in different aquatic environments and biota: a review of recent studies. Mar Pollut Bull 133:191–208

Rochman CM, Brookson C, Bikker J, Djuric N, Earn A, Bucci K, Athey S, Huntington A, McIlwraith H, Munno K, Frond HD, Kolomijeca A, Erdle L, Grbic J, Bayoumi M, Borrelle SB et al (2019) Rethinking microplastics as a diverse contaminant suite. Environ Toxicol Chem 38(4):703–711

Savelli H, Beunen R, Kalz M, Ragas A, Belleghem FV (2017) Solutions for global marine litter pollution. Curr Opin Environ Sustain 28:90–99

Siwek P, Domagala-Swiatkiewicz I, Kalisz A (2015) The influence of degradable polymer mulches on soil properties and cucumber yield. Agrochimica 59:108–123

Thompson RC, Olsen Y, Mitchell RP, Davis A, Rowland SJ, John AWG et al (2004) Lost at sea: where is all the plastic? Science 304(5672):838. https://doi.org/10.1126/science.1094559

Tokiwa Y, Pranamuda H (2001) Microbial degradation of aliphatic polyesters. In: Doi Y, Steinbuchel A (eds) Biopolymers, vol 3. Wiley-VCH, Weinheim, pp 85–103

Trevor ML (2020) Chapter 1—Introduction to plastic waste and recycling. In: Plastic waste and recycling. Academic Press, New York, pp 3–12

Tseng M, Hoang KC, Yang MK, Yang SF, Chu WS (2007) Polyester-degrading thermophilic actinomycetes isolated from different environment in Taiwan. Biodegradation 18:579–583

Turra A, Manzano AB, Dias RJS, Mahiques MM, Barbosa L, Balthazar Silva D et al (2014) Three-dimensional distribution of plastic pellets in sandy beaches: shifting paradigms. Sci Rep 4:4435. https://doi.org/10.1038/srep04435

Urbanek AK, Rymowicz W, Mirończuk AM (2018) Degradation of plastics and plastic-degrading bacteria in cold marine habitats. Appl Microbiol Biotechnol 102:7669–7678. https://doi.org/10.1007/s00253-018-9195-y

Usha R, Sangeetha T, Palaniswamy M (2011) Screening of polyethylene degrading microorganisms from garbage soil. Libyan Agric Res Center J Int 2(4):200–204

Vague M, Chan RC, Swartz NA, Mellies JL (2019) Pseudomonas isolates degrade and form biofilms on polyethylene terephthalate (PET) plastic. bioRxiv 2019:647321. https://doi.org/10.1101/647321

Wan Y, Wu C, Xue Q, Hui X (2019) Effects of plastic contamination on water evaporation and desiccation cracking in soil. Sci Total Environ 654:576–582. https://doi.org/10.1016/j.scitotenv.2018.11.123

Yu H, Zhang Y, Tan W, Zhang Z (2022) Microplastics as an emerging environmental pollutant in agricultural soils: effects on ecosystems and human health. Front Environ Sci 10:855292. https://doi.org/10.3389/fenvs.2022.855292

Zang H, Zhou J, Marshall MR, Chadwick DR, Wen Y, Jones DL (2020) Microplastics in the agroecosystem: are they an emerging threat to the plant-soil system? Soil Biol Biochem 148:107926. https://doi.org/10.1016/j.soilbio.2020.107926

Zhang M, Zhao Y, Qin X, Jia W, Chai L, Huang M et al (2019) Microplastics from mulching film is a distinct habitat for bacteria in farmland soil. Sci Total Environ 688:470–478. https://doi.org/10.1016/j.scitotenv.2019.06.108

Zhou Y, Liu X, Wang J (2020) Ecotoxicological effects of microplastics and cadmium on the earthworm *Eisenia foetida*. J Hazard Mater 392:122273

Algal-based Degradation of Microplastic: Prospects for Combating Microplastic Pollution

29

Rashidi Othman, Farah Ayuni Mohd Hatta, Nur Hanie Mohd Latif, Wan Syibrah Hanisah Wan Sulaiman, Razanah Ramya, Qurratu Aini Mat Ali, Nursyafica Nadia Johari, and Ainaa Eliah Abu Bakar

Abstract

Microalgae or microphytes are microscopic microorganisms that can be found in both freshwater and seawater. They can be described as a diverse group of unicellular and predominantly photosynthetic organisms. Eukaryotic or prokaryotic cyanobacteria (blue-green algae) can be classified as microalgae. The phycoremediation process uses microalgae to bioremediate (bioconcentrate, biotransform, sequester, assimilate, and biosorption) contaminants like heavy metals, phosphates, nitrates and the most recent microplastics from polluted waters in an eco-friendly manner without generating additional toxic compounds. It is a reliable alternative method for addressing microplastic pollution in a responsible, cost-effective, and practical manner when compared to conventional treatment methods. It also has the added benefit of lowering the carbon footprint. Several initiatives have been undertaken in the fields of immobilisation, pre-treatment techniques, genetic engineering, and other technologies to fully harness the potential of microalgae in phycoremediation and the production of

R. Othman (✉) · N. N. Johari
Herbarium Unit, Department of Landscape Architecture, Kulliyyah of Architecture and Environmental Design, International Islamic University Malaysia, Kuala Lumpur, Malaysia
e-mail: rashidi@iium.edu.my

F. A. M. Hatta · Q. A. M. Ali
Institute of Islam Hadhari, The National University of Malaysia, Bangi, Selangor, Malaysia

N. H. M. Latif · W. S. H. W. Sulaiman · A. E. A. Bakar
International Institute for Halal Research and Training, International Islamic University Malaysia, Kuala Lumpur, Malaysia

R. Ramya
Institute of the Malay World and Civilization, The National University of Malaysia, Bangi, Selangor, Malaysia

M. Shahnawaz et al. (eds.), *Microplastic Pollution*,
https://doi.org/10.1007/978-981-99-8357-5_29

value-added products (such as bioenergy and biofuel generation). Additionally, to address the ever-growing problem of water pollution in the world, particularly microplastic pollution, phycoremediation strategies must examine and develop hyperaccumulator algae species from pre-existing algal species. This integrated structure is a prime example of how to achieve numerous goals while implementing sustainable growth policies that address environmental, social, and industrial concerns, protecting natural resource, and simultaneously enhancing the quality of life for the world's growing population. Consequently, an in-depth research is necessary to analyse or produce hyperaccumulator algae species that could easily and efficiently tackle the issue of microplastic pollution.

Keywords

Biodegradation mechanism · Plastics · Phycotechnology · Microalgae · Cyanobacteria

29.1 Introduction

Global plastic production is predicted to have increased 4 times from 245 million metric tonnes in 2008–2050, accounting for 20% of global oil consumption (Tiseo 2021). According to Duru et al. (2019), thermoplastic resins accounted for approximately 67% of global plastic output in 2007, which was predicted to rise to 348 million metric tonnes in 2017. According to estimates, agricultural soils in Europe receive 125 to 850 tonnes of microplastics (per million residents) annually, totalling 44,000–300,000 and 63,000–430,000 tonnes of microplastics in North American and European farmlands, respectively (Rillig et al. 2017). According to Xia et al. (2021), Sanggou Bay's surface sediment had the highest concentration of microplastics (49.32 kg), followed by coastal waters (28.24 kg), while the bay's mouth had the lowest concentration (18.59 kg). The average amount of plastic consumed per person annually in Asia is 20 kg, and this is expected to increase. Asia consumes more plastic than any other region, accounting for nearly 30% of the world's total consumption (Hidayat et al. 2019), followed by America, Europe, and other continents. Since 2017, Malaysia has had the highest plastic pollution in the world, reflecting global trends in both the production of overall plastic garbage and the use of single-use plastics (Chen et al. 2021). Food waste is responsible for 45% of Malaysia's solid waste, followed by plastic (24%), paper (7%), glass (6%), and iron (6%) (Chua et al. 2019). The Malaysian plastic market is segmented into 7 main industries: agriculture, residential, packaging, electronics, construction, automobiles, and several sub-industries, such as medical equipment and plastic furniture (GESB 2011).

All marine environments, including seas, rivers, and freshwater bodies, are polluted with plastic on a global scale (Yonkos et al. 2014). Around 4.8–12.7 million tonnes of plastic are discharged into the sea annually, constituting 60–80% of all marine debris (Jambeck et al. 2015). Marine microplastics are mostly obtained from

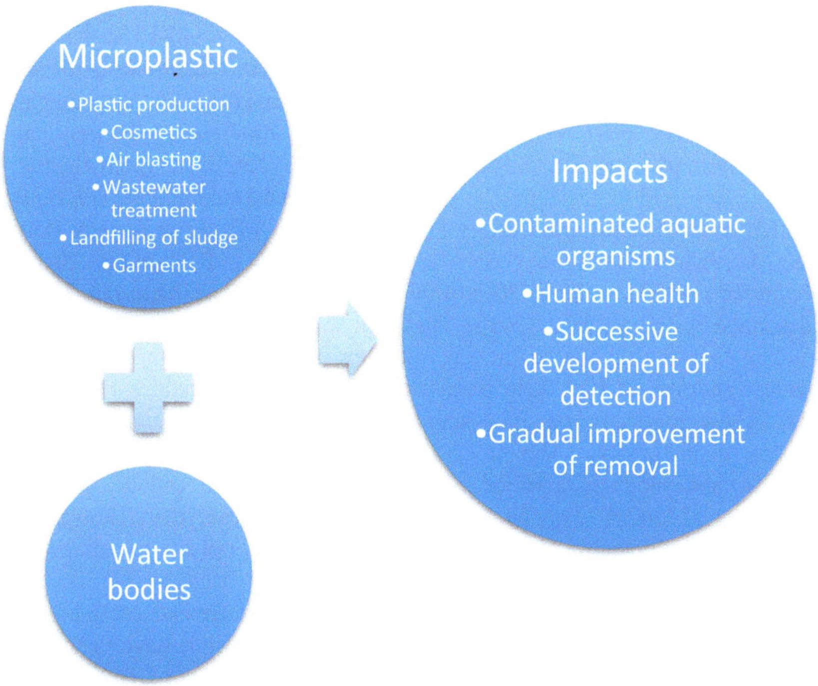

Fig. 29.1 Microplastics and their effect on the water bodies

primary and secondary sources. The primary microplastics refer to plastic particles that are added to cosmetics and personal care products, while secondary microplastics are produced through the physical, chemical, and biological degradation of large plastic items like tyre abrasion, road wear, and films (Jiang et al. 2022; Manzi et al. 2022). Large-scale plastic trash has caused white pollution in aquatic environments as well, destroying coral reefs and other marine life as well as wreaking mayhem on the environment. Figure 29.1 illustrates the effects of microplastic products that are generated from different sources like plastic manufacturing, sludge landfilling, cosmetics, wastewater treatment, air blasting, and garment removal; and discharged into the water bodies, which could affect human health, contaminate aquatic organisms, successive detection development, and progressive removal. Inhaling microplastic fibres has been linked to human lung tumours, and dispersive colours found in polyester and acrylic fibres have been linked to endocrine disruption, dermatitis, and hormonal imbalance (Anbumani and Kakkar 2018). The toxicological responses of marine species to different microplastics present a range of adverse effects, including the inhibition of cell development, increased oxidative stress, and decreased feeding rates (Guo et al. 2020).

The term "plastic" refers to a wide range of high molecular-weight organic polymers, most of which are generated from various hydrocarbon and petroleum derivatives (Ahmed et al. 2018). It can be divided into two categories: fossil- or bio-based, and non-biodegradable. The three types of bio-based plastics include modified natural polymers, synthetic bio-based polymers made from bio-based monomers, and bioplastics made from waste (Thielen 2019). On the other hand, non-biodegradable plastics are made of fossil- and bio-based polymers that are produced from hydrocarbon and petroleum derivative compounds (petrochemicals). Many plastics used in packaging, such as Polyethylene Terephthalate (PET), Polyvinyl Chloride (PVC), High-Density Polyethylene (HDPE), Low-Density Polyethylene (LDPE), Polystyrene (PS), Polypropylene (PP), Polyamide (PA), and Polyurethane (PUR), make up 90% of the world's plastic manufacturing industries, as described by Hidayat et al. (2019). These polymers comprise a variety of elements including carbon, oxygen, hydrogen, nitrogen, silicon and chloride, as well as chemical additives, which help them attain their desired qualities (Manzi et al. 2022). These materials are quite stable, and their rate of degradation prevents complete disintegration (Ahmed et al. 2018). Plastics produced from organic materials, and biodegradable or compostable plastic, are called "bioplastics" (Dijkstra et al. 2020). The bio-based and fossil-based polymers, however, are included in biodegradable plastics, depending on the level of microbial absorption and biodegradability.

Biodegradable plastics and other eco-friendly polymers are extremely popular in sectors like packaging, healthcare, and agriculture. Bioplastics have been used to wrap candies since the early nineteenth century; thus, this is not a recent development. They are expensive and derived from biological sources, so they have garnered a lot of relevance (Thiruchelvi et al. 2020). The four different types of biodegradable plastics include compostable plastics, bio-based plastics, photodegradable plastics, and biodegradable plastics (Chia et al. 2020). Bioplastics have derived a lot of relevance owing to their sustainability, and they are becoming more popular as a substitute for fossil-based plastics (Cinar et al. 2020). Bioplastics are marketed as one of the possible solutions to minimise pollution from plastic trash and petroleum-based single-use goods, and examples of "greenwashing" are associated with these emerging materials (Jankova 2018). Raw ingredients, including maize, sago, and food crops, are used to produce sustainable polymers with poor mechanical properties. Additionally, seaweeds include many natural polysaccharides that are used in pharmaceutical industries to make stabilisers, tablets, cosmetics, and bioplastic-based packaging materials (Thiruchelvi et al. 2020).

Due to the severe microplastic pollution present in the estuaries, it is recommended that plastic waste and microplastic interception devices must be set up on rivers to prevent microplastic intake. The different techniques used for decreasing plastic pollution, such as recycling plastic, incineration, chemical treatment, and landfilling, are ineffective (Chia et al. 2020). Numerous studies have investigated the abilities of various bacteria (using aerobic and anaerobic mechanisms) and microalgae (Bhuyar et al. 2019) for polymer biodegradation. The most common bacterial and fungal species employed for polymer biodegradation include *Pseudomonas aeruginosa*, *P. fluorescens*, and *Penicillium simplicissimum* (Raziyafathima et al. 2016). The monsoon season had a higher proportion of microplastics than the dry season, showing that plastic waste recycling must be intensified before the start of the rainy season (Jiang et al. 2022). Algae play a significant role in aquatic environments. During photosynthesis, they generate the highest amounts of oxygen and absorb the highest concentration of carbon dioxide (CO2) (Manzi et al. 2022). Microalgae may use their enzymes or toxins to break down plastics while absorbing plastic polymers as a source of carbon (Chia et al. 2020). Algae present a different, eco-friendly, and sustainable microplastic remediation method that combines physical and chemical characterisation techniques, biosorption, and bioaccumulation. The effectiveness of the algae-based bioremediation process is influenced by several factors, such as pH, ionic strength, temperature, contact time, and the presence of different counter-ions (Salama et al. 2019). As a significant marine trophic level producer, microalgae are essential for preserving the ecological balance of the ocean. However, the pervasive microplastics and heavy metals can inhibit the growth of microalgae cells and perhaps cause cell damage (Fu et al. 2019). To produce bioplastics, microalgae species that cannot be harvested easily, such as spirulina, are used. Seaweeds, for example, are a type of macroalgae that have more promise than the above-mentioned sources due to their large biomass, cost-effectiveness, adaptability to different environmental conditions, and year-round harvestability (Thiruchelvi et al. 2020). It was noted that the polyethylene sheets were colonised by different microalgae species like *Anabaena spiroides*, *Scenedesmus dimorphus*, and *Navicula pupula* (Manzi et al. 2022). Algae can degrade complex polymers and lack endotoxins, making them an ideal candidate for bioremediation, a technique for getting rid of microplastics (Yang et al. 2021). The process by which microplastics are degraded by cyanobacteria, and algae has not yet been fully understood.

29.2 Microplastic Pollution

When compared to other aquatic ecosystems, marine habitats are the most diverse. The contamination of marine ecology raises severe concerns that could result in serious harm to the ecosystem globally. Acidification, eutrophication, mine tailing, inappropriate sewage discharge, plastic littering, and all other harmful forms of coastal and marine habitats are the main causes of marine pollution, which is frequently induced by the discharge of land-based wastes (Nishaa et al. 2019). Wastewater treatment facilities are the main cause of these contaminations as their effluents and sludge contain concentrated microplastics from natural water sources, which then re-enter the environment (Tang 2020). The contaminants increased chemical concentrations in the environment, which could harm wildlife and people, and negatively affect the fishing, agriculture, tourism industries, and environmental health. Since 80% of the plastic pollutants end up in seas and oceans, plastic pollution significantly contributes to the negative effects that damage the marine ecosystem (Ebere et al. 2019; Howard 2019).

When the concentrations of plastic-based contaminants increased to >300 million tonnes, which seep into the sea through water runoff or are dumped in the water bodies, the problem of plastic pollution started to emerge in 2016 and has since grown into a major environmental concern (Pico et al. 2019). Given their extended lifespans and resistance to biodegradation, plastics are widely used, which indirectly increases their concentration in the environment, particularly in the seas and oceans. According to reports, the Pacific Ocean has a garbage patch covering about 620,000 square miles, and there are daily discharges of 50,000 to 15 million microplastic particles, weighing 4–12 million tonnes in the sea (Howard 2019; Pico et al. 2019). This indicates that plastic pollution is a severe issue that may have negative effects on the marine ecosystem and is predicted to outnumber fish by 2050 (Pico et al. 2019). Single-use plastic products that are still frequently utilised and discarded into the ocean are the main cause of plastic contamination. The amount of microplastics dumped into the environment has increased due to the increased usage of plastics in packaging, diverse products and raw materials. Microplastic pollution may originate from materials and products that use microplastics as an ingredient or may result from the breakdown of larger plastics (Tang 2020). The term "microplastic" refers to synthetic solid particles or polymeric matrices that originate from either primary or secondary manufacturing processes, have a variety of forms and sizes ranging from 1 to 1000 m, and are water-insoluble (Frias and Nash 2019). As reported by Bermúdez and Swarzenski (2021), microplastics are the most frequent form of plastic that is detected in the ocean. As a result of the physical and behavioural effects on aquatic species, the introduction of microplastics into the environment has unfortunately led to the toxicity of all living beings (Tang 2020). For instance, because the floating plastics resemble jellyfish and some of them smell like natural food owing to the chemicals that are emitted, many marine creatures and birds may mistakenly eat them, which may result in blockage of their digestive tracts. Additionally, some of the microplastics sink to the ocean floor, where they are consumed by corals (Howard 2019). Since microplastics include several additives and hazardous chemicals that percolate into the marine ecosystem, this will also disrupt the hormones and enzymes of the affected wildlife (Tang 2020). The aquatic ecosystem is not the destination of this microplastic waste. It gets transmitted and biomagnified up the food chain, which will undoubtedly have an impact on both plants and people.

In the future, everything related to human life, from feeder bottles and dentures to coffins, will be made of plastic. Products that are only intended to be utilised for a brief period harm the environment for many years as the waste generated from these products enters the food chain, is transferred from generation to generation, and occasionally affects the descendants of the people who had initially disposed of the waste. In the early 2000s, when the public became more aware of this fact, several scientists started investigating the effect of waste on wildlife, the environment, and eventually people. The current effects of plastic on the environment and human health are still not fully understood, but they may be extensive. Environmental consequences include the risk of ingestion (Setälä et al. 2014), wildlife entrapment, and habitat damage (Lamb et al. 2018), in addition to plastic-facilitated migration of the organisms to different ecosystems (McCormick et al. 2016; Viršek et al. 2017).

According to global data that are available, the production of plastics is increasing at the rate of 8% per year, from 21 million tonnes in the 1950s to the present production levels of >405 million tonnes. Asia accounts for 51% of the world's total production. Globally, 322 million tonnes of plastic were produced in 2014, with 58 million tonnes generated in Europe alone. In 2016, almost 380 million tonnes of plastic were produced, 40% of which was used for manufacturing disposable plastic containers. In the United States, the healthcare industry alone produces more than 1.7 million tonnes of plastic garbage annually. Around 13.2% of the total accumulated waste in the environment includes plastic-generated waste (Rizan et al. 2020).

29.3 Effect of Plastic Wastes on the Environment and Ecosystem

Plastic waste is found in every region around the world, and the sources are continuously increasing. Jambeck et al. (2015) predicted that by 2025, improper landfills in coastal nations will release over 28 million metric tonnes of plastic annually into the oceans. Additionally, as temporary sinks in the Arctic Sea ice thaw, microplastics from decades earlier that were trapped in these sinks are being released (Obbard et al. 2014). The proportions of plastic garbage in the ocean are rising, raising concerns on a global scale regarding the potential effects on the vulnerable environment. Wildlife frequently interacts with plastic waste in habitats containing higher concentrations of plastic waste. Issues like habitat loss and entrapment are well-known, but the effects of ingestion are far more complicated (Amin et al. 2020). According to estimates, 21% of seabirds and 45% of marine mammals ingest or become trapped in plastic trash (Chelsea et al. 2013). Habitat destruction can be attributed to the accumulation of plastic waste. Plastics also affect human habits. The abundance of plastic waste in the oceans shows that consumption of plastic products has significantly increased in the past few years. Recent studies show that 267 species are affected by plastic waste, while 44% of seabirds consume plastic waste (Melton 2022). It is anticipated that 99% of seabirds will have digested plastic garbage by 2050 (Bonanno 2022). Some methods of consumption that can have an impact on animal health include physical blockages of the digestive tract, biomagnification of the adsorbed poisons, and bioaccumulation of plastic particles. It is known that seabirds who eat plastic garbage die due to the blockage of their digestive tracts (Bonanno 2022). Morbidity is species-specific and affects most marine and maritime species at different ingestion quantities.

Native populations may be displaced due to invasive species and pathogens that migrate to sensitive habitats, due to the accumulation of plastic waste. Self-fertilising corals in the Netherlands, found near plastic wastes, provided evidence of microbial species migration (Póvoa et al. 2022). Microalgae and other undesirable species also proliferate in these environments (Tian et al. 2022). Plastic is a sought-after raft for animals due to its availability and sturdiness. In addition to offering a sanctuary for invasive species, it has been found that marine plastic samples strongly adsorb higher concentrations of environmental contaminants from the surrounding waters.

Plastic litter is linked to more than 75% of the chemicals on the US Environmental Protection Agency's (USEPA) list of potential hazards, including phthalates and DDT (a pesticide), which are detected at concentrations that are 100 to 1,000,000 times higher compared to those in the environment (Botterell et al. 2022). In another study, the researchers conducted a study of market fish in the United States and India and observed micro plastics or microfibers in 28% of fish, 33% of shellfish, and 55% of all the species (Rochman et al. 2015). Because synthetic clothing is so widely used in the USA, microfiber contamination was significantly more prevalent, raising questions regarding the effectiveness of wastewater treatment strategies used for their disposal. Within the range of polymers, one additive, one polymer, and two other polymers were identified as showing the largest potential effect on human health. Additionally, aquatic ecosystems have been the focus of most studies on plastic contamination, since it was considerably harder to detect and measure plastic contamination in soil. Soil is composed of several particles with a higher concentration of organic matter and a large surface area. Plastics can infiltrate the soil environment directly through biomass, atmospheric sedimentation, irrigation water, or inadvertently through the in-situ breakdown of substantial plastic chunks, such as those found in plastic mulch films. However, the field of plastic-related soil pollution has garnered a lot of scientific interest (Scheurer and Bigalke 2018; Wang et al. 2019; Chai et al. 2020; Sarker et al. 2020; Baho et al. 2021).

29.4 Effect of Plastic Wastes on the Human Health

When exposed for a lengthy period, certain polymers and the related additives are linked to endocrine disruption and carcinogenicity and can cause many cardiovascular, neurological, hepatological, and renal issues. Humans and other mammalian species are most exposed to contaminants through inhalation and ingestion. After being exposed to microplastics, particles are transferred, deposited, or excreted. Transport of micro- and nanoplastics involves the respiratory, lymphatic, and cardiovascular systems (Wright and Kelly 2017). The deposition of any material was noted in the mice's kidneys, liver, and stomach tissues (Deng et al. 2017). Inhaled polymeric particles have been associated with respiratory conditions, such as lung cancer, airway obstruction, nasal cavity cancer, respiratory diseases, and lung deposition (Prata 2018). Food products containing polymers and their related polymer additives may display neurological and psychological side effects, including abnormal maternal behaviour, neurotoxic responses, self-reported sexual dysfunction, and sex-dependent brain remodelling (Kundakovic et al. 2013). The effects of plastics on reproductive systems have been well-documented. For instance, many studies investigated the postulated adverse effects of the monomer, polymer, and plastic additives like bisphenol A, which acted as an androgen protagonist and estrogen agonist on the reproductive system, as it was seen to cause prostate cancer, ovarian cancer, breast cancer, lower the sperm count, and generally impaired the foetal development (Gao et al. 2015; Choi et al. 2022). Ingestion can also cause many harmful effects like metabolic diseases, diabetes, large bowel cancer, bladder cancer, liver disease, etc. (Guimond et al. 2022). Dermally applied polymers and their related polymer additives have also been linked to irritation without serious side effects; however, high-dose situations are rare. Additionally, consuming micro- and nano-plastics through polluted foods like honey, shellfish, sea salt, and wild-caught fish has been studied (Katsara et al. 2022; Nur et al. 2022; Wu et al. 2022), and even the excretion rates of the exposed individuals were determined (Katsara et al. 2022; Nur et al. 2022; Wu et al. 2022; Yang et al. 2022).

Plastic materials are a by-product of human activity and are not present in nature. Plastics are a commonly used material in various industries and daily life owing to their universal qualities, wide range of potential applications, and affordable production costs. Plastic materials also referred to as plastics are materials made of different chemical compounds. They are mostly composed of polymers, which refer to multi-molecular chemical compounds that are produced using the industrial polymerisation process. Plastics are synthetic organic polymers produced through the polymerisation of monomers that are derived from natural gas and petroleum sources. During the polymer production process, antioxidants, thermal stabilisers, different catalysts, and colouring agents are also added to the polymeric compounds. The plastic processing industry currently employs more than a hundred different polymers (chemically different), including Polyethylene (PE), Low-density Polyethylene (LDPE), Polypropylene (PP), and High-Density Polyethylene (HDPE), Polystyrene (PS), copolymer Styrene and Acrylonitrile (SAN), copolymer Acrylonitrile, Butadiene, and Styrene (ABS), and polyvinyl chloride (Chmielewski et al. 2022). The widespread use of plastics, which has turned into a major ecological and health issue, is the result of improper waste management of plastics (such as the dumping of PET bottles, bags, disposable bags, and disposable face masks), for instance in municipal waste dumps. Plastics are not detected easily and can be transported through the soil, water, and the atmosphere. The expenditures involved with trash management and plastic recycling, as well as the lack of social awareness, lead to a serious social issue. The management of plastic trash, which includes the burning of landfills, poses a serious threat to the environment as well as to human and animal health (Andrady 2017; Alimi et al. 2018; Rizan et al. 2020; Chmielewski et al. 2020; Walosik et al. 2021).

29.5 The Effect of Microplastics on Algae

Microplastics are small pieces (less than 5 mm in size) of plastic waste that can be produced artificially or naturally using chemical or physical processes (Liu et al. 2019). Since there is a significant increase in the concentration of microplastics in different environments, the possible risks of microplastic toxicity have been investigated in the past few years (Prata et al. 2019). In their study, Tunali et al. (2020) noted that *Chlorella vulgaris* initially exhibited a considerably higher growth inhibition rate but subsequently showed an adaptive response to microplastics, leading to an increase in the growth rate compared to the control samples. Despite this, Lyakurwa (2017) and Bergami et al. (2017) found that microplastics significantly affect the growth of algae. Additionally, a few researchers noted that the microplastic compounds could promote the growth of some algal species as they could be used as a substrate and carbon source (Yokota et al. 2017; Canniff and Hoang 2018). However, studies have indicated that microplastics can reduce the formation of Chlorophyll A and subsequently the rate of photosynthetic activity. Wu et al. (2019) examined the effects of PVC and PP microplastics at concentrations ranging from 5 to 500 mg/L on *C. pyrenoidosa* and *M. flosaquae*. It was discovered that the inhibition effect increased with an increasing microplastic concentration, with PVC having a more significant inhibitory effect on chlorophyll synthesis compared to PP (Wu et al. 2019). Additionally, Xiao et al. (2020) identified a connection between the effect of microplastics on algal growth. The researchers noted that the effect of microplastic concentration may differ based on their size, where the smaller-sized microplastics showed a greater detrimental effect compared to the larger-sized microplastics. Similar results were also reported by Hazeem et al. (2020) and Liu et al. (2019). One theory is that the adsorption capabilities of smaller microplastics at a small scale can either cause physical harm to the photosynthetic cells, as they alter the electron transport chain or obstruct the light source, thereby decreasing the light absorption (Wu et al. 2019). Although this phenomenon has been investigated in the past, it is difficult to determine the actual concentration and successive EC50 value of microplastics that could affect algae because the induced toxicity varies based on the algal species and the types and properties of available microplastics that can impact the adsorption of plastic and sunlight. Hence, additional research is required.

29.6 Algae

The algal group is diverse and encompasses several distinct taxa, ranging from primitive prokaryotic single-celled cyanobacteria to large multicellular eukaryotic algae (Tan et al. 2020). Microalgae are a type of eukaryotic organism, which have descended from the Kingdom Plantae. The two primary types of algae are macro and microalgae. In comparison to the higher and more complicated plant families, algae lack roots, stems, and leaves. Based on the pigments that are dominant in their cells, different algal strains display a variety of colours. The multicellular algae known as macroalgae (also called seaweeds) refer to a polyphyletic group of algal species that show a variety of colours like green, red, and brown. They typically inhabit the littoral zone to the depths with adequate light penetration for the photosynthesis process (Ruggiero et al. 2015; Serrano et al. 2019).

Plastic materials are a by-product of human activity and are not present in nature. Plastics are commonly used. On the other hand, microalgae are unicellular plants with vast biodiversity as more than 40,000 species have been detected after screening 200,000 and 800,000 species related to the organisms (Yin et al. 2020). Microalgae, also known as microphytes, are a heterogeneous category of unicellular autotrophs with a predominant photosynthetic apparatus. They refer to microscopic organisms that inhabit marine and freshwater ecosystems. The prokaryotic and eukaryotic cyanobacteria (blue-green algae) subgroups of microalgae have simple cell structures and can survive by consuming light, water, carbon dioxide, and nutrients such as nitrogen and phosphorus. Microalgae use these nutrients to transform them into energy for cell growth (Mutanda et al. 2020). Algae also have the characteristic attribute of maintaining chlorophyll in every cell to aid in photosynthesis (Singh and Singh 2015). Microalgae can generally be categorised into four categories, i.e. Chlorophyceae, Cyanophyceae, Bacillariophyceae, and Chrysophyceae. Microalgae from the Chlorophyceae category include organisms containing Chlorophyll A and thallus that do not differentiate into leaves, stems, or roots. Although microalgae lack the presence of complex organelles and structures, they implement the photosynthesis process using water, sunlight, and carbon dioxide. This can be attributed to the fact that they contain a photosynthetic pigment called chlorophylls (Tan et al. 2020). Three different environments—heterotrophy, autotrophy, and mixotrophy—can promote the growth of microalgae (Tan et al. 2020). Autotrophic microalgal species use light energy in addition to inorganic nutrients like carbon dioxide, nitrogen, and phosphorus to synthesise important biomass components like lipids, proteins, and carbohydrates. Both closed photobioreactors and open pond systems frequently use autotrophic cultivation. During heterotrophic cultivation, the organisms are cultivated in darkness in the presence of organic substances that can be utilised as a source of carbon and energy. Additionally, several microalgal species can thrive with glycerol and use glucose as a common carbon source. Furthermore, mixotrophic mode metabolism is described as a two-stage growth system, where Stage 1 includes a heterotrophic step that occurs in the presence of high organic carbon levels and Stage 2 is an autotrophic step that initiates with photosynthesis (is generally triggered due to low organic carbon content). Since microalgae can grow in both autotrophic and heterotrophic environments, under the right conditions (Khan et al. 2018), this technique may be effective in natural light and dark cycles.

29.7 Phycoremediation of Microplastics

An enormous amount of plastic garbage is polluting the environment, especially the marine environment, because of the rising global demand for plastic products. Since microplastics can infiltrate the food chain and result in several negative health effects, this critical problem can impact both marine and human lives. It appears that landfill, incineration, chemical treatment, and plastic recycling are not the best options for minimising plastic pollution (Wen et al. 2020). Bioplastics can be defined as polymers that are derived completely or partially from renewable or biomass sources, like food crops. They perform the same function as petroleum-based plastics (Mekonnen et al. 2013). The natural environment and the existence of various wild species have been affected due to the plastic age. The accumulation of plastic materials in aquatic environments is one issue that is becoming increasingly alarming. Microplastics are often referred to as plastic particles with a diameter <5 mm, and nanoplastics are particles with a size in the nanoscale (Steensgaard et al.

2017). Barnes et al. (2009) noted that in the past few decades, the abundance and global dispersion of microplastic fragments have dramatically increased while the average size of the plastic particles disposed in the environment has decreased. The amount of entrapped garbage that can eventually turn into marine plastic waste is mostly determined by the population density and effectiveness of the nation's waste management systems. Ordinarily, plastics account for 10% of the total waste that is discarded, but the effect is highly serious. The term "phycoremediation" refers to the process of using different types of algae to treat or remediate water pollution. As a result of this species' widespread distribution and environmental adaptation, it contains both macroalgae and microalgae, whether they are from freshwater or marine habitats. Because of their exceptional adaptogenic capacity, which allows them to thrive in a variety of conditions, microalgae are seen to have a high biodiversity, with 200,000 and 800,000 species found worldwide. This makes them ideal candidates for water treatment (Abinandan and Shanthakumar 2015; Hussain et al. 2021; Poonia et al. 2022). They can remove pollutants, including heavy and trace metal ions, pesticides, hydrocarbons, and organic and inorganic contaminants, through a variety of mechanisms, including biosorption, bioremediation, bioconcentration, and biotransformation (Phang et al. 2015; Koul and Taak 2018; Upadhyay et al. 2018). Phycoremediation sequesters toxins without generating secondary pollutants and offers tertiary bio-treatment solutions by generating biomass that can be utilised for a variety of uses, and hence, it is regarded as an environmentally beneficial and sustainable treatment strategy (Priyadharshini et al. 2021). This biological process not only aids in the removal of heavy metals, nutrients, and toxins but also enables heterotrophic aerobic bacteria to provide oxygen to mineralise the organic pollutants (Chaudhary et al. 2018). It is interesting to note that the biomass generated via sequestration contains significant fat and protein concentrations, which allows its application in animal feed, human nutrition, biofuels, aquaculture, and pharmaceuticals (Shields and Lupatsch 2012; Jais et al. 2017). Due to their fast-growing root systems, marine algae like *Ulva lactuca, Fucus vesiculosus, Gracilaria gracilis, Ascophyllum* sp., and *Sargassum* sp. are frequently utilised for rhizofiltration (Henriques et al. 2015; Koul and Taak 2018). The freshwater algal species like *Chlamydomonas*, *Chlorella,* and *Scenedesmus* genera have been used for phycoremediation because of their high capacity and adsorption mechanisms for eliminating microplastics (Priyadharshini et al. 2021).

Over the past few decades, extensive research has been done on microalgae for biotechnological uses, particularly their ability for biofuel production (Shanmugam et al. 2021). The metabolism-dependent biosorption methods for bioremediation typically comprise two steps: the attachment of pollutants to the cell surface and the passive or active transport of pollutant particles into the cells. Barone (2020) argued that the ability of plastics to serve as a source of nutrients for microorganisms is based on the oxygen content and metabolic processes of all species involved. Most of the plastic trash that is found on ocean surfaces or freshwater basins shows the presence of bacterial colonies. Cyanobacteria are generally found in nature and are involved in the ecological processes that occur in the biofilms on plastic debris (Oberbeckmann et al. 2015). The interactions between plastic waste and algal species affect the characteristics of plastic materials, such as density, biodegradation processes, and sinking behaviour. In addition, regardless of size, they may collect, control, and attach to the microplastic compounds on their surface. According to several researchers (Lucas et al. 2008; Dussud and Ghiglione 2014; Jacquin et al. 2019), the biodegradation of plastics can be summed up in four key steps:

(i) Adherence of microalgae to the plastic surfaces—This initiates the biodegradation process and changes the surface characteristics.
(ii) Biodeterioration—Algae secrete enzymes that are essential for the biodegradation of plastic.
(iii) Bio fragmentation—The plastic materials lose their mechanical stability and become fragile.
(iv) Assimilation—In the final stage, water and microbial filaments permeate the plastics, which leads to the degradation and use of plastic materials by the microbes.

As depicted in Fig. 29.2, the biodegradation of plastic materials by different algal species must be investigated further. The existing tactics included filtering, distillation, chemical degradation, and coagulation, while future strategies included microbial and enzymatic degradation (Fig. 29.2). Green algae and cyanobacteria constitute a significant proportion of phytoplankton, which are the organisms at the bottom of the trophic chain in freshwaters and play a role in the decomposition of plastic (Barone 2020). Kumar et al. (2017) presented an example of plastic bioremediation in freshwater, where the researchers collected waste PE bags from suburban water bodies and isolated diatoms, green algae, and cyanobacteria from the bags. A few researchers determined the ability of *Anabaena spiroides*, *Scenedesmus dimorphus*, and *Navicula pupula* to degrade PE wastes. It was discovered that the algae were able to more easily colonise the surfaces of LDPE sheets compared to HDPE sheets, where *A. spiroides* was seen to be the most effective microbe. Although all three of the aforementioned algae showed signs of biological LDPE degradation, *A. spiroides* showed an average of 8.18% mass loss per month (where the test set consisted of 1 cm^2 PE sheets, maintained at 27 °C), followed by *N. pupula* at 4.4%. The biodegradation of PE is more effectively accomplished by microorganisms exhibiting a higher ability to secrete ligninolytic and oxidative enzymes (Nayak and Tiwari 2011). Given that cyanobacteria depend on light sources to survive, the degradation processes implemented by these microbes are restricted to surface-accessible particles. Sarmah and Rout (2018) investigated the potential of two cyanobacterial species, i.e. *Phormidium lucidum* and *Oscillatoria subbrevis* (present in domestic sewage water) to degrade PE materials. It is believed that cyanobacterial interaction increases the surface hydrophilicity of PE materials as it generates carbonyl molecules, which is a key indicator of biodegradation. There was an increase in the carbonyl index of PE strips (1 cm^2, 20 μm thick) treated using the above cyanobacteria, which led to a 62% decrease in the crystallinity of the strips. Both the algal species utilised 3–4% of the total carbon content within 6 weeks. A 30% decrease in the weight of the pieces was noted after 42 days.

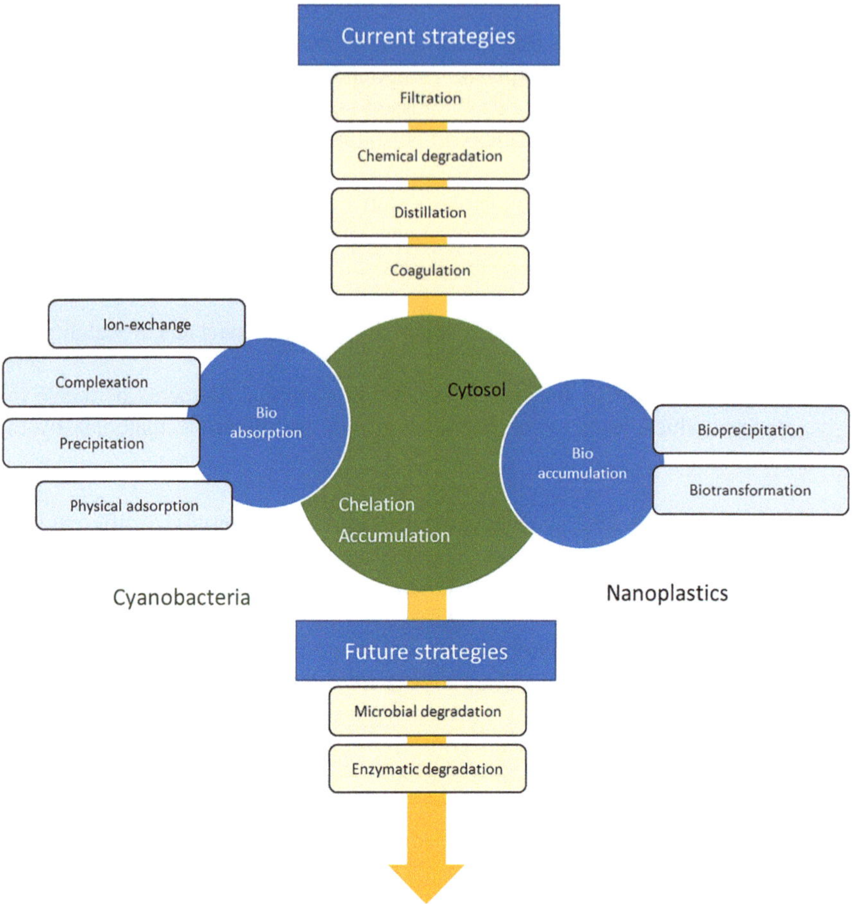

Fig. 29.2 Present and future strategies used for plastic bioremediation

Rillig et al. (2017) noted that earthworms can carry microplastic particles, such as PE beads, below the soil's surface and potentially into groundwater. The ability of *Chlorella reinhardtii*, a freshwater microalga, was studied in the PS microplastic powder solution at concentrations ranging between 0 and 100 mg/L in ultrapure water. The researchers noted lower photosynthetic activity and decreased chlorophyll fluorescence levels, which started recovering after around 10 days. The microplastic particles adhered to the microalgal surfaces, which affected the cell membrane. However, this might imply a potential cell-based strategy for lowering PS concentrations in water. Therefore, this raises the possibility that photosynthetic microorganisms, like cyanobacteria, can tolerate PS. In a different study, the researchers determined the effects of microplastic consumption by *Daphnia magna* (Canniff and Hoang 2018). Green algae, which adhere to plastic particles, are primarily consumed by zooplankton; hence, microplastics are detected in their digestive tract. *D. magna* were exposed to microplastic particles in moderately hard waters for a period of 3 weeks. In one study, the researchers noted that *D. magna* stored 0.44, 1.56, and 1.75 mg/L of microplastic particles after 5 days, when subjected to concentrations of 25, 50, and 100 mg/L. After 21 days, the values increased eight times. It is interesting to note that the reproduction and survival rates of *D. magna* were not significantly altered by the results presented by the researchers.

According to Tokiwa and Calabia (2004), several petroleum-based plastic materials are not biodegradable; which led to the production of biodegradable plastics like Poly(lactic acid) (PLA) and Poly(ε-caprolactone) (PCL). PHA-, PHB-, PLA-, and PCL-depolymerase, cutinases, esterases, ureases, proteinases, and dehydratases are a few examples of possible enzymes that could participate in the biodegradation of the above substances (Urbanek et al. 2018). Considering the biodegradation process, the discovery of cold-adapted bacterial strains displaying lipase activity may be significant. Additionally, it is anticipated that other enzymes isolated from cold settings and secreted by bacteria would exhibit biodegradable activity. Heterotrophic microorganisms belonging to the *Colvellia, Marinomonas, Pseudomonas, Pseudoalteromonas,* and *Shewanella* genera were detected in microbial strains that were recovered from the Arctic Sea ice in the Canadian Basin. A review regarding microbial ecotoxicology offered accurate observations on how contaminants and microbial communities interact with each other in relation to plastic contamination in the oceans. The study of microbial ecotoxicology focused on the effect of contaminants and microbial communities. Different populations are impacted by plastic pollution depending on whether the garbage is sunk or floating on the seafloor. The biofilms formed by the *Phormidium* and *Rivularia* genera of cyanobacteria are present near the water's surface, while the species from the Bacteroidetes (*Flavobacteriaceae*) and *Proteobacteria* genera dominate the microbiomes of the seafloor (*Rhodobacteraceae* and *Alcanivoracaceae*). The application of cyanobacteria in this field requires additional study of native pathways and associated enzymes.

Species that are known to play an important role in plastic degradation, such as *Pseudomonas, Corynebacterium, Arthrobacter, Streptomyces, Micrococcus*, and *Rhodococcus*, provide significant insights into important metabolic pathways. These serve as the foundation for expressing the potential of natural photosynthetic microorganisms. Because there are many different synthetic polymers in the market, biodegradation metabolic routes are extremely diverse. In this regard, the emphasis should be placed on model compounds for traditional and biodegradable plastics, such as PE, PS, PET, and PHA (Mintenig et al. 2017). Several metabolic pathways that were directly associated with the degradation of compounds including PE, PS, PET, and PHAs were determined. However, PE materials are not easily degraded as their structure contains balanced charges in the lengthy chains of carbon and hydrogen. Adenosine triphosphate (ATP), a critical compound for biomass synthesis and cellular growth, is also produced by the potential transformation via the tricarboxylic acid (TCA) cycle when a few products are secreted by competent wild-type or designed microbes. Some argue that the initial research on metabolically modified cyanobacteria that degrade plastic biodegradation must be conducted with the help of a culture-based technique because of the difficulties noted while dealing with natural environments and knowledge derived from earlier studies. As a result, the environmental pollution of microplastics and nanoplastics has garnered a lot of research attention. Many believed that the accumulation of plastic materials had negatively altered its appearance. Scientific reports have shown that there have been increasing concerns regarding plastic degradation since 2016. Plastic pollution in nature can be prevented or remedied promptly, which has positive effects on the environment and economy. The time needed for the anticipated quick environmental recovery from plastic pollution is not yet precisely equalled to microbe-related plastic bioremediation, which is yet understudied. However, further efforts should be taken to conduct a more thorough examination as well as to expand the limited data that are currently accessible in this situation.

29.8 Use of Algae as a Potential Agent for Microplastic Recycling

Petroleum-based synthetic polymeric plastic materials have been used in many applications. Since they have been derived from crude oil, which is a limited natural resource, they cannot be easily degraded by microbial activities in landfills. Furthermore, they release toxic compounds when incinerated. This leads to the conclusion that conventional plastic products are bad for the environment (Wang 2014; Venkatachalam and Palaniswamy 2020; Kalita et al. 2021). Recycling plastic is one way to address environmental issues. However, this procedure is hampered by numerous barriers. A viable alternative to traditional plastics is the development of bioplastics, which are at least partially biodegradable and less dependent on petroleum products (Venkatachalam and Palaniswamy 2020; Kanagesan et al. 2022). Bioplastics can be produced using a variety of natural polymers, including carbohydrates, proteins, and other molecules, such as disaccharides, sugar, and fatty acids. In the past, most bioplastics were produced primarily from PLA-based polymers and starch. Between 2014 and 2018, bioplastic production increased from 2 to 6.7 million tonnes (Cinar et al. 2020). Bioplastics derived from agricultural sources, however, compete with the demand for food and necessitate extensive crop management. In this section, the researchers have investigated the possibilities of using algae-based bioplastic materials as a substitute for traditional plastics.

Due to the high algal biomass yield compared to other biomass, easy cultivation processes, and biodegradability, algae-based bioplastic production is simple, inexpensive, and not dependent on petroleum products, with minimal effects on the food chain. Algae are significant biotic and diversified photoautotrophic organisms having a unicellular, chlorophyll-containing reproductive structure (Devadas et al. 2021). They thrive in freshwater and marine habitats and use solar energy to produce ATP (Lisha et al. 2022). Algae are characterised as a diverse class of organisms that range in size from unicellular organisms to enormous seaweeds. Algae are frequently found attached to a solid substratum or seafloor and are typically suspended in water columns, where they are known as phytoplankton. One habitat supports only one kind of algae. Microalgae, which are tiny algal colonies, must, therefore, be observed by microscopic analysis; in contrast, macroalgae are visible to the unaided eye. Microscopic plants known as microalgae can exist in both freshwater and saltwater habitats (Wang 2014). They were categorised into four groups: diatoms (Bacillariophyceae), green algae (Chlorophyceae), blue–green algae or cyanobacteria (Cyanophyceae), and golden algae (Chrysophyceae) (Devadas et al. 2021). They have a photosynthetic system that is similar to that of terrestrial plants; however, they can acquire carbon dioxide, water, and nutrients easily because of their submerged properties and basic cellular structure. This increases their capacity to convert solar energy into biomass. The primary components of microalgae are seen to be carbohydrates, lipids, and proteins (Cinar et al. 2020). Microalgae are generally cultivated for commercial activities; for instance, *Chlorella vulgaris* is utilised as a food supplement and *Haematococcus pluvialis*, a green alga, is seen to be a good astaxanthin source. Green algae and diatoms produce oils and carbohydrates.

Macroalgae, commonly referred to as seaweeds, are multicellular plants with thallus (leaf-like) forms that cling to a free-floating or hard substrate and are frequently mistaken for common aquatic plants. Based on their tissue pigmentation, the macroalgae are categorised into three categories, i.e. Rhodophytes (red), Chlorophytes (green), and Phaeophytes (brown) (Lisha et al. 2022). Studies have demonstrated the relevance of algal biomass and microalgae in biorefinery, culinary, and medicinal applications. Scientists are interested in the use of algal biomass as filler polymer preparations because of its useful environmental impacts. This method might be lucrative and likely to boost biodegradability (Kalita et al. 2021). The use of blue–green algae or cyanobacteria allows the utilisation of carbon compounds and decreases the greenhouse gas emissions from industries or power plants (Venkatachalam and Palaniswamy 2020). Algae-based bioplastics are described as bio-based polymers that are generally extracted from biomass, also referred to as agro-polymers (Wang 2014). Although most agro-polymers need to undergo chemical modification, some can be used to produce thermoplastic materials directly. They possess a few characteristics, such as hydrophilicity, rapid degradation rates, and poor mechanical qualities, particularly in moist conditions. The main technical difficulty is to properly modify the properties of these materials to overcome limitations such as water sensitivity, brittleness, and poor strength.

The main component of algae that contributes to polymer production is protein. Proteins are complicated heteropolymers that are composed of ≥20 amino acids (Wang 2014). Polymer chains are linked together by several interactions, including hydrophobic interactions, hydrogen bonding, and weak van der Waal forces. Strong covalent crosslinks can be used to further stabilise the structure once it has been folded. A vast array of biodegradable materials with differing functional qualities can be manufactured owing to the different protein building blocks and their distinctive architectures. Chain slippage is typically prevented by inter- and intra-molecular bonds, in addition to chain entanglements, which lead to the production of polymers with exceptional characteristics. In one study, Wang investigated the properties of the spirulina-extracted protein, notably, the mechanical, thermal, and morphological properties, and their probable use in bioplastic manufacturing (Wang 2014). Chlorella is often used in producing biomass-polymeric compounds, owing to their thick cell walls, higher crack resistance, and stable thermal characteristics (Cinar et al. 2020). Spirulina and chlorella are both used for making bioplastic composites because of their small cell sizes. Despite their commonalities, the various amino acid residues resulted in differing matrix behaviours (Cinar et al. 2020). *Spirulina* and *Chlorella* have been utilised to create HDPE thermoplastic blends, and the qualities of the resultant thermoplastic blends improved significantly after adding *Spirulina* (Simonic and Zemljic 2021).

Cyanobacteria display many benefits that are pertinent to the industrial sector, such as a rapid growth rate, high CO_2 utilisation, and a higher potential for genetic engineering. Polyhydroxyalkanoates (PHA) are biodegradable biopolymers that share many traits with synthetic polymers due to their large molecular weights and additional characteristics such as hydrophobicity and thermoplastic processability (Özcimenc et al. 2017). Different prokaryotic organisms can produce PHA, which are biopolyesters that can be stored within cells as insoluble granules, and used for storing carbon and energy. PHA is accumulated by two organisms, i.e. bacteria and cyanobacteria. PHA is a photoautotrophic molecule that can serve as an intracellular carbon and energy storage compound (Markl et al. 2019). PHAs are eco-friendly polymers that are utilised in many applications. They are biological materials that can be utilised in the delivery of drugs, as an implant, wound dressing, surgical suture, or other biomedical applications (Özcimenc et al. 2017). PHAs are also used for packaging, laminating, coating materials, films, and consumer goods. Poly-(hydroxybutyrate) (PHB), a homopolymer of hydroxybutyrate, is the most common type of PHA and is present in a variety of cyanobacteria, like *Aphanothece* spp., *Chlorogloea fritschii*, *Gloeothece* spp., *Gloeocapsa* spp., *Phormidium* spp., *Spirulina* spp., *Synechocystis* sp., and *Synechococcus* spp.

Marine organisms produce a wide variety of natural products such as polysaccharides. Özcimenc et al. (2017) and Lisha et al. (2022) have presented a list of natural polymers that were extracted from algal biomass, as follows:

(i) Polysaccharides—They are widely used in biomedical applications because of their biodegradability, water solubility, and functional activity. Numerous biological characteristics of sulphated polysaccharides have been identified, including immunoinflammatory, anticancer, antiviral, and anticoagulant activities. Polysaccharides are extracted from *Sargassum* sp., a type of brown seaweed.

(ii) Alginate—It is a polysaccharide that can be found in the cells of brown algal species like *Ascophyllum* sp., *Ecklonia* sp., *Lessonia* sp., *Laminaria* sp., and *Macrocystis* sp. The linear form of polyuronic acid is referred to as alginic acid. There are currently over 200 types of alginates, each with a varied concentration of α-L-guluronate and β-D-mannuronate, with a varying length of every block in different alginate sources.

(iii) Laminarin—It is the primary storage polysaccharide in brown seaweeds, accounting for 35% of the algal dry weight. It is a linear polysaccharide composed of 25–50 glucose units. It is derived from *Laminaria* sp. and *Saccharina* sp.

(iv) Fucoidans, also known as fucans, are a subclass of polysaccharides that largely consist of sulphated L-fucose along with $<10\%$ of other monosaccharides. It is primarily produced by brown seaweeds like *Ecklonia maxima* and *Sargassum muticum.*

(v) Carrageenan—It constitutes 30–75% of the algal dry weight of different species such as *Iridaea* sp., *Chondrus* sp., *Kappaphycus alvarezii*, *Eucheuma denticulatum*, and *Gigartina* sp. It is seen to be a principal component in red seaweed cell walls. They can be grouped into three main commercially-significant families—kappa, iota, and lambda—based on the number and position of the sulphate groups in recurring galactose units. The sulphate groups in carrageenan are essentially responsible for its chemical reactivity, while its variable structure accounts for its broad application.

(vi) Agar—It is a mixture of polysaccharides consisting of agarose and agaropectin and possesses many of the same structural and functional properties as carrageenan. Sulphated galactan, which is frequently formed by red algae, such as *Gracilariaceae* and *Gelidiaceae*, is the source of this heterogeneous polymer.

(vii) Ulvan—It is the main constituent of cell walls in green seaweed like *Ulva flexuosa*, *Ulva gigantina*, and *Cladophora glomerata*. Ulvans constitute about 8–29% of the algal dry weight. It is mostly composed of glucose, glucuronic acid, rhamnose, xylose, and sulphate, with trace amounts of arabinose, mannose, and galactose. There are 2 types of Ulvan, i.e. water-soluble and insoluble cellulose-like compounds. A larger ulvan concentration is found in the intercellular space and fibrillar wall, where it is uniformly distributed throughout the frond.

Unlike fossil-based plastics, microalgae-based bioplastics are produced for their easy degradability under both natural and industrial composting conditions (Simonic and Zemljic 2021). Starch-, polyhydroxyalkanoate-, PLA-, and cellulose-based plastics are some of the main types of bioplastics that are under development. PLA is a type of bioplastic material that has been investigated for its recyclability. Thermoplastic PLA is produced via the lactide ring-opening polymerisation or lactic acid condensation processes. PLA has strong mechanical strength and superior thermal characteristics in comparison to other fossil fuel-based polymers. Assimilation occurs when polymer fragments offer the microorganisms the required energy, carbon, and nutrients, where the carbon molecules in the plastic material are converted to water, CO_2, and biomass. Data showed that the biodegradability of this material, as measured by CO_2 generation, was similar to that displayed by cellulose-based thermoplastics (Simonic and Zemljic 2021). These results demonstrate the possibility of developing a unique and sustainable bioplastic material. Since algal biomass communities do not require any segregation from other waste, they are also easy to handle. Therefore, it is possible to utilise the algal biomass to develop bioplastics. The most popular technique for creating microalgae-polymer blends is compression moulding, wherein a mixture of polymers, biomass, and additives are placed into a mould and squeezed at high temperature and pressure for a brief period to generate bio-composites (Cinar et al. 2020). In some studies, the researchers use the heating and melt mixing processes at this step. The prototypes can take the form of films, dogbones, slabs, or rectangular flex bars. Another technique for creating film prototypes is solvent casting. A solvent is used to dissolve a mixture of additives, polymers, and biomass that is then cast onto surfaces and dried to create films The majority of studies focus on maximising component concentration and assessing the feasibility of the blended design (Cinar et al. 2020). It is common to employ lab-scale production techniques. However, more scalable techniques like twin-screw extrusion and injection moulding were also applied.

Algae are a great source of feedstock for manufacturing plastic or a possible recycling agent for microplastics, due to their ability to thrive in a wide range of conditions and generate a higher yield (Venkatachalam and Palaniswamy 2020). Waste stream-grown microalgae are the most affordable microalgae feedstocks that are used for a bio-refinery system. The use of algae in a biorefinery reduces production costs with a zero-waste approach (Özcimenc et al. 2017). It is possible to generate innovative bio-composite polymers by combining biomass or algal biopolymers with a different polymer. However, only 50% of algae was utilised in the manufacturing of algae-based bioplastics, and innovative progression is required to produce bioplastics made of algal composite materials (Venkatachalam and Palaniswamy 2020). High biopolymer yields can be achieved by genetically modifying specific algae species. The commercialisation of plastic made from algae exposed several technical problems that need to be resolved through biotechnological development.

29.9 Plastic Waste Treatment Strategies and the Algae Biodegradation Mechanisms

Due to the consumption of plastic and the resulting rubbish, the COVID-19 pandemic has shown higher environmental effects. This crisis has also highlighted the significance of plastic materials in day-to-day life. Single-use plastic is necessary to control the virus, even though disposability is typically seen as an environmental liability in other applications. Inadequate rubbish management has been a source of worry during the past century on a global scale. Due to the dramatic increase in COVID-19 cases worldwide, researchers have developed an evaluation technique called the Plastic Waste Footprint (PWF) to quantify plastic consumption. As a result, the energy and environmental footprints of these product systems have significantly increased (Kleme et al. 2020). New waste management challenges that arose during and after the epidemic are evaluated from the perspectives of innovative science and environmental regulations. The dramatic change in the type and quantity of waste products has emphasised the need for a waste management system that is dynamically responsive. Many methods have gained popularity for reducing plastic waste while maintaining sustainability, including recycling, pyrolysis, landfills, incineration, and bioremediation. The technique of bioremediation involves treating contaminated media, including water, soil, and subsoil materials, by altering the environment where organisms break down organic compounds, including plastic-made products (Ali et al. 2021).

The waste treatment situation includes a significant environmental load or material loss (Huysman et al. 2017). Ali et al. (2021) highlighted that polyethylene (36.3%) has the highest market percentage of all synthetic plastic types, followed by polypropylene (21%), PVC (11.8%), polyethylene terephthalate (10.2%), polyurethane (8.2%), and PS (7.6%). Due to its adaptability, excellent mechanical properties, affordability, and durability, PS is regarded as a flexible commodity in all industries. These materials are often used to make lab containers, shopping bags, food packaging materials, and power wire sheathing. These packaging materials are frequently dumped in landfills, severely polluting the environment. Sarmah and Rout (2020) assert that the disposal of synthetic polymer products in municipal solid waste and the degradation of such materials is extremely challenging and has not received enough attention. Furthermore, soil and aquatic environments commonly contain Bisphenol A, which is an important plastic component. Bisphenol A is used in the manufacture of polycarbonates, flame retardants, epoxy resins, and other products. These products are eventually disposed of in landfills, where Bisphenol A is generally present in higher concentrations in landfill leachates compared to the environmental water bodies (Hirooka et al. 2005). It is crucial to determine efficient eco-environmental solutions for treating plastic waste because conventional disposal methods have detrimental effects on the environment. Thermo-oxidative degradation, preceded by photodegradation, is used as the initial stage of spontaneous plastic breakdown that activates the energy required to initiate the incorporation of oxygen atoms into a polymer by ultraviolet radiation derived from the sun and is one potential benefit from environmental techniques used for plastic degradation (Webb et al. 2013).

Algae species have been used as bioremediation and biodegradation agents since they secrete mucilaginous extracellular polymeric compounds in sewage water for growing on synthetic substrates such as polythene surfaces. However, their growth is dependent on the enzymatic activities of intracellular and extracellular depolymerases (Sarmah and Rout 2020). Studies have demonstrated the effectiveness of synthetic biology techniques in developing environmentally safe microalgae-based microplastic breakdown processes. However, compared to other organisms, like fungi and bacteria, the significance of algae has received very little attention. However, abrupt and uncontrolled algal growth has been noted in numerous coastal places due to several environmental factors, such as climate change; specifically, eutrophication of salt water caused by farming and agriculture activities (Ali et al. 2021). Due to their ability to biosorb heavy metals and degrade harmful organic compounds, algae can help in the removal of contaminants or the treatment of wastewater (Hirooka et al. 2005). Yang et al. (2021) noted that microalgae are a good candidate as they lack endotoxins and do not need organic carbon sources to thrive under photoautotrophic conditions, as opposed to bacterial and fungal organisms that could be classified as biological pollutants owing to the endotoxins and need for a rich carbon source. In addition, algae are a commonly available, inexpensive resource that is derived from renewable marine resources that can be used to produce eco-friendly composite materials (Chiellini et al. 2008). Balaji et al. (2013) presented evidence regarding the involvement of algal species in the plastic biodegradation process (2013). The following studies have been cited regarding the viability of microplastic biodegradation processes using cyanobacteria and algae:

(i) After algal colonisation on polyethylene, surface deterioration was seen by the green alga *Scenedesmus dimorphus*, diatom *Navicula pupula*, and the blue-green alga *Anabaena spiroides*, each of which showed degradation percentages of 3.74, 8.18, and 4.44%, respectively.
(ii) The cyanobacterial species store intracellular PHB granules as the main source of carbon and energy in their cells. These PHB thermoplastics are eco-friendly, biodegradable, and biocompatible (Balaji et al. 2013; Ali et al. 2021).
(iii) Two cyanobacterial species, *Oscillatoria subbrevis* and *Phormidium lucidum*, were recognised for efficiently degrading PE compounds, and they formed 4% of their total carbon requirement. It is also a source of carbohydrates, vitamins, proteins, and antioxidants (Sarmah and Rout 2020).
(iv) When compared to medium-sized bacteria, *Chlorella fusca* was able to eliminate bisphenol A and generate many phenolic compounds, such as nitrophenol, chlorophenol, and bisphenol A (BPA; 2,2-bis (4-hidroksifenil) propana). It could also eliminate 90% of BPA under the 8:16-h light: dark conditions, which were almost as effective as the continuous light conditions (Hirooka et al. 2005).
(v) Freshwater non-toxic cyanobacterial species, such as *Phormidium lucidum and Oscillatoria subbrevis,* can colonise PE surfaces, effectively degrade LDPE without any pre-treatment, and develop on algal biofilms, under different contaminated aquatic environments, such as lakes, ponds, and wastewaters (Ali et al. 2021).
(vi) Due to its ability to bond with hydrophilic, environmentally friendly polymers like PVA, the green algae *Ulva armoricana,* isolated from the French coast, has received favourable reviews for its fibre content in composites that are created by compression moulding and casting aqueous suspensions (Chiellini et al. 2008).
(vii) The highest PE degradation rate was shown by *Anabaena spiroides* (Sarmah and Rout 2020).
(viii) Scanning electron microscopy analysis revealed that *S. dimorphus* adhered to the surface of LDPE surfaces, with minor changes (Sarmah and Rout 2020).

To date, very few researchers have investigated the effectiveness of using different algal species in pollution control. It was also seen that several fungal and bacterial species used to produce PETase, like *Ideonella sakaiensis* and *Pestalotiopsis microspore*, are not well adapted to coastal environments, which accumulate most of the plastic waste (Ali et al. 2021). Additionally, microalgae can be genetically engineered to function as a microbial cell factory that produces and secretes enzymes that degrade polymers. Both physicochemical (abiotic) and biological (biotic) processes can degrade plastic waste. According to Andrady (2017), the degradation of polymeric substances by mechanically oriented physical forces is typically the first stage in any environmental degradation process, which involves four primary mechanisms, i.e. photodegradation, hydrolysis, thermo-oxidative degradation, and biological degradation (Ali et al. 2021).

The filamentous algae could colonise the waste plastic surfaces because of environmental components that assist in algal growth, like nutrients, sunlight, and water (Ali et al. 2021). A novel strategy was proposed for degrading PE compounds with pro-oxidant additions where the microorganisms that can colonise the PE surfaces and are adhered to the material surfaces were cultured so that they could extract and use the PE oxidation products, while also growing and using a wide range of chemicals, primarily various fatty acid-like substances, as well as other functional groups including esters, alcohols, ketones, and double bond compounds (Sarmah and Rout 2020). In conclusion, the capacities of microorganisms, especially algae, in the remediation and biodegradation of plastic waste are of significant value. Additionally, many researchers have investigated the microbial ability of biodegrade plastic waste has been extensively demonstrated using several techniques, including enzymatic, spectroscopic, mechanical, thermal, and morphological techniques, including both biotic and abiotic control experiments. It is possible to isolate novel species with enhanced capacity for digesting plastic waste by studying several algae and cyanobacteria from various submerged polythene surfaces in the solid waste and sewage water disposal sites.

29.10 Microplastic Remediation Strategies

As a result of the public's growing awareness of microplastics, a category of plastics with a size smaller than 5 mm that were first discovered in 2004 and are currently being studied (Thompson et al. 2004). The maximum size of microplastics was seen to be 5 mm in most studies, but the minimum size may be 100 nm or 1 μm (Nguyen et al. 2019) because of the ambiguous barrier between MPs and nanoplastics (NPs) (Sobhani et al. 2020). Because microplastics may have a negative influence on aquatic microbes, their effects on biological treatment need to be investigated. Many researchers (Wei et al. 2019, 2020b; Zhang et al. 2020a, b) noted that microplastics, which include PET, PVC, PE, and PS, show several negative effects on aerobic and anaerobic processes, such as CH_4 and H_2 production, hydrolysis-acidification, and digestion of waste-activated sludge. An efficient way to lower the number of microplastics entering the environment is to take better precautions to lessen the amount that is released into the water. A contemporary remediation method helps in reducing the number of microplastics in water as well as degrading or eliminating any existing microplastics from the water, in addition to lowering the production and use of plastics. Several techniques can be utilised to minimise the number of microplastics, including microbial degradation, thermal treatment, physical techniques (adsorption and filtration), membrane technologies, and wastewater treatment facilities. Owing to the higher annual production and accumulation of plastics in aquatic environments, it is essential to have a better understanding of microplastics, remediation strategies, and precautionary performance.

29.10.1 Thermal Treatment

Thermal treatment is an effective technique used to degrade general wastes since it can reduce the need for landfill space and help to alleviate sanitary issues resulting from the use of waste soils in the past (Lombardi et al. 2015). This is due to the high calorific value and large production scale of plastic products that contribute to a viable source of high-density hydrogen and carbon compounds for the generation of fuel. Numerous reviews have explored the technologies used to convert plastic products into fuels, including gasification, liquefaction, pyrolysis, cracking in refinery units, and combined approaches like pyrolysis and reforming (Panda et al. 2010; Wong et al. 2015; Lopez et al. 2018; Nanda and Berruti 2021). According to Bazargan and McKay (2012), the carbon content in plastic wastes can be upgraded to value-added nanomaterials, including carbon nanotubes, which improve their physicochemical properties and their use in several applications. Recently, this technique has been applied to the conversion of microplastics. According to Tavares et al. (2018), PET was used in the co-gasification of microplastics and biomass at temperatures between 40 and 1200 °C. This co-gasification process included a variety of gasification agents, including oxygen, air, and steam. The composition of commonly used PET and biomass that produces better hydrogen yields has been demonstrated. As a result, it is claimed that the experimental settings, compositions, and ratios that need to be optimised are the factors that affect the various syngas use. Thermal pyrolysis can be used as an alternative to co-gasification and helps in converting microplastics into small molecules and value-added products. In their study, Cheng et al. (2020) asserted that high-pressure LDPE pyrolysis can be used to investigate the effect of pressure on fuel quality and identify how it may affect the distribution and product transformation processes. The fuel-like liquid products are composed of gasoline, aromatics, and isoparaffins. Thus, these liquid products contain fewer olefins and more cycloalkane compounds.

29.10.2 Microbial Degradation

This is an eco-friendly approach that can reduce costs and energy input while also using a carbon footprint. In the past, many researchers (Zheng et al. 2005; Taniguchi et al. 2019; Montazer et al. 2020) observed that bacteria use plastics as a carbon and nitrogen source to survive and proliferate. Bacterial and fungal organisms can break down polymeric compounds. Although the design does not include the use of microorganisms for microplastic degradation, they are still included based on their function.

29.10.3 Bacterial Species

Marine plastic compounds can be degraded by bacterial species (Roager and Sonnenschein 2019). The common bacterial species used for plastic degradation included several *Bacillus* sp., *Kocuria palustris*, and *Rhodococcus* sp. 36, belonging to the Actinobacteria family. However, after that, the degradation rates peaked at 30 days and lasted for a year, mainly at rates of 1–10% weight loss (Roager and Sonnenschein 2019). Additionally, the microbial consortium involved in the breakdown of plastic was identified for a single bacterium. Roager and Sonnenschein (2019) claim that bioaugmentation and acclimatisation of native microbial consortia that have been isolated from the environment are effective ways of degrading plastics.

29.10.4 Fungi

Many fungal species degrade microplastics using them as the sole source of carbon and energy, by secreting enzymes. Fungi aid in the breakdown of petroleum-based polymers (Sanchez 2020). Additionally, it has been discovered that fungi effectively grow on a wide range of plastic substrates, including Polyester Polyurethane (PPU) and polystyrene sulfonate, as well as PA, PET, LDPE, PP, HDPE, and PVC. Fungi can thrive in a variety of marine and terrestrial environments, including deep-sea and anoxic habitats, (Raghukumar 2017). They can even easily adapt to changing environmental circumstances. Natural microorganisms are not expected to break down microplastics, but they can be exploited to interact with them and build microbial communities known as plastispheres. The challenge for biochemists is that they need access to deep-sea bacteria to use the valuable resources from deep-sea microbes as a potential enzyme source (Wei et al. 2020a).

29.10.5 Physical Techniques (Adsorption and Filtration)

29.10.5.1 Adsorption on Green Algae

There has been an increased accumulation of microplastics in the aquatic environments, and this problem is more severe compared to other pollutants as they lead to harmful effects that can even lead to the death of marine birds, mammals, fish, and reptiles due to bioaccumulation and entanglement (Graham and Thompson 2009; Gregory 2009; Cole et al. 2011). Microplastics are categorised as persistent materials since they can disintegrate in different ways, based on their chemical makeup and behaviour. According to Rios et al. (2007), microplastics can easily adsorb into new habitats by absorbing a variety of aqueous pollutants on the surface. Nolte et al. (2017) observed that the PS particles (20–500 nm) were adsorbed by a type of unicellular green algae, *Pseudokirchneriella subcapitata*. This was attributed to the fact that positively charged microplastics could be easily adsorbed on negatively charged algal surfaces. The surface charge of the particle has a significant impact on the adsorption of microplastics in the algae. Martins et al. (2013) described the gelatinous properties of anionic polysaccharide substrates, where alginate increases the adherence of PS particles on seaweed surfaces. The positively charged microplastics are more effectively adsorbed to the anionic polysaccharide compounds on the algal surface.

29.10.5.2 Filtration Technologies

According to Lares et al. (2018), utilising a membrane bioreactor to remove microplastics (99.4%) is preferable to using a traditional activated sludge treatment system (98.3%). Gurung et al. (2016) claimed that using an aeration tank, for combining air with wastewater to activate microorganisms, and a sedimentation tank could help in separating the sludge and treated wastewater. For biological degradation and secondary purification, the typical activated sludge system is expected to boost activity. Additionally, many researchers (Murphy et al. 2016; Leslie et al. 2017; Mintenig et al. 2017) discussed how the comparison of processing steps and wastewater samples in the earlier works could be attributed to the observation of a narrow range of microplastic contents.

29.10.5.3 Membrane Technology

According to Lares et al. (2018) and Talvitie et al. (2017), membrane bioreactors have a higher capability to remove micro-sized plastics than basic dynamic membranes. According to Gurung et al. (2016), this approach is suitable for treating high-strength pollutants such as microplastics and polymeric debris since it is successful at eliminating a wide variety of complicated industrial wastewater. Talvitie et al. (2017) claimed that the removal of a variety of microplastics from the wastewater treatment plant discharges required the use of modern final-stage treatment technologies, including membrane bioreactors, fast sand filtration, disk filters, and dissolved air floating. The ability to remove microplastics is not solely dependent on shape but is also influenced by textile fibres, which were prevalent in both the effluents and influents during treatment and were easily eliminated. Microplastics could be easily removed from contaminated aquatic ecosystems using membrane technologies. The removal effectiveness is entirely dependent on the object's robustness, influent flux, size, and microplastic concentration. Biological processes used in addition to porous membranes help in increasing the propensity of elimination by up to 99.9%. The adsorption and filtering processes, which are typically used in conjunction with other techniques like sedimentation and biological processes, demonstrate superior efficiency for the treatment of microplastics, including wastewater. In most wastewater treatment facilities around the world, membrane bioreactors are used in tandem with other innovative physical–chemical treatments to achieve a greater capacity for waste elimination. However, it was noted that these facilities can also be used as sources of microplastics for aquatic environments.

29.10.5.4 Wastewater Treatment Plants

Although not specifically created for the removal of microplastics, each treatment step implemented in wastewater treatment facilities can assist in eliminating microplastics (Sun et al. 2019). The complete wastewater treatment process can be divided into primary, secondary, and tertiary treatments. The primary treatment strategy can individually eliminate 35%–59% and 50%–98% of microplastics. Ngo et al. (2019) observed that the sedimentation process occurring during the process could help in trapping the high-density microplastics in solid solutes during the grit removal and sedimentation steps for further sludge treatment. This process may be affected by density, which will allow the lower-density microplastics (PA, PP, and PE) to float while high-density microplastics (PET and PES) will settle down (Ngo et al. 2019). Biological treatment and clarification steps can be used to eliminate 0.2%–14% in the second stage. According to Sun et al. (2019), the bigger microplastic particles will be eliminated by allowing the accumulation of extracellular polymers from bacterial sludge flocs in the clearing tank. In the third step, the presence of microplastics can be reduced to 0.2–2% by applying several procedures. According to Talvitie et al. (2017), the tertiary treatment includes fast sandpaper, a membrane bioreactor, a disk filter, and dissolved air floatation. Thus, the membrane bioreactor shows a maximum removal efficiency of 99.9%. Poerio et al. (2019) stated that membrane bioreactors are more effective than other membrane processes, such as reverse osmosis, dynamic membrane technology, and ultrafiltration. The presence of microplastics will result in toxic oxidative stress, which will lead to the accumulation of a variety of hazardous compounds, including additives, heavy metals, and persistent organic pollutants (POPs), which slow down the enzymatic activity or diminish cell viability (Zhang et al. 2020c).

29.11 Conclusions

Even though extensive production and use of plastics and synthetic organic polymers initially started in 1950, it is now unthinkable to envision a world without them. Due to the extensive use of fossil fuels and increasing, uncontrollable global disposal of conventional, non-biodegradable plastics, researchers are under pressure to develop efficient methods to biodegrade plastic wastes and find alternatives to plastics to prevent marine plastic pollution. There is a need to prevent the extinction of some aquatic species since marine plastic pollution can harm and devastate marine life and its ecosystem. Legislation is needed to restrict the application of throwaway plastic products and microplastics to dramatically reduce the production and discharge of microplastics used for primary and secondary products. Secondly, methods for intercepting plastic and microplastic wastes must be implemented on the rivers to prevent their entry into the ocean, given the significant microplastic pollution detected in the estuaries. Thirdly, all local communities and national governments need to take a proactive approach to solve this common environmental problem of waste pollution by establishing viable policies and rules that will prevent microplastics from affecting the marine ecosystem. As a result, the removal of microplastic using algae may be conceivable with the application of an algomics technique, and extensive analysis of various enzymes is involved in the microplastic breakdown. It is fascinating to note that algae may grow on plastic surfaces, secrete plastic-degrading enzymes, and use plastic polymers as a source of energy and carbon for algal growth. Traditional plastics can be replaced by inexpensive, environmentally responsible microalgae-based bioplastics. Applying abiotic factors including salt, nutrient restriction, temperature, phytohormones, and light to encourage algae growth and enzyme activity may be able to boost microplastic decomposition. Many researchers are investigating the role of algae in the degradation of bioplastics, which could lead to a sizeable and sustainable production of bioplastics from algal biomass. Very few researchers have studied plastic degradation and bioplastic production by different algae species. This review presents an overview of different plastics and microplastics. This review also defines the social, ecological, and economic effects of plastic and microplastic pollution on cyanobacteria and algae growth. Then, the plastic biodegradation mechanisms used by different algal species have also been discussed. Thereafter, the biodegradation mechanism was presented, and the algal species involved in plastic pollution were identified. Then, the effect of algae on microplastics has been discussed. Finally, the potential of algal species as a source of bioplastics is thoroughly addressed.

References

Abinandan S, Shanthakumar S (2015) Challenges and opportunities in application of microalgae (Chlorophyta) for wastewater treatment: a review. Renew Sust Energ Rev 52:123–132

Ahmed T, Shahid M, Azeem F, Rasul I, Shah AA, Noman M, Hameed A, Manzoor N, Manzoor I, Muhammad S (2018) Biodegradation of plastics: current scenario and future prospects for environmental safety. Environ Sci Pollut Res 25(8):7287–7298

Ali SS, Elsamahy T, Al-Tohamy R, Zhu D, Mahmoud YAG, Koutra E, Metwally MA, Kornaros M, Sun J (2021) Plastic wastes biodegradation: mechanisms, challenges and future prospects. Sci Total Environ 780:146590

Alimi OS, Farner Budarz J, Hernandez LM, Tufenkji N (2018) Microplastics and nanoplastics in aquatic environments: aggregation, deposition, and enhanced contaminant transport. Environ Sci Technol 52(4):1704–1724

Amin RM, Sohaimi ES, Anuar ST, Bachok Z (2020) Microplastic ingestion by zooplankton in Terengganu coastal waters, southern South China Sea. Mar Pollut Bull 150:110616

Anbumani S, Kakkar P (2018) Ecotoxicological effects of microplastics on biota: a review. Environ Sci Pollut Res 25(15):14373–14396

Andrady AL (2017) The plastic in microplastics: a review. Mar Pollut Bull 119(1):12–22

Baho DL, Bundschuh M, Futter MN (2021) Microplastics in terrestrial ecosystems: moving beyond the state of the art to minimize the risk of ecological surprise. Glob Chang Biol 27(17): 3969–3986

Balaji S, Gopi K, Muthuvelan B (2013) A review on production of poly β hydroxybutyrates from cyanobacteria for the production of bio plastics. Algal Res 2(3):278–285

Barnes DK, Galgani F, Thompson RC, Barlaz M (2009) Accumulation and fragmentation of plastic debris in global environments. Philos Trans R Soc B Biol Sci 364:1985–1998

Barone GD (2020) Microalgae/cyanobacteria in biodegradation of plastics. In Encyclopedia. https://encyclopedia.pub/entry/5947

Bazargan A, McKay G (2012) A review—synthesis of carbon nanotubes from plastic wastes. Chem Eng J 195:377–391

Bergami E, Pugnalini S, Vannuccini ML, Manfra L, Faleri C, Savorelli F, Corsi I (2017) Long-term toxicity of surface-charged polystyrene nanoplastics to marine planktonic species Dunaliella tertiolecta and Artemia franciscana. Aquat Toxicol 189:159–169

Bermúdez JR, Swarzenski PW (2021) A microplastic size classification scheme aligned with universal plankton survey methods. MethodsX 8:101516

Bhuyar P, Muniyasamy S, Govindan N (2019) Biodegradation of plastic waste by using microalgae and their toxins. J Biotechnol Biomater 8:55

Bonanno G (2022) Marine-protected areas and plastic pollution. In: Plastic pollution and marine conservation. Academic Press, pp 249–273

Botterell ZL, Bergmann M, Hildebrandt N, Krumpen T, Steinke M, Thompson RC, Lindeque PK (2022) Microplastic ingestion in zooplankton from the Fram Strait in the Arctic. Sci Total Environ 831:154886

Canniff PM, Hoang TC (2018) Microplastic ingestion by Daphnia magna and its enhancement on algal growth. Sci Total Environ 633:500–507

Chai B, Wei Q, She Y, Lu G, Dang Z, Yin H (2020) Soil microplastic pollution in an e-waste dismantling zone of China. Waste Manag 118:291–301

Chaudhary R, Dikshit AK, Tong YW (2018) Carbon-dioxide biofixation and phycoremediation of municipal wastewater using Chlorella vulgaris and Scenedesmus obliquus. Environ Sci Pollut Res 25(21):20399–20,406

Chelsea MR, Mark AB, Benjamin SH, Brian TH, Eunha H, Hrissi KK, Richard CT (2013) Policy: classify plastic waste as hazardous. Nature 494(7436):169–171

Chen HL, Nath TK, Chong S, Foo V, Gibbins C, Lechner AM (2021) The plastic waste problem in Malaysia: management, recycling and disposal of local and global plastic waste. SN Appl Sci 3(4):1–15

Cheng L, Gu J, Wang Y, Zhang J, Yuan H, Chen Y (2020) Polyethylene high-pressure pyrolysis: better product distribution and process mechanism analysis. Chem Eng J 385:123866

Chia WY, Ying Tang DY, Khoo KS, Kay Lup AN, Chew KW (2020) Nature's fight against plastic pollution: algae for plastic biodegradation and bioplastics production. Environ Sci Ecotechnol 4: 100065

Chiellini E, Cinelli P, Ilieva VI, Martera M (2008) Biodegradable thermoplastic composites based on polyvinyl alcohol and algae. Biomacromolecules 9:1007–1013

Chmielewski J, Zeber-Dzikowska I, Luszczki JJ, Szajner J, Bartyzel M, Dziechciaz M, Chmielowiec B, Gworek B, Wojtowicz B (2020) Release of pollutants into the environment as a result of landfill fires and their impact on human health in the context of health education. Przemysl Chemiczny 99(8):1149–1154

Chmielewski J, Luszczki JJ, Zeber-Dzikowska I, Dziechciaz M, Czarny-Dzialak M, Gworek B, Wojciechowska M (2022) Microplastic in the environment. The role of education in raising social awareness of the handling of plastic waste. J Elem 27(1):31–45

Choi JY, Lee J, Huh DA, Moon KW (2022) Urinary bisphenol concentrations and its association with metabolic disorders in the US and Korean populations. Environ Pollut 295:118679

Chua HS, Bashir MJK, Tan KT, Chua HS (2019) A sustainable pyrolysis technology for the treatment of municipal solid waste in Malaysia. AIP Conf Proc 2124:020016

Cinar SO, Chong ZK, Kucuker MA, Wieczorek N, Cengiz U, Kuchta K (2020) Bioplastic production from microalgae: a review. Int J Environ Res Public Health 17(11):1–21

Cole M, Lindeque P, Halsband C, Galloway TS (2011) Microplastics as contaminants in the marine environment: a review. Mar Pollut Bull 62(12):2588–2597

Deng Y, Zhang Y, Lemos B, Ren H (2017) Tissue accumulation of microplastics in mice and biomarker responses suggest widespread health risks of exposure. Sci Rep 7(1):1–10

Devadas VV, Khoo KS, Chia WY, Chew KW, Munawaroh HSH, Lam MK, Lim JW, Ho YC, Lee KT, Show PL (2021) Algae biopolymer towards sustainable circular economy. Bioresour Technol 325:124702

Dijkstra H, van Beukering P, Brouwer R (2020) Business models and sustainable plastic management: a systematic review of the literature. J Clean Prod 258:120967

Duru RU, Ikpeama EE, Ibekwe JA (2019) Challenges and prospects of plastic waste management in Nigeria. Waste Disposal Sustain Energy 1(2):117–126

Dussud C, Ghiglione JF (2014) Bacterial degradation of synthetic plastics. CIESM Workshop Monogr 46:43–48

Ebere EC, Wirnkor VA, Ngozi VE (2019) Uptake of microplastics by plant: a reason to worry or to be happy? World Scientific News 131:256–267

Frias JP, Nash R (2019) Microplastics: finding a consensus on the definition. Mar Pollut Bull 138: 145–147

Fu D, Zhang Q, Fan Z, Qi H, Wang Z, Peng L (2019) Aged microplastics polyvinyl chloride interact with copper and cause oxidative stress towards microalgae Chlorella vulgaris. Aquat Toxicol 216:105319

Gao H, Yang BJ, Li N, Feng LM, Shi XY, Zhao WH, Liu SJ (2015) Bisphenol A and hormone-associated cancers: current progress and perspectives. Medicine 94(1):e211

GESB (2011) A study on plastic management in peninsular Malaysia. National Solid Waste Management Department Ministry of Housing and Local Government Malaysia, pp 1–282

Graham ER, Thompson JT (2009) Deposit and suspension-feeding sea cucumbers (Echinodermata) ingest plastic fragments. J Exp Mar Biol Ecol 368(1):22–29

Gregory MR (2009) Environmental implications of plastic debris in marine settings entanglement, ingestion, smothering, hangers-on, hitch-hiking and alien invasions. Philos Trans R Soc Biol Sci 364:2013–2025

Guimond AJ, Zevon ES, Tucker-Seeley RD, Giovannucci EL, Trudel-Fitzgerald C, Kubzansky LD (2022) Childhood socioeconomic status, healthy lifestyle, and colon cancer risk in a cohort of US women. Prev Med 107097

Guo Y, Ma W, Li J, Liu W, Qi P, Ye Y, Guo B, Zhang J, Qu C (2020) Effects of microplastics on growth, phenanthrene stress, and lipid accumulation in a diatom, Phaeodactylum tricornutum. Environ Pollut 257:113628

Gurung K, Ncibi MC, Fontmorin J-M, Sarkka H, Sillanpaa M (2016) Incorporating submerged MBR in conventional activated sludge process for municipal wastewater treatment: a feasibility and performance assessment. J Membr Sci Technol 6(3):1000158

Hazeem LJ, Yesilay G, Bououdina M, Perna S, Cetin D, Suludere Z, Barras A, Boukherroub R (2020) Investigation of the toxic effects of different polystyrene micro-and nanoplastics on microalgae Chlorella vulgaris by analysis of cell viability, pigment content, oxidative stress and ultrastructural changes. Mar Pollut Bull 156:111278

Henriques B, Rocha LS, Lopes CB, Figueira P, Monteiro RJ, Duarte ADC, Pardal MA, Pereira E (2015) Study on bioaccumulation and biosorption of mercury by living marine macroalgae: prospecting for a new remediation biotechnology applied to saline waters. Chem Eng J 281: 759–770

Hidayat YA, Kiranamahsa S, Zamal MA (2019) A study of plastic waste management effectiveness in Indonesia industries. AIMS Energy 7(3):350–370

Hirooka T, Nagase H, Uchida K, Hiroshige Y, Ehara Y, Nishikawa JI, Nishihara T, Miyamoto K, Hirata Z (2005) Biodegradation of bisphenol A and disappearance of its estrogenic activity by the green alga Chlorella fusca var. vacuolata. Environ Toxicol Chem 24(8):1896–1901

Howard J (2019) Marine pollution, explained: a wide range of pollution-from plastic pollution to light pollution-affects marine ecosystems. https://www.nationalgeographic.com/environment/article/critical-issues-marine-pollution. Accessed 27 Aug 2022

Hussain F, Shah SZ, Ahmad H, Abubshait SA, Abubshait HA, Laref A, Iqbal M (2021) Microalgae an ecofriendly and sustainable wastewater treatment option: biomass application in biofuel and bio-fertilizer production. A review. Renew Sust Energ Rev 137:110603

Huysman S, De Schaepmeester J, Ragaert K, Dewulf J, De Meester S (2017) Performance indicators for a circular economy: a case study on post-industrial plastic waste. Resour Conserv Recycl 120:46–54

Jacquin J, Cheng J, Odobel C, Pandin C, Conan P, Pujo-Pay M, Barbe V, Meistertzheim AL, Ghiglione JF (2019) Microbial ecotoxicology of marine plastic debris: a review on colonization and biodegradation by the "plastisphere". Front Microbiol 10:865

Jais NM, Mohamed RMSR, Al-Gheethi AA, Hashim MA (2017) The dual roles of phycoremediation of wet market wastewater for nutrients and heavy metals removal and microalgae biomass production. Clean Techn Environ Policy 19(1):37–52

Jambeck JR, Geyer R, Wilcox C, Siegler TR, Perryman M, Andrady A, Narayan R, Law KL (2015) Plastic waste inputs from land into the ocean. Science 347:768–771

Jankova V (2018) Bioplastics: opportunities and challenges. Central European University

Jiang Y, Yang F, Hassan Kazmi SSU, Zhao Y, Chen M, Wang J (2022) A review of microplastic pollution in seawater, sediments and organisms of the Chinese coastal and marginal seas. Chemosphere 286:131677

Kalita NK, Damare NA, Hazarika D, Bhagabati P, Kalamdhad A, Katiyar V (2021) Biodegradation and characterization study of compostable PLA bioplastic containing algae biomass as potential degradation accelerator. Environ Challenges 3:1–10

Kanagesan K, Abdulla R, Derman E, Sabullah MK, Govindan N, Gansau JA (2022) A sustainable approach to green algal bioplastics production from brown seaweeds of Sabah, Malaysia. J King Saud Univ Sci 34(7):102268

Katsara K, Kenanakis G, Alissandrakis E, Papadakis VM (2022) Honey quality and microplastic migration from food packaging: a potential threat for consumer health? Microplastics 1(3): 406–427

Khan MI, Shin JH, Kim JD (2018) The promising future of microalgae: current status, challenges, and optimization of a sustainable and renewable industry for biofuels, feed, and other products. Microb Cell Factories 17:36

Kleme J, Van Fan Y, Tan RR, Jiang P (2020) Minimising the present and future plastic waste, energy and environmental footprints related to COVID-19. Renew Sust Energ Rev 127:109883

Koul B, Taak P (2018) Biotechnological strategies for effective remediation of polluted soils. Springer

Kumar RV, Kanna GR, Elumalai S (2017) Biodegradation of polyethylene by green photosynthetic microalgae. J Bioremed Biodegr 8(1):1–8

Kundakovic M, Gudsnuk K, Franks B, Madrid J, Miller RL, Perera FP, Champagne FA (2013) Sex-specific epigenetic disruption and behavioral changes following low-dose in utero bisphenol A exposure. Proc Natl Acad Sci 110(24):9956–9961

Lamb JB, Willis BL, Fiorenza EA, Couch CS, Howard R, Rader DN, True JD, Kelly LA, Ahmad A, Jompa J, Harvell CD (2018) Plastic waste is associated with disease on coral reefs. Science 359(6374):460–462

Lares M, Ncibi MC, Sillanpaa M, Sillanpaa M (2018) Occurrence, identification and removal of microplastic particles and fibers in conventional activated sludge process and advanced MBR technology. Water Res 133:236–246

Leslie H, Brandsma S, Van Velzen M, Vethaak A (2017) Microplastics en route: field measurements in the Dutch river delta and Amsterdam canals, wastewater treatment plants, North Sea sediments and biota. Environ Int 101:133–142

Lisha VS, Kothale RS, Sidharth S, Kandasubramanian B (2022) A critical review on employing algae as a feed for polycarbohydrate synthesis. Carbohydrate Polymer Technol Appl 4:100242

Liu G, Jiang R, You J, Muir DC, Zeng EY (2019) Microplastic impacts on microalgae growth: effects of size and humic acid. Environ Sci Technol 54(3):1782–1789

Lombardi L, Carnevale E, Corti A (2015) A review of technologies and performances of thermal treatment systems for energy recovery from waste. Waste Manag 37:26–44
Lopez G, Artetxe M, Amutio M, Alvarez J, Bilbao J, Olazar M (2018) Recent advances in the gasification of waste plastics. A critical overview. Renew Sustain Energy Rev 82:576–596
Lucas N, Bienaime C, Belloy C, Queneudec M, Silvestre F, Nava-Sauced J-E (2008) Polymer biodegradation: mechanisms and estimation techniques—a review. Chemosphere 73:429–442
Lyakurwa DK (2017) Uptake and effects of microplastic particles in selected marine microalgae species; Oxyrrhis marina and Rhodomonas baltica. (Master's thesis, Retrieved from SINTEF)
Manzi HP, Abou-Shanab RA, Jeon BH, Wang J, Salama ES (2022) Algae: a frontline photosynthetic organism in the microplastic catastrophe. Trends Plant Sci 27:1159
Markl E, Grünbichler H, Lackner M (2019) Cyanobacteria for PHB bioplastics production: a review. In Algae. IntechOpen. doi: https://doi.org/10.5772/intechopen.81536
Martins MJF, Mota CF, Pearson GA (2013) Sex-biased gene expression in the brown alga *Fucus vesiculosus*. BMC Genomics 14(1):294–326
McCormick AR, Hoellein TJ, London MG, Hittie J, Scott JW, Kelly JJ (2016) Microplastic in surface waters of urban rivers: concentration, sources, and associated bacterial assemblages. Ecosphere 7(11):e01556
Mekonnen T, Mussone P, Khalil H, Bressler D (2013) Progress in bio-based plastics and plasticizing modifications. J Mater Chem A 1:13379
Melton C (2022) Development and evaluation of novel corrosion sensing & protective coatings based on polymeric powders for marine environments (Doctoral dissertation, University of Leicester)
Mintenig S, Int-Veen I, Loder MG, Primpke S, Gerdts G (2017) Identification of microplastic in effluents of waste water treatment plants using focal plane array-based micro-Fourier-transform infrared imaging. Water Res 108:365–372
Montazer Z, Najafi MBH, Levin DB (2020) Challenges with verifying microbial degradation of polyethylene. Polymers 12:123
Murphy F, Ewins C, Carbonnier F, Quinn B (2016) Wastewater treatment works (WwTW) as a source of microplastics in the aquatic environment. Environ Sci Technol 50(11):5800–5808
Mutanda T, Naidoo D, Bwapwa JK, Anandraj A (2020) Biotechnological applications of microalgal oleaginous compounds: current trends on microalgal bioprocessing of products. Front Energy Res 8(598803):1–21
Nanda S, Berruti F (2021) Thermochemical conversion of plastic waste to fuels: a review. Environ Chem Lett 19:123–148
Nayak P, Tiwari A (2011) Biodegradation of polythene and plastic by the help of microbial tools: a recent approach. Int J Biomed Adv Res 2:344–355
Ngo PL, Pramanik BK, Shah K, Roychand R (2019) Pathway, classification and removal efficiency of microplastics in wastewater treatment plants. Environ Pollut 255:113326
Nguyen B, Claveau-Mallet D, Herandez LM, Xu EG, Farner JM, Tufenkji N (2019) Separation and analysis of microplastics and nanoplastics in complex environmental samples. Acc Chem Res 52:858–866
Nishaa LJL, Arthanareeswaranb G, Poonguzhalia TV, Mohanc TA, Valentinaa J (2019) Phycoremediation of hydrocarbon using two marine seaweeds from the Bay of Bengal coast of India. Desalin Water Treat 156:378–386
Nolte TM, Hartmann NB, Kleijn JM, Garnæs J, van de Meent D, Hendriks AJ, Baun A (2017) The toxicity of plastic nanoparticles to green algae as influenced by surface modification, medium hardness and cellular adsorption. Aquat Toxicol 183:11–20
Nur AAU, Hossain MB, Banik P, Choudhury TR, Liba SI, Umamaheswari S, Yu J (2022) Microplastic contamination in processed and unprocessed sea salts from a South Asian country and potential risk assessment. Chemosphere 308:136395
Obbard RW, Sadri S, Wong YQ, Khitun AA, Baker I, Thompson RC (2014) Global warming releases microplastic legacy frozen in Arctic Sea ice. Earth's Future 2(6):315–320

Oberbeckmann S, Loeder MG, Gerdts G, Osborn AM (2015) Spatial and seasonal variation in diversity and structure of microbial biofilms on marine plastics in northern European waters. FEMS Microbiol Ecol 90:478–492
Özcimenc D, İnan B, Morkoç O, Efe A (2017) A review on algal biopolymers. J Chem Eng Res Updates 4:7–14
Panda AK, Singh RK, Mishra DK (2010) Thermolysis of waste plastics to liquid fuel: a suitable method for plastic waste management and manufacture of value-added products—a world prospective. Renew Sust Energ Rev 14:233–248
Phang SM, Chu WL, Rabiei R (2015) Phycoremediation. In: The algae world. Springer, Dordrecht, pp 357–389
Pico Y, Alfarhan A, Barcelo D (2019) Nano-and microplastic analysis: focus on their occurrence in freshwater ecosystems and remediation technologies. TrAC Trends Anal Chem 113:409–425
Poerio T, Piacentini E, Mazzei R (2019) Membrane processes for microplastic removal. Molecules 24:4148
Poonia AK, Kajla S, Koul B, Panwar JS (2022) Algae: the high potential resource for biofuel production. In: An integration of phycoremediation processes in wastewater treatment. Elsevier, pp. 155–176.
Póvoa AA, de Araújo FV, Skinner LF (2022) Macroorganisms fouled in marine anthropogenic litter (rafting) around a tropical bay in the Southwest Atlantic. Mar Pollut Bull 175:113347
Prata JC (2018) Airborne microplastics: consequences to human health? Environ Pollut 234:115–126
Prata JC, da Costa JP, Lopes I, Duarte AC, Rocha-Santos T (2019) Effects of microplastics on microalgae populations: a critical review. Sci Total Environ 665:400–405
Priyadharshini SD, Babu PS, Manikandan S, Subbaiya R, Govarthanan M, Karmegam N (2021) Phycoremediation of wastewater for pollutant removal: a green approach to environmental protection and long-term remediation. Environ Pollut 290:117989
Raghukumar S (2017) Fungi in coastal and oceanic marine ecosystems. Springer
Raziyafathima MPPK, Praseetha PK, Rimal IRS (2016) Microbial degradation of plastic waste: a review. Chem Biol Sci 4:231–242
Rillig MC, Ziersch L, Hempel S (2017) Microplastic transport in soil by earthworms. Sci Rep 7(1): 1–6
Rios LM, Moore C, Jones PR (2007) Persistent organic pollutants carried by synthetic polymers in the ocean environment. Mar Pollut Bull 54(8):1230–1237
Rizan C, Mortimer F, Stancliffe R, Bhutta MF (2020) Plastics in healthcare: time for a re-evaluation. J R Soc Med 113(2):49–53
Roager L, Sonnenschein EC (2019) Bacterial candidates for colonization and degradation of marine plastic debris. Environ Sci Technol 53:11,636–11,643
Rochman CM, Tahir A, Williams SL, Baxa DV, Lam R, Miller JT, Teh SJ (2015) Anthropogenic debris in seafood: plastic debris and fibers from textiles in fish and bivalves sold for human consumption. Sci Rep 5(1):1–10
Ruggiero MA, Gordon DP, Orrell TM, Bailly N, Bourgoin T, Brusca RC, Smith TC, Guiry MD, Kirket PM (2015) A higher level classification of all living organisms. PLoS One 10(6): e0130114
Salama ES, Roh HS, Dev S, Khan MA, Abou-Shanab RAI, Chang SW, Jeon BH (2019) Algae as a green technology for heavy metals removal from various wastewater. World J Microbiol Biotechnol 35:75
Sanchez C (2020) Fungal potential for the degradation of petroleum-based polymers: an overview of macro- and microplastics biodegradation. Biotechnol Adv 40:107501
Sarker A, Deepo DM, Nandi R, Rana J, Islam S, Rahman S et al (2020) A review of microplastics pollution in the soil and terrestrial ecosystems: a global and Bangladesh perspective. Sci Total Environ 733:139296

Sarmah P, Rout J (2018) Efficient biodegradation of low-density polyethylene by cyanobacteria isolated from submerged polyethylene surface in domestic sewage water. Environ Sci Pollut Res 25:33,508–33,520

Sarmah P, Rout J (2020) Role of algae and cyanobacteria in bioremediation: prospects in polyethylene biodegradation. Adv Cyanobacterial Biol:329–346

Scheurer M, Bigalke M (2018) Microplastics in Swiss floodplain soils. Environ Sci Technol 52(6): 3591–3598

Serrano, O., Kelleway, J. J., Lovelock, C., & Lavery, P. S. (2019). Conservation of blue carbon ecosystems for climate change mitigation and adaptation. In Coastal wetlands. Elsevier, pp. 965–996

Setälä O, Fleming-Lehtinen V, Lehtiniemi M (2014) Ingestion and transfer of microplastics in the planktonic food web. Environ Pollut 185:77–83

Shanmugam S, Hari A, Kumar D, Rajendran K, Mathimani T, Atabani A, Brindhadevi K, Pugazhendhi A (2021) Recent developments and strategies in genome engineering and integrated fermentation approaches for biobutanol production from microalgae. Fuel 285: 119052

Shields R, Lupatsch I (2012) Algae for aquaculture and animal feeds. J Anim Sci 21(1):23–37

Simonic M, Zemljic F (2021) Production of bioplastic material from algal biomass. Chem Ind Chem Eng Q 27(1):79–84

Singh SP, Singh P (2015) Effect of temperature and light on the growth of algae species: a review. Renew Sust Energ Rev 50:431–444

Sobhani Z, Zhang X, Gibson C, Naidu R, Megharaj M, Fang C (2020) Identification and visualisation of microplastics/nanoplastics by Raman imaging (i): down to 100 nm. Water Res 174:115658

Steensgaard IM, Syberg K, Rist S, Hartmann NB, Boldrin A, Hansen SF (2017) From macro- to microplastics—analysis of EU regulation along the life cycle of plastic bags. Environ Pollut 224:289–299

Sun J, Dai X, Wang Q, van Loosdrecht MC, Ni BJ (2019) Microplastics in wastewater treatment plants: detection, occurrence and removal. Water Res 152:21–37

Talvitie J, Mikola A, Koistinen A, Setala O (2017) Solutions to microplastic pollution removal of microplastics from wastewater effluent with advanced wastewater treatment technologies. Water Res 123:401–407

Tan JS, Lee SY, Chew KW, Lam MK, Lim JW, Ho SH, Show PL (2020) A review on microalgae cultivation and harvesting, and their biomass extraction processing using ionic liquids. Bioengineered 11(1):116–129

Tang KHD (2020) Effects of microplastics on agriculture: a mini-review. Asian J Environ Ecol 13(1):1–9

Taniguchi I, Yoshida S, Hiraga K, Miyamoto K, Kimura Y, Oda K (2019) Biodegradation of PET: current status and application aspects. ACS Catal 9(4089):4105

Tavares R, Ramos A, Rouboa A (2018) Microplastics thermal treatment by polyethylene terephthalate-biomass gasification. Energy Convers Manag 162:118–131

Thielen M (2019) Bioplastics: plants and crops raw materials products. Fachagentur für Nachwachsende Rohstoffe Ev, German

Thiruchelvi R, Das A, Sikdar E (2020) Bioplastics as better alternative to petro plastic. Mater Today Proc 37:1634–1639

Thompson RC, Olsen Y, Mitchell RP, Davis A, Rowland SJ, John AW, McGonigle D, Russell AE (2004) Lost at sea: where is all the plastic? Science 304(5672):838

Tian Y, Yang Z, Yu X, Jia Z, Rosso M, Dedman S, Wang J (2022) Can we quantify the aquatic environmental plastic load from aquaculture? Water Res 219:118551

Tiseo I (2021) Plastics industry—statistics & facts. Statista. https://www.statista.com/topics/5266/plastics-industry/#topicHeader__wrapper

Tokiwa Y, Calabia BP (2004) Review degradation of microbial polyesters. Biotechnol Lett 26: 1181–1189

Tunali M, Uzoefuna EN, Tunali MM, Yenigun O (2020) Effect of microplastics and microplastic-metal combinations on growth and chlorophyll a concentration of *Chlorella vulgaris*. Sci Total Environ 743:140479

Upadhyay AK, Singh R, Singh DP (2018) Phycotechnological approaches toward wastewater management. In: Bhargava RM, Chowdhary P (eds) Emerging and eco-friendly approaches for waste management. Springer, Singapore, pp 423–435

Urbanek AK, Rymowicz W, Miro'nczuk AM (2018) Degradation of plastics and plastic-degrading bacteria in cold marine habitats. Appl Microbiol Biotechnol 102:7669–7678

Venkatachalam H, Palaniswamy R (2020) Bioplastic world: a review. J Adv Sci Res 11(3):43–53

Viršek MK, Lovšin MN, Koren Š, Kržan A, Peterlin M (2017) Microplastics as a vector for the transport of the bacterial fish pathogen species *Aeromonas salmonicida*. Mar Pollut Bull 125(1–2):301–309

Walosik A, Szajner J, Pawlas K, Zeber-Dzikowska I (2021) Municipal waste and related health risks. J Elem 26(3):573–582

Wang K (2014) Bio-plastic potential of spirulina microalgae (master of science). University of Georgia

Wang J, Liu X, Li Y, Powell T, Wang X, Wang G, Zhang P (2019) Microplastics as contaminants in the soil environment: a mini-review. Sci Total Environ 691:848–857

Webb HK, Arnott J, Crawford RJ, Ivanova EP (2013) Plastic degradation and its environmental implications with special reference to poly(ethylene terephthalate). Polymers 5(December):1–18

Wei W, Huang QS, Sun J, Dai X, Ni BJ (2019) Revealing the mechanisms of polyethylene microplastics affecting anaerobic digestion of waste activated sludge. Environ Sci Technol 53:9604–9613

Wei R, Tiso T, Bertling J, O'Connor K, Blank LM, Bornscheuer UT (2020a) Possibilities and limitations of biotechnological plastic degradation and recycling. Nat Catal 3:867–871

Wei W, Hao Q, Chen Z, Bao T, Ni BJ (2020b) Polystyrene nano plastics reshape the anaerobic granular sludge for recovering methane from wastewater. Water Res 182:116041

Wen YC, Doris YYT, Kuan SK, Andrew NKL, Kit WC (2020) Nature's fight against plastic pollution: algae for plastic biodegradation and bioplastics production. Environ Sci Ecotechnol 4(100065):1–10

Wong SL, Ngadi N, Abdullah TAT, Inuwa IM (2015) Current state and future prospects of plastic waste as source of fuel: a review. Renew Sust Energ Rev 50:1167–1180

Wright SL, Kelly FJ (2017) Plastic and human health: a micro issue? Environ Sci Technol 51(12):6634–6647

Wu Y, Guo P, Zhang X, Zhang Y, Xie S, Deng J (2019) Effect of microplastics exposure on the photosynthesis system of freshwater algae. J Hazard Mater 374:219–227

Wu D, Feng Y, Wang R, Jiang J, Guan Q, Yang X et al (2022) Pigment microparticles and microplastics found in human thrombi based on Raman spectral evidence. J Adv Res 49:141–150

Xia B, Sui Q, Sun X, Zhu L, Wang R, Cai M, Chen B, Qu K (2021) Microplastic pollution in surface seawater of Sanggou Bay, China: occurrence, source and inventory. Mar Pollut Bull 162:111899

Xiao Y, Jiang X, Liao Y, Zhao W, Zhao P, Li M (2020) Adverse physiological and molecular level effects of polystyrene microplastics on freshwater microalgae. Chemosphere 255:126914

Yang X, Wang H, Zhang L, Kong L, Chen Y, He Q, Li L, Grossart HP, Ju F (2021) Marine algae facilitate transfer of microplastics and associated pollutants into food webs. Sci Total Environ 787:147535

Yang L, Liu Z, Peng Z, Song P, Zhou J, Wang L, Dong Q (2022) Exposure to DEHP is potential to increase the risk of overactive bladder, evidence from NHANES 2003–2008. Environ Sci Pollut Res 29:1–9

Yin ZL, Zhu S, Li T, Hu R, Chu F, Mo D, Hu C, Liu, Li B (2020) A comprehensive review on cultivation and harvesting of microalgae for biodiesel production: environmental pollution control and future directions. Bioresour Technol 301:122804

Yokota K, Waterfield H, Hastings C, Davidson E, Kwietniewski E, Wells B (2017) Finding the missing piece of the aquatic plastic pollution puzzle: interaction between primary producers and microplastics. Limnol Oceanogr Lett 2(4):91–104

Yonkos LT, Friedel EA, Perez-Reyes AC, Ghosal S, Arthur CD (2014) Microplastics in four estuarine rivers in the Chesapeake Bay, USA. Environ Sci Technol 48(24):14,195–14,202

Zhang YT, Wei W, Sun J, Xu Q, Ni BJ (2020a) Long-term effects of polyvinyl chloride microplastics on anaerobic granular sludge for recovering methane from wastewater. Environ Sci Technol 54:9662–9671

Zhang YT, Wei W, Huang QS, Wang C, Wang Y, Ni BJ (2020b) Insights into the microbial response of anaerobic granular sludge during long-term exposure to polyethylene terephthalate microplastics. Water Res 179:115898

Zhang X, Chen J, Li J (2020c) The removal of microplastics in the wastewater treatment process and their potential impact on anaerobic digestion due to pollutants association. Chemosphere 251:126360

Zheng Y, Yanful EK, Bassi AS (2005) A review of plastic waste biodegradation. Crit Rev Biotechnol 25:243–250

Recent Trends to Address Microplastic Waste at the Global Level

30

Pradnya P. Nadkarni

Abstract

Microplastics are a nuisance to marine and terrestrial ecosystems. These tiny particles of plastic can also originate from the degeneration of large plastic units. Being nonbiodegradable, they have the ability to stay for hundreds and thousands of years. They even pollute the sand on the shores and alter its texture. They are easily consumed by sea creatures leaving their presence inside them because of their tiny size of less than 5 mm. Various governmental, nongovernmental, and private sector agencies are taking efforts to deal with the problem of microplastic in ecosystems. This chapter makes an attempt toward understanding the global initiatives for the reduction of microplastic in the natural ecosystems.

Keywords

Microplastics · Effects of microplastic pollution · The marine and terrestrial ecosystem · Global initiatives for reduction of microplastic

30.1 Introduction

Plastics have been posing an issue to the health of the ecosystem and human beings for a long time. The first synthetic plastic, Bakelite, was produced in 1907, ushering in a new era for the global plastics industry (Ritchie and Roser 2018). However, it was not until 1950s that plastic was produced and used in larger magnitudes. The annual production of plastic has increased by about 230 times. The packaging industry is the main source of plastic trash, and it has increased its contributions as

P. P. Nadkarni (✉)
R.A. Podar College of Commerce & Economics (Autonomous), Mumbai, Maharashtra, India

M. Shahnawaz et al. (eds.), *Microplastic Pollution*,
https://doi.org/10.1007/978-981-99-8357-5_30

the world has moved away from reusable to single-use goods (Geyer et al. 2017). As a result, the contribution of plastics in municipal solid waste (by weight) amplified from about 1% in 1960 to more than 10% in 2005. This shift was mainly seen in middle- and high-income countries (Jambeck et al. 2015). Differences are also seen at the global level with regard to the daily per capita generation of plastic waste. Countries like Kuwait, Guyana, Germany, The Netherlands, Ireland, and the USA generate plastic waste more than ten times than countries like India, Tanzania, Mozambique, and Bangladesh (Ritchie and Roser 2018).

Conventionally used plastics are nonbiodegradable. Thus, they do not degrade but rather build up in landfills and other natural areas like oceans (Barnes et al. 2009). The plastic waste found in terrestrial ecosystems can be traced and collected more conveniently than in aquatic ecosystems. On the land, it can be managed through incineration, which is a commonly used technique. However, the aquatic ecosystems, because of their more dynamic environment, allow for a long-term biophysical breakdown of plastics (Thompson et al. 2004). As long as they are present on the earth, they cause esthetic problems if in the open ecosystem. If inside the living systems, they may interfere with the metabolism, reproduction, etc. of the organisms.

Apart from the plastic types and their chemical composition, the size of the plastics in the waste can also be used to classify it: macroplastics (plastics more than 20 mm in diameter) and microplastics (plastics which are less than 5 mm in diameter). Among these two types, microplastics are documented to be the major pollutants in natural ecosystems. These microplastics are either primary, which are produced for direct use as tiny-sized particles by the design itself, or secondary microplastics, which are formed as a result of the degradation of macroplastics (Alabi et al. 2019). The major issues related to microplastics are faced because of the difficulty in their monitoring and their cumulative physical and chemical effects on the environment and human health, which is mainly due to their higher volume-to-surface area ratio (Science for Environmental Policy 2011). Inadequate waste management and indiscriminate dumping are the major routes of entry of microplastics into the marine environment (Pettipas et al. 2016). Various efforts are being taken globally in order to tackle the issue of microplastic generation and accumulation. International agencies, individual nations, and NGOs are working in their own capacities to reduce the passage of microplastics from the sources to various pockets of ecosystems. Despite these efforts, no concrete effects have been observed in regard to the improvement of terrestrial and marine ecosystems being affected by microplastics. In fact, it is observed that the amount of microplastics released in the natural ecosystems is increasing each year (Ritchie and Roser 2018). The treaties and conventions released by international agencies have been observed to have established partial or no improvement agencies working toward microplastic reduction and do not have a common thread between them.

30.1.1 The Definition of Microplastics

Microplastics, as the name suggests, are tiny particles of plastics, which are usually less than 5 mm in size. They are much smaller than the regular plastic materials which are found around, which makes them more detrimental to ecosystems and human health. Even though the use of plastics on a regular basis started in the 1950s, it was not until the 1970s that the presence of microplastics was first recognized. In today's times, the abundance of microplastics has allowed them to be present in extreme locations, such as Mount Everest and the Mariana Trench (An et al. 2020).

30.1.2 The Scale of the Problem of Microplastic Pollution

The small particle size of microplastics allows them to be more movable in nature, facilitating easy movement through the ecosystem. They can be easily carried to places over long distances because of the action of wind, water, etc. and can get distributed widely across terrestrial ecosystems, rivers, lakes, seas, oceans, estuaries, and coastal regions. Their presence is also found in deep seas and as well as polar ecosystems (Chen et al. 2021).

The tiny size of microplastics also allows for easy accumulations of the same into the systems of living organisms. Microplastics have been detected in more than 100 species, especially aquatic organisms, probably because of the more even distribution of microplastics in the aquatic environment. Since fishes, mussels, shrimps, and crabs are among the most staple diet around the world, this microplastic contamination of aquatic species poses a serious threat to human health at the global level (Chen et al. 2021).

30.1.3 The Need for Global Action

Geographic and political boundaries seem to be irrelevant when it comes to the distribution of microplastics around the globe. They have crossed all the boundaries and surpassed all barriers to mark their presence in natural ecosystems. Though they are abundantly found in the aquatic ecosystem, their origin is still terrestrial activities with multiple sources. One notable issue with plastics is that the region where they are found may not necessarily be the region of their origin as well. In other words, the ease with which they can move around makes it possible that the source and destination of certain microplastic pollution can be continents apart from each other. Thus, the issue of microplastic pollution is not limited to a certain region, a nation, or a continent. A severely microplastic-polluted region can be the result of a source many miles away. Rules and regulations imposed by one nation are not enough to control global microplastic pollution. Thus, there is a growing need for all nations to come together to combat this problem.

30.2 The Sources of Microplastic Waste

There is not just one source of microplastic pollution. In fact, every piece of plastic, large or small, can eventually lead to microplastic pollution in some or the other way during its life cycle.

Thus, the number of microplastic sources is in direct proportion to the total amount of plastic manufactured, utilized, and disposed all over the world. This makes identification of the sources of microplastics even more complicated and difficult (Prata et al. 2019).

The size of the microplastic particles is the main criterion for their classification. Because of this, they are divided into two groups: primary microplastic sources and secondary microplastic sources.

30.2.1 Primary Sources

Primary microplastic sources are all activities that lead to the direct manufacture and production of small plastic particles, mainly for commercial purposes. The primary microplastics are tiny pieces of plastic, usually less than 5 mm, which are intentionally manufactured to be part of various commercial materials like cosmetics, other personal care products, paints, and medicines and are also used in air blasting technology (Wikipedia). Personal care products and cosmetics such as exfoliating cleansers and scrubs contain tiny beads of plastic with the promise of a better cleansing effect. Through the toilet runoff, these microbeads end up in the seas and oceans (Vince and Hardesty 2017). Microplastics are also used in the medical field in the form of tiny plastic particles which act as vectors to carry drugs. Although this facilitates an easy pathway for drugs to reach various body parts, it becomes a major source of microplastics. In air blasting technology, microplastics are used in the form of scrubbers made from acrylic, melamine, or polyester. To get rid of rust and outdated paint, they are blasted at engines, machinery, or boat hulls. Over a period, they lose their cutting power and become contaminated with heavy metals, on which they are used for blasting, such as cadmium, chromium, and lead (National Geographic Society 2022). Other sources of primary microplastics are sewage sludge as well as plastic running tracks in schools and artificial turf.

30.2.2 Secondary Sources

Secondary sources of microplastics are those which originate from large plastic materials via physical, chemical, or biophysical degradation. Such degradation leads to a breakdown of bigger plastic material into finer, smaller plastic particles, thus releasing secondary microplastics into the environment. Photo-oxidation due to exposure to sunlight also makes plastic brittle and prone to easy breakdown. Physical abrasions also lead to the disintegration of larger plastics into smaller microplastics. The wear and tear of fishing nets is also another major reason for the entry of secondary microplastics into the marine ecosystem.

30.3 The Impact of Microplastic Waste

Microplastics can be carried by ocean currents and other biological vectors such as fish, which harbor microplastics in other areas of the ocean. Not only the tiny size but also the chemical composition of plastics poses a threat to human health. The research in the field of understanding the impact of microplastic pollution on the physiology of human and aquatic organisms is still in the primary stage and thus requires further inspection. However, it is common knowledge that chemicals released from plastics have an adverse impact on the health and development of living organisms including human beings.

30.3.1 Ecological Impacts

Microplastics can enter the ecological environment through runoff from surfaces and wastewater released from wastewater treatment plants, fishery and aquaculture, the household and industrial water being dumped into aquatic ecosystems, and atmospheric emission coming in contact with the ocean (Talvitie et al. 2016). Being tiny in size, they can be conveniently ingested by aquatic organisms. Once inside the organisms, microplastic accumulation can be detrimental to tissues, organs, and the intestinal tract (Hoffmann et al. 2020). Various contaminants from the environment also get easily adsorbed on the surface of microplastics, thus carrying these contaminants into the systems of marine organisms (Khalid et al. 2021). Research has proven the presence of microplastics in plankton, fish, and seabirds (Xie et al. 2020). Microplastics and the adsorbed chemical contaminants can have the following impacts on marine organisms.

30.3.1.1 Oxidative Stress

Oxidative stress caused by microplastics leads to damage to the cell and cell membrane structure. It results in the decrease of permeability of the cell membrane. It reduces the cell growth rate, thus affecting the physiological processes of the cells (Mao et al. 2022).

30.3.1.2 Neurological Damage

Various organisms which were exposed to microplastics were observed to have shown inhibition of acetylcholinesterase (AchE). It can reduce the neuronal network function of the organism (Keefer et al. 2001).

30.3.1.3 Endocrine Disruption

During the manufacturing, synthesis, and processing of plastics in modern industrial production, many organic as well as inorganic components are added to enhance the efficiency of the final plastic product. These components (e.g., phthalates, polybrominated diphenyl ethers, bisphenol A, etc.) give plastic properties such as flame retardation, stabilization, and resistance to fungus. These substances are known to affect development and reproduction, further leading to endocrine-disrupting effects (Hahladakis et al. 2018). Phthalates and bisphenol A have the ability to damage larval development among amphibians by destroying their thyroid function (Oehlmann et al. 2009). Phthalates are also responsible for endocrine disruption by the destruction of the signaling pathway which targets the receptor for nuclear hormone (Lang et al. 2008; Kim et al. 2002).

30.3.1.4 Immune Injury

Microplastics can alter the immune response of organisms by entering their immune system after being accumulated and transferred to tissues and organs. They can cause stress in fish by interfering with their natural defense mechanism. It leads to a decrease in the phagocytic activity of immune cells and cell viability. It can also cause damage to the lysosome membrane (Sharifinia et al. 2020). Exposure to microplastics has adverse effects on the release of cytokines and also causes changes in inflammatory responses (Hu and Palić 2020).

Apart from these effects, microplastics can also become a vector for various eukaryotic and invasive prokaryotic species over long distances in the marine environment (Khalid et al. 2021).

30.3.2 Health Impacts

Microplastics have shown their occurrence in the air, drinking water, and food, things human beings require for their survival. The human respiratory system can be affected by suspended microplastic particles in the air. Microplastics can enter the human body through the intake of contaminated food and water (Dris et al. 2015). Research is still going on in the field of understanding the possible impacts of microplastics on the human body, but it is very much evident that they have a negative impact on overall human health (Wright and Kelly 2017).

There are three ways in which microplastics can enter the human body.

30.3.2.1 Gastric Exposure

Research has found that microplastics can be ingested via food and drinks (Lehner et al. 2019). For example, analysis of human stool samples has shown the presence of microplastic particles (Ge et al. 2018). Nanoparticles of plastics may interact with various molecules in the gastrointestinal tract like proteins, lipids, nucleic acids, carbohydrates, water, and ions, causing changes in their rate and ability of absorption (Maria Fernanda et al. 2022).

30.3.2.2 Pulmonary Exposure

The fact that microplastics can penetrate the lungs leads to the theory that they may damage the respiratory system by increasing the carcinogenic rate (Prata 2018).

Synthetic textiles can be a source of indoor microplastic pollution. Contaminated aerosols from ocean waves or airborne fertilizer particles from sewage treatment plants can cause outdoor microplastic pollution. Inhalation of these nanoparticles of microplastics can cause chemical and particle toxicity and they can also act like vectors by introducing pathogens and parasites in the respiratory tract (Yee et al. 2021).

30.3.2.3 Dermal Exposure

Cosmetic products like face and body scrubs used on the skin contain polyethylene and thus are the major sources of microplastic exposure through the skin. Drugs that are administered through the skin along with nanocarriers can also be another reason for the entry of microplastics through the skin (Yee et al. 2021).

30.3.3 Economic Impacts

The economic effects of microplastic contamination are less obvious but just as detrimental as the ecological and human health effects. The plastic waste management strategies employed by many developed and high-income countries have resulted in the transfer of plastic waste to low-income countries increasing their economic burden. It may have had a positive impact on the circular economy of the low-income groups in these developing nations, but nonetheless, it has caused a significant increase in the health issues of the people and has adversely affected the economic activities which are dependent on the ecosystem services. This has threatened the well-being of people and reduced their ability to live a sustainable lifestyle (Marine Litter Vital Graphics 2016).

30.4 Recent Trends in Addressing Microplastic Waste at a Global Level

Since microplastic pollution is a borderless issue, various governments, NGOs and local authorities have addressed it globally. The effectiveness of the strategies adopted by these agents depends upon the seriousness of enforcement, the willingness of the participants, and the socio-economic status of the society.

30.4.1 International Agreements and Conventions

International agreements and conventions signed by the participant nations have been looked at as an effective methodology to tackle many social, environmental, and economic issues, mainly for their universal nature and applicability.

30.4.1.1 The United Nations Environment Assembly

The United Nations Environment Assembly (UNEA) has been actively delivering guidelines with regard to implementing new strategies, identifying possible loopholes and suggesting methodologies to overcome plastic pollution at the international level (Walker 2021). It has conducted three important sessions in the same regard and the most recent one was conducted in December 2017. It has also established a committee in order to formulate response strategies so that plastic pollution can be tackled through individual government policies and voluntary initiatives taken by individual groups (Usman et al. 2020).

Another initiative taken by the United Nations (UN) is the formation of the United Nations Convention on the Law of the Sea (UNCLOS). It has been lauded as being the "Constitution for the Ocean" (Hu and Palić 2020). It has a consideration for all human marine activities, which affect the overall health of the marine environment, as it was discussed in the resolutions of the UNEA. As compared with other conventions on the protection of the marine environment, UNCLOS has more comprehensive provisions and also considers the conflict of interest between nations when it comes to the protection of marine litter. In addition to Part XII of UNCLOS, which covers unique provisions for safeguarding and protecting the marine environment, it also has several other provisions which encompass all types of sources which cause marine pollution. It also assesses the detrimental effects of marine pollution and how it can be effectively controlled. The breakthrough provision in UNCLOS is the identification and regulation of terrestrial pollution sources since they are the major contributors toward marine microplastic pollution (Dris et al. 2015).

The "Clean Seas" campaign of UNEP is devoted to ending marine pollution. It is one of the most impactful strategies which connects individuals, civil society groups, manufacturers, and policymakers at the global level in order to reduce marine litter. They have highlighted the issue of marine pollution due to microplastics. Their tie-up with various agencies worldwide has helped in creating awareness and identifying the subtle sources of marine microplastic pollution, which usually go unnoticed. They have partnered with the Secretariat of the World Health Organization Framework Convention on Tobacco Control (WHO FCTC) to reduce cigarette consumption at the global level through consumers, policymakers, and business houses. The filters of cigarettes contain microplastics called cellulose acetate fibers, which release along with heavy metals and many other chemicals when the cigarettes break down under water (Kershaw et al. 2011). By emphasizing the adverse impact of cigarettes on humans, they are drawing attention toward their contribution to marine pollution as well, since they are one of the most common marine litters on the beaches worldwide (Clean Seas Campaign, UNEP 2019).

UNEP has also participated in the UN Alliance for Sustainable Fashion in order to create awareness against the use of synthetic fabrics containing plastics such as polyester, acrylic, and nylon which causes around 9% of the marine microplastic pollution (UN Alliance for Sustainable Fashion). UNEP also tied hands with the Ellen McArthur Foundation to act globally for achieving a circular economy for plastics. It has 500 stakeholders which include plastic producers, financial institutions, and governments in order to reach the goal of reuse and regeneration of plastics by 2025 (Inside the Clean Seas Campaign, UNEP 2022). *UNEP is also in the process of creating a digital platform to encourage action to prevent marine plastic litter through the formation of the Global Partnership on Marine Litter (Global Partnership on Plastic Pollution and Marine Litter, UNEP).*

30.4.1.2 The Basel Convention

The Basel Convention has been brought into action for controlling the transboundary movement of hazardous waste across nations. The original scope of this convention was not meant for plastic waste. However, recently, the convention has focused more on the prevention and control of pollution rather than compensation for the damage that happens after pollution (Sulaiman and Razman 2010).

In May 2019, the convention adopted a resolution named "Plastic Waste Amendments", which directly binds the nations for the prevention and treatment of microplastics for the first time. Several provisions and amendments have been made to identify the hazardous waste, the rights and responsibilities of the participants, and the mechanism to implement the same has been made under this resolution (U.S. Department of States 2019). The Basel Convention also addresses the issues faced by developing countries where hazardous waste, including plastics, is deposited.

Thus, the Basel Convention is regarded as the only entity that specifically includes microplastics in its scope, whereas other conventions mention it only at limited levels (Li 2022). One of the most important changes introduced by this resolution was the control of plastic waste. Two important measures were considered under the same: (a) source control—to make an effort toward effectively managing the plastic waste in the country of its origin and to control the production of plastics and (b) approval for the cross-border transfer of plastic waste. This amendment considers plastic waste under special consideration and includes specific types of plastic waste, for example, used bottles of beverages, kitchen garbage, etc. This provision makes it mandatory for the exporting country to give prior notice and the importing country to give consent for the same before the export is attempted (Li 2022).

The Basel Convention also aims to reduce persistent organic waste by encouraging manufacturers to diminish the environmental impact of their products at every stage of its life cycle and by creating awareness about the hazardous nature of the chemicals produced by persistent organic pollutants (Magulova 2012). This convention plays an important role in preventing and controlling marine microplastic pollution and thus becomes the only international agreement with legal strength especially when dealing with plastic waste (Li 2022).

30.4.1.3 The Stockholm Convention

Considering the serious damage caused by persistent organic pollutants to human health, the Stockholm Convention further contributes to protecting human beings from such substances. However, this convention only controls the production and circulation of hazardous plastic products, having no significant impact on the limitation and control of the generation of plastic waste (Zhang et al. 2022).

30.4.2 Government Policies and Initiatives

The governments have made efforts for microplastic reduction in the form of various laws and policy enforcement.

30.4.2.1 Plastic Bans and Restrictions

Internationally, many governments have enforced different strategies to keep a check on the utilization and distribution of plastic in many forms. Imposing a ban on plastic materials contributing to secondary microplastic pollution (such as single-use plastic bags) and/or direct use of primary microplastics (such as cosmetics containing microbeads) are used as effective measures.

Single-use plastic mainly comprises thin sheets of plastic, which are suitable only for short-term use and thus end up in the trash soon after its use. Many countries have taken decisive steps toward restricting the manufacture, storage, distribution, and use of such single-use plastic in order to control their entry into the ecosystems. This includes banning plastic bag sales, charges on plastic bags, and imposing taxes on the sellers of plastic bags. In the US A, Washington D.C. was one of the first cities to impose taxes on plastic use. The revenue generated was utilized for the driver clean-up drive. They furthered the efforts by donating reusable bags to the poor and aged community in the city (Schnurr et al. 2018). San Francisco has implemented a policy on charging extra fees on single-use bags. Some other states in the USA like Seattle have imposed a complete ban on the use of single-use plastic bags. On the other hand, states like Boston have implemented a combination of tax and ban both. As a result of these efforts, a reduction in plastic use was seen. In South America, Chile, Columbia, Brazil, and Argentina have banned the usage of single-use plastic, businesses that distribute plastic bags. The government of Chile has also come up with the beach clean-up drives during peak vacation periods (Poortinga et al. 2013).

Canada has imposed a ban on all plastic shopping bags. The government of Canada listed single-use plastic items as toxic substances as mentioned in Schedule 1 of the Environmental Protection Act, 1999 in Canada. This allowed them to prohibit single-use plastic items. The prohibition is applicable for the following six classes of single-use plastic items which are harmful: plastic checkout bags, straws, stir sticks, six-pack rings, cutlery, and food containers made from plastics which are difficult to recycle (Schnurr et al. 2018).

Australia has banned all types of plastic bags with a single use. The general public was urged to bring reusable bags for shopping. In 2008, South Australia implemented the "Zero Waste" program resulting in the reduction of 400 million bags annually (Block 2009). New Zealand has banned the single-use plastic having a thickness of less than 70 microns.

China has introduced a strict ban on the circulation of plastic bags with a single-use lifespan in the stores in the country. China has also enforced a national program for marine waste monitoring and microplastic monitoring since 2016. China participated in 2019 in the process of the amendment of the *Basel Convention* appendix, which encouraged the management of marine plastic wastes (Yang et al. 2021). Other Southeast Asian countries like Malaysia have banned nonbiodegradable plastic waste, which is poorly enforced.

India has executed a ban on the single-use plastic ban, charging on plastic bags in a phased manner. The enforcement of these policies is weak due to a lack of jurisdiction (DasGupta and Shaw 2013).

Many African countries including Nigeria, Benin, Kenya, Ghana, Togo, Morocco, Rwanda, and South Africa have contributed to their share of reducing plastic pollution by enforcing a partial and complete prohibition of nonbiodegradable plastic bags. Out of the countries mentioned above, only Kenya and South Africa have been able to successfully ban the plastic bags. Other countries have witnessed poor enforcement (Schnurr et al. 2018; Block 2009).

In comparison with the laws for the ban on single-use plastic, there are few jurisdictions for the ban of microbeads. Only a few countries have taken steps toward the ban and control of the use of microbeads and have implemented jurisdiction for the same (Plastic in cosmetics, UNEP 2015).

Canada has included microbeads as toxic substances under the Canadian Environmental Protection Act. They have furthered their efforts by imposing a national ban on the sale and use of microbeads. It has also enforced a ban on toiletries and cosmetics which included some exfoliating personal care products containing microbeads from stores. The province of Ontario passed a law to ban the manufacture of microbeads in 2015 (Walker and Xanthos 2018).

The use of microbeads in cosmetics was first banned in the Netherlands in 2016. They have also banned the import, production, and sale of microbeads in washable cosmetics (Schnurr et al. 2018).

The United Kingdom has implemented the Environmental Protection (Beads) Regulations (England) Act in 2017, which bans the use of not only nonbiodegradable microbeads but also the biodegradable ones. The reason stated for this was that even the biodegradable microbeads require specific state of conditions to be able to decompose. Such conditions may not be present in all ecosystems, especially the deep marine water which may not allow the degradation to happen (Anonymous 2015).

The National Reforms and Development Commission in China has declared the country's policy to describe their ban on microbeads. Chinese legislation bans the manufacture of new cosmetics containing microbeads since 2020. Banning the sale of existing cosmetics containing microbeads was also planned to be implemented before 31 December 2022 (Yang et al. 2021).

The Microbeads Free Water Acts (2015) of the USA has banned the sale of cosmetics which have microbeads since 2017, promoting the use of biodegradable options such as polylactatide (PLA) and polyhydroxyalkanoates (PHA) (Usman et al. 2020).

China has implemented the Law on Prevention and Control of Environmental Pollution by Solid Wastes (LPCEPSW) for the regulation of sites of waste disposal and prohibition on the plastic waste dumping in rivers, lakes, and reservoirs (Usman et al. 2020).

30.4.2.2 Extended Producer Responsibility (EPR) Schemes

Extended Producer Responsibility (EPR) policy involves the responsibility taken by producers toward environmental protection from hazardous substances, including plastics. Under this policy, the producers are expected to take financial and physical responsibility for treating their products after they become waste, provide incentives to reduce waste at the source, and manufacture the product in such a manner that it causes minimum harm to the environment. This can be practiced by the producers by helping in precollecting the waste products before they end up in the trash, segregation of the same, and aiding in their management (Anonymous 2019).

UNEA also recognizes the contribution of the private sector in preventing and minimizing marine litter and microplastics. According to them, this can be done through sharing information, creating awareness, developing innovative ecofriendly technologies, and using human resources in clean-up activities. A cooperation between the private sector and governmental bodies is required for the same (Marine litter and microplastics, UNEP 2018).

UNEA holds not only the producers but also the retailers, distributors, and packaging industries responsible. They are expected to provide information about what impact their products can have on the environment and human health and promote innovative approaches to reduce the impact (Addressing single-use plastic products pollution, UNEP 2021).

30.4.2.3 Recycling and Waste Management Programs

The measures adopted by the nations for waste management can have long-term impacts on the potency of waste and how it affects the surroundings. Thus, many nations have come up with strategies for waste management to be implemented in order to diminish the effect of waste, including plastic waste. The state of California in the USA has a mandatory strategy named the National Pollutant Discharge Elimination System for promoting the installation of meshes to filter out plastics in the primary stages of waste management (Bahadori and Smith 2016). The recycling and waste reduction bill which was introduced in Australia in 2020 is another example which gives a flowchart for the management and recycling of waste in the entire nation. This law also bans exporting waste materials including plastic, glass, and paper, encouraging their recycling within the country (Anonymous 2020).

30.4.3 Initiatives by Industries

The success of the waste management program taken by governments toward the reduction of plastic waste will remain limited if similar initiatives are not taken by industries. Industries make up a large contribution to plastic waste generation. Thus, a moral, legal, and financial responsibility also lies upon them in order to lessen their environmental effect. This can be achieved through replacing plastic with other ecofriendly materials (e.g., glass or paper), fundamental design changes to reduce the amount of plastic used in manufacturing, controlling the number of additives, polymers, and mixtures which allows better recycling, and extending the durability of the product by allowing repair and reuse. It also includes not using some types of single-use plastics (Prata et al. 2019).

30.4.3.1 Development of Biodegradable and Compostable Plastics

Biodegradable plastics are the ones which fulfill the requirements of conventional plastics and at the same time have the ability to be degraded in the environment by natural processes in a sound manner. Biodegradable plastics are manufactured through complex engineering processes. They have the ability to replace most of the plastics conventionally used in various processes (Shen et al. 2020; Moshood et al. 2022). However, currently, the rate of substitution of conventional plastic with biodegradable plastic is still low (Shen et al. 2020).

Large market expansion is seen due to the increasing demand for biodegradable plastic, especially in developing nations like India, Malaysia, China, and Brazil (Moshood et al. 2022). Various nations have their own certification standards to identify the quality of the biodegradable plastic products made by the industries. For example, according to the TUV Austria database, the number of certified products made from biodegradable plastic can be summarized as follows: 1463 are industrially compostable, 739 are home compostable, 23 can be composted by the natural soil environment, 1 product can be composted in fresh water and no products could be listed under the category of marine compostable. Biodegradable Plastics Institute (BPI) of the USA has over 8000 certified products which are industrially compostable (Ghosh and Jones 2021).

30.4.3.2 Sustainable Packaging and Product Design

The packaging industry contributes to approximately half of the plastic waste generated globally (Walther et al. 2020). The objective behind packaging a product is to ensure safety, easy handling, and information provision to customers (Chen et al. 2013). Considering the growing concerns around the environmental problem created by plastic waste, various packaging industries and their partners and distributors have brought up strategies for replacing plastic in packaging with more ecofriendly options, as discussed below.

30.4.3.2.1 Paper and Paperboard

There is a huge scope for paper and paperboard to replace plastic when it comes to packaging products like food and beverages, furniture, construction materials, electrical appliances, etc. (Cela and Kaneko 2011). Paper and paperboard are superior to plastic in terms of sustainability and cost-effectiveness. However, their use is limited because of being less resistant to water, moisture, and chemicals. They also possess lower strength toward heavy goods (Chen et al. 2013).

30.4.3.2.2 Glass

Glass was a commonly used packaging material before plastics took over. Glass proves to be a safer option for packaging because it is chemical resistant, it can be made into any desired shape, and it retains the quality of products, especially food and medicines. Hence, the food and pharmaceutical industries still use it extensively. However, its use is limited for being fragile, heavier in weight, prone to breakage, and nonthermal resistant (Ibrahim et al. 2022).

Metal: Metals like tin, aluminum, lead, chromium, etc. are commonly used for packaging. Among them, aluminum is the most widely used metal for packaging, in the form of cans, boxes, or foils. This is attributed to its properties like cheap availability, less weight, flexibility, ease of recycling, and resistance to heat (Ibrahim et al. 2022).

30.4.3.3 Voluntary Commitments and Partnerships

Apart from the jurisdictions and mandatory policies enforced by governments, many plastic industries have willingly joined hands to combat the issues caused by microplastics in the natural environment. Operation Clean Sweep is one such international initiative, formed with an intention of handling plastic pellets, flakes, and powders with utmost care so that they do not end up in rivers or seas. Through this initiative, the plastic industries pledge to exercise the most efficient methodologies in order to reduce plastic pellet loss so as to protect the marine environment (Moore et al. 2005). Many textile industries have come together to produce synthetic fabrics which shed fewer plastic microfibers and therefore are ecofriendlier (Carney Almroth et al. 2018).

30.5 Challenges and Future Directions

The goal of reducing the entry of microplastics into the natural environment faces many challenges. Numerous jurisdictional drawbacks, a lack of implementation strategies, and the fundamental omnipresence of plastics throughout the globe are limiting the success of these efforts. The tiny size of microplastic particles and the complexity of their source and origin are some of the challenges faced by the existing efforts which the nations, industries, and NGOs are taking for their control.

30.5.1 Implementation and Enforcement Challenges

Many developing countries have no statutory laws or regulations for controlling plastic and microplastic discharge into the marine environment. In a few developing countries, the laws are inadequate and poorly enforced (Onyena et al. 2022; Vince and Hardesty 2018). The decentralized and fragmented nature of international collaboration makes it difficult to form and implement uniform laws for all nations to combat microplastic pollution (Thompson 2015).

The gap between the directives and actual actions taken by governments hampers communication and collaborative efforts required for plastic waste control within a nation (Stoett 2019).

In addition, most of the policies on the single-use plastic ban were lifted during the COVID-19 pandemic where the use of single-use masks, PPE kits, and gloves was encouraged (Silva et al. 2020).

30.5.2 Knowledge Gaps and Research Needs

A lack of significant amount of data in the field of the toxic effects of microplastics and measures to control the same is seen in most developing countries (Jeyasanta et al. 2020). No sufficient amount of information is available regarding human exposure levels of nanoparticles of microplastics through food (Dang et al. 2022). There is not enough research to examine the effect of microplastics on the health of human beings (O'Neill and Lawler 2021). There is no sufficient understanding and documentation of how particles are transported vertically at water levels and near the ocean floor. In addition, the presence of complex oceanic phenomena such as eddies, turbulence, and fronts makes the study of marine microplastics difficult to understand (Derraik 2002; Jensen et al. 2022).

30.5.3 The Role of Consumers and Public Awareness

In many nations, there are no clear guidelines on how the public can be effectively involved in controlling microplastic pollution. On many occasions, the policies for microplastic ban face interventions which leave people in confusion. There is also a limited level of awareness about the threats posed by microplastics among the public, especially among the rural and coastal communities (Onyena et al. 2022; Adam et al. 2020).

There is a general lack of awareness about the issues caused by single-use plastic. Especially after the COVID-19 pandemic, the increased use of PPE kits has resulted in unmanageable dumping of plastic waste worldwide (Silva et al. 2020).

30.6 Conclusion

Microplastic pollution poses a severe threat to the environment, human health, and economies worldwide. This issue has garnered attention from international agencies, governments, and independent parties, leading to various measures to combat it. However, the success of these policies varies depending on a country's development status, public awareness, and socio-economic factors. To ensure effective change, practical solutions must be discussed and implemented. Strengthening existing laws and regularly reviewing plastic-producing industries are vital steps. Creating a common platform for open dialogue and collaborative actions across communities can significantly contribute to tackling microplastic pollution. Implementing policies should involve giving ample time to raise awareness and prepare the public for their implications. Encouraging research on microplastic dispersal, types, and their discovery at the source is crucial. Improved collaboration between nations can help to keep oceans cleaner with better waste management facilities in ports and harbors for recycling fishing nets. A holistic and empathetic approach is needed to address ecological problems. Nations should prioritize the well-being of all species over short-term political gains. Industries must prioritize global well-being over monetary gains, and the public can contribute by adopting ecofriendly and sustainable lifestyles. Combining efforts and committing honestly to a microplastic-free ecosystem will pave the way for a cleaner, healthier planet for generations to come.

References

Adam I, Walker TR, Bezerra JC, Clayton A (2020) Policies to reduce single-use plastic marine pollution in West Africa. Mar Policy 116:103928

Alabi OA, Ologbonjaye KI, Awosolu O, Alalade OE (2019) Public and environmental health effects of plastic wastes disposal: a review. J Toxicol Risk Assess 5:021. https://doi.org/10.23937/2572-4061.1510021

Alison Watson, a report on 'Addressing Single-use Plastic Product Pollution' published 30 June 2021, ISBN: 978-92-807-3841-4, https://wedocs.unep.org/bitstream/handle/20.500.11822/35109/ASUP.pdf?sequence=3&isAllowed=y

An L, Liu Q, Deng Y, Wu W, Gao Y, Ling W (2020) Sources of microplastic in the environment. In: He D, Luo Y (eds) Microplastics in terrestrial environments. The handbook of environmental chemistry, vol 95. Springer, Cham. https://doi.org/10.1007/698_2020_449

Anonymous (2015) What are microbeads and why would canada ban them? The Economist. https://www.economist.com/the-economist-explains/2015/08/03/what-are-microbeads-and-why-would-canada-ban-them Accessed 19 June 2023

Anonymous (2019) UNEP CHW.13/INF/8—Draft Practical Manuals on extended producer responsibility and financing system for environmentally sound management. https://www.brsmeas.org/Decisionmaking/COPsandExCOPs/2019COPs/MeetingDocuments/tabid/7832/ctl/Download/mid/21391/language/en-US/Default.aspx?id=367&ObjID=21132 Accessed 13 May 2023

Anonymous (2020) Australia: Recycling and Waste Reduction Act 2020, No. 119. https://www.legislation.gov.au/Details/C2021C00501. Accessed 19 June 2023

Bahadori A, Smith ST (2016) Dictionary of environmental engineering and wastewater treatment. Springer, Cham. https://doi.org/10.1007/978-3-319-26261-1_14

Barnes DK, Galgani F, Thompson RC, Barlaz M (2009) Accumulation and fragmentation of plastic debris in global environments. Philos Trans R Soc Lond B Biol Sci 364(1526):1985–1998

Block B (2009) China reports 66-percent drop in plastic bag use. Eye on Earth, World watch Institute. https://www.thefreelibrary.com/China+reports+66-percent+drop+in+plastic+bag+use.-a0207123518. Accessed 15 May 2023

Cárdenas-Alcaide MF, Godínez-Alemán JA, González-González RB, Iqbal HMN, Parra-Saldívar R (2022) Environmental impact and mitigation of micro(nano)plastics pollution using green catalytic tools and green analytical methods. Green Anal Chem 3:100031. https://doi.org/10.1016/j.greeac.2022.100031. ISSN 2772-5774. https://www.sciencedirect.com/science/article/pii/S2772577422000301

Carney Almroth BM, Åström L, Roslund S, Petersson H, Johansson M, Persson NK (2018) Quantifying shedding of synthetic fibers from textiles; a source of microplastics released into the environment. Environ Sci Pollut Res 25:1191–1199

Cela E, Kaneko S (2011) Determining the effectiveness of the Danish packaging tax policy: the case of paper and paperboard packaging imports. Resour Conserv Recycl 55(9–10):836–841

Chen W, Wang X, Tao Q, Wang J, Zheng Z, Wang X (2013) Lotus-like paper/paperboard packaging prepared with nano-modified overprint varnish. Appl Surf Sci 266:319–325

Chen G, Li Y, Wang J (2021) Occurrence and ecological impact of microplastics in aquaculture ecosystems. Chemosphere 274:129989

Dang F, Wang Q, Huang Y, Wang Y, Xing B (2022) Key knowledge gaps for one health approach to mitigate nanoplastic risks. Eco-Environ Health 1(1):11–22

DasGupta R, Shaw R (2013) Changing perspectives of mangrove management in India–An analytical overview. Ocean Coastal Manag 80:107–118

Derraik JG (2002) The pollution of the marine environment by plastic debris: a review. Mar Pollut Bull 44(9):842–852

Dris R, Gasperi J, Rocher V et al (2015) Microplastic contamination in an urban area: a case study in Greater Paris. Environ Chem 12:592–599

Ge H, Yan Y, Wu D, Huang Y, Tian F (2018) Potential role of LINC00996 in colorectal cancer: a study based on data mining and bioinformatics. Onco Targets Ther 14:4845–4855

Geyer R, Jambeck JR, Law KL (2017) Production, use, and fate of all plastics ever made. Sci Adv 3(7):e1700782

Ghosh K, Jones BH (2021) Roadmap to biodegradable plastics—current state and research needs. ACS Sustain Chem Eng 9(18):6170–6187. https://doi.org/10.1021/acssuschemeng.1c00801

Hahladakis JN, Velis CA, Weber R, Iacovidou E, Purnell P (2018) An overview of chemical additives present in plastics: migration, release, fate and environmental impact during their use, disposal and recycling. J Hazard Mater 344:179–199

Hoffmann L, Eggers SL, Allhusen E, Katlein C, Peeken I (2020) Interactions between the ice algae Fragillariopsis cylindrus and microplastics in sea ice. Environ Int 139:105697

Hu M, Palić D (2020) Micro- and nano-plastics activation of oxidative and inflammatory adverse outcome pathways. Redox Biol 37:101620

Ibrahim ID, Hamam Y, Sadiku ER, Ndambuki JM, Kupolati WK, Jamiru T, Eze AA, Snyman J (2022) Need for sustainable packaging: an overview. Polymers 14(20):4430. https://doi.org/10.3390/polym14204430

Jambeck JR, Geyer R, Wilcox C, Siegler TR, Perryman M, Andrady A, Narayan R, Law KL (2015) Plastic waste inputs from land into the ocean. Science 347:768–771

Jensen S, Grøsvik BE, Halsband C, Halldórsson HP, Leslie HA, Gunnlaugsdóttir H, Guls HD, Vorkamp K, Granberg ME, Sigurðsson V, Jörundsdóttir HÓ (2022) Understanding microplastic pollution in the Nordic marine environment–knowledge gaps and suggested approaches. Microplast Nanoplast 2(1):1–9

Jeyasanta KI, Sathish N, Patterson J, Edward JP (2020) Macro-, meso-and microplastic debris in the beaches of Tuticorin district, southeast coast of India. Mar Pollut Bull 154:111055

Keefer EW, Norton SJ, Boyle NAJ, Talesa V, Gross GW (2001) Acute toxicity screening of novel AChE inhibitors using neuronal networks on microelectrode arrays. Neurotoxicology 22:3–12

Kershaw P, Katsuhiko S, Lee S, Samseth J, Woodring D (2011) Plastic debris in the ocean. UNEP Year Book, pp. 20–33. https://citeseerx.ist.psu.edu/viewdoc/download;jsessionid=03F07686A4BFC15E8676A0D9A580E3ED?doi=10.1.1.691.5994&rep=rep1&type=pdf. Accessed 13 June 2023

Khalid N, Aqeel M, Noman A, Hashem M, Mostafa YS, Alhaithloul HA, Alghanem SM (2021) Linking effects of microplastics to ecological impacts in marine environments. Chemosphere 264:128541

Kim E, Kim J, Lee S (2002) Inhibition of oocyte development in Japanese medaka (Oryzias latipes) exposed to di-2-ethylhexyl phthalate. Environ Int 28:359–365

Lang IA, Galloway TS, Scarlett A, Henley WE, Depledge M, Wallace RB et al (2008) Association of urinary bisphenol a concentration with medical disorders and laboratory abnormalities in adults. J Am Med Assoc 300:1303–1310

Lehner R, Weder C, Petri-Fink A, Rothen-Rutishauser B (2019) Emergence of nanoplastic in the environment and possible impact on human health. Environ Sci Technol 53:1748–1765

Li Y (2022) Scientific uncertainty of marine microplastic pollution and the dilemma of future international unified legislation. Int J Environ Res Public Health 19(24):16394. https://doi.org/10.3390/ijerph192416394

Magulova K (2012) Stockholm convention on persistent organic pollutants: triggering, streamlining and catalyzing global scientific exchange. Pollut Res 3:366–368

Mao X, Xu Y, Cheng Z, Yang Y, Guan Z, Jiang L, Tang K (2022) The impact of microplastic pollution on ecological environment: a review. Front Biosci (Landmark Ed) 27(2):46

Moore CJ, Lattin GL, Zellers AF (2005) Measuring the effectiveness of voluntary plastic industry efforts: AMRF'S analysis of operation clean sweep. In Proceedings of the plastic debris rivers to sea conference, Algalita Marine Research Foundation, Long Beach, CA

Moshood TD, Nawanir G, Mahmud F, Mohamad F, Ahmad MH, AbdulGhani A (2022) Biodegradable plastic applications towards sustainability: a recent innovations in the green product. Clean Eng Technol 6:100404

National Geographic, article on 'Plastic Detective' Imogen Napper traces pollution to the source, published October 21, 2022, https://www.nationalgeographic.com/impact/article/imogen-napper-explorer-story

O'Neill SM, Lawler J (2021) Knowledge gaps on micro and nanoplastics and human health: a critical review. Case Stud Chem Environ Eng 3:100091

Oehlmann J, Schulte-Oehlmann U, Kloas W, Jagnytsch O, Lutz I, Kusk KO et al (2009) A critical analysis of the biological impacts of plasticizers on wildlife. Philos Trans R Soc Lond B Biol Sci 364:2047–2062

Onyena AP, Aniche DC, Ogbolu BO, Rakib MRJ, Uddin J, Walker TR (2022) Governance strategies for mitigating microplastic pollution in the marine environment: a review. Microplastics 1(1):15–46. https://doi.org/10.3390/microplastics1010003

Pettipas S, Bernier M, Walker TR (2016) A Canadian policy framework to mitigate plastic marine pollution. Mar Policy 68:117–122

Poortinga W, Whitmarsh L, Suffolk C (2013) The introduction of a single-use carrier bag charge in Wales: attitude change and behavioural spillover effects. J Environ Psychol 36:240–247

Prata JC (2018) Airborne microplastics: consequences to human health? Environ Pollut 234:115–126

Prata JC, Silva ALP, da Costa JP, Mouneyrac C, Walker TR, Duarte AC, Rocha-Santos T (2019) Solutions and integrated strategies for the control and mitigation of plastic and microplastic pollution. Int J Environ Res Public Health 16(13):2411. https://doi.org/10.3390/ijerph16132411

Ritchie H, Roser M (2018) "Plastic Pollution". Published online at OurWorldInData.org. 'https://ourworldindata.org/plastic-pollution' (online resource)

Schnurr RE, Alboiu V, Chaudhary M, Corbett RA, Quanz ME, Sankar K, Srain HS, Thavarajah V, Xanthos D, Walker TR (2018) Reducing marine pollution from single-use plastics (SUPs): a review. Mar Pollut Bull 137:157–171

Science for Environmental Policy (SEP) (2011) Plastic waste: ecological and human health impact, in depth report, pp 1–44

Sharifinia M, Bahmanbeigloo ZA, Keshavarzifard M, Khanjani MH, Lyons BP (2020) Microplastic pollution as a grand challenge in marine research: a closer look at their adverse impacts on the immune and reproductive systems. Ecotoxicol Environ Saf 204:111109

Shen M, Song B, Zeng G, Zhang Y, Huang W, Wen X, Tang W (2020) Are biodegradable plastics a promising solution to solve the global plastic pollution? Environ Pollut 263:114469

Silva ALP, Prata JC, Walker TR, Campos D, Duarte AC, Soares AM, Barcelò D, Rocha-Santos T (2020) Rethinking and optimising plastic waste management under COVID-19 pandemic:

policy solutions based on redesign and reduction of single-use plastics and personal protective equipment. Sci Total Environ 742:140565
Stoett P (2019) Global ecopolitics: crisis, governance, and justice, 2nd edn. University of Toronto Press, Toronto, ON
Sulaiman AB, Razman MR (2010) Study on international environmental law and governance: focusing on Basel convention and role of precautionary principle. J Food Agric Environ 8: 1216–1219
Talvitie J, Mikola A, Setälä O, Heinonen M, Koistinen A (2016) How well is microlitter purified from wastewater?—a detailed study on the stepwise removal of microlitter in a tertiary level wastewater treatment plant. Water Res 109:164–172
Thompson RC (2015) Microplastics in the marine environment: sources, consequences and solution. In: Bergmann M, Gutow L, Klages M (eds) Marine anthropogenic litter. Springer, Cham, pp 185–200
Thompson RC, Olsen Y, Mitchell RP, Davis A, Rowland SJ, John AW, McGonigle D, Russell AE (2004) Lost at sea: where is all the plastic? Science 304(5672):838. https://doi.org/10.1126/science.1094559
U.S. Department of States (2019). https://www.state.gov/key-topics-office-of-environmental-quality-and-transboundary-issues/basel-convention-on-hazardous-wastes/. Accessed 15 June 2022
United Nations Environment Assembly of the United Nations Environment Programme, UNEP/AHEG/2018/1/INF/3, report of first meeting 'Combating marine plastic litter and microplastics: an assessment of the effectiveness of relevant international, regional and subregional governance strategies and approaches – a summary for policymakers' 8 May 2018, https://apps1.unep.org/resolutions/uploads/unep_aheg_2018_inf3_summary_assessment_en_rev.pdf
United Nations Environment Programme (2015) Plastics in cosmetics: are we polluting the environment through our personal care (Factsheet). https://wedocs.unep.org/20.500.11822/21754
United Nations Environment Programme, & Grid-Arendal (2016) Marine litter: Vital graphics. https://wedocs.unep.org/20.500.11822/9798
United Nations Environment Programme, Mid-term Evaluation of the UN Environment "Clean Seas Campaign", October 2019, p 76. https://wedocs.unep.org/bitstream/handle/20.500.11822/30929/01882_2019_mte_unep_em_global_clean_seas_campaign.pdf?sequence=1&isAllowed=y
United Nations Environment Programme, an article 'Inside the clean seas campaign against microplastics' published on 17 February 2022, https://www.unep.org/news-and-stories/story/inside-clean-seas-campaign-against-microplastics
Usman S, Abdull Razis AF, Shaari K, Amal MNA, Saad MZ, Mat Isa N, Nazarudin MF, Zulkifli SZ, Sutra J, Ibrahim MA (2020) Microplastics pollution as an invisible potential threat to food safety and security, policy challenges and the way forward. Int J Environ Res Public Health 17(24):9591. https://doi.org/10.3390/ijerph17249591
Vince J, Hardesty BD (2017) Plastic pollution challenges in marine and coastal environments: from local to global governance. Restor Ecol 25(1):123–128
Vince J, Hardesty BD (2018) Governance solutions to the tragedy of the commons that marine plastics have become. Front Mar Sci 5:214
Walker TR (2021) (Micro)plastics and the UN sustainable development goals. Curr Opin Green Sustain Chem 30:100497. https://doi.org/10.1016/j.cogsc.2021.100497. https://www.sciencedirect.com/science/article/pii/S2452223621000535, ISSN: 2452-2236
Walker TR, Xanthos D (2018) A call for Canada to move toward zero plastic waste by reducing and recycling single-use plastics. Resour Conserv Recycl 133:99–100
Walther BA, Kusui T, Yen N, Hu CS, Lee H (2020) Plastic pollution in East Asia: macroplastics and microplastics in the aquatic environment and mitigation efforts by various actors. In: Plastic aquatic environ—part I. Springer Nature, Cham, pp 353–403
Wright SL, Kelly FJ (2017) Plastic and human health: a micro issue? Environ Sci Technol 51:6634–6647

Xie YS, Zhang QK, Zhang LS, Zhong S, Liu JY, Hou XY et al (2020) Impact of a sewage treatment plant on the accumulation of microplastics in freshwater organisms in the Lijiang River of the Guilin urban section. Environ Sci 41:4999–5007

Yang Y, Chen L, Xue L (2021) Looking for a Chinese solution to global problems: the situation and countermeasures of marine plastic waste and microplastics pollution governance system in China. Chin J Popul Resour Environ 19(4):352–357

Yee MS-L, Hii L-W, Looi CK, Lim W-M, Wong S-F, Kok Y-Y, Tan B-K, Wong C-Y, Leong C-O (2021) Impact of microplastics and nanoplastics on human health. Nanomaterials 11(2):496. https://doi.org/10.3390/nano11020496

Zhang Y, Peng Y, Xu S, Zhang S, Zhou G, Yang J, Li H, Zhang J (2022) Distribution characteristics of microplastics in urban rivers in Chengdu city: the influence of land-use type and population and related suggestions. Sci Total Environ 846:157411

MIX
Papier aus verantwortungsvollen Quellen
Paper from responsible sources
FSC® C105338

If you have any concerns about our products,
you can contact us on
ProductSafety@springernature.com

In case Publisher is established outside the EU,
the EU authorized representative is:
Springer Nature Customer Service Center GmbH
Europaplatz 3, 69115 Heidelberg, Germany

Printed by Libri Plureos GmbH
in Hamburg, Germany